Contamination
of
Groundwaters

Advances in Environmental Science

Contamination
of
Groundwaters

Edited by

Domy C. Adriano
Professor of Environmental Soil Science,
University of Georgia Savannah River Ecology Laboratory

Alex K. Iskandar
Senior Scientist, Department of the Army Cold Regions
Research and Engineering Lab

Ishwar P. Murarka
Program Manager, Electric Power Research Institute

CRC Press
Taylor & Francis Group
Boca Raton London New York

CRC Press is an imprint of the
Taylor & Francis Group, an **informa** business

First published 1999 by St. Lucie Press

Published 2019 by CRC Press
Taylor & Francis Group
6000 Broken Sound Parkway NW, Suite 300
Boca Raton, FL 33487-2742

© 1999 by Taylor & Francis Group, LLC
CRC Press is an imprint of Taylor & Francis Group, an Informa business

First issued in paperback 2019

No claim to original U.S. Government works

ISBN-13: 978-0-367-44952-0 (pbk)
ISBN-13: 978-0-905927-44-2 (hbk)

Visit the Taylor & Francis Web site at
http://www.taylorandfrancis.com

and the CRC Press Web site at
http://www.crcpress.com

Library of Congress Cataloging-in-Publication Data

Catalog record is available from the Library of Congress

Contents

Preface

The series was initiated in 1986 with Prof. Dr. W. Salomons of the Institute for Soil Fertility of the Netherlands and Dr. Domy C Adriano of the University of Georgia's Savannah River Ecology Laboratory of the United States as editors. The strategy for the series was to adopt a thematic approach to address critical, contemporary environmental issues. The first five volumes were on acidic precipitation, which is still viewed today as one of the most controversial environmental issues and as an integral part of the global change issue. Adverse consequences from acidic precipitation have become obvious during the last two decades or so, as manifested in tree die back in Europe and North America, as well as acidification of numerous lakes on these two continents.

The topic of groundwater contamination is also a critical environmental issue because of the importance of groundwater supply and quality in sustainable development. On a global scale, the concern is scarcity of potable water and irrigation water for agricultural lands. In many countries, concerns also include the groundwater quality due to pollution from anthropogenic chemicals. Tainted groundwaters can pose threats to public health and may limit the adequate use of this resource by the current and future generations. Because contaminated groundwaters in deep aquifers can be impossible or impractical to remediate to restore their quality, it is necessary that chemicals should be handled properly and disposed of rationally.

Groundwaters may be subject to contamination by inorganic and organic chemicals due to human activities. The sources of chemicals are many and varied, especially since new generation chemicals are ever being produced. In the USA alone, about 500,000 potentially hazardous chemicals are in existence. When soils are contaminated by these chemicals the contaminants may eventually move and reach the groundwaters, and once in the groundwaters they are usually persistent.

In this volume, the chapters contributed by international experts are divided into two categories. In **Section 1: Methodology and Modeling** – there are nine chapters from international contributors dealing with both inorganic and organic contaminants including those from agricultural operations. In **Section 2: Case Studies** – there are scenarios for contamination with both inorganic and organic chemicals including agriculturally-related constituents, such as the nitrates.

It is not the intent of the editors and the members of the editorial board to have a comprehensive topic on groundwater contamination. However, it is hoped that with this timely volume addressing various aspects of this issue that this volume could be a welcome addition to the literature and should enhance our knowledge of the most recent information on methodology, modeling, and real world situations.

Domy C. Adriano
Alex K. Iskandar
Ishwar P. Murarka

Contamination of Groundwaters

Editorial Board

Series Editor

Domy C. Adriano University of Georgia SC 29802, USA

Associate Editors

SECTION 1
Methodology and Modeling

1   Hydrogeochemical modeling to predict subsurface transport

Dave A. McIntosh
Environment Division, Electric Power Research Institute, Palo Alto, California, USA

Abstract

Hydrologic and geochemical processes must be given due consideration when evaluating the adequacy of waste disposal facility design for meeting regulatory criteria. Assessing the engineering performance and the alternate design options for a waste disposal site requires a capability for estimating leachate flux and chemical composition, groundwater travel times, and potential geochemical reactions. Because hydrologic and geochemical processes are occurring simultaneously in a heterogeneous subsurface setting, one must resort to computer codes to encompass the complexity of the problem. Therefore, it has become common for regulators and the regulated community to use models for simulating chemical transport and determining regulatory compliance. This chapter describes in some detail the hydrologic processes and geochemical reactions that act to determine the fate of chemicals in the groundwater environment. In addition, the mathematical relationships that are used to represent these processes are presented along with some of the numerical techniques that may be used to arrive at a solution. The Electric Power Research Institute (EPRI) hydrogeochemical code FASTCHEM description focuses on the linkages among the various modules and provide the pre and post-processing possibilities. Finally, the code is applied to a subsurface contamination problem involving several chemical species, solid phases, adsorbents, and four soil strata.

Introduction

The abundant groundwater resource in the United States is of relatively good quality and supplies twenty five percent of all water used in the country. In addition, groundwater serves to maintain many ecosystems that support fish and wildlife. Nationally the groundwater contamination problem has not reached crisis proportions. However, there have been numerous instances of water supply well closures and public health alerts resulting from the appearance of organic and inorganic chemicals in groundwater (Pye, 1983).

To effect a groundwater protection strategy, both the federal and state governments have implemented legislation to curtail activities that may result in the degradation of groundwater quality. Federal statutes restricting underground injection of wastes and human activities in the recharge zones of vulnerable groundwater systems have been enacted in the Safe Drinking Water Act to ensure the quality of drinking water from aquifers. Control of the generation and disposal of hazardous waste together with the regulation of leaking underground storage tanks are addressed by the Resource Conservation and Recovery Act.

Management of the groundwater resource requires knowledge of the quantity and quality of available water and a capability to assess the impact of management decisions related to groundwater withdrawals and surface landuse restrictions. Groundwater transport and fate models provide a convenient assessment tool that integrates our present understanding of hydrologic and geochemical processes. In this chapter the rudiments of such models will be covered and a specific application of a model to a groundwater contamination problem will be presented.

Value of hydrogeochemical codes

Consideration of the simultaneous occurrence of the processes of fluid flow, advective and diffusive transport, and chemical reactions in a heterogeneous media with a complex geometrical configuration demands the use of a computer code to comprehend or predict the distribution of chemicals in groundwater. Mathematical representations of the hydraulic and geochemical processes governing the transport and fate of chemicals in the subsurface are coded for solution on a digital computer and used for physicochemical simulations. After proper calibration, such codes can be used to model the present state of a groundwater system or to examine the response of the system to a specified stress. Hydrogeochemical codes can be applied to address issues of well-interference, contamination from the operation of a waste disposal facility, and the remediation of existing groundwater contamination. Given these capabilities, models can aid the regulator in the development of groundwater quality and quantity criteria, and guide the water resource manager in selecting among various management strategies to achieve regulatory compliance.

The simulation of an existing groundwater contamination condition using a hydrogeochemical code serves to validate the user's conceptual model of the system. The identification of the dominant processes at work, the values of hydrologic and geochemical parameters, and the nature of the boundary and initial conditions can all be checked through the simulation of a groundwater system in its existing condition. This understanding and proper conceptualization of the existing state of affairs is crucial for directing future data acquisition efforts for environmental impact assessment and may help determine model appropriateness.

Regulators in their efforts to ensure that the disposal of hazardous substances do not pose a risk to human health or the environment may resort to computer codes. The longterm risk of a land-disposal facility may be established in terms of the probability that a chemical concentration is not exceeded at a downgradient compliance location. To assess the fate of chemicals released from a disposal site and their migration to a monitoring point require the consideration of hydrogeochemical processes and hence the use of a model. Incorporation of such a model into a risk framework, provides the regulator a method for estimating the probability distribution of the concentration at the receptor location.

From the regulated community standpoint these modeling approaches can be used to back-calculate the effluent concentrations and release rates that are permissible at the disposal site to ensure that the receptor concentration meets water quality standards. Consequently, leachate pretreatment systems or facility liners can be designed to achieve a certain facility performance.

Under circumstances where a contaminant plume has developed and the regulation requires cleanup to specified standards, the application of a hydrogeochemical model could assist in developing remediation strategies and assessing the performance of alternate remediation options. Also, the model could be included as an algorithm in a non-linear programming routine to provide optimum designs for pumping in pump-and-treat systems.

Hydrologic and geochemical processes

The transport and fate of chemicals released in the groundwater environment are dictated by hydrologic processes for their migration in the subsurface and by geochemical and biological processes for their attenuation and transformation. Because of the limited importance of biotransformation in determining the fate of inorganic chemicals (the focus of this chapter), the hydrologic and geochemical processes related to inorganic transport will be elucidated here.

Advective and dispersive transport

The rationale for decomposing hydrologic transport into advection and dispersion is forced by the realization that a detailed hydraulic characterization of the subsurface necessary to produce a well defined velocity field is impractical. Consequently, the mean movement of a mass of contaminant is captured by advection and deviations from the mean are characterized with a dispersion term.

Advection and dispersion are extremely important in chemical transport and fate modeling because they represent the mechanism whereby the contaminant is brought into contact with the ambient fluid and minerals. Additionally, these processes influence the extent of the contaminated zone and the magnitude of any remediation efforts that may be necessary. Therefore, the application of hydrogeochemical computer codes for contaminant impact predictions and for evaluating management options require an accurate representation of these processes to ensure reliable results.

Advection

Advection represents the process whereby dissolved or suspended substances are carried along by the moving ground water undergoing Darcy flow. If the subsances of interest were a tracer, then by definition the advective component of its motion would be the Darcy velocity of the groundwater.

Dispersion

Generally, dispersion describes the spreading of a solute about its center of mass as it advects (travels) through the subsurface. Fundamental local dispersion, the kind of dispersion that would occur in a perfectly homogeneous porous medium, is due to heterogeneities below the Darcy scale. Heterogeneities on scales above the Darcy scale are responsible for the intertwining of flow paths and the potential for increased mixing that occurs as the solute migrates through the porous medium. Dispersion caused by heterogeneities above the Darcy scale is sometimes called macrodispersion and is not understood as well as local dispersion. The macrodispersion process is driven by velocity variations about the mean Darcy velocity, and if so-called Fickian conditions are reached it can be expressed in terms of gradients in solute concentrations. Velocity variations above the Darcy scale facilitate the transport of solute in complex patterns through the porous medium thereby providing an opportunity for local dispersion and diffusion to mix zones of unequal concentrations. However, there is disagreement about when the above Darcy scale velocity variations should be treated as a dispersion process, which obeys the mathematics of diffusion, and when the variations should be measured and included explicity as part of the advective field.

Molecular diffusion

Mixing brought about by molecular agitation is ultimately the process by which contaminant mass is integrated with ambient fluids. Advective and dispersive transport simply facilitate the opportunity for molecular diffusion. Therefore, under conditions where advective and dispersive transport are small, molecular diffusion can play a large role in redistributing contaminants in the subsurface. However, under all circumstances it is the only process that causes molecular scale mixing.

Geochemical processes

During their transport by advection, dispersion and diffusion, chemicals are brought into contact with various components of the ambient environment where geochemical processes intervene to determine the fate of the reactive chemicals. The mobilization of chemicals through dissolution and desorption or their attenuation through precipitation and sorption reactions are keys to understanding the level and extent of contamination. Therefore, efforts to simulate subsurface contamination distributions require the representation of geochemical as well as hydrologic processes in the computer code.

Precipitation-dissolution

Precipitation-dissolution is the process by which a chemical enters the aqueous phase or leaves to form a solid phase. The occurrence of trace metals in many solid phases of soil is an important influence on the aqueous concentration of these metals. The dissolution of unstable solid phases and the precipitation of stable phases work in concert to bound the amount of various substances in solution. The partitioning of metals between mobile and immobile forms is fundamental to the accurate simulation of a contaminant plume evolution. Factors that affect the precipitation-dissolution reactions are pH, type and concentration of complexing inorganic ligands, oxidation state of mineral components, temperature, and pressure.

Complexation

Similar to precipitation-dissolution reactions, complexing ligands react with aqueous metal ions to form soluble complexes or insoluble phases. Some important complexing ligands to be considered in groundwater contamination problems are chloride, sulfate, bicarbonate, and sulfide species. The formation of metal ligand precipitates can alter the distribution of trace metals in the aqueous phase of a plume. As well, the formation of soluble complexes may act to dissolve other insoluble phases resulting in an increase in aqueous metal concentration. Therefore, complexation reactions are an essential inclusion in computer codes that simulate groundwater geochemistry. The degree to which these reactions occur depends on the metal ion concentration, inorganic anion concentration, the concentration of dissolved oxygen, and pH (Rai and Zachara, 1984).

Sorption-desorption
Sorption represents another geochemical process whereby dissolved contaminants are partitioned between the aqueous and solid phases in the subsurface. The sorption process is induced by chemical reactions that result in binding of chemicals to surfaces, chemicals being incorporated into solid interiors, and charged ions being attracted to the surface of solids. Common adsorbents present in the subsurface are hydrous metal oxides, clays, sands, organic matter and humic coatings on surfaces. Clay, because of its large surface area and net negative charge, has a great affinity for positively charged ions. The replacement of positively charged ions residing on mineral surfaces is one mechanism by which contaminants can be removed from solution and bonded to surfaces. The ability of surfaces to sorb chemicals is controlled by groundwater pH. Many sorption reactions are completely or partially reversible. In situations where this is the case, sorption may be reversed when aqueous concentrations are reduced.

Oxidation-reduction
Oxidation-reduction (redox) reactions involve the transfer of electrons from one atom to another as contrasted with reactions that involve the combination or separation of ions or molecules. The state of oxidation in an aqueous environment as measured by the redox potential indicates the oxidation state of multivalent metal ions (*e.g.*, Mn, Fe, Cu, Se, Cr) and indicates the potential redox changes in available and competitive ligands that could in turn affect the aqueous concentration of trace metals. Consequently, the significance of redox reactions to hydrogeochemical process modeling is evident.

Mathematical framework for modeling

Mathematical expressions that represent the hydrologic and geochemical processes are developed in this section. These equations, their numerical approximation, and subsequent solution, represent the foundation of computer models. The equations that evolve are based on fundamental laws of physics and chemistry, *i.e.*, conservation of mass and momentum, and the law of chemical equilibrium expressed by the constancy of the mass action expression.

Groundwater flow equations
The conservation of mass and momentum are the fundamental equations governing

groundwater flow in a porous medium. Equations of state that give density and viscosity as functions of pressure, temperature, and contaminant concentration provide the coupling between hydrogeology and geochemistry. Here, though, we consider the flow and geochemistry as uncoupled. Conservation of momentum is introduced to the computation of groundwater flow via Darcy's equation for both saturated and partially saturated conditions. Darcy's equation for saturated conditions is written in tensor form using the Einstein summation convention, as

$$q_i = - K_{ij} \frac{\partial h}{\partial x_j} \qquad\qquad i,j = 1,2,3 \qquad\qquad (1)$$

for unsaturated conditions, as modified by Richard, as

$$q_i = - \left(K_{ij} k_r (\theta) \frac{\partial}{\partial x_j}, \left(\psi(\theta) + x_3 \right) \right) \qquad\qquad i,j = 1,2,3 \qquad\qquad (2)$$

where q_i = Darcy velocity, K_{ij} = saturated hydraulic conductivity tensor, h = hydraulic head, $k_r (\theta)$ = relative permeability with respect to water phase, $\psi(\theta)$ = pressure head, x_j = spatial coordinates, and x_3 = vertical coordinate. Both $k_r (\theta)$ and $\Psi(\theta)$ are related to the soil moisture content, (q). Consequently, the use of these equations for computing groundwater flows requires that the characteristic curves k_r (θ) and $\psi(\theta)$ be available for input.

The principle of conservation of mass is invoked along with Darcy's law to analyze flow through porous media. Simply stated, the conservation principle calls for the mass inflow rate to be equal to the mass outflow rate plus the change of storage with time. In differential equation form, the conservation equation can be represented as

$$\frac{\partial(\theta \rho)}{\partial t} = \frac{\partial(\rho q_i)}{\partial x_i} \qquad\qquad i = 1,2,3 \qquad\qquad (3)$$

where ρ = mass density of water, which cancels for incompressible or slightly compressible flow.

The incorporation of Darcy's representation of q_i into the conservation of mass equation produces the groundwater flow equation:

$$\frac{\partial(\theta\rho)}{\partial t} = -\frac{\partial}{\partial x_i}\left[\rho K_{ij}\frac{\partial h}{\partial x_j}\right] \qquad\qquad i,j = 1,2,3 \qquad\qquad (4)$$

and for soil moisture transport in the unsaturated zone , the incorporation of the unsaturated form of Darcy's law into equation (3) yields

$$\frac{\partial(\theta\rho)}{\partial t} = -\frac{\partial}{\partial x_i}\left[\rho K_{ij}k(\theta)\right](\frac{\partial}{\partial x_j}(\Psi(\theta) + x_3)) \qquad\qquad (5)$$

which is sometimes referred to as the Richard equation.

In order to incorporate the pumping (or injection) of wells in the aquifer system, a source/sink term must be added to the groundwater mass balance equation. The mass flow rate of the source/sink is given by

source/sink $= \rho q_s$

where q_s = volumetric flow rate per unit porous medium volume of the source/sink.

For a slightly compressible fluid, such as water, the transport equation (with source/sink terms) may be written as

$$\eta\frac{\partial\psi}{\partial t} = \frac{\partial}{\partial x_i}\left[K_{ij}k_r(\theta)\frac{\partial(\psi + x_3)}{\partial x_j}\right] + q_s \qquad\qquad (6)$$

with

$$\eta = S_w S_s + \phi\frac{dS_w}{d\psi} \qquad\qquad (7)$$

where S_w = water phase saturation, ϕ = effective porosity, and the specific storage coefficient $S_s = \rho g\,(\psi\beta + \alpha),\beta$ and α and represent the coefficient of compressibility of the fluid and the granular skeleton respectively. Note that $\theta = \phi S_w$.

For strictly saturated groundwater flow, the governing equation may be rewritten in the form

$$\frac{\partial}{\partial x_i}\left(K_{ij}H\frac{\partial h}{\partial x_j}\right) = H\eta\frac{\partial h}{\partial t} - q_s \qquad\qquad (8)$$

where H is equal to the saturated thickness of the aquifer for areal flow analyses and equal to unity for cross-sectional analyses. Since $S_w = 1$, a constant, for saturated flow, the coefficient h for this case becomes

$$\eta = S_s \qquad \text{for a confined aquifer}$$

$$\eta = \frac{S_y}{H} + S_s \qquad \text{for an unconfined aquifer}$$

where Sy is the specific yield of the unconfined aquifer.

A thorough development of these equations governing fluid flow in the subsurface can be found in the textbooks of Freeze and Cherry (1979), Bear (1979) and deMarsily (1986).

Geochemical equilibrium computations
The subsurface can be characterized as a dynamic geochemical system with the simultaneous occurrence of many reactions involving the solid, aqueous, and gaseous phases. The complexation, solubility and adsorption reactions that are incorporated in geochemical models will be discussed here. The mathematical development of the geochemistry problem follows closely the work of Hostetler (1989).

Complexation
Aqueous complexation reactions involving charged complexes and ion pairs can be represented by the relation

$$S_i \rightarrow b_{i1}P_1 + b_{i2}P_2 + \cdots + b_{iN}P_N \qquad \text{for a confined aquifer}$$

or in more concise notation

$$S_i \rightarrow \sum_{j=1}^{N} b_{ij}P_j \qquad \text{for an unconfined aquifer} \qquad (9)$$

where S_i = chemical formula of species i, b_{ij} = stoichiometric reaction coefficient (moles of P_j required to form one mole of S_i), P_j = chemical formula of component j. The mass action equation associated with the above aqueous speciation reaction is

$$K_i = -\frac{a_{si}}{\prod_{j=1}^{N} a_{pj}^{b_{ij}}} \qquad (10)$$

where K_i is the equilibrium constant, and a_{si} and a_{pj} are the activities of species i and components j respectively. The relation between activities and concentrations will be developed in a later section. Suffice to state at this point that activities represent effective thermochemical concentrations. The mathematical formulation utilizing the equilibrium constant approach was presented by Brinkley, 1946, 1947 and Kandiner and Brinkley 1950.

Solubility

Similarly, solubility reactions that describe the partitioning of mass between the aqueous and solid phases can be written as

$$D_k \rightarrow b_{ki}P_1 + b_{k2}P_2 \cdots + b_{kN}P_N$$

or in more concise notation

$$D_k \rightarrow \sum_{j=1}^{N} b_{k_j} P_j \tag{11}$$

where D_k = chemical formula of the precipitated solid phase k, b_{kj} = stoichiometric reaction coefficient that indicates the number of moles of P_j required to form one mole of the solid phase D_k. The associated mass-action expression for this equilibrium solubility reaction is

$$K_k = -\frac{a_{Dk}}{\displaystyle\prod_{j=1}^{N} a_{pj}^{b_{k_j}}} \tag{12}$$

where a_{Dk} is the activity of the kth solid phase which equals unity.

Adsorption

To calculate the transfer of mass between aqueous phase components and the absorbent, a surface complexation model can be formulated to represent surface-aqueous interactions. This formulation could be used to mimic both ion exchange and specific ion adsorption reactions.

The surface-complexation reaction used to represent adsorption is written as

$$A_m \rightarrow \sum_{j=1}^{N} b_{mj}P_j + b_{ms}z_s \tag{13}$$

where A_m = chemical formula of the mth adsorbate, b_{mj} = stoichiometric reaction coefficient indicative of the number of moles of the aqueous component P_j required to form one mole of adsorbate A_m, b_{ms} = stoichometric reaction coefficient that represents the number of moles of z_s adsorption site that are required for the formation of A_m.

The mass-action relation for the adsorption reaction is

$$K_m = \frac{a_{Am}}{\prod\limits_{j=1}^{N} a_{pj}^{b_{mj}} \cdot a_{zs}^{b_{ms}}} \tag{14}$$

where a_{Am} is the activity of the mth adsorbate defined as the ratio of the number of moles of adsorbate to moles of surface sites, and a_{zs} is the activity of the adsorption sites computed as the ratio of the number of moles of free sites to the total number of moles of surface sites.

Mass balance
The conservation of mass for the chemical components P_j in a system in which the geochemical processes of aqueous speciation, precipitation-dissolution, and adsorption-desorption are active is given by

$$T_j = n_{pj} + \sum_i b_{ij} n_{si} + \sum_k b_{kj} n_{Dk} + \sum_m b_{mj} n_{Am} \tag{15}$$

where T_j is the total number of moles of uncomplexed component j, n_j the second term is the mass (moles) of component j in the complex i, the third term represents the mass of component j in the solid phases k, and the last term is the amount of component j in each of the adsorbates. The terms n_{si}, n_{Dk} and n_{Am} represent the mass of specie i, the mass of solids, and mass of adsorbate.

The mass balance equation for the total number of moles of surface sites is:

$$T_{js} = n_{js} + \sum_m b_{mj} n_{Am} \tag{16}$$

where n_{js} = moles of unoccupied surface sites and n_{Am} = moles of sites occupied by adsorbates.

Activity coefficient
The activity of the aqueous species appearing in the mass-action expressions are calculated from the relation

$$a_{si} = \gamma_{si} \left[\frac{1000 \, n_{si}}{(\text{Mol. Wt})_{H_2O} \cdot n_{H_2O}} \right] \qquad (17)$$

where γ_{si} is the activity coefficient of the species i, the term in brackets is the expression for the molality of species i, and n_{H_2O} is the number of moles of water. Similarly the activity of component P_j can be expressed as:

$$a_{pj} = \gamma_{pj} \left[\frac{1000 \, n_{pj}}{(\text{Mol. Wt})_{H_2O} \cdot n_{H_2O}} \right] \qquad (18)$$

The use of an activity rather than a concentration is required to maintain the equilibrium constant independent of concentration. The Davies (1962) equation can be used to compute the activity coefficient for charged specie as follows:

$$\log \gamma_{si} = - A Z_i^2 \left[\frac{\sqrt{I}}{(1 + \sqrt{I})} - 0.3I \right] \qquad (19)$$

where I is the ionic strength of the solution, Z_i is the change or valence of the ith ion, and A is the Debye-Huckel parameter (tabulated in the FASTCHEM database). The ionic strength of the solution is estimated from the relation

$$I = 0.5 \left(\sum_i m_i Z_i^2 \right) + 0.5 \left(\sum_j m_j Z_j^2 \right) \qquad (20)$$

where m_i and m_j are the aqueous concentration of complexes and components respectively.

The activity coefficient of neutrally charged species is calculated by

$$\log \gamma_{si} = 0.1I \qquad (21)$$

The activity of water is computed using an approximation of the Raoults' law expression:

$$a_{H_2O} = 1 - 0.017\left(\sum_i m_i + \sum_j m_j\right) \qquad (22)$$

For the adsorbed species, the activity of the mth adsorbate is defined as

$$a_{Am} = \gamma_{Am}\, n_{Am} \qquad (23)$$

where n_{Am} is the number of moles of adsorbate and γm is the reciprocal of the number of moles of surface sites.

Advective-dispersion equation

Chemical transport in the subsurface environment can be described using the advection-dispersion equation. Represented in this equation are the processes of advection, Fickian dispersion, retardation (owing to adsorption), first-order decay, and source/sink terms. The equation, written in two-dimensional form below, is a conservation of mass representation for a single chemical species:

$$\frac{\partial}{\partial x_i}\left(D_{ij}\frac{\partial c}{\partial x_j}\right) - \frac{\partial}{\partial x_i}\left(V_i\, c\right) - \kappa R c = R\frac{\partial c}{\partial t} - \frac{C_s w}{\theta} \qquad (24)$$

where D_{ij} are components of dispersion, V_i are components of the velocity vector, R is the retardation factor, k is the first-order decay rate, C_s is the chemical concentration of the source/sink and W is the flow rate of the source/sink.

In contrast to the groundwater flow equations and geochemical equilibrium relations developed earlier in this chapter, the advection-dispersion-retardation-decay formulation assumes knowledge of the velocity components and handles geochemical attenuation through the processes of first-order decay and linear partitioning.

Solution techniques

Both analytical and numerical solution techniques are available for solving the transport and mass balance equations presented in the previous section. Some of the techniques commonly used in attaining a solution are covered in this section.

Analytical method for advection-dispersion equations

Analytical solutions to the advection-dispersion equation have been achieved for groundwater systems that can be characterized as homogeneous, and isotropic, with constant velocity and dispersivity. In addition, simple boundary conditions, and system definitions are required to make an analytical solution feasible. However, as a result of the simplifications necessary to achieve a solution to the advectiondispersion equation, the results are usually used for preliminary assessments of contamination problems. The MYGRT model (Summers, *et al.*, 1990) developed for the Electric Power Research Institute represents a two-dimensional version of the advection-disperson relation (equation (24)) for a finite boundary source.

Finite-element method for flow equations
The finite-element method provides a means of discretizing the groundwater system and using interpolation functions (basis functions) to provide a piecewise continuous approximation of the dependent variable over the problem domain. The problem domain is divided into elements with nodes located in the center of each element. Rectangular and triangular elements are used to represent the domain of flow.

The trial function is comprised of the product of the basis functions and the model values of pressure head. The trial function, therefore, represents a piecewise continuous approximation of the dependent variable over the problem domain. This approximation of the dependent variable is substituted into the groundwater flow equation and integrated over the problem domain. The governing system of partial differential equations for flow is thus transformed into a system of ordinary differential equations that must be integrated with respect to time. An implicit time-stepping scheme is utilized and the result is a highly non-linear system of algebraic equations. Both backward and central difference time-stepping is performed depending on whether the model point is in the saturated or unsaturated zone. As a consequence of the non-linearity of the resulting equations an iterative method (Picard or Newton-Raphson) is required to obtain a solution at each time step. The iterative scheme produces a linearized system of algebraic equations (described in the next section) that is solved using a symmetric banded matrix solver. Boundary conditions are incorporated through the boundary integral term. The method allows arbitrarily shaped elements with arbitrary orientation thereby allowing accurate representation of subsurface features and external boundary features. More detailed discussion and development of the finite-element approach can be found in the work of Zienkiewiz and Cheung (1965).

Numerical method for mass-balance expression
The conservation of mass for the chemical components P_j is given by the expression (equation (15))

$$T_j = n_{pj} + \sum_i b_{ij} n_{si} + \sum_k b_{kj} n_{Dk} + \sum_m b_{mj} n_{Am} \tag{24}$$

where T_j is the total mass of component j expressed in terms of the uncomplexed amount of component j, the amount of species i, the amount of the kth solid phase, the amount on m^{th} adsorbate, and the known stoichiometry (b_{ij}, b_{kj} and b_{mj}). The mass of component n_{si} is related to the activity a_{si} through equation (17) and a_{si} is related to a_{pj} through the mass-action expression (equation (10)) and finally a_{pj} may be expressed in terms of n_{pj} using equation (18). In a similar fashion n_{Am} can be expressed in terms of n_{pj} and n_{js} using equations (23), (14) and (18). The resulting system of equations would be non-linear if the stoichiometric coefficients (exponents in the mass-action expression) are greater than unity, which is normally the case. The mass acton expression for the solubility reactions (equation (12)) may be rearranged as

$$1 = K_k \prod_{j=1}^{N} a_{pj}^{b_{kj}} \qquad (25)$$

and the mass balance for the adsoption surfaces (equation (16)) can be combined with the mass-action expression for adsorption (equation (14)) to give

$$T_{js} = n_{js} + \sum_m b_{mj} K_m \prod_{j=1}^{N} a_{pj}^{b_{mj}} \cdot a_{zs}^{b_{ms}} \qquad (26)$$

Again a_{pj} can be related to n_{pj} through equation (18) and a_{zs} is related to n_{js}. Therefore, the problem at hand becomes one of computing values of n_{pj}, n_{Dk}, and n_{js} that satisfies equations (15), (25) and (26).

Several numerical techniques exist for solving systems of non-linear, algebraic equations, however, the Newton-Raphson method will be highlighted here because it is utilized in achieving a solution to the transport equation as well.

The mass-balance expression is reformulated in terms of residuals Y, such that

$$Y_j = T_j - \left[n_{js} + \sum_i b_{ij} n_{si} + \sum_k b_{kj} n_{Dk} + \sum_m b_{mj} n_{Am} \right] \qquad (27)$$

$$Y_k = K_k \prod_{j=1}^{N} a_{pj}^{b_{kj}} - 1 \qquad (28)$$

$$Y_{js} = n_{js} \sum_m b_{mj} K_m \prod_{j=1}^{N} a_{pj}^{b_{mj}} \cdot a_{zs}^{b_{ms}} \qquad (29)$$

The goal of the iterative technique for accomplishing a solution is to make the residuals approach zero. The Newton-Raphson procedure for achieving this goal is based on the formula

$$n^{\text{revised}} = n^{\text{old}} + \Delta n \tag{30}$$

$$Y^{\text{revised}} = Y^{\text{old}} + \frac{\partial Y}{\partial n} \Delta n = 0 \tag{31}$$

To initiate the process, trial estimates of the independent variables n_{js}, n_{pj} and n_{Dk} are provided. This allows the computation of values for Y^{old} and the derivative of Y with respect to n. Consequently, equation (31) is linearized and can be solved to provide estimates of Δn. These values of Δn are used in equation (30) to revise n^{old} and the process repeated until the residuals Y meet some acceptable tolerance. The Newton-Raphson approach along with other numerical techniques can be found in the book by Hildebrand (1956).

Once the speciation problem is solved in this fashion, the masses of solids specified by the user are checked to determine if dissolution occurs and the saturation indices of potential solids are checked to determine if precipitation occurs. If either condition is present, the stable phase assemblage is redefined and the Newton-Raphson calculations repeated. This procedure continues until neither precipitation nor dissolution is indicated, *i.e.*, a stable phase assemblage is found, along with a set of equilibrium aqueous phase species.

FASTCHEM EPRI hydrogeochemical code

The FASTCHEM computer code (developed by Battelle, Pacific Northwest Laboratory under EPRI Contract RP24852) is designed to simulate equilibrium geochemistry, advection, and dispersion in the saturated and unsaturated subsurface environments. The hydrologic flow component of the code performs the computation of steady-state velocities and pressures. The output from the hydrologic flow component serves as input to the module that develops the two-dimensional streamtubes that define the paths for the movement and interaction of leachate with ambient soils and pore water. The advective and dispersive transport processes are simulated using a Markov procedure (Moore and Yackel, 1974) that provides the flexibility of handling Fickian as well as non-Fickian representations of macrodispersion. Following the displacement of solute resulting from the Markov transport, the equilibrium geochemistry module simulates aqueous speciation, solubility reactions involving solids, and adsorption-desorption of aqueous constituents on solid surfaces. This sequential coupling of transport and geochemistry is done at each time step for the stipulated duration of the simulation.

Hydrologic modules

The hydrologic modules in the FASTCHEM Code are used to generate the pathways of flow in the subsurface. This is accomplished by first solving the flow equations to generate a pressure field (EFLOW) and subsequently using the pressure field data with Darcy's equations to produce detailed patterns of velocity. The velocity field is used in a particle-tracking routine to describe flow pathways (stream-tubes).

Flow module EFLOW

Advective transport in the saturated and unsaturated zones is simulated using a two-dimensional, finite-element model. Spatial variations in the soil hydraulic properties can be accommodated by the code.

In simulating two-dimensional, saturated-unsaturated flow conditions, EFLOW is capable of handling seepage faces, infiltration, evaporation, plant root moisture extraction and pumped (or injection) wells. In simulations that address saturated, unconfined groundwater flow in the areal plane, ETUBE accounts for storage conversion, and vertical recharge from precipitation. Some of the assumptions made in developing EFLOW are: (1) the water flow is governed by Darcy's law, (2) the water is considered homogeneous (no variations in density, viscosity, *etc.*) and slightly compressible, (3) hysteresis in the hydraulic conductivity relations are negligible, and (4) only single-phase flow (water) is considered. Details on the structure of the code and the numerical solution technique can be found in the EPRI Report EA5870, Volume 2.

Streamtube generation module, ETUBE

The pathlines created with this module result from the tracking of waterparcel movements that are produced by the steady-state flow field generated using the hydrologic flow module. Acceleration and velocity terms are determined from the spatial gradients of pressure (or hydraulic head) and are obtained from the output of the hydrologic flow module. In the process of taking discrete steps to create the pathline, ETUBE keeps track of longitudinal position, moisture content, pore-water velocity and the cross sectional area of the stream tube being created. Subsequently, in the coupling module EICM, groundwater routed through the streamtube is advected, dispersed, and reacted with the soils and ambient groundwater.

All of the assumptions associated with the EFLOW code hold true for ETUBE. In addition, ETUBE is applicable to steady-state flows only.

Parent geochemical module, ECHEM

To enhance the computational efficiency of the coupled code EICM, a separate geochemistry code is used to develop a site-specific thermodynamic database. The relevant components of the site-specific database are determined by considering a range of pH-Eh and water saturations for each geochemical domain, and boundary and initial conditions. All aqueous species and minerals that may be important to the site assessment are identified by ECHEM. The resulting site-specific thermodynamic database is used by for geochemical computation in the coupling code EICM. The processes represented in ECHEM, the numerical schemes used, and the types of geochemical problems that can be addressed are all presented in the EPRI Report EA5870, Volume 4.

Hydrogeochemical coupling code, EICM

The coupled code utilizes a two-step process for performing the hydrologic transport (advection and dispersion) and geochemical transformation (speciation, precipitation-dissolution, and adsorption) of the dissolved chemicals. The information on advection is generated by the EFLOW/ETUBE modules and made available to the EICM module. The dispersion and geochemical processes are coded in the hydrogeochemical coupling module.

The advective velocity along a given streamtube is calculated by ETUBE and supplied to EICM, while the dispersion processes are simulated using the Markov method. During each time step fluid-parcels are distributed from each point within the streamtube to downstream locations in accordance with the Darcy velocity within a tube and a specified probability distribution function (pdf) for displacement about the mean. The appropriate Darcy velocity represents the advective component and the spread about the mean is a reflection of the dispersive contribution to transport. It is worth noting that the selection of a Gaussian probability distribution function to characterize dispersion is akin to simulating Fickian dispersion under the Markov approach.

The simulation of aqueous speciation, precipitation-dissolution, and adsorption-desorption in EICM are based on the principles of equilibrium chemical thermodynamics. The geochemical segment sets up a mass balance equation for each chemical component represented in the problem. Mass action expressions are then substituted into the mass balance equations for each component. This results in a set of non-linear simultaneous equations which is solved using a Newton-Raphson iteration technique as discussed in above.

The hydrogeochemical coupling is accomplished by the sequential execution of the Markov transport segment and the geochemistry segment, once per time step. This two-step coupling is illustrated in Figure 1. During the transport phase, all solutes are displaced in accordance with the applicable tube velocity and selected pdf. This transport step is followed by the execution of the geochemistry segment to simulate the solution and soil reactions in each volume element of the streamtube. Hostetler (1989) presents all the details pertaining to the structure and verification of the EICM code.

FASTCHEM code structure

The relationship among the various modules that make up the FASTCHEM code is depicted in Figure 2. The modules are linked through the exchange of output files form one module to serve as input for another. The manipulation and management of these inter-module file transfers are achieved through the use of the preprocessor.

The input file for EFLOW (containing the hydrologic parameters and boundary conditions) is developed on the PC-based preprocessor and transferred to the mainframe for execution. The output files from EFLOW containing information on the system geometry, the finite-element mesh, and the pressure distribution are used as input to ETUBE. Along with this information from EFLOW, the user must specify the starting location of the stream-tubes desired. This set of data is shipped to the mainframe for execution by ETUBE. The output file from ETUBE containing information on bin volume and length, material type encountered at each location,

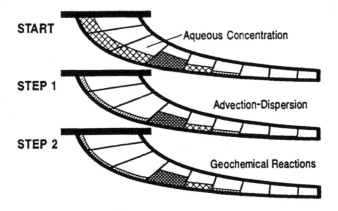

Figure 1 *The coupling of hydrology and geochemistry by EICM.*

water content, bulk density, and time step is returned to the preprocessor to eventually serve as input to EICM.

The input file for ECHEM (containing the geochemical constituents of the boundary and initial conditions) is created on the preprocessor and sent to the mainframe for processing by ECHEM. The ECHEM output files with information on the site-specific thermochemical database (a subset of the complete database) and equilibrated concentrations of the constituents of the geochemical initial and boundary conditions are returned to the PC for use with EICM.

Finally, the ECHEM and ETUBE outputs are combined with user supplied information on the dispersion characteristics and uploaded to the mainframe for processing by the coupled code EICM.

Pre and post-processing

Linkages among the modules of FASTCHEM are facilitated by an IBM-PC based workstation that host the pre and post-processing codes. The workstation codes (ECPR 1-6, and EGRF) assist the user in development of a hydrogeochemical conceptual model of the groundwater system to be analyzed. Because the EFLOW, ETUBE, ECHEM and EICM modules perform their computationally intensive calculations on a mainframe computer, a communication link between the PC and mainframe is required. By developing input data files on the PCbased workstation, mainframe time is optimized.

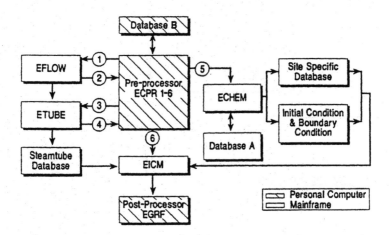

Figure 2 *FASTCHEM structure-relation among modules.*

FASTCHEM pre-processor

The preprocessor is composed of six modules that provide a facility for developing input files for use with EFLOW, ETUBE, ECHEM and EICM. In addition, the pre-processor is used to preview, in graphical form, the output from EFLOW and ETUBE.

The pre-processor features a multilevel design that caters to users of different levels of expertise in hydrology and geochemistry, and handles problems of varying levels of complexity. Three levels (0, 1, 2) of detail can be specified for hydrology and geochemistry. Because of the two-step method used for coupling hydrology and geochemistry, different levels of detail for hydrology and geochemistry can be matched to arrive at a final hydrogeochemical characterization of the problem to be solved.

Level O Hydrology requires that the user specifies a single flow velocity, defines the length of the associated streamtube, and the computational parameters of bin length and time step. Consequently, this results in a conceptual model for hydrology that is one-dimensional, steady-state flow along a stream-tube This simplified specification of the hydrologic characteristics negates the use of EFLOW and ETUBE.

Level 1 Hydrology requires the user to create input files for submission to EFLOW and ETUBE on the mainframe. The conceptual model that can be developed at this level is depicted in Figure 3 for a vertical crosssection. The soil hydraulic characteristics are treated as homogeneous and the user selects from four soil types provided by the preprocessor. The input file for EFLOW and ETUBE to address more complex problems at Level 2 must be developed outside of the pre-processor

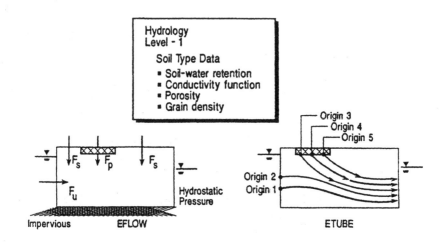

Figure 3 *Conceptual model for hydrology level 1.*

using a text editor. Level 2 problems include multiple soil types, seepage faces, pumping, irregular geometries, and surface evaporation.

Level O Geochemistry involves the development of a conceptual model that represents all geochemical reactions with a constant distribution coefficient (K_d) within the stream-tube. The user provides a value of K_d appropriate for the chemical component of concern. In addition, the user must specify the concentration of the chemical component entering the streamtube and the initial concentration within the tube. In pursuing the problem with Level 0 Geochemistry, the user is not required to execute the parent geochemical module ECHEM.

For Level 1 Geochemistry the user is provided a choice of the groundwater chemical composition that establishes the aqueous phase initial condition. As well, the user selects the boundary condition from a library of leachate compositions that are stored in the pre-processor and can be reviewed by the user prior to selection. The selection of the initial and boundary conditions determines the thermochemical database required and the nature of the speciation/dissolution/adsorption conceptualization of the problem. Again, because of the pre-specification of geochemical initial and boundary conditions, execution of the ECHEM module is not required.

Level 2 Geochemistry requires that the user first run ECHEM to equilibrate the measured chemical composition of the initial and boundary conditions, and produce a site-specific thermochemical database. Here the aqueous chemical composition and the relevant minerals for defining the initial and boundary conditions are provided by the user (and not selected form a library). In addition to developing information for

Level 2 Geochemistry, the pre-processor for ECHEM allows the user to develop input files for stand-alone geochemical equilibration problems.

The use of the FASTCHEM code at various levels for hydrology and

Table 1.*Chemical composition of pond water and ambient groundwater.*

Component	Pond water (ppm)	Groundwater (ppm)
pH	4.1	6.0
Al	8.3	0.2
Cu	2.0	2.0
Fe	1.85	0
B	0.6	0.6
K	11.54	2.74
Si	26.0	14.33
Mn	48.5	0.14
Ca	304.6	6.6
SO_4	137.0	4.03
Ni	0.34	0
Zn	0.39	0.03
Mg	39.26	2.43
Na	8.51	6.44
Cl	3.10	2.98
CO_3	1.0	40.0

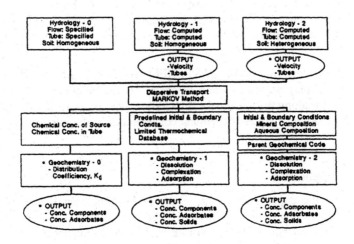

Figure 4 *Multi-level uses of the FASTCHEM code.*

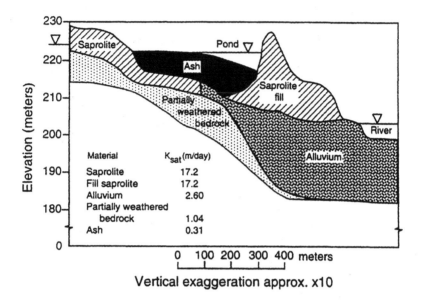

Figure 5 *Geological cross-section of disposal site.*

geochemistry is summarized in Figure 4. This figure highlights the many combinations of hydrology and geochemistry levels available to the user. The selection of a particular combination will depend on data availability, appropriateness of assumptions, and emphases of the study.

Post-processing
The post-processor residing on the FASTCHEM workstation provides the capacity for graphically displaying the variation of chemical concentrations of components, sorbates and solids along the streamtube. In addition, influent and effluent plots show the temporal variation of aqueous chemical concentration at the inlet or outlet of the streamtube.

FASTCHEM application

To demonstrate the capabilities of the code, a two-dimensional, heterogeneous, chemical transport problem was investigated using FASTCHEM. The site investigation was prompted by the presence of a waste disposal pond located approximately 350m from an adjacent river. The waste sluiced to the pond consisted mainly of fly ash, bottom ash, and pyrites from a 400 MW coalfired power plant. Because the pond is unlined, the site assessment focused on the potential for ash

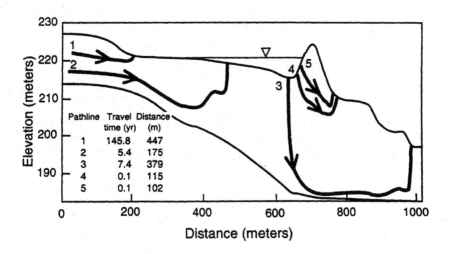

Figure 6 *Pathlines from ETUBE output.*

leachate to enter the subsurface and eventually migrate to the river. In addition, the configuration of the pond with its water surface 22 m above that of the river led one to suspect the existence of a seepage face along the dike built to create the pond. Consequently, this pathway of leachate migration to the river was also investigated in the study.

Hydrologic conditions
The geologic composition of the vertical cross-section to be analyzed is shown in Figure 5. The hydrologic attributes of these four deposits alluvium, saprolite, partially weathered bedrock, ash are necessary to proceed with the flow analysis in FASTCHEM. The soil hydraulic properties obtained from laboratory and field permeameter tests were used in this application.

A hydrostatic head was applied as the upgradient hydraulic boundary condition, the downstream boundary was characterized by the hydrostatic head in the river, and the surface boundary was described by the elevation of water in the pond. The bottom of the cross-section, as well as the non-pond areas at the surface, are treated as no-flow boundaries. However, one should recognize that this no-flow stipulation does not prohibit the calculation of seepage faces along these boundaries. Because a steady state solution will be pursued, initial hydraulic conditions are immaterial.

Geochemical conditions

The chemical composition of the pond water and the ambient groundwater were measured from samples of pond pore water and an up-gradient sampling well. The compositions are listed in Table 1. The elevated concentrations of iron, sulfate,

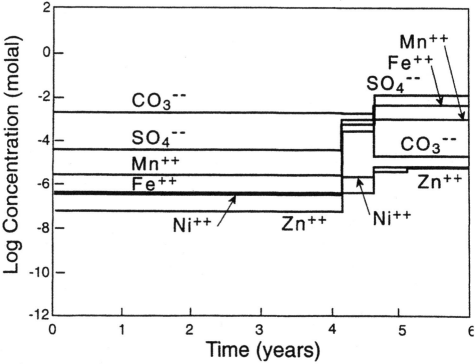

Figure 7 *Concentration of aqueous chemical components at river .*

manganese, calcium and magnesium represented the source of concern for the pond's impact on the subsurface environment and adjacent river.

The initial geochemical conditions in the system are achieved by sampling the many geochemical regimes to establish characteristics of the four material types identified. The results of these field and laboratory efforts are the source of the data used in defining the initial conditions.

Results and discussion

The hydrogeochemical assessment of the threat posed by the presence of the ash disposal facility was performed using the FASTCHEM Code with Hydrology Level 2 and Geochemistry Level 2.

The finite-element grid used to capture the two dimensional heterogeneity of the problem was composed of 947 triangular elements and nodes. The arrangement of elements are geared to capturing the various soil types with a more dense array at the bottom of the pond where difficulties arose during preliminary runs. The tubes simulated are shown in Figure 6. These results indicate the presence of two seepage zones at the surface. The major seep present on the down-gradient face of the dike accounted for 96% of the flow leaving the pond-a significant result in and of itself. The amount of pond leachate making its way to the river via the sub-surface is approximately 4% of the total flow leaving.

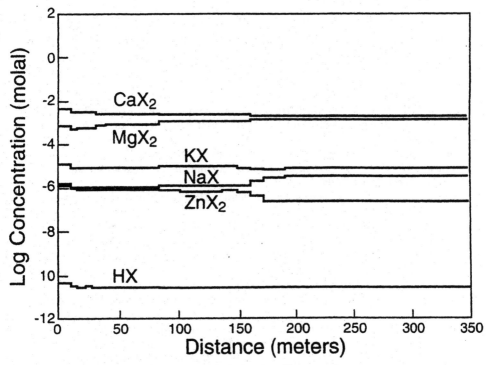

Figure 8 *Concentrations of adsorbents along pathline no. 3.*

The equilibrated initial and boundary conditions, the site specific database, and the streamtube database were submitted as inputs to EICM for an evaluation of streamtube 3. The results of this effort shown in Figures 7 and 8 indicate that the components Fe^{2+} and Mn^{2+} were affected by the advection-disperson process only. On the other hand, the concentrations of K^+, Mg^{2+}, Ca^{2+}, and SO_4^{2-} are all modified by the geochemical processes of adsorpton and precipitation-dissolution. The surface complexaton reactions involving clays and hydrated ferric oxide (HFO) dominated the distribution of Ca^{2+}, Mg^{2+}, K^+, H^+ and SO_4^{2-}. The Ca^{2+}, Mg^{2+}, K^+ and H^+ were found to displace the Mg^{2+} and Na^+ on the clays (see Figure 8). The replacement of H^+ on the HFO sites was done by H_2SO_4 in the acidic leachate.

Applications of FASTCHEM to four case studies were performed by Kitanidis and Freyberg(1992) and Narasimhan and Apps(1992). In the latter publication, FASTCHEM results were compared with results from the DYNAMIX code to ascertain the importance of redox reactions.

Conclusion

The mathematical basis for equilibrium hydrogeochemical modeling was presented along with descriptions of some numerical techniques for achieving a solution to the governing equations. The incorporation of the mathematical framework and numerical techniques to develop the FASTCHEM hydrogeochemical code was highlighted. Finally, the application of FASTCHEM demonstrated the hydrological and geochemical insights that can be gained by using a coupled code to perform a water quality impact assessment.

References

Battelle and GeoTrans.1989. *FASTCHEM Package: User's Guide to EFLOW.* The Groundwater Flow, EA5870, Vol. 2. Electric Power Research Institute, Palo Alto, California

Bear, J. 1979. *Hydraulics of Groundwater.* McGrawHill, New York.

Brinkley, S.R. 1946. Notes on the Condition of Equilibrium for Systems of Many Constituents. *J. Chem. Phys.*, 14, 563564.

Brinkley, S.R. 1947. Calculation of the Equilibrium Composition of Systems of Many Constituents. *J. Chem. Phys.*, 15, 107110.

Criscenti, L.J., Kemner, M. L., Erickson, R.L., Hostetler, C.J. and Morrey, J.R. 1946. *The FASTCHEM Package Workstation: Integration of Pre- and Post-processing Functions*, EA5871, Electric Power Research INstitute, Palo Alto, California. Davies, C.W. 1962). Ion Association, Butterworths, Washington, D.C.

deMarsily, G. 1986. *Quantitative Hydrogeololgy.* Academic Press, Orlando, Florida.

Freeze, R.A. and Cherry, J.A. 1979. *Groundwater.* PrenticeHall Englewood Cliffs, New Jersey.

Hildebrand, F.H. 1956. *Introduction to Numerical Analysis.* McGrawHill, New York.

Hostetler, C.J., Erikson, R.L., Fruchter, J.S. and Kincaid, C.T. 1989. *Overview of FASTCHEM Package: Application to CHemical Transport Problems*, EA5870, Vol. 1., Electric Power Research Institute, Palo Alto, California

Kandiner, H.J. and Brinkley, S.R. 1950. Calculation of Complex Equilibrium Relations. *Ind. Eng. Chem.*, 42, 850855.

Kincaid, C.T. *FASTCHEM Package: User's GUide to ETUBE 1989. The Pathline and Streamtube Database Code*, EA5870, Vol. 3., Electric Power Research Institute, Palo Alto, California.

Krupka, K.M., Erikson, R.L., Mattigod, S.V., Schramke, J.A., Cowan, C.E., Eary, L.E., Morrey, J.R., Schmidt, R.L. and Zachara, J.M/ 1989. *Thermochemical Data Used by the FASTCHEM Package*, EA5872, Electric Power Research Institute, Palo Alto, California.

Moore, D.S. and Yackel, J.W. 1974. Applicable RFinite Mathematics, Houghton Mifflin Co., Boston, Massachusetts.

Morrey, J.R. 1974. *FASTCHEM Package: User's Guide to ECHEM, The Equilibrium Geochemistry Code*, EA5870, Vol. 4., Electric Power Research Institute, Palo Alto, California.

Pye, V.I., Patrick, R. and Quarles, J. 1983. *Groundwater Contamination in the United States.* University of Pennsylvania Press, Philadelphia.

Rai, D. and Zachara, J.M. 1984. *Chemical Attenuation Rates, Coefficients and Constants in Leachate Migration*, EA3356,V1, Electric Power Research Institute, Palo Alto, California.

Summers, K.V., Gherini, S.A., Lang, M.M., Ungs, M.J. and Wilkinson, K.J. 1989. *MYGRT Code Version 2.0: An IBM Code for Simulating Migration of Organic and Inorganic Chemicals in Groundwater*, EN6531, Electric Power Research Institute, Palo Alto, California.

Zienkiewicz, O.C. and Cheung, Y.K. 1965. Finite Elements in the Solution of Field Problems. *The Engineer*, London.

2   Transport of reactive solute in soil and groundwater

Sjoerd E.A.T.M. van der Zee

*Department of Soil Science and Plant Nutrition, Agricultural University
Wageningen, Dreijenplein 10,6703 HB Wageningen, The Netherlands*

Abstract

*The intrinsic heterogeneity of natural porous media, with regard to physical and
chemical properties, gives rise to complicated transport behaviour of reactive
chemicals. For an adequate description of transport, first the scale of interest needs
to be defined. At this scale, the transport equations are formulated and smaller scale
heterogeneity needs to be treated adequately, e.g. by averaging, but it is not taken
explicitly into account. Small scale heterogeneity may give rise to e.g. Freundlich
type adsorption that is locally not in equilibrium. For a one-dimensional flow
domain, this may induce traveling wave type of displacement, provided certain
conditions are met. These conditions are indicated. Spatial variability of properties
and boundary conditions in the horizontal direction in the vadose zone may be dealt
with using either impulse-response functions or stochastic methods. Use of both
approaches (until 1990) is addresses. In particular, spatial variability with regard to
both physical and chemical properties-appears to be important. Chemical
heterogeneity appears to be relatively underexposed in the literature. Different types
of heterogeneity may give rise to fingering phenomena in the unsaturated zone. For
reactive solutes the persistence of fingers as well as the travel time distributions
originating form this type of flow pattern may be of main importance with regard to
solute leaching into groundwater. In groundwater, the main direction of flow is
generally parallel to stratification. The loss of solute by diffusion into low
permeable layers depends on dispersive properties of different layers, as well as
differences in retarded velocities. Different recent advances for perfectly stratified
media and transport parallel to layering indicate that transversal losses may be
significant. Whereas profound advances have been made in solute transport
modeling, both theoretical and experimental research concerning the interaction of
(bio)chemical aspects on transport behaviour in natural porous media is needed.*

Introduction

To describe the fate of contaminants that enter the soil or groundwater, much use has been made of mathematical models. Most modeling was based on the distinction of two modes of displacement, *i.e.*, convective and diffusive/dispersive transport. For reacting compounds the displacement is furthermore subject to chemical interactions (adsorption, precipitation, complexation), uptake and transformations by biota, and transfer between the liquid and the gas phase of the system. Mathematical modeling has resulted in many models for different situations. However, most models have in common that the usually adopted continuum approach is only appropriate when the intrinsic heterogeneity of the porous medium is of sufficiently small scale. When this is not the case, heterogeneity must be accounted for explicitly. In this chapter transport modeling in heterogeneous soils and aquifers is reviewed. Because of the rapid developments in this area, any pretense of completeness would be false. In view of the limited space available, I discuss flow (water movement) only in so far as this is necessary for understanding transport (solute movement). For reviews giving more emphasis to flow I refer to Van der Heijde *et al.* 1985), Nielsen *et al.* (1986) and Van Genuchten and Jury (1987). Gas phase transport was omitted completely. Because details regarding the two-component transport equation and multicomponent transport were provided by Abriola (1987) and Yeh and Tripathi (1989), respectively, they were not reviewed here. For a comprehensive text with regard to stochastic modeling reference can be made to Dagan (1989). Although for heterogeneous transport in groundwater it is somewhat arbitrary to deal with deterministic models while omitting stochastic approaches, I ignored the latter, except in the special case of one-dimensional transport in the vadose zone.

Basic considerations

The continuum approach

To describe flow and transport in porous media, which consist of solid matrix particles separated by pores, usually the continuum concept is used. A continuum is an imaginary medium with the same average characteristics at each point in space. Thus, individual pores and particles are not taken into account because this would imply an enormous effort with respect to modeling and because the necessary data required for modeling pore geometry and dimensions generally are not available. Also, this approach would lead to a too detailed description. By representing the medium with a continuum, a description is obtained that is accurate and sufficiently detailed for our purpose. Small scale heterogeneity that may represent for instance pore size variations or particle roughness is then averaged in the model. This averaging may be done making wrong assumptions. This, in fact, is one of the causes for discrepancies between modeling and experimental results.

Essential for applicability of the continuum concept is that a representative elementary volume (REV) can be distinguished that has the average characteristics of the real porous medium. This REV is much smaller that the dimensions of the system of prime interest. This system contains a large number of REV's. In some cases a continuum cannot be constructed for the medium at hand. This may occur for instance for fractured rock or macroporous soil. When the channels (fractures, macropores) are of similar dimensions as the system it may be necessary to account for them deterministically. Then the governing transport equations for the channel are formulated separately from the porous matrix. Both sets of transport equations are coupled at the specified channel/matrix interface. For the latter case the REV that represents pore-scale variability is of the same dimensions as the porous medium considered. This invalidates the assumption of one continuum.

To replace a porous medium by a continuum requires the averaging of the properties within an REV to macroscopic parameters. Much attention has been given to this averaging procedure (Bear, 1979; Baveye and Sposito, 1984; Cushman, 1984,1986; Quintard and Whitaker, 1988). An appealing illustration with respect to the size of the REV was given by Bear (1972) for porosity and bulk density of a granular medium. In essence the porosity shows a large variation when a volume of the scale of the pore diameter is moved over small distances through the medium. When the volume is increased to a .larger scale this property becomes almost independent of the exact position in the porous medium, and the larger volume could be chosen for the REV. With a further increase in volume, larger scale heterogeneity (*e.g.* due to differences in strata) may be captured in the volume and the property may change significantly when the volume is moved around slightly. The accuracy of calculations and the system's dimensions determine whether this larger scale heterogeneity is averaged in an REV of even larger scale, or modeled as a separate sub-domain. A literature review of recent advances in macroscopic averaging is beyond the scope of this paper. While omitting mathematical details, it is, nevertheless necessary to give some of the background of theoretical work done by Cushman (1984, 1985, 1986, 1987), Baveye and Sposito (1984), and Quintard and Whitaker (1988).

The concept of heterogeneity is intimately coupled with the concept of scale. Modeling efforts are usually constrained to a limited range of scales, and so are measurements. Heterogeneity at scales smaller than the resolution of modeling or measurement is not present explicitly. Heterogeneity at larger scales is recognized as such. Hence, Cushman (1986) distinguished implicit and explicit heterogeneity, respectively. An illustration when modeling flow is the implicit heterogeneity, due to pore size and geometry variability, and the larger scale explicit heterogeneity due to cracks in clayey soils. When input data are gathered using a permeameter, the positioning of the device with respect to cracks controls the measurement. This is not the case for the positioning with respect to individual (micro)pores (Nielsen *et al.*, 1983). Nevertheless, implicit heterogeneity is relevant. One manifestation of implicit heterogeneity is well known in transient flow modeling, when the resolution of flow modeling is not attempted at the pore-scale. Heterogeneity leads macroscopically to the hysteresis phenomenon. The heterogeneity causing hysteresis is "effectively" filtered out of most models and has to be re-introduced again, often on an *ad hoc* basis. Another well known example is mechanic (convective) dispersion. On the macroscopic (continuum) scale of the convection-dispersion equation (CDE), microscopic variations in the fluid velocity are not taken into account explicitly. These variations, however, give rise to enhanced front spreading. To account for these observations convective dispersion is introduced as an additional process into the transport equation (and usually lumped with the diffusion process). Thus, microscopic heterogeneity affects the mathematical nature of the transport equation (when molecular diffusion is left out of consideration). Of some interest is the process-dependent nature of heterogeneity as observed by Quintard and Whitaker (1988) who remark that a porous medium may be homogeneous regarding molecular diffusion, but heterogeneous with regard to convective dispersion.

Whereas implicit heterogeneity is filtered out in the measurement or model, the explicit heterogeneity must be characterized by large numbers of measurements or a fine modeling resolution. This may induce uncertainty when the number of measurements (or nodal points in the computations) is limited. The relationships between scale, heterogeneity, and measurement were discussed in a series of papers by Cushman. A good starting point for reading his rather abstract work is his 1987-paper, that gives examples and an expansion of earlier developments.

Before considering the effect of heterogeneity on transport, a proper definition of this concept is needed. As follows intuitively scale plays a central role. Equally important is the question which features are essential to the process, and which level of detail is needed and possible. The latter aspect usually sets the scale of measurement or modeling. In the porous media-continuum approach the REV-scale implies averaging at the microscopic pore scale and the incorporation of liquid/solid interfaces into the macroscopic equations. Sometimes, this is followed by successive averaging (now always over a liquid/solid continuum) in order to consider larger scales of heterogeneity. Arriving at the scale of prime interest, the system's homogeneity can be evaluated using the definition given by Quintard and Whitaker (1988): "A porous medium is homogeneous with respect to a given process and a given averaging volume when the effective transport coefficients in the volume-averaged transport equations are independent of position". Otherwise it is heterogeneous. Note that this definition involves besides the process also the averaging volume. Instead of a specific averaging volume, weighing functions may be used. Whitaker (1986c) showed that weighing functions, which may differ for each dependent variable just like the REV, do not change the general form of the macroscopic equations but may affect the macroscopic transport coefficients. A theoretical analysis of volume averaging was given by Quintard and Whitaker (1988). A readable account of statistical aspects was given by Dagan (1986). In the latter work a definition is given of the heterogeneity scale in terms of the linear integral scale, which is defined by an integral of the covariance function of a variable. This variable may be of a geometrical or a physical nature.

In summary, we may expect the porous medium of interest (soil, aquifer, ground water basin) to exhibit heterogeneity on different scales. Starting with the pore scale level, we obtain a homogeneous medium by averaging to a larger scale. On a still larger scale, heterogeneity may appear due to lamination patterns. Increasing the scale, lamination heterogeneity may be averaged (and yield anisotropic behaviour), and we may observe larger scale depositional and scouring structures. These structures may consist of lenses with gravel, sand, loam or clay material, which obviously control the hydraulic properties of these structures. At still larger scales we may encounter large-scale lenses and practically continuous layering. Besides the effect of these structures on the transmissive properties, also the difference in chemical reactivity between different materials is well known. The effect of such heterogeneities on transport may therefore be enhanced in the displacement of reacting solute. Clay, primary minerals, metal (hydr)oxides, and organic matter may be equally important with respect to displacement, depending on the solute of interest. A positive correlation between the water transmissive and solute sorbing properties may therefore be found equally well as a negative correlation, depending on the relative occurrence of these solid phase compounds in differently textured strata.

Similar differences in heterogeneity with increasing scale may be observed in the water unsaturated soil. Adopting for the moment a one dimensional vertical flow pattern in the vadose zone different types of heterogeneity may be recognized. Many soils have developed in unconsolidated, horizontally layered material. Because flow is more or less vertical, the main flow direction is perpendicular to layering. This is one of the differences with the situation commonly encountered in groundwater. Moreover, since soil forming processes are significantly affected by the direction of flow, soil horizons are also oriented perpendicular to the mean direction of flow. Two common types of heterogeneity may furthermore be mentioned. Soils containing significant amounts of clay, organic matter, or metal(hydr)oxides may exhibit aggregation. In that case, two distinct pore size distributions (bimodal distributions) control flow and transport. Also, on a somewhat larger scale, macropores may be due to the action of soil biota (*e.g.*. earth worm or root channels) and caused by cracking in case of shrinking soils. Such macropores also affect the response of the soil with respect to flow and transport. Again, besides the differences in water transmission between different flow domains, the chemistry may be different too. An example is the difference in composition and reactivity between different soil horizons.

Averaging microscopic chemical surface heterogeneity
For reacting solutes we may assume that we deal with microscopic heterogeneity both of physical and of chemical properties. Different solid phase minerals with different chemical behaviour are present in soil. For homogeneous mineral particles we may also encounter "non-uniformity". The parameters regulating chemical interactions are not single valued (uniform) but vary sufficiently to be distributed around some mean value. This distribution may have a single modus. However, a bimodal distribution is possible, for instance with clay colloids, that exhibit different characteristics for the planar and the edge sides of the particle.

Non-uniformity of the mineral phase surface is usually of too small a scale to be accounted for in transport modeling, since we do not consider the pore-level processes explicitly. For a macroscopic (system-level) description we therefore need to average the chemistry also. Consider a relatively simple situation, that is relevant for the monocomponent approach adopted here. We assume that the specific interaction between a dissolved species (C) and a surface site (S), with the intrinsic affinity coefficient ($K_{j,i}$) is described by the following reaction equation.

$$S_j^y + C_i^z \underset{\rightarrow}{\leftarrow} S_j^y C_i^z; \qquad K_{j,i} \qquad (1)$$

The equilibrium is defined on the basis of K and the concentration of C at the interface, and C is equal to the bulk concentration multiplied by the Boltzmann factor. The saturation $(\theta_{j,i})$ of sites j with solute i can be given with a Langmuir equation

$$\theta_{j,i} = \frac{K_{j,i}[C_i^z]}{1 + K_{j,i}[C_i^z]} \tag{2}$$

To derive an expression for the surface saturation (Θ_i) by species i for the entire reactive surface we need to add all local (j) saturations weighed with the fractional occurrence of sites j (with coefficient $K_{j,i}$) with respect to the total number of sites. When the probability density function (PDF) for K (or $\log K$) is continuous this yields

$$\Theta_i = \int \theta_{j,i} f_{\log K} \, d \log K \tag{3}$$

In terms of moment theory (Aris, 1959) this can be recognized as the expectation value of the surface coverage. Analytical solutions are feasible for a few special cases, assuming that the surface potential is uniform. Then all sites experience the same overall potential. Omitting details that were presented elsewhere (Kinniburgh *et al.*, 1983; Sips, 1948, 1950; Toth, 1974) the result for three frequency functions (*f*) is given. For a symmetric nearly Gaussian function as given by Sposito (1984) (his equation 6.38) for monocomponent adsorption the Langmuir-Freundlich equation is obtained.

$$\Theta(c) = \frac{(Kc)^m}{1 + (Kc)^m} \tag{4}$$

(Sips, 1948, 1950; Van Riemsdijk *et al.*, 1986). For a right hand side exponential PDF, integrating equation(3) results in

$$\Theta(c) = \left(\frac{Kc}{1 + Kc} \right)^m \tag{5}$$

For a left hand side widened quasi-Gaussian PDF the result is (Van Riemsdijk *et al.*, 1986)

$$\Theta(c) = \frac{Kc}{\{1 + (Kc)^m\}^{1/m}} \tag{6}$$

The parameter m (Sposito's β) represents the degree of microscopic heterogeneity, and its value is restricted to $0 < m < 1$. When $m=1$ the surface is uniform with respect to the reaction affinity coefficient (K), and the above three cases reduce to the Langmuir equation. When m is not equal to 1 and Kc $\ll$ 1, equations similar to the Freundlich-Van Bemmelen equation are obtained (except)

$$\Theta(c) = K c^m \tag{7}$$

for equation(6) which yields a linear equation). In view of the discussion by Sposito (1981) it is worthwhile to note that the Freundlich coefficient in the equation is commonly expressed in terms of $q = Q\,\Theta$ and implicitly accounts for an adsorption maximum (Q). In case of multicomponent adsorption also the behaviour of other competing ions in solution is intrinsic in K. The averaging procedure presented here for one specifically adsorbing solute may be extended to more complicated situations. Such analyses are beyond the scope of this illustration, and reference is made to Van Riemsdijk *et al.* (1986). Other averaging procedures were discussed by Harmsen (1982).

The transport equation for a macroscopically homogeneous medium
During the past few decades much theoretical work has been devoted to the development of the macroscopic transport equation. Rigorous derivations have been provided by Bear (1979, 1987) and Quintard and Whitaker (1988), among others, and will not be reproduced here. For a macroscopically homogeneous porous medium the transport equation is given by Bear (1972)

$$\frac{\partial \theta c}{\partial t} = -\nabla.[\theta v c - \theta D.\nabla(c)] \tag{8}$$

where (θ) is the volumetric water fraction, c is the concentration, t is time, and v is the interstitial flow velocity. Source terms were omitted. Much work has been devoted in the past to describe and understand the dispersion tensor (D). One contribution to this tensor is due to molecular diffusion in the porous medium. This contribution to displacement can be described with Fick's law. In the dispersion tensor, molecular diffusion is partly accounted for explicitly. Partly, diffusion acts on the second process of mechanical or convective dispersion which accounts for variations in the velocity of the fluid (water) in the porous medium pores at the microscopic level. This second contribution is not modeled explicitly but is implicit in the coefficient of mechanical or convective dispersion. Thus, we obtain (Bear, 1979, pp. 232-233)

$$D = D_M + D_{dis} \tag{9}$$

Both contributions to D in equation (9) depend on medium properties such as tortuosity. Furthermore, D_{dis} depends on pore geometry and length and diffusion between streamlines at the microscopic level. Such effects are lumped into the (macroscopically defined) dispersivity (L). The dispersivity is given by a fourth rank tensor, which may be characterized by two parameters in case of isotropy: the longitudinal and the transversal dispersivities. For more complicated cases consult Bear (1979, p.234) or the literature referenced in that work. Of main interest is that when one of the principal coordinates is directed in the mean flow direction at the macroscopical level, the dispersion coefficient is given by

$$D_{11} = L_{11}\bar{v}; \quad D_{22} = L_{22}\bar{v}; \quad D_{33} = L_{33}\bar{v} \tag{10}$$

for an isotropic medium. In (10) the direction of flow is given by '11'. Non-diagonal terms of the 3x3 mechanical dispersion matrix are zero for that case. An important practical consequence of equation (10) is therefore that mechanical dispersion is dependent on the average flow velocity whereas that is not the case for the molecular diffusion contribution of D. This dependency enables one to assess the parameters of equations (9) and (10) by variation of the flow velocity.

Transport in homogeneous media

The mass balance equation for transport (8) has been the basis of many studies on transport in porous media. Most work, using data obtained from column leaching experiments in the laboratory, considered one-dimensional transport. Early examples in the soil literature were provided by Nielsen and Biggar (1961, 1962) and Biggar and Nielsen (1962). Analytical solutions could often be taken from the existing literature on diffusion (Crank, 1956) and heat conduction (Carslaw and Jaeger, 1959). Because most solutes of interest reacted with the solid phase, the incorporation of reaction terms was attempted at an early stage (Van der Molen, 1956). Because of the simplicity to account for linear adsorption which affects the solutions for non-reactive transport only by a constant, mostly linear adsorption has been assumed. Because adsorption chemistry is essentially a multicomponent process (Abriola, 1987), the assumption of linear adsorption is only valid in special cases. Examples where this is the case are when the solute adsorbs with high specificity, or when the concentrations of the solute of interest are small with regard to the concentrations of other competing solutes (if Kc is small).

For some solutes, such as nitrate, pesticides, and chlorinated hydrocarbons, (bio)chemical transformation processes must be taken into account. These were commonly described by zeroth or first order irreversible rate processes. The latter was also used to describe radio-active decay. Mathematical tools (*e.g.*. Laplace transform methods) of use for non-reactive transport can also be used if zeroth or first order loss rates are added to the transport equation. Consequently, a large number of analytical solutions have been derived or adapted from existing literature to model transport with linear adsorption and with the above mentioned rate processes. A compilation was given by Van Genuchten and Alves (1982), and Javandel *et al.* (1984). Solutions for multidimensional transport were provided by Bear (1979), and Van der Heijde *et al.* (1985).

Most experimental effort concerned non-reacting solutes to assess the dispersivities that characterize the transport behaviour when the macroscopic flow field is well known. At the laboratory scale many experiments have been performed. The data collected by Perkins and Johnston (1963), Pfannkuch (1963), Klotz and Moser (1974) and Shamir and Harleman (1966) supported theoretical analyses as reviewed by Bear (1979) and Bolt (1982), that the longitudinal dispersivity is for granular media of the order of the grain size. Transversal dispersivities may be one or several orders of magnitude smaller. (De Josselin de Jong, personal communication, 1987).

With growing field data of transport, it became clear that quite different values of dispersivities were needed to describe transport in the field. The dispersivities appeared to be scale (or time) dependent. As will be discussed below, the large values of field scale dispersivities (which may be orders of magnitudes larger than those measured at the laboratory scale) may arise from matching the spatial or temporal behaviour of solute plumes with models that assume homogeneity. Because soils and aquifers are heterogeneous, it became necessary to leave the concept of homogeneity and to develop models that take heterogeneity or spatial variability explicitly into account.

Transport in heterogeneous soil systems

Soil differs from groundwater reservoirs in several ways. In this review, soil is defined as the porous medium in the water unsaturated (vadose) zone above ground water. The hydraulic head is usually negative, when the free water table is taken as the zero hydraulic head reference. At the field scale the boundaries for flow are the soil surface and the freatic water table. Depending on the flow boundaries, flow will be on average vertically downward in areas with net-recharge or vertically upward, in dry regions with shallow groundwater. Increasing the resolution in time or space by looking at smaller time or spatial scales, the flow process becomes more complicated. Transient flow with varying directions and more complicated flow patterns than the one dimensional case may be observed. Significant horizontal flow components may arise due to layering or flow to roots, rootchannels and cracks. On a yet smaller scale flow around aggregates may be observed. To take such complicated three-dimensional transient flow patterns into account requires numerical solutions. Exceptions to this general rule are well-defined systems that exhibit symmetry, such as considered by Clothier and Elrick (1985). In this section I consider several types of heterogeneity relevant for the unsaturated soil. The three-dimensionality of flow in soil is not taken into account. Instead vertical one-dimensional flow is assumed.

Mathematical formulation of one-dimensional flow and transport
The mathematical analysis of flow in soil is usually based on the solution of the one-dimensional Richard's equation, which reads

$$C(H)\frac{\partial H}{\partial t} - \frac{\partial}{\partial z}\left[K_w(H)\frac{\partial H}{\partial z}\right] + \frac{\partial K_w(H)}{\partial z} = 0 \qquad (11)$$

where H is the pressure head, C is the slope of the soil water retention curve, θ (H), K_w (H) is the hydraulic conductivity and z is the vertical spatial coordinate taken positive downward.

To solve equation(11) we need to provide initial and boundary conditions in terms of water potential or volumetric water fraction, as well as the retention curve $(H(\theta))$ and the $K(\theta)$- relationship. Defining the saturation (Θ)

$$\Theta = \frac{\theta - \theta_r}{\theta_s - \theta_r} \tag{12}$$

where θ_r and θ_s denote the irreducible and saaturated volumetric water fraction, respectively, two successful models may be used. The relationship of Brooks and Corey (1964) is given by

$$\Theta = \left(\frac{H_a}{H}\right)^{\beta 1}; \qquad \frac{K_w(H)}{K_s} = \left(\frac{H_a}{H}\right)^{\beta 2} = \Theta^{\beta} \tag{13}$$

where $\beta = \beta_2 / \beta_1$. The values for β_1 and β_2 are in the range of 0.25-0.5 and 2.0-3.5, respectively. The value of β is approximately 7.2 and has been shown to be rather constant for different soils (Bresler and Dagan, 1979). An alternative is the retention curve relationship suggested by Van Genuchten (1980)

$$\Theta = [1 + |bH|^n]^{-m} \tag{14}$$

When we assume that $m = (n-1)/n$, we can combine (14) with the conductivity model given by Mualem (1976) in closed form. This yields

$$K_w(\Theta) = K_s \Theta^{\varepsilon} \{1 - (1 - \Theta^{1/m})^m\}^2 \tag{15}$$

The parameter (ε) in equation(15) equals 0.5 for many soils (Mualem, 1976).

The solution of the above set of equations has been the subject of a large number of studies in the area of soil physics. It is beyond the scope of this review to summarize the main results. Some interesting studies that combine transient flow with transport for macroscopically homogeneous columns were reviewed by Van der Zee and Van Riemsdijk (1990). Of interest may be that the transient nature of flow on the breakthrough curve could be dealt with elegantly by considering the discharge from columns as the time variable rather than time itself (Wierenga, 1977). The effect of hysteresis, neglected in (13)-(15) appeared to be important in the work by Jones and Watson (1987) and Curtis *et al.* (1987).

To emphasize solute transport, let us assume that flow is steady. In that case the solution of the flow equation yields the flow velocity and water content, which serve as input for the transport equation. For a macroscopically homogeneous soil column with one-dimensional flow the transport equation is given by

$$\frac{\partial \theta c}{\partial t} = \frac{\partial}{\partial z}\left(\theta D \frac{\partial c}{\partial z}\right) - \frac{\partial \theta v c}{\partial z} \tag{16}$$

When θ and v are constant (16) becomes after division by (θ)

$$\frac{\partial c}{\partial t} = D \frac{\partial^2 c}{\partial z^2} - v \frac{\partial c}{\partial z} \tag{17}$$

This diffusion equation is also called the Convection Dispersion Equation (CDE). Of particular interest has been the value of the diffusion/dispersion coefficient, D. This coefficient accounts for diffusion in the liquid filled soil pores, as well as for the microscopic heterogeneity that is not accounted for explicitly (in a deterministic fashion) in the macroscopic equation. The contribution of dispersion to D, is given by D_{dis} in the equation (9) and describes the additional mixing caused by Poisseuille type flow in individual pores. Hence, D_{dis} describes the flow distribution within pores, differences in local velocities for differently sized pores, and effects arising from the tortuosity of paths of solute molecules in the pore system. It was found experimentally for large columns that dispersion can be described similarly as molecular diffusion. The molecular diffusion coefficient in soil

$$D_M = \alpha D_M^o \tag{18}$$

is the coefficient under standard conditions in water, corrected by a factor (α) to account for tortuosity effects. Even if molecular diffusion were insignificant, microscopic heterogeneity requires that a dispersion term as in equations (16) or (17) is taken into account. Although microscopic heterogeneity is not modeled deterministically it affects the mathematical formulation of the transport process. The scale of microscopic heterogeneity is important and quantified with the dispersion coeffient, D_{dis}. As mentioned above, this coefficient is related to medium properties and the flow velocity (see equation(10))

$$D_{dis} = L_D v; \qquad L_D = L_{11} \tag{19}$$

However, the dispersivity of a heterogeneous medium reflects the characteristic scale of the heterogeneity as well as the physical transport processes involved. Unless this is realized, fitted L_D -values are little more than fitting parameters.

Effect of small-scale heterogeneity on transport
Much of the experimental work on transport in the vadose zone considered non-reacting or linearly reacting solutes displaced in homogeneous columns. The linear adsorption equation on a mass basis and the linear retardation factor (R_l) are given by

$$s = K_d c; \qquad R_l = 1 + \frac{\rho}{\theta} K_d \qquad\qquad (20)$$

respectively. Using equation (20) the transport equation for steady state flow is

$$R_l \frac{\partial c}{\partial t} = D \frac{\partial^2 c}{\partial z^2} - v \frac{\partial c}{\partial z} \qquad\qquad (21)$$

When the assumption of linear adsorption is not valid equation (21) may yield poor results. Some examples were discussed by Bolt (1982) in relation to ion exchange. Equation (21) may also be inadequate to describe experimental data when the assumption of local equilibrium sorption (assumed in equation (21)) is invalid. Sorption as estimated from batch experiments involving large contact times may then lead to an overestimation of R-values; both first breakthrough sooner than expected and considerable breakthrough tailing may result, that is not described by equation (21). To account for the time dependence of sorption the two-component models were developed. These models are also called two-site models in case two ensembles of sorption sites were postulated, or dual-porosity or mobile/immobile models in case two pore regions were assumed. In general two conceptual regions for solute storage are considered with different (chemical or mass transfer) kinetics and different storage capacities. Mathematically, the various two-component models are similar (De Smedt and Wierenga,1979; Nkedi-Kizza *et al.*, 1984). Reviews of the two-component model with linear sorption in both regions given by Van Genuchten and Cleary (1982), Nielsen *et al.* (1986), and Van der Zee and Van Riemsdijk (1990). These reviews also mention similar models that specify the geometry and size (distribution) for the regions in the macroscopically homogeneous flow domain.

A recent development that also illustrates the behaviour of the two-component models was given by Van der Zee (1990a). In this paper one region has a linear sorption equation and is locally at equilibrium, while the second region is not at equilibrium and has a non-linear sorption equation (either Langmuir or Freundlich equation). Using for the second region, a Freundlich equation on a volumetric basis $\theta = (\rho) s$

$$q = K c^n; \qquad 0 < n < 1 \qquad\qquad (22)$$

in some aspects a different transport behaviour as compared to the linear two-component model is found. The two-component model is given by

$$\dot{R_l} \frac{\partial c}{\partial t} = D \frac{\partial^2 c}{\partial z^2} - v \frac{\partial c}{\partial z} - \frac{1}{\theta} \frac{\partial q}{\partial t} \qquad\qquad (23a)$$

$$\frac{\partial q}{\partial t} = k_r [f(c) - q] \tag{23b}$$

The only difference between the models of Van Genuchten and Wierenga (1976) and Van der Zee (1990a) is that $f(c)$ is linear in the former and described by equation (22) in the latter case (for the analytical solutions). For simplicity, set $R_1 = 1$, although this was not done in my original paper. As was recently proven by Van Duijn and Knabner (1991) a travelling wave type of displacement is found when (22) holds and the feed concentration (c_0 is larger than the initial resident concentration (c_i). Such travelling waves arise due to the opposing actions of non-linear sorption and dispersion on the front shape. Non-linearity leads to steepening of the front whereas dispersion tends to flatten fronts. When both effects are opposite but equally large, a front develops with a constant shape and a constant velocity through a (macroscopically) homogeneous column. Analytical solutions for the shape and position of the front were given for the travelling wave in case of non-equilibrium (Van der Zee, 1990a), thereby generalizing the results of Van der Zee and Van Riemsdijk (1987), and of Bolt (1982) for similar (exchange) problems. Introducing the dimensionless variables

$$Z = \frac{z}{L}; \quad T = \frac{vt}{L}; \quad P = \frac{vL}{D}; \quad Da = \frac{k_r L}{v} \tag{24}$$

travelling wave solutions are found after transformation according to

$$\eta = Z - \alpha T, \tag{25}$$

In equation (25), α is the inverse retardation factor for the non-equilibrium sorption process. The transformed problem

$$\frac{dc}{d\eta} = \frac{1}{\alpha}\frac{dc}{d\eta} - \frac{1}{\alpha P}\frac{d^2 c}{d\eta^2} - \frac{1}{\theta}\frac{dq}{d\eta} \tag{26a}$$

$$\frac{dq}{d\eta} = -\frac{Da}{\alpha}[f(c) - q] \tag{26b}$$

was solved after specifying $f(c)$, for the infinite system. As an example the Freundlich front shape is given:

$$c(\eta) = c_0 \left\{ 1 - \exp\left[\frac{(1-n)Kc_0^{n-1}\alpha}{\theta A}(\eta - \eta^*) \right] \right\}^{\frac{1}{1-n}} \tag{27a}$$

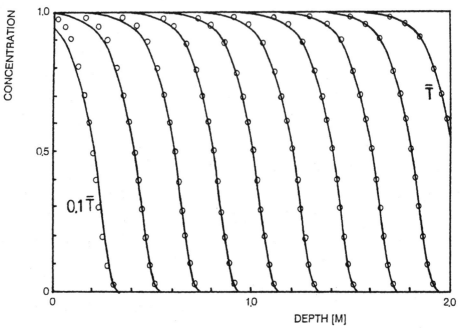

Figure 1 *Fronts for cadmium for 10 times using parameter values $v = 1$ myr^{-1}, $\theta = 0.45$, $\rho = 1,350$ kg m^{-3}, $L = 2$ m, $k_r =$ infinity, $K = 0.00636$ $m^3 mol^{-1}$, $n = 0.65$, $c_0 = 0.02$ mol m^{-3}, $c_i = 0$. Different from the case given by Van der Zee (1990) the dispersivity $L_D = 0.03$ m instead of 0.12 m. Shown are numerically obtained fronts (circles) and the analytical travelling wave solution (solid line). Curves for 0.1 t till 1.0 E with E is mean breakthrough time (left to right), and concentrations are divided by c_0.*

$$A = P^{-1} + \frac{\alpha(1 - \alpha)}{D\alpha} \qquad (27b)$$

The parameter A represents an apparent inverse Peclet number, that accounts for the non-equilibrium effects. The similarity with the linear two-component model is that the apparent dispersion coefficient found by Valocchi (1985) is equal to the apparent coefficient resulting from A. The front thickness that can be calculated with (27) is for larger flow domains smaller than for the case of linear sorption (for which (27) does not hold, since no travelling wave develops in that case) except when the front has moved small distances (Van der Zee, 1990a). The front shapes shown by Van der Zee (1990a) illustrate the steep fronts that develop as non-linearity becomes more profound. Whereas (27) holds in theory for infinite times, Figure 1 shows that it gives a good approximation for depths greater than about 10-20 times A. In terms of the effective dispersivity that accounts for non-equilibrium effects, the travelling

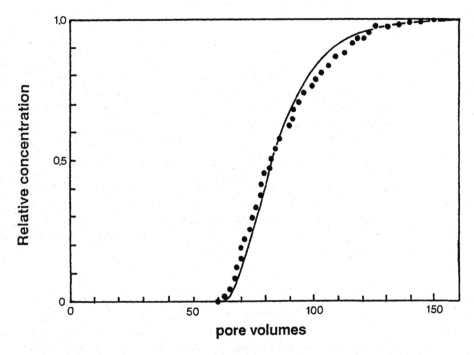

Figure 2 *Predicted (line) and measured (points) breakthrough curves for cadmium. Concentration relative with respect to c_0 (see Figure 1). Parameter values: as Figure 1, but $v = 333 \ m \ yr^{-1} \ L_D = 1 \ mm, \ k_r = 1{,}000 \ yr^{-1} \ L = 11 \ cm, \ c_0 = 0.018 \ mol \ m^{-3}$.*

wave solution describes numerically obtained fronts well for depths (z) given approximately by

$$z > 4\sqrt{L_D} \qquad\qquad (28)$$

When a solute adsorbs according to the Freundlich equation, as may be the case for heavy metals like cadmium (Harmsen, 1977; Christensen, 1981; Van der Zee and Van Riemsdijk, 1987; Boekhold *et al.*, 1990), the breakthrough curve may be described using equation (27) subject to some additional assumptions. Simultaneous fitting of the analytical equation with k_r as the only adaptable parameter for different c_0 -values, gave results as shown in Figure 2. The evaluation of the analytical approximation is much simpler than numerical computations, and therefore, worthwhile as a first analysis.

Of interest is the behaviour for the zero initial concentration assumed to derive equation (27). Zero concentrations appear to be found at finite distances (equal to η^*) from the mean front. If instead of equation (23b) the non-equilibrium process is described by

$$\frac{\partial c}{\partial t} = k'_r \left[c - \left(\frac{q}{K}\right)^{1/n} \right]$$

(23c)

as is also done frequently, a different behaviour results as was shown by Van Duijn and Knabner (1991). Equation (23c) can not be derived from equation (23b). When adsorption is given by the linear or the Langmuir equation, (23b) and (23c) are equivalent, *c.f.* Boesten (1986). Although both (23b) and (23c) have the Freundlich equation for the LEA-limit, Van Duijn and Knabner (1991) found that for equation (23c) a zero concentration is located at infinite distance downstream of the mean front (*i.e.*, at $(\eta)\infty$. This shows that equation (23c), conceptually related to micropore diffusion, and equation (23b), conceptually related to pore-surface diffusion, are mathematically essentially different and should never be used as freely exchangeable (Van Duijn *et al.*, 1993).

Effect of layered soil on transport

The presence of different soil layers and horizons is one of the best documented types of soil heterogeneity. These variations are one of the foundations of soil mapping. With respect to flow and transport research, the effect of layering has remained underexposed. With regard to transport it is of immediate interest whether soil layering affects breakthrough. Intuitively, it is expected to have large effects when layering influences the one-dimensionality of flow. A special case where this may occur is discussed below. When flow remains one-dimensionally vertical, which suggests perfect horizontal stratification, it is of interest whether layering order affects breakthrough at the groundwater level. Analyses of this aspect are scarce. An early analytical and experimental treatment was provided by Shamir and Harleman (1966, 1967) who used a system's analysis approach. They made the assumption that different layers are independent with regard to solute travel time. Thus, in essence they postulated the presence of different mixed reservoirs (the different layers) set in series. For such reservoirs the mean travel time and the travel time variance (related to the dispersion coefficient) are additive. Each layer's response serves then as the boundary condition for the downstream layer. This allowed a solution based on convolution, as used also by Gureghian and Jansen (1985) for a three-member radionuclide decay chain. Shamir and Harleman (1966) observed that the order of layering did not affect breakthrough significantly.

This interesting result was further elaborated upon by Barry and Parker (1986) who used various analytical approaches. They assumed that the flow velocity, the dispersion coefficient and the linear retardation factor differed for each layer. When for a two-layered system an "equivalent" single layer with the same mean and variance of the travel time distribution exists, then the "equivalent" moments (Aris, 1958) are easy to derive. For the two-layer system the moments can be assessed using the flux concentration Laplace domain solution (Barry and Parker, 1986). The two-layer system has by definition the same first and second moment as the

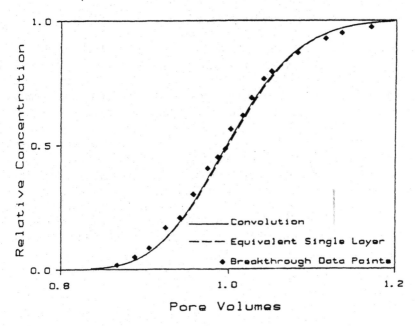

Figure 3 *Agreement between theoretical and measured breakthrough curves for the case layering order is unimportant (Reproduced from Barry and Parker, 1987, who used data of Shamir and Harleman, 1966).*

equivalent layer and, if the third (skewness) moment is also the same, an equivalent layer is virtually indistinguishable from the two-layer system. A third solution used by Barry and Parker (1986) was based on convolution of the two layers, and was expected to give good results when an "equivalent" single layer exists. Thus, when the "equivalent" layer exists, the exact (Laplace) and the convolution result are similar. In that case they observe that the order of layering does not affect the breakthrough curve significantly. Equivalent parameters for the layered soil can then be assessed using the expressions given by Barry and Parker (1986). By consecutive lumping (first two layers, the result with a third layer, *etc.*) a multilayered system can be treated in the same manner. They gave a simple guideline to assess whether the three approaches will give the same response. According to this guideline the equivalent layer peclet number should be approximately equal to the sum of the peclet numbers of the two layers. When one layer is relatively long compared with the other one, layering order is unimportant. An example for which layering order is unimportant was given by Shamir and Harleman (1986), see Figure 3. Interesting is also the extension given by Barry and Parker (1986) using the convolution approach to incorporate the two-component transport equation. With experimental results of the two-layer problem of Panigatti (1970) where immobile water was observed, they showed that in that case layering order was also not important.

A numerical analysis of transport in multilayered soils was provided by Selim *et al.* (1977). Their observation that layering order did not affect results for breakthrough can be understood with the analytical analysis of Barry and Parker (1986). Puzzling, however, is that Selim *et al.* (1977) concluded that layering order is also unimportant for Freundlich adsorption ($n = 0.7$). I attribute this result, no doubt accurate for their results, to the small peclet number assumed for the non-linear layer, that prevents non-linearity effects to be clearly manifested. I will use a hypothetical result to show that layering order should have an effect. Consider, when LEA is valid, two layers with equal L_D-values of which one sorbs linearly while the other one sorbs non-linearly. From the travelling wave discussion of the previous section (3.2) and Van der Zee (1990) (who used a similar n-value), a steep travelling wave is expected when the non-linear layer has a thickness of say $16/P$ (with P the peclet number). When this steep front enters the subsoil, which is non-reacting, we may approximate the subsoil front as having an upstream boundary condition described by a step change. For the small dispersivity assumed for the first layer, the front in the subsoil is given by

$$\bar{c} = \frac{c - c_i}{c_0 - c_i} = 0.5 \, erfc\left[\frac{\bar{z} - v\bar{t}}{2\sqrt{D\bar{t}}}\right] \tag{29}$$

where time and depth refer to the residence time and the thickness of the linear layer. The thickness of the front described by (29) is defined as the δ_l distance between $\bar{c} = 0.16$ and $\bar{c} = 0.84$. This thickness equals $\delta_l = 2\sqrt{(2 \, L \, L_D)}$ and is about 11 times the dispersivity when $L = v\bar{t} = 0.5$ m and L_D is 3 cm. When we reverse the layering order, a front as given by equation (29) arrives at the interface between the two layers and enters the non-linear subsoil layer. In view of the results obtained by Van der Zee (1990a) the travelling wave has developed by the time the mean front leaves the subsoil layer. That instead of the case of Van der Zee (1990a) the input condition for the non-linear layer is not a step change of concentration does not affect the development of the travelling wave (Van Duijn and Knabner, 1990, Van Duijn *et al.*, 1993). Hence, for this reversed case the front thickness is controlled by the non-linear downstream layer and significantly smaller than 11 times the dispersivity. When parameters are chosen such, that $\delta_l = 30$ cm for a front depth of 58cm, the front thickness that would be found in case of non-linear sorption ($n = 0.66$ as used by Selim *et al.* (1977)) would be only 19 cm. Observe that the linear front is still flattening as the front gets deeper (Bosma and Van der Zee, 1992).

In conclusion, it is unlikely that the reversal of the layering order does not affect breakthrough in case of non-linear sorption. For these cases the analyses of Shamir and Harleman (1967) and Barry and Parker (1986) do not hold. Generalizing the above results, the mean breakthrough time for a multilayered system is found by summation of the residence times in each layer, *i.e.*,

$$t_n = \sum_{i=1}^{n} R_i L_i / v_i \qquad (30)$$

with i the layer number. When the last layer exhibits non-linear sorption and is thick enough, it controls the shape of the front at the end of the last layer. If it is thick enough and exhibits linear sorption, while the $n-1$ th layer is non-linear to such an extent that it yield almost a step front, the response is controlled by equation (29), with appropriate rescaling of z and time as indicated in (29) and by Bosma and Van der Zee (1992). When neither constraint is met, a simple analytical prediction may not be feasible.

Transport in spatially variable soils
When transport models as discussed in the previous sections are used to evaluate transport in the field, predictions yield often poor results (Beek, 1979). This is even the case when use was made of dispersion coefficients adapted for aggregation. One of the causes for the discrepancies between model and experiments for the field situation is the difference in heterogeneity between columns in the laboratory and field soils. Not only the level of heterogeneity differs for these two systems, but also the scale of heterogeneity. In most cases, the dispersion coefficient and other parameters, as obtained in the laboratory reflect only pore scale or soil sample scale heterogeneity. However, in the field or regional scale, larger scale heterogeneity is found as may be mapped in soil maps. In fact, much of the literature to be reviewed in this section suggests that in many cases the effect of pore or sample scale heterogeneity is secondary with respect to field scale heterogeneity. Moreover, it is often of a too fine resolution to be taken explicitly into account. For the field scale, the transport process therefore has to be appropriately averaged. One of the main working hypotheses, so far, has been that such averaging has to be done for the transport in many imaginary columns, that are parallel and non-interacting with each other. To deal with such averaging when only the mean response of the unsaturated soil system is of importance (*i.e.*, the areally averaged process is of importance, instead of the process as a function of the horizontal coordinates x and y) use was made of stochastic theory. An extension to uncertainty of predictions was given by Destouni (1992). Basic to this theory is that the assumptions of stationarity (parameters such as mean and variance are considered to hold for all positions) and ergodicity are invoked. A spatially variable stochastic process $G(x,y,z)$ is characterized by a probability density function (PDF) at each position (x,y,z). When measuring, of course only one realization is obtained. Stationarity then implies that the PDF is independent of the position (x,y,z). Ergodicity implies that the statistics derived from the one ensemble for the entire domain of interest adequatly characterize the statistics (mean, variance, *etc.*) for each PDF at each point (x,y,z). An exception is where different subdomains are distinguished. Thus, three different layers may be distinguished in a model, and stationarity and ergodicity are assumed to hold for each subdomain to be characterized by a single PDF.

Coarse-textured soil: 01

silt	(%)	8.0
organic matter	(%)	0.5
bulk density	(g/cm³)	1.52
M50	(μm)	140

a

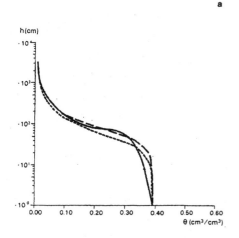

K(cm/day)

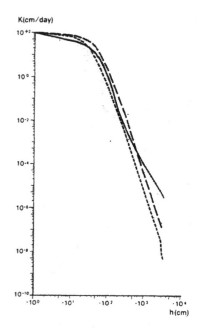

h(cm)

Medium-textured soil: B8

clay	(%)	12.8
organic matter	(%)	3.5
bulk density	(g/cm³)	1.39

b

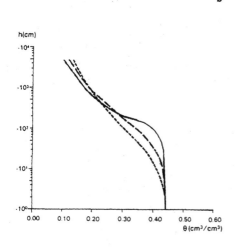

K(cm/day)

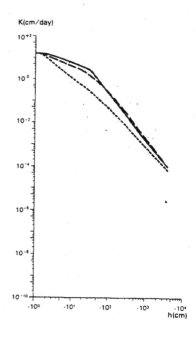

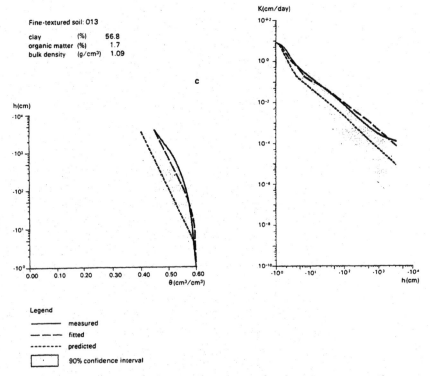

Figure 4 *Scaled retention and conductivity curves for three important soil texture groups of The Netherlands (Reproduced from Wösten and Van Genuchten, 1988).*

A good starting point for the present discussion may be the work by Nielsen *et al.* (1973) and Biggar and Nielsen (1976). They measured flow and transport in a field at a large number of locations. They established that the flow rate differed significantly at different locations. Also the dispersion coefficient varied as a function of horizontal position (x,y). Both flow rate and dispersion coefficient appeared to be reasonably lognormally distributed. This observation can be related with the fundamental work by Miller and Miller (1955, 1956), who developed the scaling theory of similar media. Specifically, the scaling parameter seemed to be lognormally distributed.

The similar media theory shows that when inert soil particles and their configuration and geometry are the same for two different ensembles, whereas the absolute size of pores and particles differs, we deal with similar media which have a different characteristic length scale (size λ). When we denote one ensemble with the subscript m (for mean) and the other with the subscript i, the magnification factor is given by

$$\alpha_i = \frac{\lambda_i}{\lambda_m} \tag{31}$$

The factor α is called the scaling parameter. Assuming Poisseuille flow, Miller and Miller (1955, 1956) showed that the retention curve, H (θ) and conductivity curve, $K(\theta)$ can be scaled at designated (constant) θ . In particular::

$$H_i = H_m / \alpha_i \tag{32}$$

$$K_i = \alpha_i^2 K_m \tag{33}$$

Thus, the pressure head and the conductivity can be directly found from the scaling parameter and mean (subscript m) values of H and K, provided the saturation is the same. It reduces the number of parameters needed for characterizing soil hydraulic behaviour, when soils can be scaled. Consequently, much effort has been given to scale soil hydraulic behaviour, usually leading to a lognormally distributed scaling parameter. When this parameter is lognormally distributed, the head and conductivity are also lognormally distributed at a particular volumetric moisture fraction or, moisture saturation, θ . Interesting papers in this respect were provided by Warrick *et al.* (1977), Nielsen *et al.* (1983), Russo and Bresler (1980), Wösten and Van Genuchten (1988), and Hopmans (1987). The publications on this matter showed the use of scaling with respect to compressing the information of hydraulic behaviour into a small number of parameters (see Figure 4). As Hopmans (1987) showed, scaling according to different procedures may yield different results. Of course, different measurement techniques also yield different results.

It may be clear that when the scaling parameter is distributed due to variability in the horizontal plane in the field, also flow and recharge may be distributed. This is immediately realized for ponding conditions at saturated conditions. Then the flow rate differs from place to place depending on the saturated conductivity and the local scaling parameter. To my knowledge Dagan and Bresler (1979) and Bresler and Dagan (1979) were the first to quantify the effect of spatially variable unsaturated flow on solute transport. For their analysis they used scaling theory, allowing the scaling parameter and recharge to vary randomly in space *(x,y)*. The scaling parameter was considered to be lognormally distributed, while recharge was rectangularly distributed. Ponding conditions were allowed as well as unsaturated flow elsewhere. In their stochastic approach, Dagan and Bresler evaluated the expected value of the concentration as a function of depth and time. They postulated that in many cases this average behaviour is of more direct interest that the local response, *i.e.,* the concentration front at a particular position *(x,y)*. Calculating the concentration front depth at each position would reveal too much information, but also require much more with regard to the data necessary for model input. Pore scale dispersion was neglected by Dagan and Bresler (1979), and a solution was obtained for the first moment of the local concentration, *i.e.,*

$$< \bar{c}(z,t) > = \int \bar{c} f(z,t;\bar{c}) d\bar{c} \tag{34}$$

assuming steady flow. In equation (34) the averaged concentration (normalized to one) at depth (z) and time (*t*) is defined. The PDF of the concentration is given by *f*. Using the chain rule (34) can be recast into

$$<\bar{c}(z,t)> = [\bar{c}Pr(\bar{c})]_0^1 - \int Pr(\bar{c})d\bar{c} = 1 - \int Pr(\bar{c})d\bar{c} \qquad (35)$$

as zero pore scale dispersion was assumed. This expression equals the probability that the front has passed a particular depth, *i.e.*,

$$<\bar{c}(z,t)> = 1 - Pr\left[v \leq \frac{z}{t} \right] = Pr[vt > z] \qquad (36)$$

The distribution function Pr(c) in equation (36) gives the probability that *c* is lower than a particular value. Calculations showed that spatial variability may result in a profound " field scale dispersion ". Field scale averaged fronts may significantly deviate from Fickian behaviour. Therefore, a "field-equivalent" column that can be described by the CDE with average parameter values and that reproduces the transport in a spatially variable field, may not always be found. In later work, Bresler and Dagan (1983) showed that pore scale dispersion is often of secondary importance.

Following the pioneering work of Bresler and Dagan (1979, 1981, 1983), others likewise analyzed the effect of spatially variable flow on non-interacting solute transport (Amoozegar-Fard *et al.*, 1982, Persaud *et al.*, 1985). They did not relate the PDF of v with scaling theory and used Monte Carlo simulations to illustrate convective-dispersive transport in heterogeneous fields. Their conclusions agreed with those of Bresler and Dagan.

Of more interest are studies proceeding along different lines or extending the already developed theory. In this respect the Transfer Function Model (TFM) approach of Jury (1982) is worth mentioning. His working hypothesis was that complicated processes in real soil environments are incompletely understood or often hard to parametrize. Obviously, an incomplete description limits the possibilities of a deterministic modelling approach. Instead, Jury (1982) argued that the overall behaviour may be adequatly characterized by measuring the travel time density function. The PDF of travel times (or residence times) accounts implicitly for all processes occurring in the system. The approach is a special case of the impulse response systems analysis for linear systems as known from hydrology (*e.g.*, the Instantaneous Unit Hydrograph, IUH, concept) and used for solute transport in chemical engineering (Levenspiel, 1972). Jury's (1982) initial system was characterized by a lognormal travel time PDF, stationary flow, input at the soil surface and losses at a particular depth for a non-reacting solute. By characterizing the PDF for simple initial and boundary conditions the impulse-response function is obtained for the system, from which the response for more complicated situations can be derived by convolution. A significant advantage of this line of work is that it considers the flux rate at the surface and the breakthrough of flux concentrations at a depth of interest. For many situations the breakthrough of fluxes is more interesting than the changes of resident concentrations at a given depth. A disadvantage is that the PDF may be depth-dependent, when various layers have different hydraulic or dispersive behaviour. Extrapolation of a measured PDF to different depths than used to assess the function may then lead to errors. With regard to providing a description of transport in the field or in a column, the approach could yield good results (Jury *et al.*, 1982).

The similarities between the TFM-approach of Jury and the work by Bresler and Dagan became more apparent in later work (Jury, 1983), where the TFM was related to soil properties. In the last paper, the effect of spatially variable sorption (accounted for via the retardation factor) was illustrated, and shown to be potentially significant. To my knowledge, Jury (1983) was the first considering spatial variable flow as well as sorption. The case of sorption with a single valued (non-random) retardation factor is a trivial extension of stochastic models for non-reacting solutes.

The TFM-approach was considered in more detail by Jury *et al.*(1986). Denoting solute input times in the transport volume by primes and solute lifetimes without primes, both τ and τ' were assumed jointly distributed, with a PDF denoted with $f(\tau, \tau')$ Both time variables were assumed to be positive and the joint distribution function is subject to normalization to one. Then the (conditional) PDF that t lies in $\tau_1 < \tau < \tau_1 + d\tau$ for a parcel of solute that first entered the transport volume at τ' equals

$$g(\tau \mid \tau') = \frac{f(\tau, \tau')}{Q_\epsilon(\tau')} \tag{37}$$

The probability that a parcel injected at τ' in the range $<0, t>$ did not yet leave (the first time) at time (t), implies that the lifetime is in $<t, \infty>$, *i.e.*,

$$Pr[\tau' + \tau > t \mid \tau' < t] = \int_0^t \int_{t-\tau}^\infty f(\tau, \tau') d\tau d\tau' \tag{38}$$

This equation can be recast into

$$Pr[\tau' + \tau > t \mid \tau < t] = \int_0^t Q_\epsilon(\tau') d\tau' - \int_0^t \int_0^{t-\tau'} g(\tau \mid \tau') Q_\epsilon(\tau') d\tau d\tau' \tag{39}$$

where the fractional injection rate is

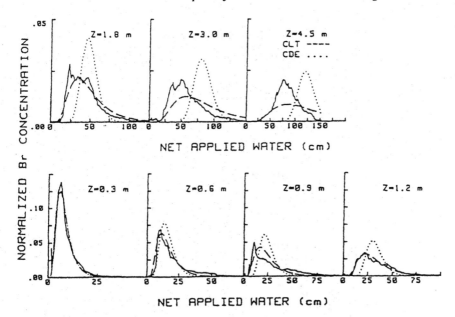

Figure 5 *Fitted and predicted bromide breakthrough curves for different depths, using the transfer function (CLT) and the convection-dispersion (CDE) models (Reproduced from Butters et al., 1989). Whereas clearly the CDE has poor predictive capabilities, the CLT performs better although agreement with the data is limited for depth significantly differing from the 0.3 m depth used to assess parameter values.*

$$Q_\varepsilon(\tau') = \int_0^\infty f(\tau, \tau')d\tau \qquad (40)$$

In the right hand side of equation (39) the output is subtracted from the fraction of mass injected up till time t. Taking the time derivative yields the net accumulation rate, where the output rate given by

$$Q_{out}(t) = \int_0^t g(t - \tau' \mid \tau')Q_\varepsilon(\tau')d\tau' \qquad (41)$$

In the last equation, $g(.)$ describes all processes such as convection, hydrodynamic dispersion and chemical, radio active or (bio)chemical transformations. The function θ_ε gives the feed boundary condition. Consequently, all process information is captured in $g(.)$. For a conservative solute and steady state

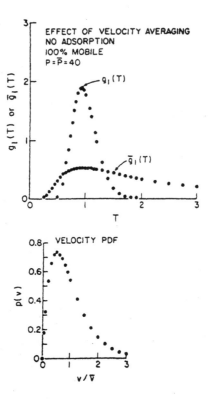

Figure 6. *Travel time probability density function, $\bar{g}$, (top) obtained from equation (43) and the gamma-distributed flow velocity (bottom), showing the large effect on the travel time pdf (Reproduced from Sposito et al., 1986). Also shown is (g), the pdf for zero variability of the velocity in the field.*

flow (at conductivity $K(\theta)$ and τ' statistically independent of τ the TFM by Jury (1982) can be derived as a special case:

$$c(L,t) = \int_0^\infty f[K(\theta)t - K(\theta)\tau']c(0,\tau')d\tau' \qquad (42)$$

Several examples (such as transport of chloride, *E. coli* bacteria) using equation (41) were illustrated by White *et al.* (1986). They showed that a lognormal distribution was less appropriate for some situations. Recently, Butters and Jury (1989) compared the deterministic CDE with the TFM approach (Figure 5). The latter model appeared to yield best results. A special case of the TFM is the two-component CDE or dual porosity model, as was shown for the one-dimensional case by Sposito *et al.* (1986). The consistency between the two models for the three-dimensional case, and the assumptions needed to derive one from the other, was later discussed by Rinaldo *et al.* (1989). In particular, Sposito *et al.* (1986) related the Laplace transformed travel time PDF with the two component CDE. For special cases the inversion can be done analytically, but in general it has to be done numerically. A special case considered by Sposito *et al.* (1986), is the classical CDE which has the travel time PDF given by

$$g_L(t,v) = \frac{L}{(4\pi D t^3)^{1/2}} \exp\left[-\left(\frac{L - vt}{2\sqrt{Dt}}\right)^2 \right] \qquad (43)$$

Assuming equation (43) to hold and that the pore water velocity varies randomly in the field, *e.g.* lognormally, an analytical solution for the field averaged travel time PDF (similar as done earlier for the field averaged concentration) is difficult to obtain. However, when the flow velocity is not distributed lognormally but has a gamma distribution, which may have a similar shape as the lognormal PDF, the averaging for the field may be performed analytically (see Figure 6). As was shown before this averaging may give rise to significant field *vs* scale dispersion.

The work reviewed so far for spatially variable transport in the water unsaturated zone considered mostly non-reacting solutes. In practice most solutes react with the soil matrix or are subject to transformation or decay. Because variability of the retardation factor affects the solute velocity or residence time in soil similarly as variability of v, it may be expected that it similarly affects the field averaged response. Based on the conceptual approach of Dagan and Bresler (1979) and Jury (1982, 1983), the transport of phosphate (P) in a spatially variable field was studied by Van der Zee and Van Riemsdijk (1986). Both the net input (A_P) of P (due to differences in flow, and manure or fertilizer application) and P-sorption (F_p) were assumed to be random and normally distributed. Local front spreading due to dispersion was neglected in view of sorption non-linearity. The PDF of A_P was assessed by measuring resident P for a large number of positions. The same was done for the retention capacity, by relating F_p with the amount of amorphous (oxalate extractable) metaloxides (Fe, Al). Such oxides often control P-sorption. Due to sorption non-linearity, periods of zero-input do not affect the P-front (total-P, dissolved plus sorbed) significantly. Therefore, realistic P-boundary conditions do not disturb the schematized problem. Since the P-front is controlled by P-input and P-chemistry, the degree of P-saturation for a column of length L can be assessed by dividing A_P with F_P. When both A and F are spatially variable, as experimental evidence showed, the moment of first breakthrough at level z=L should not be estimated using mean values of A_P and F_P. Instead, the field averaged P-saturation profile may be calculated with the theory of Dagan and Bresler (1979) (with small adaptations), taking the experimental PDF's of A_P and F_P into account.

As was shown by Van der Zee and Van Riemsdijk (1986), significant field scale

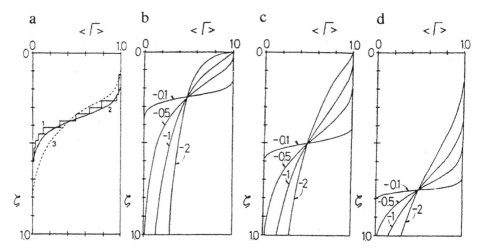

Figure 7 *Field averaged phosphate saturation profiles for the field case of Van der Zee and Van Riemsdijk (1986), assuming locally piston flow (Figure 7a). Dimensionless saturation <T> is the phosphate storage (A_P/F_P) relative to its maximum and averaged for the field and ζ is a dimensionless depth. Curve 1 (discrete): experimental data; Curve 2 (solid smooth line): predicted curve for partial correlation coefficient between A_P and F_D of 0.6; Curve 3 (dashed): predicted in the absence of correlation. Figures. 7b-7d: Field average fronts for lognormally distributed local front penetration depth (Y) with median depths of ζ = 0.25, 0.50, and 0.75, respectively, and variation coefficients of Y as indicated. (Reproduced from Van der Zee and Van Riemsdijk, 1987)*

dispersion may result when A_P and F_P are uncorrelated. For reasons given in the original paper, the experimental and calculated field averaged P-profiles agreed well when a partial correlation between A_P and F_P was assumed as is shown in Figure 7. Observe that the front found with median parameter values (A_P, F_P) would be situated at about a dimensionless depth (or P-saturation) of 0.35. Using approximations as given by Van der Zee and Boesten (1991), an analytical treatment is also possible for the assumed normal PDF's. A small variation by assuming lognormal PDF's for A and F (rationalized by the small differences between lognormal and normal PDF's for small variation coefficients), was treated by Van der Zee et al. (1988), and may also be solved analytically. In the original paper several constraining assumptions were made. Most important were neglecting pore scale dispersion and neglecting the P-precipitation kinetics, which were discussed by Van der Zee et al. (1989). When such simplifications are not made, the fronts shown in Figure 8 may be obtained for parameter values that were in close agreement with those of Van der Zee and Van Riemsdijk (1986), and with a pore scale dispersivity of 3 cm. In Figure 8a, the fronts for adsorbed, precipitated, total, and dissolved P are shown using mean values for the retention capacity (F (t $\rightarrow \infty$)) of the topsoil (till z = 0.4 m, with capacity of 50 mmol kg^{-1} and the subsoil (z>0.4 m, with capacity of 10 mmol kg^{-1}). Before and

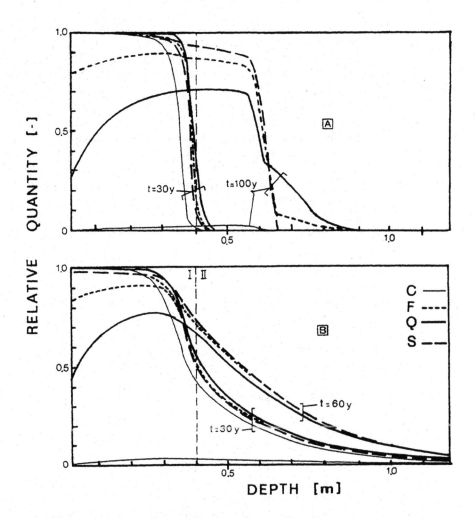

Figure 8 *Field average fronts for phosphate for zero variability (A) and finite variability of the phosphate retention capacity (B). Shown are concentration $\bar{c}$ = $\bar{c}/c_0$), adsorbed amount ($\bar{Q}$ =Q/Q_m) precipitated amount ($\bar{S}$ = S/S_m), and total-P ($\bar{F}$ = F/F_m) for two times indicated, with topsoil metal content $\hat{M}$ = 100 mmol kg^{-1} and soil $\hat{M}$ = 10 mmol kg^{-1}. P-input discontinued after 35 years, $F_m = 0.5\ \hat{M}$ $Q_m = 0.135\ \hat{M}$ and $S_M = F_m$-Q_m. Curves obtained with model of Van der Zee et al. (1989), for (B) with statistics as given by Van der Zee and Van Riemsdijk (1986).*

after discontinuation of P-input (at time t = 35 yrs) relatively well defined (thin) downstream fronts are observed. Taking randomness of retention capacities into account (Figure 8b) yields for both times a significantly dispersed field average downstream front. The effects of spatial variability are small in the initially P-saturated topsoil. Observe that for the case considered, leaching of phosphate at a hazardous level occurs in Figure 8b, whereas it does not occur in Figure 8a. A more detailed discussion of the shown trends was given by Van der Zee (1990b) and Van der Zee and De Haan (1990).

An extension was given by Van der Zee and Van Riemsdijk (1987) for the related case of non-linear Freundlich adsorption of heavy metals (Cu, Zn, Cd) with a relatively large retardation factor at prevailing feed concentrations. Randomness of flow, via the PDF of the scaling parameter and the conductivity function, and of the retardation factor was taken into account. In both cases a lognormal PDF was assumed and justified. This resulted in a simple analytical expression for the field average concentration (or solute saturation) front, *i.e.,*

$$< \bar{c}(z,t) > = \frac{1}{2}\left\{ 1 - erf\left[\frac{\ln(z) - m_{lnY}}{s_{lnY}\sqrt{2}} \right] \right\}$$

(44)

where Y is the dimensionless front penetration depth ($Y = vt/RL$). The general behaviour is shown in Figure 7b for three median penetration depths (*i.e.* found from v, t, R, and L by using mean values) and four variation coefficients of Y. The statistics of R followed from the PDF of the random organic carbon fraction (oc) and the pH, which appeared to be lognormally and normally distributed, respectively. For the case considered, randomness of R dominated field scale dispersion, and randomness of flow could be neglected without much loss of accuracy. To deal with correlation between flow and sorption the functional relationship given by Jury (1983) was used

$$R = \exp(m_{lnR})\left[\frac{\exp(m_{lnv})}{v} \right]^{\kappa}$$

(45)

In principle both positive and negative correlations are possible. Small differences in soil surface level may channel flow towards lower parts of the field. This may be accompanied with organic matter erosion at places where overland flow occurs and organic matter accumulation where water infiltrates (Boekhold *et al.,* 1990). For that case, a positive correlation may be expected between mean water fluxes and heavy metal and pesticide sorption, as the latter depend on organic matter content. When such spatially variable flow causes differences in leaching of iron, a negative correlation may be expected for phosphate. Phosphate retardation may be affected significantly by the amount of amorphous iron-oxides in soil.

Transport of solute in spatially variable fields was also considered by Destouni and Cvetkovic (1989) and Cvetkovic and Destouni (1989). An important difference with previous approaches of Dagan, Bresler, Van der Zee and Van Riemsdijk, was that they evaluated the effect of spatial variability of groundwater depth, conductivity, saturated moisture content and recharge (infiltration) on the field averaged solute mass flux ($<j_S>$) rather than the resident concentration front. For a

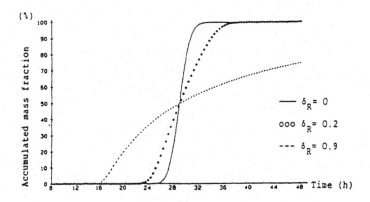

Figure 9 *Field average mass breakthrough curves for indicated variability of recharge ∂_P. (Reproduced from Destouni and Cvetkovic, 1989).*

single semi-infinite column subjected to an instantaneous injection of mass (m), the local mass flux j_S is given by

$$j_s = \frac{mL}{\sqrt{4\pi Dt^3}} \exp\left[-\left(\frac{L-vt}{2\sqrt{Dt}}\right)^2\right] \qquad (46)$$

(Kreft and Zuber, 1978). One of the results of Destouni and Cvetkovic (1989), where only infiltration is spatially variable is given in Figure 9. Their working hypothesis was that the field averaged flux concentration is usually of more interest (for ground water quality predictions) than are resident concentrations. The field averaged flux concentration is given by ($<j_s>/<q>$), where $<q>$ is the field averaged specific discharge. When flow is spatially variable, it is clear that the probability that at (x,y) a front has reached depth $z=L$ depends among others on the velocity at (x,y). Hence, the resident concentration and v are correlated, *i.e.*, the PDF of $c(z,t)$ and of v are jointly distributed with (time-dependent) correlation coefficient (see below). Most effect on the field averaged behaviour had spatial variability of infiltration and unsaturated zone thickness. In particular, when the saturated volumetric water fraction and conductivity were perfectly correlated, their joint variability compensated for unsaturated conditions. For saturated flow, randomness of the saturated conductivity is important. An example where flow variability was detected, for saturated (ponding) conditions was given by Biggar and Nielsen (1976). Cvetkovic and Destouni (1989) studied breakthrough for an instantaneous resident injection boundary, and compared flux and resident concentrations. The probability of finding a pulse at $z=L$ is positively correlated with the flow velocity for times shorter that the mean breakthrough time of the field. For this reason, field averaged flux concentrations that take correlation between the PDF of $c(L,t)$ and the PDF of v into account, are expected to be larger than field averaged resident concentrations,

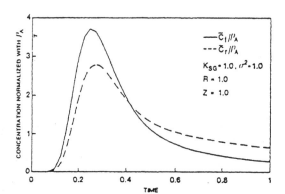

Figure 10 *Comparison of field average breakthrough based on flux ($\bar{c}_f$) and resident ($\bar{c}_r$) concentrations in case of flow variability (Reproduced from Cvetkovic and Destouni, 1989).*

that do not account for such correlation. Indeed, Cvetkovic and Destouni (1989) and Destouni and Cvetkovic (1991) observed this behaviour as shown in Figure 10. On the other hand, for times larger than the field average residence time, the PDF's of v and $c(L,t)$ are negatively correlated, as the probability of finding solute at $z=L$ is larger for 'columns' with a small than with a large velocity v. Figure 10 shows that in that case field average flux concentrations are smaller than resident concentrations. To evaluate leaching losses, the field averaged flux concentrations is of more interest than resident concentrations. However, when solute residence times are large, and the flow rate at a particular position (x,y) is affected by tillage and may therefore change significantly from year to year, such temporal changes may effectively mask the difference between $<c(L,t)>$ and $<j_s(z,t)>/<\theta>$. A useful extension was given by Destouni (1992a) by looking at the prediction uncertainty associated with field scale heterogeneity. The effects of vertical heterogeneity and of transient flow on this modeling approach were examined recently by Destouni (1992b, 1991).

A useful extension to model transport on a still larger scale was given by Rinaldo and Marani (1987) for a basin or watershed scale. They related the solute yield at a particular downstream position such as a measurement point in a river, with the hydrological response. The solute discharge

$$Q_s(t) = \int_{-\infty}^{t} G_s(t,t')i(t')dt' \tag{47}$$

where Q_s is the mean concentration in the discharged water at time t, G_s the (instantaneous unit) mass response function (MRF) and $i(t')$ gives the rainfall intensity at time $t' < t$. For the hydrological response a similar expression as equation (47) can be given. A pocket of water entering the transport volume at t'

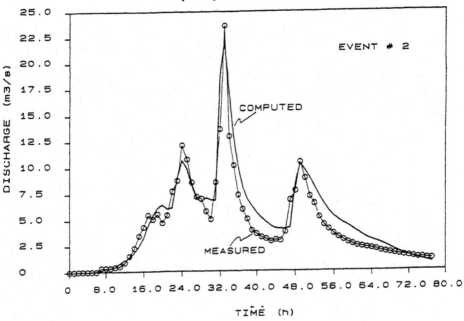

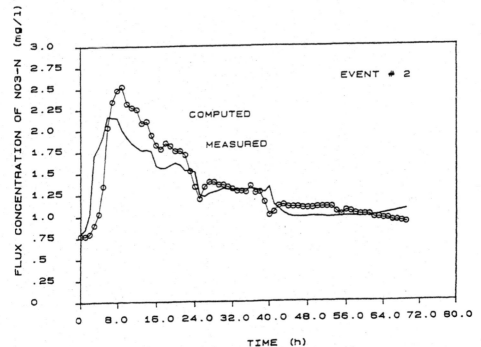

Figure 11 *Predicted hydrologic responce (top) and mass output of nitrate (bottom), using the MRF-approach with parameters obtained from a different event (i.e., event # 1) (Reproduced from Rinaldo and Marani, 1987).*

stays in different particular states according to different residence time PDF's for each state. These PDF's are assumed to be statistically independent. Essential assumptions made by Rinaldo and Marani (1987) were that the solute transfer across transport volume boundaries (*i.e.*, from mobile to immobile phases and vice versa) can be described by a first order mass transfer expression at the megascopic scale and is only dependent on the contact times. Furthermore the chain of events is ergodic. The assumed megascopic first order mass transfer does not imply that at the smaller (*e.g.* streamtube) scale also mobile and immobile zones have to be postulated. The background of the MRF-approach was to give a predictive model with easily assessible model parameters. Jury's (1982) TFM-approach requires that prior to predictions, the adaptable parameters of the residence time distribution are experimentally assessed. Predictions are thus possible only for depths and solutes experimentally evaluated, which is a disadvantage. It would require a large effort to characterize a field in this way. From the management point of view it is clearly more interesting to characterize at once an entire watershed (rather than a column or field) for future predictions, if this does not involve more work than characterizing a soil column. A basic result obtained by Rinaldo and Marani (1987) enables a simple watershed characterization. Their MRF decouples the hydrological response from chemistry affected terms because it can be written as

$$G_s = c'(t - t', t') f[i(t'), t - t'] \qquad (48)$$

Hence, the MRF is the product of the hydrological response $f(.)$ (the IUH) and c'), which is the solution of the megascopic mass transfer rate expression and accounts for (solute specific!) physico-chemical interactions. The IUH may be measured at the observation point if this has not yet been done, and the same may be done for the solute yield. From this the form of the solute function c prime can be assessed, and used for predictions of future events. An example of predictions is given in Figure 11, where parameters were obtained from other events. The excellent result is promising for this line of approach that directly considers the scale of prime interest. Further elaborations were presented by Rinaldo (1988) and Rinaldo *et al.* (1989). It was shown that equation (47) and the 3-D megascale two-component transport equation are consistent. The solute function c' was shown to be a resident concentration if the equilibrium concentration in the transport volume depends on the contact time with the immobile phase rather than the spatial position. Then, the characteristic times of transport and of mass transfer are of comparable magnitude. The flux concentration may be found by appropriate weighing (using the hydrologic response) of the MRF and can explain hysteresis observed in concentration and water time series at watershed outlets. Such hysteresis is commonly observed.

Preferential displacement in soils

Besides preferent flow through fractures and macropores (for different examples see Van Genuchten *et al.* (1990)) channeling may occur also in unstructured soils . Due to different causes an initially planar wetting or concentration front may displace

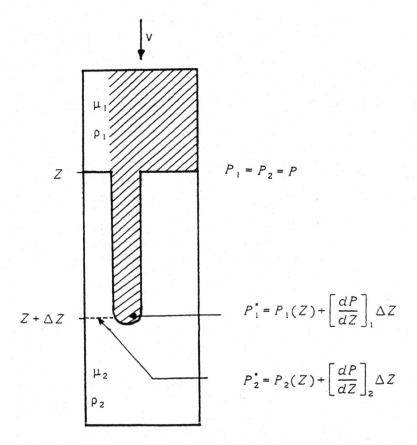

Figure 12 *Illustration of flow instability for fluid 1 displacing one-dimensionally fluid 2. If $P_1^* > P_2^*$ the flow is unstable.*

downward non-uniformly and exhibit fingering. In column experiments Rose and Passioura (1971) found that due to differences in packing between the center of columns and near the walls preferential flow at the walls occurred. Other conditions, leading to unstable flow, are entrapped air and air pressures that limit infiltration (Raats, 1984, Stroosnijder, 1976), changes in the hydraulic conductivity as a function of depth (Parlange and Hill, 1976, Raats, 1973, Hillel and Baker, 1988), poor wettability of upper soil horizons or layers (Van Ommen *et al.*, 1988,1989, Hendrickx *et al.*, 1988), and gradients in the density and/or viscosity of the fluid in the direction of flow (Saffman and Taylor, 1958, Rose and Passioura, 1971, Homsy, 1987). The effects of such fingering phenomena may be large for the fate of contaminants, erosion and irrigation efficiency. Due to the bypassing of part of the reacting soil matrix the filtering function of the soil compartment may be significantly reduced. Then for instance pesticides may be leached into groundwater before a significant decay has taken place. In case of a hazard of wind erosion the non-wetted surface may be more readily eroded. Additionally, surface runoff due to a limited infiltration capacity may invoke erosion, by overland flow. It is therefore important to be able to quantify fingering phenomena and to understand when exactly this process is important.

One of the first papers to address unstable flow was Hill (1952) who considered the displacement of one resident fluid by another fluid in a packed column. The mean density, viscosity and permeability for each fluid may be different. On the microscopic scale flow may show small variations even when the packing was done carefully. The imposed planar interface between the fluids, oriented perpendicular to the mean direction of flow, may on the microscopic level become perturbed. Such perturbations may grow or dampen out, depending on the situation. Based on Darcy's law, Hill (1952) and Saffman and Taylor (1958) gave a criterion regarding the stability of displacement. Considering the pressures in the tip of and next to the instabilities, fingers were assumed to grow when the pressure in the tip is larger than in the surrounding resident fluid (Figure 12). The criterion for the critical mean velocity above which unstable displacement occurs was given by

$$v_c = \frac{\bar{g}(\rho_2 - \rho_1)}{\left[\frac{\mu_2}{\kappa_2} - \frac{\mu_1}{\kappa_1} \right]} \qquad (49)$$

As (49) shows, instability may be gravity as well as viscosity driven. Experimental work by Hill (1952) was qualitatively in agreement with equation (49). Effects not taken into account in equation (49) such as surface tension (for immiscible displacement) and dispersion (for miscible displacement) effects may affect flow and finger dimensions but do not stabilize flow that is unstable according to equation (49). A special phenomenon observed by Saffman and Taylor (1958) and called shielding refers to the situation that one finger grows at the expense of other fingers because it has the largest pressure drop at the tip. In that case, one dominant finger advances and captures a growing portion of the discharge of the displacing fluid. A disadvantage of neglecting dispersion is that small perturbations seem least stable, *i.e.,* smaller fingers predominate. This was shown by Saffman and Taylor (1958) to be not the case. Considering immiscible displacement in Hele-Shaw models Chuoke *et al.* (1959) presented a linear stability analysis that accounted for surface tension effects. They found that large surface tensions stabilize small scale perturbations. The fastest growing perturbation is of finite size due to surface tension effects. For miscible displacement it appears difficult to take the effects of dispersion (similar as surface tension effects for immiscible displacement (Yortsos, 1987)) rigorously into account. Experimentally Slobod and Thomas (1963) and Perkins et al (1965) showed that the smallest scale perturbations are indeed stabilized for miscible displacement.

An interesting situation with unstable flow is wetting front instability. In case of a vertically moving wetting front in the absence of capillarity and air entrapment (Raats, 1984; Stroosnijder, 1976) we may use equation (49). Because the viscosity and density of air are much smaller than for the infiltrating water, equation (49) simplifies to

$$v_c \theta < \frac{\kappa \rho_u \tilde{g}}{\mu_w} = K_s \qquad (50)$$

This shows that unstable flow may result when flow is less than the saturated hydraulic conductivity. Indeed, when infiltration occurs from a lower permeability layer into a higher permeability layer and when the upper layer is saturated while the lower is not, finger formation can be observed. Typical finger diameters are equal to a few centimeters (Hill and Parlange, 1972). These observations were an incentive to analyze unstable wetting fronts mathematically (Raats, 1973; Philip, 1975a,b). Of particular interest is the analysis by Parlange and Hill (1976). They considered a cylindrical finger with $\theta = \theta_s$ in a system with uniform initial volumetric moisture fraction, θ_i. The variation of θ perpendicular to the finger surface was assumed to be abrupt, *i.e.*, the front is sharp. Based on mass conservation they related the velocity of a stabilized front with that of the perturbed front.

The linear stability analysis of Saffman and Taylor (1958), was extended by considering a continuous pressure across the front by Chu and Parlange (1962). They showed that the fastest growing perturbation has a finger radius given by

$$r_f \sim \pi S^2 / \{(\theta_s - \theta_i)[K_s - v(\theta_s - \theta_i)]\} \qquad (51)$$

$$S^2 = 2(\theta_s - \theta_i) \int_{\theta_i}^{\theta_s} \tilde{D} d\theta \qquad (52)$$

As Parlange and Hill (1958) noted, equation (51) suggests that the finger radius decreases as the flow rate that is imposed decreases. Earlier measurements revealed, though, that a change in the flow rate did not decrease the radius. Instead the number of fingers decreased. Merging of fingers to perturbations moving at maximum speed ($K_s/ \theta_s - \theta_i$ for a rate smaller than half the saturated conductivity was postulated, with a merged dominant radius of fingers equal to

$$r_f^m = \frac{2\pi S^2}{K_s(\theta_s - \theta_i)} \qquad (53)$$

Important is the dependency of r on S, as it predicts large radii in fine-textured soils where capillarity effects dominate over gravity effects. As initial moisture fractions increase, the entrapment of air in the larger pores will cause flow to occur mainly in the finer pores. The soil then behaves as a finer textured soil, which explains the effect of initial moisture in equation (53). Both latter effects are in disagreement with Philip's analysis (1975a,b).

An extension of the theory given by Parlange and Hill (1976) was given by Glass *et al.* (1989a). They presented a dimensional analysis of fingering taking the following parameters into account: (1) porous medium properties (parameters describing the grain/pore size distribution and the microscopic length scale), (2) fluid properties (surface tension, viscosity, density, and contact angles between fluid, gas and solid phase), (3) hydraulic properties (conductivity, retention curve, diffusivity, sorptivity), (4) hydraulic heterogeneity (type, level, and scale), (5) macroscopic length scale (flow domain dimensions), (6) initial and boundary conditions for flow.

Consideration of the flow domain dimensions (point 5, above) was necessary because the scale of the measurement device affects the flow process. Decreasing the measurement scale may prevent the formation of fingers at that scale. However, this does not mean that at a larger scale, fingers will not form. By appropriate scaling the Richards' equation was made dimensionless by Glass *et al.* (1989a)

$$\frac{\partial \theta_*}{\partial t_*} = \nabla_* . D_*(\theta_*) \nabla_* \theta_* - N \frac{\partial K_*(\theta_*)}{\partial z_*} \tag{54}$$

The resulting general equation for the finger diameter is

$$r_f = \frac{\sigma}{\rho \tilde{g} M K_{s*}(\theta_s - \theta_i)} \frac{S_*^2}{S_*^2} f_{dF}(R_F) \tag{55}$$

when use is made of the similar media concept (Miller and Miller, 1955, 1956). Likewise an expression can be found for the finger velocity, v

$$v = \frac{\rho \tilde{g} M^2}{\mu} \frac{K_{s*}}{\theta_s - \theta_i} f_{vF}(R_F) \tag{56}$$

Of interest is an immediate consequence of equations (55)-(56), *i.e.*, when the average flux into the finger divided by M^2 is constant, the product of r and M is constant. Hence, when M decreases the fingers will have a larger cross sectional area. Then also the ratio v/M^2 is constant.

For the macroscopic system the flux-conductivity ratio was introduced given by

$$R_s = \frac{q_s}{K_s} \tag{57}$$

which is related to the flux-conductivity ratio defined using the average flux through the fingers $\overline{R}_F$

$$\overline{R}_F = \beta R_s; \qquad \beta = A_s / n \overline{A}_f \tag{58}$$

When it is assumed that the dominant finger size found with linear stability analyses is valid for the fully developed finger and the mean finger size is adequatly estimated with such analyses, then expressions for the finger diameter may be given explicitly as a function of R_S

On the finger scale Glass *et al.* (1989a) note that an increase in r with increasing flow rate through the finger is not supported by experimental evidence of Hill and Parlange (1972). This difference is attributed to heterogeneity in packing, which leads to merging, and meandering of fingers in the columns of Hill and Parlange. In their own experiments, Glass *et al.* (1989b) found that the stability analysis of Parlange and Hill (1976) gave a good approximation for the finger diameter. The relation predicted by linear theory

$$\beta = 1 / \overline{R}_F \qquad (59)$$

was found to be satisfied for $R_S > 0.05$

This relationship implies

$$R_S = \overline{R}_F^2 \qquad (60)$$

The results obtained by Glass *et al.* (1989a,b) show that finger diameter increases when pore and particle size decrease, *i.e.*, R_S increases and other parameters are held constant.

The systematic analysis of Glass *et al.* (1989a,b) provides insight to a possibly important phenomenon in field soils. Although field soils may be expected to be heterogeneous (and even a slight degree of heterogeneity as in experiments by Hill and Parlange (1972) may significantly affect fingering) their results consider not only an artefact. For understanding, a homogeneous system had to be considered first. Whether their analysis holds for 3-D media with a chamber thickness exceeding finger dimensions remains questionable. Another important constraint (the medium was initially dry) was relaxed by Glass *et al.* (1989c). They showed with new experimental techniques that fingers may persist over different infiltration cycles, which was explained with hysteresis effects.

As mentioned, other causes may lead to preferential flow exhibiting finger morphology. The various causes such as topsoil hydrophobicity, non-uniform infiltration, *etc.*, may not be (mathematically or physically) related with instability. Small differences in height (*e.g.* gilgay soils), channeling of water by plants towards or away from the stem, and poor wettability of organic layers in the topsoil may result in fingering patterns of displacement. An example of such patterns explained by poor wettability is given in Figure 13 (Van Ommen *et al.*, 1989). Unknown at this stage is whether such fingers are also persistent. With respect to contaminant leaching into groundwater, persistence may be one of the most important aspects of fingering phenomena. For some chemicals it is feasible that incidental fingers (one event) may lead to severe bypass of the vadose zone volume. For events that involve small infiltrated water quantities, or for reacting chemicals, it is less likely that one event will lead to enhanced leaching rates at the phreatic level. Finger persistence

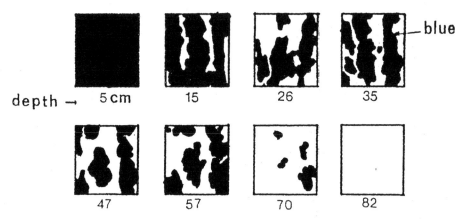

Figure 13 *Evidence of preferential flow in a coarse textured soil (Reproduced from Van Ommen et al., 1989). The dark patches were active in transferring iodide - marked water and show a complicated three-dimensional and non-uniform flow pattern. Depths in cm as indicated.*

controls the effective degree of bypass after N infiltration cycles, when leaching at the phreatic level starts. This can be easily demonstrated assuming one infiltration (or application) event is normally distributed, with mean m and variance s^2, respectively, for say heavy metal input. For persistent fingers and neglecting desorption phenomena for simplicity, the distributed front penetration depth is related with the cumulative input (Van der Zee and Van Riemsdijk, 1986) and has after N events the variation coefficent $\eta_f = s / m$), which is the same as for the individual event. When fingers are non-persistent and consecutive cycles are assumed statistically independent, the variation coefficient after the N-th cycle has become instead $\eta_f = s / (m \sqrt{N})$.

For $N=1$ both variation coefficients are the same, and therefore the degree of bypass is the same. For $N>1$, the variation coefficient for persistent fingers is larger than for non-persistent ones. The fingering pattern is stabilized towards a horizontally planar front in case of non-persistence. In this latter case the degree of bypass decreases slowly to zero. Besides the physics causing fingering, also other factors (*e.g.* tillage) may affect whether fingers are persistent. Obviously, for reacting solutes a second cause of fingering patters may arise besides flow variability, *i.e.*, regular short range variations in the retardation factor. In principle such variations are superposed on flow variability.

Another question that needs to be addressed is how flow regime and other environmental factors affect finger size and their spacing in natural media, as this is required for predictions of solute leaching. Even when such information is available, it is probably impossible to predict exactly where a finger will develop when it is caused by flow instability. For predictions of the location, detailed information and modeling on a much smaller scale than of environmental interest would be necessary, and this requirement is likely to be prohibitive. With information of the dependency of finger size, spacing and discharge as a function of different factors it does seem feasible to describe flow and bypass in probabilistic terms. In that case at least part of the variability of flow attributed to the scaling parameter will be attributed to the fingering process. This points out some conceptual uncertainties with regard to current stochastic flow modeling. Whereas in saturated aquifers scaling theory can be used to describe the variability of flow due to variations in the saturated hydraulic conductivity, use of this theory is questionable for unsaturated flow. Observed (lognormal) PDF's describing flow may be due to fingering phenomena or spatially variable infiltration. Though scaling theory may adequately characterize the observed variability, it is conceptually difficult to envision a single field as consisting of many different vertical tubes, each with a slightly different texture, unless the soil parent material significantly varies horizontally. Rather, it seems that different causes of flow variability are scaleable. It would be interesting to see whether wetting front instability may yield a lognormal flow rate distribution, within acceptable accuracy. Whereas much progress has been made regarding flow and transport in the vadose zone, we are still confronted with many challenging conceptual, theoretical and experimental gaps in our knowledge.

Transport in stratified aquifers

Transport in groundwater differs in several respects with transport in soils. One difference is of course the water pressure, which is predominantly negative in unsaturated soil and positive in saturated groundwater. A second important difference is the mean direction of flow. Whereas in soils this direction is usually vertical (downward or upward), the direction for groundwater flow is predominantly horizontal. Often an important difference is that in soils flow is perpendicular to the horizontal stratification, whereas in groundwater it is usually parallel to the main direction of stratification, at least for sedimentary deposits, emphasized in this section. Such stratification may be perfect, meaning that we find continuous layers. These layers may be either parallel or have a small inclination angle, with the mean flow direction. Often layers are not continuous, and we find lenses with different materials as in cross-bedding or cross-lamination structures (Allen, 1970). Both situations are similar in that different materials have the longest primary axes in the plane of mean flow. The dimensions of lenses and layers may vary irregularly with location. This may result in a complicated pattern of conductivities which affects transport. Furthermore, a related pattern of chemical properties may be expected that may or may not be correlated with the conductivity pattern . In view of the irregular patterns that may be found for the conductivity, many stochastic studies have been presented during the past decade. Stochastic studies usually emphasized a relatively small variability of properties with regard to the scale of transport, and are complemented by a deterministic approach that describes profound differences between a limited number of well defined sub domains. In the latter case assumptions of statistical homogeneity, invoked for the stochastic approaches, are not needed for the entire flow domain of interest. For the sake of brevity, deterministic studies that concern a stratified medium are taken into account here, and the stochastic approach is ignored.

As a starting point for the discussion, let us consider an aquifer with a thickness (*H*) of say 10-40 m, and extending in the horizontal direction of kilometers. Whereas in graphic model output the vertical scale is often exaggerated, the predominant dimension is the horizontal one. We deal in two dimensions with what resembles a horizontal column. For the three-dimensional situation we expect a "pan-cake" shape: large horizontal dimensions but a shallow depth. It is therefore quite understandable that in many simplifications made for aquifer transport modeling the variations of conductivity, retardation and the resulting concentration distribution as a function of depth (z) are ignored. Such approximations are in fact based on depth-averaged properties such as the concentration distribution:

$$\bar{c}_H(x,y,t) = \frac{1}{H} \int_0^H c(x,y,z,t)\,dz$$

$$(62)$$

Considering a planar source at particular *x* and/or a planar sink (*e.g.* a well screen) at another *x*-position, both extending along the direction *y*, we then obtain a one-dimensional transport problem. When the source and sink do not extend infinitely along *y*, we obtain problems, that may be radially symmetric. For such depth-averaged problems solutions that are found for the CDE assume often mean parameter values. However, as was illustrated by Sudicky *et al.* (1983) and Gelhar (1986) the longitudinal dispersivity appears to grow with time (or distance) in field scale experiments. This suggests that the aquifer properties change in the horizontal direction or that the dispersivity L_{11} should increase with time or displaced distance. This finding suggests that the CDE as such is inadequate for modeling field scale transport when no experimental evidence of gradual horizontal changes exists, because the longitudinal dispersivity is supposed to be an aquifer characteristic. As was already discussed , the apparent inadequacy of the CDE is due to the scale of the flow domains, the heterogeneity as well as the multidimensionality of transport, as is shown in the sequel.

When an aquifer is situated between two confining layers that do not allow vertical recharge or loss of water for the aquifer, the mean flow and transport directions are horizontal. For different geometries analytical solutions are available to describe transport. For such a description the use of macroscopic parameters, where macroscopic defines the minimum size of the medium for which the CDE holds, may give a poor prediction of measured transport. At (megascopic) aquifer scale, a good description may be obtained with megascopic average parameter values. Due to heterogeneity of the aquifer, these parameter values may differ significantly from those at the macroscopic level. An example was given by Bear (1977) for the megascopic dispersivity in the longitudinal direction A_{11}, when the hydraulic conductivity varies in the vertical direction. This dispersivity differs from the macroscopic dispersivity L_{11} obtained from column experiments, which is usually much smaller. Apparently, going from the column to the aquifer scale, the dispersivity increases from L_{11} to A_{11} . Clearly, the differences in flow velocity at the aquifer scale are not taken into account in pore scale dispersion for granular media (Bear, 1979). An *ad hoc* approach, based on measuring and fitting to solutions of the CDE, has limited value when the rate of increase of the dispersivity is yet unknown, as it requires measurement for each distance (or time) of interest. Provided the granular medium dispersivity at the macroscopic scale is well known, by using relationships as given by Pfannkuch (1963) or Moser and Klotz (1974), and when the variations of the flow field in the aquifer are known, we can discount for such heterogeneity effects deterministically. In view of the limited knowledge one usually has for the conductivity field of the aquifer, such deterministic studies use abstracted vizualizations of the aquifer.

Due to the differences in conductivity we may expects several effects on the transport process. In the high conductivity zones the fronts move more rapidly in the x-direction than in the low conductivity zones. This induces transversal (vertical) concentration gradients, and necessarily transversal loss of solute from the high conductivity zones caused by diffusion. Due to such solute losses, the mean displacement of injected mass will have a velocity intermediate to the flow velocities in the high and the low conductivity zones (disregarding retardation due to chemical interactions). The loss of solute also leads to enhanced longitudinal dispersion for the entire aquifer, as longitudinal gradients are caused by the transversal variations.

A simple case to describe the effects of transversal solute losses into aquitards (confining layers with zero flow) is a system comprising of two semi-infinitely thick slabs of different porous materials. Flow for this case is parallel to the direction of the interface (x). The mathematical formulation is given by the transport equations in both regions and the initial and boundary conditions that follow.

$$c(x,z,t) = 0 \qquad t \leq 0 \qquad -\infty < z < \infty \qquad 0 < x < \infty \qquad (63)$$

$$c(0,z,t) = c_0(t) \qquad t > 0 \qquad -\infty < z < \infty \qquad (64)$$

where equation (63) is the initial condition. The transport equation for the most permeable region (subscript 1) is given by

$$R_1 \frac{\partial c}{\partial t} = D_{11}^1 \frac{\partial^2 c}{\partial x^2} + D_{22}^1 \frac{\partial^2 c}{\partial z^2} - v_1 \frac{\partial c}{\partial x} \qquad x > 0 \qquad z > 0 \qquad 65)$$

For the least permeable layer (subscript 2) we have

$$R_2 \frac{\partial c}{\partial t} = D_{11}^2 \frac{\partial^2 c}{\partial x^2} + D_{22}^2 \frac{\partial^2 c}{\partial z^2} - v_2 \frac{\partial c}{\partial x} \qquad x > 0 \qquad z < 0 \qquad (66)$$

Hence, we do not postulate yet that the confining layers have zero flow. At the interface ($z=0$) we have two conditions that have to hold

$$\lim_{z \downarrow 0} c(x,z,t) = \lim_{z \uparrow 0} c(x,z,t) \qquad (67)$$

$$\lim_{z \downarrow 0} n_1 D_{22}^1 \frac{\partial c}{\partial z}(x,z,t) = \lim_{z \uparrow 0} n_2 D_{22}^2 \frac{\partial c}{\partial z}(x,z,t) \qquad (68)$$

where equation (67) gives continuity of concentration and equation(68) gives the continuity of the transversal dispersion fluxes at both sides of the interface. We assume homogeneity at the macroscopic level, for which equations (65) and (66) are assumed to be valid, for each region $j = 1$ and 2. Furthermore, linear sorption is accounted for through use of the linear retardation factors (R). The dispersion coefficients are defined by L_T = transversal dispersivity and L_L= longitudinal dispersivity

$$D^j_{22} = L_{j,T} v_j + D_m \qquad D^j_{11} = L_{j,L} v_j + D_m \qquad (69)$$

As was stated previously, the mean flow direction at the macroscopic level is in the x-direction.

The diffusional loss into impermeable $v_2 = 0$ confining or interbedded layers was studied by Gillham *et al.* (1984) to illustrate the effect on transport. They did not account for longitudinal dispersion in the permeable zone (with number 1), while transversal dispersion in this zone was assumed to be infinitely large. The latter assumption has two consequences. Because it ensures perfect mixing and zero transversal concentration gradients in region 1, (68) can not hold. Also the loss of solute from region 1 is at its maximum as the interface is optimally exposed to solute. Flow in the less permeable zone was assumed to be negligible. Their analytical solution differed from the fracture solution of Tang *et al.* (1981) as Gillham *et al.* did not account for adsorption and a pulse type solute input was assumed. Gillham *et al.* (1984) found that the mobile zone initially has a sharp pulse of solute, that rapidly smoothes. The resulting concentration slug exhibited considerable tailing. The cause is that the less permeable zone releases solute once the main slug has passed a particular position. The resulting skewed concentration distribution (in the direction x) is likely to yield a large dispersivity when it is fitted with the one-dimensional CDE (for depth averaged properties of zone 1). In view of the assumed zero longitudinal dispersion coefficient, this demonstrates that the transport equation (CDE) is not appropriate for their system, as its fitting result does not agree with the parameter input. This does not imply that the equation is invalid, as it may give a good description provided the non-uniformity of flow is taken into account for the entire system (*i.e.*, regions 1 and 2) as discussed by Güven *et al.* (1986) and Sudicky and Gillham (1986). Güven *et al.* (1986) mentioned that macroscopic dispersivities obtained from megascale experiments were in good agreement with laboratory measurements when variability of flow is taken into account (see also Killey and Moltyaner (1984) and Moltyaner and Killey (1984)). Therefore, equations based on the CDE may yield good results. This agrees with the remarks made by Gillham *et al.* (1984) who intended to say (Sudicky and Gillham, 1986) that the scale dependency of dispersion in the one-dimensional CDE (a deviation from the concepts that state, that the dispersivity is a medium characteristic which is constant) originates from inappropriate averaging of parameters over the vertical direction. Flow variability in the vertical direction may cause a growth of the dispersivity from macroscale to megascale values. Whether such a grow is limited or not, depends on the medium.

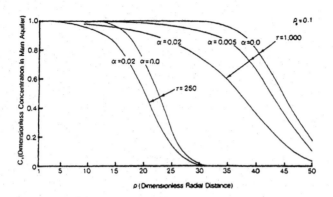

Figure 14 *Front shapes for different times and values of parameter* α *as a function of radial distance (Reproduced from Chen 1985).*

A similar case was studied by Chen (1985) who considered a fully penetrating injection well in a radially symmetric flow field. Chen (1985) cites several papers that did not account for the effect of diffusive losses into the confining layers (which were assumed completely inactive). The analytical and numerical solutions (the latter based on a numerical inversion of Laplace domain solutions, according to Stehfest (1970)) for small and intermediate to large time periods, which were given by Chen (1985), took the diffusive loss into confining layers $v_2 = 0$ into consideration. Chen (1985) assumed a finite longitudinal dispersivity similar to the fracture model of Tang *et al.* (1981), and likewise assumed perfect mixing in the aquifer (or fracture). Observe that his assumption 7 (p.1070) is not quite correctly phrased, as transversal dispersion was not neglected in the aquifer but was assumed to be infinitely large. Due to the absence of vertical concentration gradients in the aquifer, it was possible to average over the vertical direction. This leads to an artificial condition at the interface between regions 1 and 2 as continuity of fluxes could not be maintained. After all, a zero gradient in the aquifer implies zero transversal flux towards the interface, whereas this flux is not zero at the interface on the side of the low permeability zone.

For some of the cases considered by Chen (1985) the effect of diffusive losses may be profound, as is shown in Figure 14, even for non-reacting solutes. The dimensionless parameter α relates the parameters that govern transversal losses with the parameters that control longitudinal convection, and increases when transversal losses become greater. The loss of solute is in principle estimated too large because infinite mixing is assumed. If diffusive losses at the interface are completely neglected the solution of Bear (1972) is found. Though linear adsorption was not incorporated in the model by Chen (1985), this may be done straightforwardly, due to the perfectly mixed region 1 (if this region were not perfectly mixed, sorption would affect the interface conditions (71)-(72) and incorporation would become less simple). In a related analysis, dedicated to fracture flow, Chen (1986) extended his

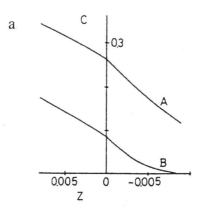

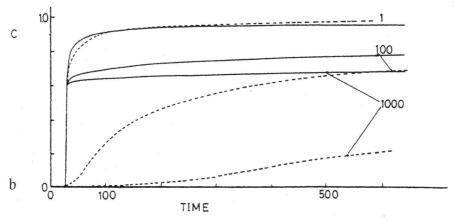

Figure 15 *Effect of transversal dispersive losses on the concentration in the permeable region (z > 0) for 15a: Concentration fronts perdendicular to the interface for (A): $R_2 = 100$ and (B): $R_2 = 1,000$, at $x = 0.5$ m and $t = 150$ days and Figure 15b: breakthrough curves for different retardation factors in region 2 indicated at the curves found analytically with finite transversal dispersivity, solid lines and for infinite transversal dispersivity (dashed lines) (Reproduced from Van Duijn and Van der Zee, 1986).*

earlier analysis with linear adsorption and first order decay. Also the source was allowed to decay with time. Rather obviously, the effect of a decaying source was negligible when the decay rate was much smaller that the period of flow considered. Furthermore, it appeared that the effects of (macroscopic) longitudinal dispersion decrease as transport proceeds. Due to the diffusional losses to confining layers the megascopic longitudinal dispersion (which take these losses into account) of course does not diminish. As steady state is reached for the transport process the effects of retardation appear to vanish. Because retardation occurs in the time derivative terms of equations (65) and (66) this could also be expected.

Whereas in these studies an impermeable confining layer was assumed, Chen (1989) also took the effects of leakage through the aquitard into consideration. In the aquifer, flow was strictly horizontal and radially symmetric while in the aquitard flow was strictly vertical. Constraints were presented for the validity of neglecting longitudinal dispersion in the aquifer and for the validity of neglecting transversal dispersion in the aquitard. The effects of leakage are two-fold. First, the flow velocity decreases more rapidly as the distance from the injection well increases. Secondly, the amount of solute lost into the aquitard is larger than in case of purely diffusive losses. The overall leakage effect is a slower front displacement of the aquifer front. For small leakage factors the correspondence with earlier results (Chen, 1986) appeared to be good.

Assuming that macroscopic longitudinal dispersion becomes of little importance for large enough times, the problem given by equations (63) - (69) was analyzed by Van Duijn and Van der Zee (1986), setting the longitudinal dispersion coefficients equal to zero. Different from the previously cited papers, the transversal dispersion coefficient was not assumed to be infinite and the proper interface conditions were accounted for. Furthermore, flow as well as linear adsorption in both regions were allowed. Constraining, without loss of generality, $v_2 / R_2 <= v_1 / R_1$ analytical approximations were presented for the solute loss from region 1 to region 2, for the interface concentration, and for the breakthrough curve in region 1 (the aquifer). A profound advantage of this work is the simplicity of the found analytical expressions, which duplicated numerically obtained results well. Only first breakthrough was not described well, due to neglecting longitudinal dispersion. It was shown that in particular for significant retardation factors in the low permeable region the results for infinite and for finite transversal mixing may become large, as shown in Figure 15. When the transversal dispersivity becomes smaller than in the situation considered by Van Duijn and Van der Zee (1986), such differences may become even more profound. A disadvantage of the solutions given by Van Duijn and Van der Zee (1986) is the limited practical use for layers of finite thickness (see their equations (65) and (67)). This is due to the semi-infiniteness of the two regions, assumed in the analytical approach.

A less constrained solution was developed by Sudicky *et al.* (1985), who allowed the permeable region to be of finite thickness. Their exact analytical solution is more complicated than those given by Van Duijn and Van der Zee (1986), but applicable as long as the diffusion front in the impermeable region does not reach its boundaries (their solutions were also developed for a semi-infinite impermeable region 2, and $v_2 = 0$. In Figure 16 some of their results are shown and compared with the solutions of Tang *et al.* (1981). This figure suggests that indeed longitudinal dispersion in region 1 soon becomes of secondary importance, namely when the transversal loss of solute starts to dominate the megascopic process. The solution with a finite transversal dispersion coefficient yields a better agreement with measurements than in case of infinite transversal mixing. Interesting is the large effect of diffusion into region 2 compared to the case where this region is inaccessible by diffusion. Apparently, low permeable regions are still important for the megascopic transport process. Relevant is the conclusion that transversal solute losses cause a time- or

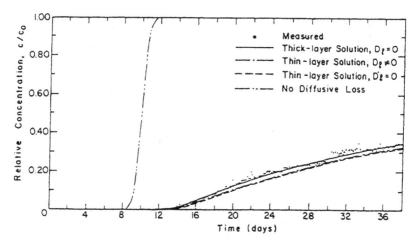

Figure 16 *Breakthrough curves for fracture models assuming no diffusive loss to confining strata (dash-double dot), Zero D_{11} and infinite D_{22}-values (dashed), finite D_{11} and infinite D_{22}-values (dash-dot) and infinite D_{22} and zero D_{11}-values (solid) (Reproduced from Sudicky et al. 1985). Points were measured.*

distance-dependent megascopic longitudinal dispersion coefficient. It is an illustration that the one-dimensional CDE is not in question (as it is used to obtain the results), but that variability of properties may necessitate that more dimensions or subdomains be taken into account.

In a second paper, Starr *et al.* (1985) incorporated linear adsorption (via the linear retardation factors for the regions 1 and 2) in the model developed by Sudicky et al (1985). As they considered reactive radionuclides, first order decay was also added to the transport equations for both regions and for the source boundary condition. The solution provided, again assumed that longitudinal (macroscopic) dispersion is negligible and that transversal dispersion is infinitely large in region 1. They mentioned that the model with finite transversal mixing as developed by Sudicky *et al.* (1985) was also modified to account for adsorption. Because the effect of transversal dispersion in region 1 was insignificant, this solution was not given. With data they showed that transversal solute loss into a low permeable adsorbing layer causes larger first arrival times (due to retardation in region 1), reduced peak concentrations and a more asymetric breakthrough (due to transversal loss into a reacting layer 2). Due to backward diffusion into region 1 the tailing of breakthrough is more severe in the absence of sorption. Although with the simple model a good fit could be obtained, some parameter values changed significantly for different flow rates in region 1 while they were not supposed to. For instance, the fitted value of R depended sensitively on the flow velocity. Using independently assessed parameter values likewise resulted in discrepancies between simulation and experiment that cast doubt with regard to the accuracy of the formulation of adsorption. In view of the results of Moreno *et al.*(1985) for solute transport in fractures (their Figure 11) it is also possible that longitudinal dispersion in the permeable region may still have some effects for the cases of Starr *et al.* (1985).

A situation that resembles the one studied earlier by Van Duijn and Van der Zee (1986) was analyzed by Leij and Dane (1990). Their aim was to develop solutions for different feed boundary conditions, that may be used for the assessment of longitudinal and transversal dispersivities. Their flow domain of two semi-infinite slabs separated by an interface in the direction of flow resembled the experimental set-up of Shamir and Harleman (1966). Likewise, and different from the case of Van Duijn and Van der Zee (1986), the interface is a plane of symmetry and does not separate two different materials. Both the longitudinal and the transversal dispersivities were assumed to be finite, which is an extension of the work of Van Duijn and Van der Zee (1986). Adsorption was not accounted for, but because of the symmetry and the resulting interface conditions, it can be easily incorporated. With regard to transport in stratified aquifers, their solution is only of interest because of the similarities with other models. As the flow domain is homogeneous, it does not exhibit the essential phenomena of the other studies. Compared to the work by Sudicky *et al.* (1985) it is considerably more constrained. The only attempt to provide a solution for the two-layer problem that accounts for longitudinal as well as transversal dispersion, for the correct interface conditions and for a finite thickness of each layer was given by Al-Niami and Rushton (1979). Unfortunately, Veling (1989) showed that the reasoning in the derivation is flawed, leading to an incorrect result. A successful solution of the full problem is unknown to me.

Several interesting cases of transport in a perfectly stratified aquifer were studied by Güven *et al.* (1984) using Aris' (1956)-moment theory. For non-reacting solutes they evaluated the megascopic dispersivity and concentration distributions for four vertical profiles (*i.e.*, transverse to the mean flow direction) of the conductivity. The macroscopic transversal dispersion coefficient was either constant or dependent on the flow velocity. For a parabolic conductivity profile expressions were obtained for the short term and the long term megascopic dispersivity as well as for the transient regime. A solution was also given for the concentration profile for this case. Similar results were presented for a step change in conductivity for a two layer medium, a linearly changing conductivity as a function of depth and a cosine-form conductivity profile. However, for some of the cases considered, the transient regime solutions for the megascopic dispersivity and the concentration distribution could not be derived (for a step profile, a linear or a cosine-profile with transversal macroscopic dispersion coefficient being velocity dependent). It is interesting to observe that for their cases where the transient regime solutions could be derived, the megascopic dispersivity increases linearly for short times (starting at the macroscopic longitudinal dispersivity value) and approach the large time limit exponentially. An important conclusion is that the contribution of trends in the conductivity or velocity) profile to limiting $t \rightarrow$ infinity megascopic dispersivities may be larger than contributions due to random components of conductivity or velocity variability. An extension for a step change was given by Valocchi (1989) who derived an expression for the limiting dispersion coefficient for a series of strata with different water velocities, dispersivities, adsorption coefficients and reaction rates.

With respect to inverse modeling it was illustrated by Bullivant and O'Sullivan (1989), that experimental data of one field experiment (considered was fractured rock) can be fitted to many of the above mentioned models. Notably for complicated systems, but also for more simple systems where important information is not available, the degrees of freedom in the proposed models (from a Taylor's dispersion model for a single fracture to models for multifractured media with first-order mass transfer or diffusive losses in geometrically well defined subdomains) is large enough to get an acceptable fitting accuracy. By an acceptable fitting accuracy it is never possible, therefore, to conclude that the mechanisms assumed in the used model are (the most) important. Such conclusions require a large number of experiments where the relevant parameters or variables are being varied significantly.

The papers cited in this section concern geometrically well defined systems. Sometimes analytical solutions can be derived for such systems for the longitudinal dispersion coefficient and the rate of approach to the limiting (megascale) value, if such a limit exists. Such asymptotic values depend on the degree of heterogeneity, as expressed in the variances of relevant properties, and the characteristic scales involved. When the geometry is less well defined, when interfaces are less abrupt, or when many different materials are present, we deal with systems that are too complicated for deterministic modeling. A stochastic approach may be feasible then. Although no details are provided here of stochastic groundwater modeling, it may be appropriate to suggest some papers for introductory reading. Useful discussions were given by Freeze (1975), Dagan (1986), Black and Freyberg (1987) and Cvetkovic and Shapiro (1989).

Conclusions

In this chapter developments concerning reactive transport in soils and in groundwater (aquifers) were reviewed. As from the environmental point of view natural porous media are of most concern, emphasis was given to some effects of the intrinsic heterogeneity of such media on transport. Heterogeneity of the transport volume being dependent on the scale (both with regard to the degree of variability and with regard to the characteristic scales), modeling attempts should emphasize those phenomena that control transport at the scale of interest.

For soil columns the heterogeneity that controls transport may be associated with microscopic phenomena that give rise to adsorption non-linearity, pore scale dispersion, and mass transfer into aggregates. These microscopic effects may influence the effect of larger scale heterogeneity (*i.e.*, layering) on transport. This was illustrated with the dependence of breakthrough on layering order the peclet numbers due to sorption (non-linearity of each individual layer.

At a larger scale, where the unsaturated zone of an entire field is considered, horizontal differences of properties further complicate transport modeling. Such (spatial) variability may be found with respect to the hydraulic properties of soil, with regard to the chemical or biological properties, as well as for initial and boundary conditions. In much of the cited literature, smaller scale heterogeneity was disregarded as a secondary effect in field scale modeling using a stochastic framework. With regard to spatially variable flow modeling, questions were raised whether scaling theory is more than an effective way to describe heterogeneity observed. A good argument can be given in favour of spatial variability of initial and boundary conditions rather than randomness of the scaling parameter (which is perhaps no more than a fitting parameter). Several suggestions for additional research were given earlier and are not repeated here.

At the aquifer scale a number of deterministic analyses were reviewed that concerned stratified aquifers. These analyses illustrated that for saturated ground water the transversal dispersional losses into confining layers or interbedded lenses with zero or insignificant flow is important. Hence, even on a megascopic scale, the aquifer solute balance is significantly affected by processes that occur on a much smaller scale (*i.e.*, local transversal dispersion at the interface of different strata). Even when lenses or layers are sufficiently impermeable to prevent flow, solute storage in these subdomains may be profound. This may cause an increase of the longitudinal dispersion coefficient from its (local) macroscopic value to a megascopic (depth averaged) value. In the specific case of parallel, continuous layers in the plane of flow with zero dispersion across their interfaces a limited megascopic longitudinal dispersion coefficient may not be found. Deterministic studies as cited complement stochastic analyses for megascopically homogeneous flow domains. In the grey zone of megascopically heterogeneous flow domains, to which neither deterministic nor stochastic analyses are dedicated, an additional effort is needed. In such cases, besides significant trends also important random components need to be accounted for.

Despite significant advances, both conceptually and with regard to modeling, further efforts are needed to understand transport phenomena at the field scale. To move one step beyond and be able to predict transport will require the availability of data with regard to parameters of interest. Such data are very limited with regard to the parameters needed for sorption modeling. Besides data on the variability of sorption parameters also the correlation with hydraulic properties needs to be known, as most solutes of interest react with the solid phase. In this context it may be worth mentioning that most theoretical developments are dedicated to non-reactive solutes, which characterize the flow of water. From the environmental point of view, reacting solutes are usually more relevant yet unfortunately their behaviour in natural media has remained underexposed.

Acknowledgements

This work has been partly funded during 1990 by the Commission of the European Communities via the STEP-Research Program, Directorate for Science, Research and Development, contract number CT-900031.

References

Abriola, L.M. 1987. Modeling contaminant transport in the subsurface: an interdisciplinary challenge. *Reviews of Geophysics,* 25(2), 125-134.

Allen, J.R.L. 1970. *Physical processes of sedimentation.* George Allen and Unwin, London.

Amoozegar-Fard, A., Nielsen, D.R. and Warrick, A.W. 1982. Soil solute concentration distributions for spatially varying pore water velocities and apparent dispersion coefficients. *Soil Sci. Soc. Am.,* 46, 3-9.

Al-Niami, A.N.S. and Rushton, K.R. 1979. Dispersion in stratified porous media: Analytical solutions. *Water Resour. Res.,* 15(5), 1044-1048.

Aris, R. 1956. On the dispersion of a solute in a fluid flowing through a tube. *Proc. R. Soc. London, Ser,.* A. 235, 67-77.

Aris, R. 1958. On the dispersion of kinematic waves. *Proc. Royal Soc. London,* A, 245, 268-277.

Barry, D.A. and Parker, J.C. 1987. Approximations for solute transport through porous media with flow transverse to layering. *Transp. Porous Media,* 2, 65-82.

Baveije, P. and Sposito, G. 1984. The operational significance of the continuum hypothesis in the theory of water movement through soils and aquifers. *Water Resour. Res.,* 20, 521-530.

Bear, J. 1972. *Dynamics of fluids in porous media.* Elsevier, New York.

Bear, J. 1977. On the aquifer's integrated balance equation. *Adv. Water Resour.,* 1(1), 15-23.

Bear, J. 1979. *Hydraulics of groundwater.* McGraw-Hill Inc, New York, 569 pp.

Beek, J. 1979. Phosphate retention by soil in relation to waste disposal, Ph.D.-thesis, Agric. Univ. Wageningen.

Biggar, J.W. and Nielsen, D.R. 1962. Miscible displacement II, Behaviour of tracers. *Soil Sci.Soc.Am.Proc.* 26, 125-128, 1962.

Biggar, J.W. and Nielsen, D.R. 1976. Spatial variability of leaching characteristics of a field soil. *Water Resour.Res.,* 12, 78-84.

Black, T.C. and Freijberg, D.L. 1987. Stochastic modeling of vertically averaged concentration uncertainty in a perfectly stratified aquifer. *Water Resour. Res.,* 23, 997-1004.

Boekhold, A.E., van der Zee, S.E.A.T.M. and de Haan, F.A.M. 1990, Prediction of cadmium accumulation in a heterogeneous soil using a scaled sorption model, In: Kovar, K. (ed.), *Calibration and Reliability in groundwater modeling,* pp. 211-220. IAHS publ. 195, IAHS-press, Wallingford.

Boekhold, A.E., van der Zee, S.E.A.T.M. and de Haan, F.A.M. 1991. Spatial patterns of cadmium contents related to soil heterogeneity. *Water, Air, Soil Pollution* 57, 479-488.

Boesten, J.J.T.I. 1986. Behaviour of herbicides in soil: simulation and experimental assessment, Ph.D.thesis, Agric. Univ. Wageningen.

Bolt, G.H. 1982. Movement of solutes in soil: Principles of adsorption/exchanges chromatography. In: Bolt, G.H. (ed.), *Soil Chemistry B.Physico-Chemical models,* pp.285-348. Elsevier, New York.

Bresler, E. and Dagan, G. 1979. Solute dispersion in unsaturated heterogeneous soil at field scale, II Applications. *Soil Sci.Soc.Am.J.,* 43, 467-472.

Bresler, E. and Dagan, G. 1981. Convective and pore scale dispersive solute transport in unsaturated heterogeneous fields. *Water Resour. Res.,* 17, 1685-1693.

Bresler, E. and Dagan, G. 1983. Unsaturated flow in spatially variable fields, 3, Solute transport models and their application to two fields. *Water Resour. Res.,* 19, 429-435.

Brooks, R.H. and Corey, A.T. 1964. *Hydraulic properties of porous media, Hydrol. Paper,* No. 3, Civil Eng. Dept., Colorado State Univ., Fort Collins, CO.

Bullivant, D.P. and O'Sullivan, M.J. 1989. Matching a field tracer test with some simple models. *Water Resour. Res.,* 25, 1879-1892.

Butters, G.L. and Jury, W.A. 1989. Field scale transport of bromide in an unsaturated soil, 2, Dispersion modeling. *Water Resour. Res.*, 25, 1583-1589.

Carslaw, H.S. and Jaeger, J.D. 1959. *Conduction of heat in solids*, 2nd edn. Oxford University Press, London.

Chen, C.S. 1985. Analytical and approximate solutions to radial dispersion from an injection well to a geological unit with simultaneous diffusion into adjacent strata. *Water Resour. Res.*, 21(8) 1069-1076.

Chen, C.S. 1986. Solutions for a radionuclide transport from an injection well into a single fracture in a porous formation. *Water Resour. Res.*, 22(4) 508-518.

Chen, C.S. 1989. Solutions approximating solute transport in a leaky aquifer receiving wastewater injection. *Water Resour. Res.*, 25(1), 61-72.

Christensen, T.H. 1980. *Cadmium sorption onto two mineral soils*. Report, Dept. Sanit.Eng., Techn. Univ. Denmark, Lyngby.

Chu, B.T. and Parlange, J.Y. 1962. On the stability of laminar flame. *J. Mécan.*, 1, 293-312.

Chuoke, R.L., van Meurs, P. and van der Poel, C. 1959. The stability of slow immiscible, viscous liquid-liquid displacements in porous media. *Trans.Am.Inst.Min.Eng.* ,216, 188-194.

Clothier, B.E. and Elrick, D.E. 1985. Solute dispersion during axisymmetric three-dimensional unsaturated water flow. *Soil.Sci.Soc.Am.J.*, 49, 552-556.

Crank, J. 1956. *The mathematics of diffusion*. Clarendon Press, Oxford.

Curtis, A.A., Watson, K.K. and Jones, M.J. 1987. The numerical analysis of water and solute movement in scale heterogeneous profiles. *Transp. Porous Media*, 2, 479-496.

Cushman, J.H. 1984. On unifying the concepts of scale, instrumentationand stochastics in the development of multiphase transport theory. *Water Resour. Res.*, 20, 1668-1676.

Cushman, J.H. 1985. Multiphase transport based on compact distributions. *Acta Appl.Math.*, 3, 239-254.

Cushman, J.H. 1986. On measurement, scale and scaling. *Water Resour. Res.*, 22(2), 129-134.

Cushman, J.H. 1987. Stochastic filtering of multiphase transport phenomena. *Transp. Porous Media*, 2, 425-453.

Cvetkovic, V. and Shapiro, A. 1989. Solute advection in stratified formations. *Water Resour. Res.*, 25(6), 1283-1289.

Cvetkovic, V. and Destouni, G. 1989. Comparison between resident and flux-averaged concentration models for field-scale solute transport in the unsaturated zone. In: Kobus and Kinzelbach, (eds),*Transport in Groundwater,* pp.245-250. Balkema, Rotterdam.

Dagan, G. and Bresler, E. 1979. Solute dispersion in unsaturated heterogeneous soil at field scale, I. *Theory, Soil Sci. Soc. Am.J.*, 43, 461-467.

Dagan, G. 1986. Statistical theory of groundwater flow and transport: pore to laboratory, laboratory to formation and formation to regional scale. *Water Resour. Res.*, 22, 120S - 134 S.

Dagan, G. 1989. *Flow and Transport in Porous Formations*. Springer Verlag, Berling, 465 pp.

De Smedt, F. and Wierenga, P.J. 1979. A generalized solution for solute flow in soils with mobile and immobile water. *Water Resour. Res.*, 15, 1137-1141.

Destouni, G. and Cvetkovic, V. 1989. The effect of heterogeneity on large scale solute transport in the unsaturated zone. *Nordic Hydrology*, 20, 43-52.

Destouni, G. 1991. Applicability of the steady state flow assumption for solute advection in field soils. *Water Resour. Res.*, 27, 2129-2140.

Destouni, G. and Cvetkovic, V. 1991. Field scale mass arrival of sorptive solute into the groundwater. *Water Resour. Res.*, 27, 1315-1325.

Destouni, G. 1992a. Prediction uncertainty in solute flux through heterogeneous soils. *Water Resour. Res.*, 28, 793-801.

Destouni,G. 1992b. The effect of vertical soil heterogeneity on field scale solute flux. *Water Resour. Res.*, 28, 1303-1309.

Freeze, R.A. 1975. A stochastic-conceptual analysis of one-dimensional groundwater flow in non-uniform homogeneous media. *Water Resour. Res.*, 11, 725-741.

Gelhar, L. 1986. Stochastic subsurface hydrology from theory to applications. *Water Resour. Res.*, 22, 135 S-145 S.

Gillham, R.W., Sudicky, E.A., Cherry J.A. and Frind, E.O. 1984. An advection-diffusion concept for solute transport in heterogeneous unconsolidated geological deposits. *Water Resour. Res.*, 20, 369-378.

Glass, R.J., Parlange, J.-Y. and Steenhuis, T.S. 1989a. Wetting front instability, 1, Theoretical discussion and dimensional analysis. *Water Resour. Res.*, 25 (6), 1187-1194.

Glass, R.J., Steenhuis, T.S. and Parlange, J.-Y. 1989c. Mechanism for finger persistence in homogeneous, unsaturated porous media: Theory and verification. *Soil Sci.*, 148(1), 60-70

Glass, R.J.,Steenhuis, T.S. and Parlange, J.-Y. 1989b. Wetting front instability, 2, Experimental determination of relationships between system parameters and two-dimensional unstable flow field behaviour in initially dry porous media. *Water Resour. Res.*, 25(6), 1195-1207.

Gureghian, A.B. and Jansen, G. 1985. One-dimensional analytical solutions for the migration of a three-member radionuclide decay chain in a multilayered geologic medium. *Water Resour. Res.*, 21, 733-742.

Güven, O., Molz, F.J. and Melville, J.G. 1984. An analysis of dispersion in a stratified aquifer, *Water Resour. Res.*, 20, 1337-1354.

Güven, O., Molz, F.J. and Melville, J.G. Comment on "An advection-diffusion concept for solute transport in heterogeneous unconsolidated geological deposits" by Gillham *et al.* *Water Resour. Res.*, 22, 89-91.

Harmsen, K. 1977. Behaviour of heavy metals in soils, Ph.D. thesis, Agric. Univ. Wageningen

Harmsen, K. 1982. Theories of cation adsorption by soil constituents: Discrete site models. In: Bolt, G.H. (ed.), *Soil chemistry B, Physico-Chemical Models*, 2nd edn, pp. 77-139. Elsevier, Amsterdam.

Hendrickx, J.M.H., Dekker, L.W., van Zuilen, E.J. and Boersma, O.H. 1988. Water and solute movement through a water repellent sand soil with grass cover, In: Wierenga, P.J. and Bachelet, D. (eds), *Int. Conf. and Workshop Validation Flow and Transport Models for the Unsaturated Zone*, pp.131-146. NMSU report, 88-SS-04.

Hill, D.E. and Parlange, J.-Y. 1972. Wetting front instability in layered soils. *Soil Sci. Soc. Am. Proc.*, 36, 697-702.

Hill, S. 1952. Channeling in packed columns. *Chem. Eng. Sci.*,1, 247-253.

Hillel, D. and Baker, R.S.1988. A descriptive theory of fingering during infiltration into layered soils. *Soil Sci.*, 146, 51-56.

Homsy, G.M. 1987. Viscous fingering in porous media. *Ann. Rev. Fluid Mech.* 19, 271-311.

Hopmans, J.W. 1987. A comparison of various methods to scale soil hydraulic properties. *J. Hydrol.*, 93, 241-256.

Javandel, I., Doughty, C. and Tsang, C.F. 1984. groundwater transport: Handbook of mathematical models. *Water Resour. Monograph Series*, 10, 228 pp. AGU, Wash. DC.

Jones, M.J. and Watson, K.K. 1987. Effect of soil water hysteresis on solute movement during intermittent leaching. *Water Resour. Res.*, 23, 1251-1256.

Jury, W.A. 1982. Simulation of solute transport using a transfer function model. *Water Resour. Res.*, 18, 363-368.

Jury, W.A. 1983. Chemical transport modeling: Current approaches and unresolved problems. In: *Chemical Mobility and reactivity in soil systems*. ASA-SSSA, Madison, WI.

Jury, W.A., Sposito, G. and White, R.E. 1986. A transfer function model of solute transport through soil, 1, Fundamental concepts. *Water Resour.Res.*, 22, 243-247.

Jury, W.A., Stolzy, L.H. and Shouse, P. 1982. A field test of the transfer function model for predicting solute transport. *Water Resour. Res.*, 18, 369-375.

Killey, R.W.D. and Moltyaner, G.L. 1984. Field studies of dispersion in porous media: Methods, 1984 Spring Ann. Meet. AGU, Cincinatti, OH, May 14-17, 1984. In Güven *et al.*, 1986.

Kinniburgh, D.G., Barker, J.A. and Whitfield, M. 1983. *J. Colloid Interface Sci.*, 95, 370.

Klotz, D. and Moser, H. 1974. Hydrodynamic dispersions as aquifer characteristic. In: *Int.Atom.Energy Ag.Symp.Isotope Techn. in groundwater Hydrol*, II, 341-355.

Kreft, A. and Zuber, A. 1978. On the physical meaning of the dispersion equation and its solutions for different initial and boundary conditions. *Chem. Eng. Sci.*, 33, 1471-1480.

Leij, F.J. and Dane, J.H. 1990. Analytical solutions of the one-dimensional advection equation and two-or three dimensional dispersion equation. *Water Resour. Res.*, 26, 1475-1482.

Levenspiel, O. 1972. *Chemical reaction engineering*, 2nd edn. J. Wiley & Sons, New York, 578 pp.

Miller, E.E. and Miller, R.D. 1955. Theory of capillary flow: I. Practical implications. *Soil Sci. Soc. Am. Proc.*,19, 267-271.

Miller, E.E. and Miller, R.D. 1956. Physical theory for capillary flow phenomena *J. Appl. Phys.*, 27, 324-332.

Moltyaner, G.L. and Killey, R.W.D. 1984. *Field studies of dispersion in porous media: Analysis of experimental data*, Spring. Ann. Meet AGU, Cincinatti OH, May 14-17, 1984.

Moreno, L., Neretnieks, I. and Eriksen, T. 1985. Analysis of some laboratory tracer runs in natural fissures. *Water Resour. Res.*, 21, 951-958.

Mualem, Y. 1976. A new model for predicting the hydraulic conductivity of unsaturated porous media. *Water Resour. Res.*, 12, 513-522.

Nkedi-Kizza, P., Biggar, J.W., Selim, H.M. van Genuchten, M. Th., Wierenga, P.J., Davidson, J.M. and Nielsen, D.R. 1984. On the equivalence of two conceptual models for describing ion exchange during transport through an aggregated oxisol. *Water Resour. Res.*,20, 1123-1130.

Nielsen, D.R. and Biggar, J.W. 1961. Miscible displacement in soils, I, Experimental information. *Soil Sci.Soc.Am.Proc.*, 25, 1-5.

Nielsen, D.R. and Biggar, J.W. 1962. Miscible displacement in soils, II, Theoretical considerations. *Soil Sci.Soc.Am.Proc.*, 26, 216-221.

Nielsen, D.R., Biggar, J.W. and Erh, K.T. Spatial variability of field measured soil-water properties. *Hilgardia*, 42, 215-221.

Nielsen, D.R., van Genuchten, M.Th. and Biggar, J.W. 1986. Water flow and solute transport processes in the unsaturated zone. *Water Resour. Res.*, 22(9), 89 S - 108 S.

Nielsen, D.R., Wierenga, P.J. and Biggar, J.W. 1983. Spatial soil variability and mass transfers from agricultural soils. In: *Chemical mobility and reactivity in soil systems*, Chap. 5, pp.65-77. ASA-SSA, Madison, WI.

Quintard, M. and Whitaker, S. 1988. Two-phase flow in heterogeneous porous media: The method of large scale averaging. *Transp. Porous Media,* 3, 357-413.

Parker, J.C. and van Genuchten, M.Th. 1984. *Determining transport parameters from laboratory and field tracer experiments,* Bull. 84-3, 96 pp., Va.Agric.Exp.Stn., Blacksburg.

Parlange, J.-Y. and Hill, D.E. 1976. Theoretical analysis of wetting front instability in soils. *Soil Sci.,* 122, 236-239.

Perkins, T.K. and Johnston, O.C. 1963. A review of diffusion and dispersion in porous media. *Soc.Petr.Eng.J.,* 3, 70-84.

Perkins, T.K., Johnston, O.C. and Hoffman, R.N. 1965. Mechanics of viscous fingering in miscible systems. *Soc. Petrol. Eng. J.,* 5, 301-317.

Persaud, N., Giraldez, J.V. and Chang, A.C. 1985. Monte-Carlo simulation of non-interacting solute transport in a spatially heterogeneous soil. *Soil Sci. Soc. Am. J.,* 49, 562-568.

Pfannkuch, H.O. 1963. Contribution á l'etude des deplacements de fluides miscibles dans un milieu poreux. *Rev.Inst.Fr.Pet.,* 18, 215-270.

Philip, J.R. 1975a. Stability analysis of infiltration. *Soil Sci. Soc. Am. Proc.,* 39, 1042-1049.

Philip, J.R. 1975b. The growth of disturbances in unstable infiltration flows. *Soil Sci. Soc. Am. Proc.,* 39, 1049-1053.

Raats, P.A.C. 1984. Tracing parcels of water and solutes in unsaturated zones. In: Yaron, B., Dagan, G. and Goldshmid, J. (eds.), *Pollutants in porous media,* Ecological Studies, 47, pp.4-16. Springer, Berlin.

Raats, P.A.C.1973. Unstable wetting fronts in uniform and non-uniform soils. *Soil Sci. Soc. Am. Proc.,* 37, 681-685.

Rinaldo, A. 1988. Solute lifetime distributions in soils and the two-component convection-dispersion equation. In: *Proc.Int.Conf. and Workshop on the Validation of Flow and Transport Models for the Unsaturated Zone,* pp.340-345. May 23-26, 1988 Ruidoso, N.M., Report 88-SS-04, Dept. Agron.Horticult., NMSU, Las Cruces, N.M.

Rinaldo, A., Marani, A. and Bellin, A. 1989. On mass response functions. *Water Resour. Res.,* 25, 1603-1617.

Rinaldo, A. and Marani, A. 1987. Basin scale model of solute transport. *Water Resour. Res.,* 23, 2107-2118.

Rose, D.A. and Passioura, J.B. 1971. The analysis of experiments on hydrodynamic dispersion. *Soil Sci.,* 111, 252-257.

Russo, D. and Bresler, E. 1980. Scaling soil hydraulic properties of a heterogeneous field. *Soil Sci. Soc. Am. J.,* 44, 681-684.

Saffman, P.G. and Taylor, G. 1958. The penetration of a fluid into a porous medium or Hele-Shaw cell containing a more viscous liquid. *Proc. Roy. Soc. London,* A, 245, 312-331.

Selim, H.M., Davidson, J.M. and Rao, P.S.C. 1977. Transport of reactive solutes through multilayered soils. *Soil Sci. Soc. Am. J.,* 41, 3-10.

Shamir, U.Y. and Harleman, D.R.F. 1966. *Numerical and analytical solutions of dispersion problems in homogeneous and layered aquifers,* Techn. Rep. 89, Hydrodynamics Lab., Mass. Inst. Technol. Cambridge, Mass., 206 pp.

Shamir, U.Y. and Harleman, D.R.F. 1967. Dispersion in layered porous media. *Proc. Am. Soc. Civil Eng. Hydr. Div.,* 93, 236-260.

Sips, R. 1948. On the structure of a catalyst surface, I. *J. Chem.Phys.,* 16, 490-495

Sips, R. 1950. On the structure of a catalyst surface, II. *J. Chem.Phys.,* 18, 1024-1026.

Slobod, R.L. and Thomas, R.A. 1963. Effect of transverse diffusion on fingering in miscible-phase displacement. *Soc. Petrol. Eng. J.*, 3, 9-13.

Sposito, G. 1984. *The surface chemistry of soils.* Oxford Univ. Press, New York.

Sposito, G., White, R.E., Darrah, P.R. and Jury, W.A. 1986. A transfer function model of solute transport through soil, 3, The convection-dispersion equation. *Water Resour. Res.*, 22, 255-262.

Starr, R.C., Gillham, R.W. and Sudicky, E.A. 1985. Experimental investigation of solute transport in stratified porous media, 2, The reactive case. *Water Resour. Res.*, 21, 1043-1050.

Stehfest, H. 1970. Numerical inversion of laplace transforms-algorithm 368. *Commun. Assoc. Comput. Mach,* 13(1), 47-49

Stroosnijder, L. 1976. Infiltratie en herverdeling van water in grond, Ph.D. thesis, Agric. Univ. Wageningen, Pudoc, Wageningen.

Sudicky, E.A. and Gillham, R.W. 1986. Reply (to comments by Güven *et al.* 1986), *Water Resour. Res.*, 22, 93-94.

Sudicky, E.A., Cherry, J.A. and Frind, E.O. 1983. Migration of contaminants in groundwater at a land fill: a case study, 4, A natural gradient dispersion test. *J. Hydrol.*, 63, 81-108.

Sudicky, E.A., Gillham, R.W. and Frind, E.O. 1985. Experimental investigation of solute transport in stratified porous media, 1, The non-reactive case. *Water Resour. Res.*, 21, 1035-1041.

Tang, D.H., Frind, E.O. and Sudicky, E.A. 1981. Contaminant transport in fractured porous media: Analytical solution for a single fracture. *Water Resour. Res.*, 17, 555-564.

Toth, J.W. Rudzinkski, Waksmundzky, A., Jaroniec, M. and Sokolowsky, S. 1974. *Acta Chim. Hung.* 82, 11.

Valocchi, A.J. 1985. Validity of the local equilibrium assumption for modeling sorbing solute transport through homogeneous soils. *Water Resour. Res.*, 21, 808-820,

Van der Heijde, P., Bachmat, Y., Bredehoeft, J., Andrews, B., Holtz, D. and Sabastian, S.(eds.), 1985. Groundwater management: The use of numerical models. *Geophys. Monograph Series,* 5, 180 pp, AGU, Wash. DC.

Van der Molen, W.H. 1956. Desalinization of saline soils as a column process. *Soil Sci.,* 81, 19-27.

Van der Zee, S.E.A.T.M. 1990a. Analytical travelling wave solutions for transport with non-linear and non-equilibrium adsorption. *Water Resour. Res.*, 26(10), 2563-2578.

Van der Zee, S.E.A.T.M. 1990b. Analysis of solute redistribution in a heterogeneous field. *Water Resour. Res.*, 26, 273-278.

Van der Zee, S.E.A.T.M. and Boesten, J.J.T.I. 1991. Effect of soil heterogeneity on pesticide leaching to groundwater. *Water Resour. Res.*, 27, 3051-3063.

Van der Zee, S.E.A.T.M. and van Riemsdijk, W.H. 1986. Transport of phosphate in a heterogeneous field. *Transp. Porous Media,* 1, 339-359.

Van der Zee, S.E.A.T.M. and van Riemsdijk, W.H. 1987. Transport of reactive solute in spatially variable soil systems. *Water Resour. Res.*, 23, 2059-2069

Van der Zee, S.E.A.T.M. and van Riemsdijk, W.H. 1991. Transport of reactive solutes in soil. *Adv. Porous Media* (in press).

Van der Zee, S.E.A.T.M., Leus, F. and Louer, M. 1989. Prediction of phosphate transport in small columns with an approximate sorption kinetics model. *Water Resour.,* 25, 1353-1365.

Van der Zee, S.E.A.T.M., van Riemsdijk, W.H. and de Haan, F.A.M. 1988. Transport of heavy metals and phosphate in heterogeneous soils. In: Wolf, K., van den Brink, W.J. and Colon, F.J. (eds.),*Contaminated Soil '88*, pp.23-32. Kluwer.

Van der Zee, S.E.A.T.M. and de Haan, F.A.M. 1990. Vulnerability of heterogeneous soil systems to pollution. *Proc. Environ. Contamination, 4th International Conf.,*pp. 274-276. Barcelona, October, 1990, CEP Consult. Ltd.

Van Duijn, C.J. and Knabner, P. 1991. *Solute transport in porous media with equilibrium and non-equilibrium multiple site adsorption: Travelling waves,* Reine Angew. Math., 415, 1-49.

Van Duijn, C.J. and van der Zee, S.E.A.T.M. 1986. Solute transport parallel to an interface separating two different porous materials. *Water Resour. Res.,* 22, 1779-1789

Van Duijn, C.J., Knabner, P. and van der Zee, S.E.A.T.M. 1993. Travelling waves during the transport of reactive solute in porous media: Combination of Langmuir and Freundlich isotherms. *Advances in Water Resources,* 16, 97-105

Van Genuchten, M.Th. 1980. A closed-form equation for predicting the hydraulic conductivity of unsaturated soils. *Soil Sci. Soc.Am.J.,* 44, 892-898.

Van Genuchten, M.Th. and Wierenga, P.J. 1976. Mass transfer studies in sorbing porous media, I, Analytical solutions. *Soil Sci.Soc.Am.J.,* 40, 473-480.

Van Genuchten, M.Th. and Cleary, R.W. 1982. Movement of solutes in soil: Computer simulated and laboratory results. In: Bolt, G.H. (ed.), *Soil Chemistry B Physico-Chemicalmodels,* pp. 349-386. Elsevier, New York.

Van Genuchten, M. Th. and Alves, W.J. 1982. Analytical solutions of the one-dimensional convective-dispersive solute transport equation. *USDA Techn. Bull.* no. 1661, 151 pp.

Van Genuchten, M.Th., Ralston, D.E. and Germann, P.F. 1990. Transport of water and solutes in macropors. *Geoderma,* 46, 1-3, 296 pp.

Van Genuchten, M.Th. and Jury, W.A. 1987. Progress in unsaturated flow and transport modeling. *Reviews of Geophysics,* 25(2),135-140.

Van Ommen, H.C., Dekker, L.W., Dijksma, R., Hulshof, J. and van der Molen, W.H. 1988a. A new technique for evaluating the presence of preferential flow paths in non-structured soils. *Soil Sci. Soc. Am. J.,* 55, 1192-1193

Van Ommen, H.C., Dijksma, R., Hendrickx, J.M.H., Dekker, L.W., Hulshof, J. and van den Heuvel, M. 1989. Experimental assessment of preferential flow paths in a field soil. *J. Hydrol.,* 105, 253-262

Van Riemsdijk, W.H.,Bolt, G.H., Koopal, L.K. and Blaakmeer, J. 1986. Electrolyte adsorption on heterogeneous surfaces: adsorption models. *J.Coll.Interf.Sci.,* 109, 219-229.

Van Riemsdijk, W.H., Koopal, L.K. and de Wit, J.C.M. 1987. Heterogeneity and electrolyte adsorption: intrinsic and electrostatic effects. *Neth.J.Agric.Sci.,* 35, 241-257.

Veling, E.J.M. 1989. Comment on "Dispersion in stratified porous media: Analytical solutions" by Al-Niami, A.N.S. and Rushton, K.R. *Water Resour. Res.,* 25 (9), 2081-2082.

Valocchi, A.J. 1989. Spatial moment analysis of the transport of kinetically adsorbing solutes through stratified aquifers. *Water Resour. Res.,* 25 (2), 273-280.

Warrick, A.W., Mullen, G.J. and Nielsen, D.R. 1977. Scaling field-measured soil hydraulic properties using a similar media concept. *Water Resour. Res.,* 13, 355-362.

White, R.E., Dyson, J.S. Haigh, R.A., Jury, W.A. and Sposito, G. A transfer function model of solute transport through soil, 2, Illustrative applications. *Water Resour. Res.,* 22, 248-254.

Wierenga, P.J. 1977. Solute distribution profiles computed with steady state and transient water movement models. *Soil Sci.Soc.Am.J.,* 41, 1050-1055.

Wösten, J.H.M. and van Genuchten, M.Th. 1988. Using texture and other soil properties to predict the unsaturated hydraulic functions. *Soil Sci. Soc. Am.J.,* 52, 1762-1770.

Yeh, G.T. and Triphathi, V.S. 1989. A critical evaluation of recent developments in hydrogeochemical transport models of reactive multichemical components. *Water Resour. Res.*, 12 (1), 93-108.

Notation

A apparent inverse peclet number
A_p phosphate input rate
A_S cross-sectional area of system (L^2)
$\overline{A}_F$ average cross-sectional finger area (L^2)
A_{ii} megascopic dispersivity (L)
b parameter
c concentration $(M\ L^{-3})$
C_i^z concentration of species (i) with valency (z) $(M\ L^{-3})$
$c_i,\ c_0$ initial resident and feed concentration $(M\ L^{-3})$
$\overline{c}$ relative concentration
$\overline{c}_H$ relative concentration averaged over depth H
D diffusion/dispersion coefficient $(L^2\ T^{-1})$
D_M molecular diffusion coefficient $(L^2\ T^{-1})$
D_{dis} convective dispersion coefficient $(L^2\ T^{-1})$
D_{ii} convective dispersion coefficient in the principal directions
$\quad$ $(i = 1$ longitudinal; $i = 2,3$ transversal$)$ $(L^2\ T^{-1})$
Da Dahmkohler number
D_* dimensionless diffusivity
D diffusivity $(L^2\ T^{-1})$
F_P phosphate sorption capacity
$\overline{F}_P$ relative amount sorbed phosphate
f probability density function (pdf)
$f(c)$ adsorption isotherm function
f_{dF} function
g probability density function (pdf)
g_L pdf of travel times for depth L
G_S mass response function
$\overline{g}$ gravity accelleration (LT^{-2})
H Heavyside step function, or aquifer thickness
h pressure head (L)
j_S solute mass flux $(M\ T^{-1})$
$K_{j,i}$ affinity parameter for species i and surface site j $(L^3\ M^{-1})$
K affinity parameter $(L^3 M^{-1})$
K_d linear adsorption coefficient $(L^3\ M^{-1})$
K_* conductivity relative to conductivity in finger
K_S saturated hydraulic conductivity $(L\ T^{-1})$
K_w hydraulic conductivity $(L\ T^{-1})$
K_m reference hydraulic conductivity $(L\ T^{-1})$
$K_r K_r$ rate parameters (T^{-1})

L_{ii}, L_D dispersivities (L) (see D_{ii})

L reference distance

M microscopic length scale (L)

m injected mass (M)

m_x mean of random variable

n parameter

N dimensionless gravity-capillarity ratio

P dimensionless variable (peclet number)

Pr probability, distribution function

Q adsorption maximum (M L^{-3})

Q_E fractional injection rate

Q_{out} fractional output rate

Q_S mean concentration in discharged water (M L^{-3})

q adsorbed amount (M L^{-3})

Q relative adsorbed amount phosphate

R, R_i retardation factor (in region i)

R_L linear retardation factor

R_F flux conductivity ratio

R_S average ratio of flux through fingers over conductivity

R_S system flux conductivity ratio

r_f finger radius (L)

r_f^m merged dominant finger radius (L)

S relative amount precipitated phosphate

S sorptivity (L T$^{-0.5}$)

S_* dimensionless sorptivity

s adsorbed amount (M M^{-1})

s_X standard deviation of X

T dimensionless time

t time (T)

$\bar{t}$ residence time in linearly adsorbing layer

t_* dimensionless time

u velocity (L T^{-1})

v_i velocity (L T^{-1}) in layer i

v_c critical velocity (L T^{-1})

x distance (L)

y distance (L)

Y mensionless front penetration depth

z distance ,depth (L)

Z dimensionless depth

z depth in linearly adsorbing layer (L)

z_* dimensionless depth

α_i, α_m scaling parameter with reference α_m

α dimensionless wave (front) velocity

β_t parameters($i = 1,2,3$)

β parameter, ratio of system surface area to wet finger surface area

δ, δ_l front thickness, linear front thickness, parameters

ϵ parameter

κ parameter (equation (45))

κ_i permeability for fluid i (equation(50)

λ_i ,λ_m scaling length and scaling length reference (L)

μ_i viscosity (M L^{-1} T^{-1})

μ parameteri

η transformed coordinate with reference η^*

η_F, $\overline{\eta}_F$ variation coefficient for persistent/non-persistent fingers

ρ dry bulk density (M L^{-3})

ρ_i density of fluid i (M L^{-3})

σ surface tension (M T^{-2})

$\theta_{\varphi,j,i}$ saturation of surface sites j with species i

Θ_t saturation of surface with species

θ volumetric water fraction

Θ water saturation

θ_r,θ_s irreducible and saturated waterfraction

θ_* dimensionless volumetric moisture fraction

τ solute life time (T)

τ' solute input time (T)

3   Modeling of diffuse-source nitrate transport beneath grazed leguminous pastures

Peter J. Dillon[1] and William E. Kelly[2]

[1]*Centre for Groundwater Studies, CSIRO Division of Water Resources, PMB2 Glen Osmond, SA, 5064, Australia*
[2] *Dept of Civil Engineering, University of Nebraska, Lincoln, Nebraska, USA*

Abstract

Nitrate leached from grazed leguminous pastures has contaminated groundwater in several areas of Australia, New Zealand and temperate regions of several other countries. This paper focuses on those pastures which do not receive nitrogen fertiliser or slurry applications, where fixation of atmospheric nitrogen by clovers is the dominant nitrogen input. Pasture residue and livestock waste products mineralise and are available for uptake by the pasture and for leaching. The non-uniform distribution of livestock waste on a pasture in space and time leads to nitrate concentrations which locally exceed the uptake/attenuation capacity.

The irregular pattern of leaching of nitrate complicates experiments at paddock scale (10 to 40 ha). Observations at pasture plots subject to applied nitrogen loads and models of nitrogen transformation and leaching processes have been applied to estimate the nitrate leached to groundwater beneath pastures at paddock-scale. These models have also been used to predict the effects of changes in pasture management on the leaching of nitrate. Furthermore the fate of nitrate in groundwater may be simulated to forecast groundwater nitrate concentrations under alternative land management scenarios. These methods may have applications in other cases where diffuse sources of contamination are in fact a collocation of point sources.

Introduction

Nitrate contamination of groundwater in New Zealand and Australia is a widespread problem. Nitrate concentrations in many areas exceed the World Health Organization's drinking water guideline. The principal cause of contamination in temperate regions of these countries is the leaching of nitrate beneath grazed leguminous pastures. Unlike those pastures in Europe and North America from which nitrate is known to be leaching to groundwater, nitrogen fertilisers are generally not used, and atmospheric nitrogen fixed by clovers is the major nitrogen source (Dillon *et al.*, 1989). The proportion of these pastures which are irrigated is small but increasing.

Control of diffuse-source contamination will be essential to preserve groundwater supplies of drinking water in a number of areas. The effects of pasture management practices have been crudely assessed using simple models and further experimental work is currently underway. It is likely that pasture management practices which retain nutrients for uptake by the pasture grasses and clovers, will simultaneously improve pasture yields and reduce the leaching of nitrate.

An understanding of the nitrogen budget of the grazed leguminous pasture is needed, and to date this has largely hinged on plot data and models. Livestock wastes, and urine in particular, which are distributed very irregularly across the pasture surface, mineralise and make a significant contribution to the pool of nitrogen available for uptake by plants or leaching. This spatial variability is particularly important for leaching as this only occurs when the pasture uptake capacity is locally exceeded. Field experimentation and estimates of mean nitrate concentrations leached from pastures are complicated by this spatial variability. This has lead to approaches described in this chapter to account for these effects, and to simulate the nitrate leaching under changes in pasture management.

The consequences of these changes on groundwater quality in an underlying aquifer is also addressed. This relies on an analytical model to describe solute transport on a vertical slice through the aquifer at a regional scale over a period of several hundred years. The timescale of changes in groundwater quality due to changes in pasture management can be identified to assist with integrated land and water planning and management. The limitations of these approaches and requirements for further research are also identified.

Models of leaching of nitrate

Reviews of models of leaching of solutes in soils have been presented by Tanji (1982), Adiscott and Wagenet (1985), Dillon (1988a), Vachaud *et al.* (1990), Wagenet *et al.* (1990) and Bogardi *et al.* (1990). At least 12 different models have been developed to consider the leaching of nitrate through soils taking account of biochemical transformations between nitrogen species. A comparison of predictive performance of four of these models is described by Rijtema and Kroes (1991). The results show that models provided a reasonable agreement with reality.

The common challenges of models are considered to be (Vachaud *et al.*, 1990):

(1) adequate definition of the microbiological transformations;
(2) how best to deal with preferential water flow and transport;
(3) how best to account for effects of spatial and temporal variability of soil hydraulic parameters on solute transport;
(4) how to efficiently couple multi-species geochemical submodels with unsaturated-saturated flow models;
(5) how to improve field methods for estimating vadose zone transport parameters, and the scale at which these apply, and

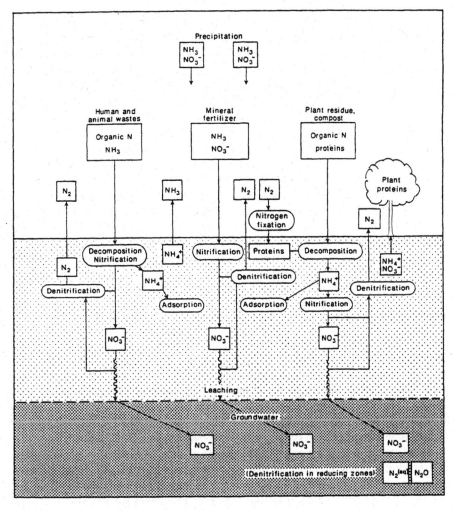

Figure 1 *Sources, movement and reactions of nitrogen in soils and groundwater (from Madison and Brunett, 1985).*

(6) how to predict the long term consequences of short-term management decisions.

Some of these issues have been discussed at length. For example Woolhiser *et al.* (1990) considered the effects of spatial and temporal variability on the ultimate form of the model, talking account of stochastic and monte-carlo approaches and drawing on scaling theory and sensitivity analysis methods.

Van der Heijde and Prickett (1990), Wagenet *et al.* (1990) and de Willigen *et al.* (1990) categorised models by use. Research models contained numerous parameters to describe the environmental variables which influence leaching in order to better understand the transformation and leaching processes. Other model categories include action-agency models and farm/extension models which rely on a number of assumptions or approximations and require less parameters to be defined. Such models are commonly used due to the significantly reduced effort required for model calibration and validation.

A trade off between mechanistic approaches and more empirical methods depends on the availability of data for model calibration. Beven and Jakeman (1990) argue that a stochastic model formulation provides a way of rationalising the number of parameters, and that model structure as well as parameter estimation error contribute to sources of predictive uncertainty. Use of models for regulatory purposes depends on an ability to estimate the prediction errors. A quality assurance policy, together with adequate data used correctly, an appropriate model, and properly posed problem and boundary conditions will assist in the use of the results of modeling for land and water resources management (Shaw and Falco,1990).

Nitrogen budget of grazed leguminous pastures

The nitrogen fluxes and transformations in a pasture system are illustrated in a diagram of Madison and Brunett (1985) (Figure 1). Descriptions of these processes are provided in a monograph edited by Stevenson (1982) and in other chapters of this book. This diagram shows the sequence of processes by which pasture residues and livestock wastes are mineralised and nitrified to produce nitrate which may be leached to groundwater. This mineralised nitrogen may be taken up through the roots of pasture plants and return to the pool of organic nitrogen for further cycling.

A net loss of nitrogen occurs in the protein of milk, wool and/or meat produced in a grazing operation. Gaseous losses of nitrogen occur through volatilisation of ammonia and denitrification. A net increase in organic nitrogen (40 to 50 kg ha^{-1} yr^{-1}) has also been observed in legume pastures of Southern Australia over 40 years (Russell 1960, Simpson and Stobbs, 1981). Hence the assumption of a closed system in equilibrium must be disgarded.

A crude model was written to describe the nitrogen budget, and especialy to estimate leaching, beneath grazed leguminous pastures in South Australia, Dillon (1988a). The model named NITWIT (NITrogen and Water In Transit) was used to determine a field (paddock) -scale estimate of the nitrogen budget (Figure 2). This revealed a net nitrogen turnover of 153 kg N ha^{-1} resulting in animal products (38%), gaseous losses (30%), increase in organic nitrogen storage (24%) and leaching (8%). Within this system there is considerable cycling through the soil inorganic nitrogen store with a turnover of almost 300 kg N ha^{-1} yr^{-1}.

This budget was determined by applying the model independently for a range of loadings of animal waste. The proportion of paddock area with each loading rate was determined from frequency distributions based on experimental evidence. The mean concentration and mass of nitrate leached from the paddock was calculated by

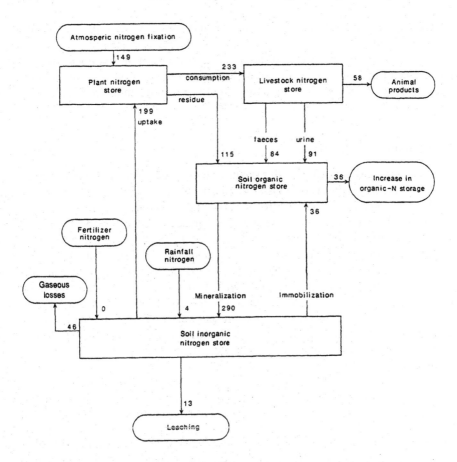

Figure 2 *Nitrogen flows modeled by NITWIT. Rectangular boxes represent nitrogen stores. Rounded boxes are net inputs and outputs of a non-closed cycle. Numbers next to arrows are estimates of mean annual nitrogen flows in kg N Ha^{-1} for the case of beef grazing pastures on aeolianite soils.*

Table 1 *Effects of extremes in pasture management on aquifer nitrogen load, nitrate-N concentration of recharge, and net annual increase in soil organic nitrogen storage (by NITWIT model).*

Pasture management	N load kg N ha^{-1} yr^{-1}	NO$_3$-N mg L^{-1}	Increase in organic N kg N ha^{-1} yr^{-1}
a. no harvest	15	7.4	124
b. harvest, no return	3	1.4	-7
c. livestock	13	6.2	36

aggregating the results of the model runs weighted by their corresponding paddock areas. The more novel features of this procedure are described below.

It is interesting to discover that the irregular distribution of livestock waste over the surface of the paddock results in the leaching of nitrate from pastures which are nitrogen-deficient in spring. If the nitrate could be retained in the root zone pasture growth would be enhanced.

Distribution of livestock wastes

Livestock play a significant part in the nitrogen cycle of pastures. They are thought to provide the major means of transfer of clover nitrogen to grass plants (MacLusky, 1960). More importantly grazing reduces shading of clovers by grasses and allows clovers to compete for the available sunlight Snaydon (1981). Defoliation therefore directly affects the legume content of pastures. Trampling by livestock in the vicinity of dairies and troughs also compacts soil and denudes it of cover (Till, 1981).

The leaching of nitrate is strongly influenced by the spatial pattern in which livestock redistribute nitrogen over a pasture. Ryden *et al.* (1984) found the amount of nitrate leached below a grazed pasture in Berkshire, UK was more than five times that leached below a comparable cut pasture and exceeded losses normally observed from arable land. Ball (1982), Quin (1982), Rijtema (1986) and Close and Woods (1986) among others have noted that urine patches may be the source of most nitrogen leached from grazed pastures.

Studies of the degree of non-uniformity of excreta distribution have been made by MacLusky (1960) for dairy cows, Petersen *et al.* (1956) for beef cattle, and Hilder (1966) for sheep. Simpson and Stobbs (1981) provided a brief review of the literature. In addition Brockington (1972) developed a model for pasture contamination by grazing cattle and the effect on palatable pasture available for consumption. There have also been studies (*e.g.* Dwyer, 1961) on the size and temporal frequency of discrete urination and defaecation products.

Petersen *et al.* (1956) made observations and fitted hypothetical distribution functions to data on the number of faeces in each 3m x 3m grid square in a North Carolina pasture grazed by beef cattle. A reanalysis of Petersen's data, changing the scale of interest from 9 sq m to the average area of a dung pat, resulted in selecting a negative binomial distribution with a constant nonuniformity factor (k=2) for all cases involving cattle (Dillon, 1988a).

This distribution did not fit Hilder's (1966) data for pastures grazed by sheep. Hilder found that about one third of excreta were concentrated on less than 5% of the paddock area. The tendency of sheep to camp in one or two locations resulted in a most irregular distribution of nutrients. This appears to fit a Poisson distribution which is applied to load rate - weighted subareas of the paddock (Dillon, 1988a). The Poisson distribution is appropriate if the tendency to find excreta at any position in a subarea is uniform over that subarea.

Two assumptions are made to simplify the model:

(1) That urine and faeces follow the same general pattern of distribution. (Generally faeces were observed in experiments). The mean excretal density for urine is

taken as the mean excretal density for faeces scaled up by the ratio of the area of urine to area of faeces contaminated by one animal per day.

(2) That the long term leaching of nitrogen from faeces is negligible. It is known that mineralisation of nitrogen in faeces occurs slowly over periods between one and five years. According to MacLusky (1960) it takes less than 4 years at normal stock densities for cattle faeces to affect 100% of the pasture area. Hence the effect of uneven distribution of faeces over a paddock is expected to be insignificant for leaching and result instead in variability of grass yield.

By comparison, the urea in urine hydrolyses rapidly to ammonium compounds which may subsequently volatilise (as ammonia) or be bacterially converted to nitrate. The rates of these reactions depend on soil moisture content, temperature and pH (Nelson, 1982; Aslying, 1986). The timing of urine application is therefore very important for leaching. In winter, ammonia volatilisation is slow and with the rapid passage of water through the root zone, the opportunity for uptake is reduced resulting in leaching of nitrate. In summer the conditions are reversed.

The typical water and nitrogen load applied to the area of soil covered by a single urination is about 10 mm and 260 kg N ha^{-1}.

Application of the NITWIT model

Kolenbrander (1981) described the two prerequisites for leaching of nitrogen as the occurrence of groundwater recharge and the presence of nitrogen in a mobile form in the soil at the time of year when recharge occurs. Leaching predictions in the area

Table 2 *Effect of number of urine applications on aquifer nitrogen load, recharge nitrate-N concentration, and net increases in soil organic nitrogen storage (by NITWIT model).*

No. urine applications	Sub-area fraction	Nitrogen load (kg N ha^{-1} yr^{-1})	Nitrate-N concentration (mg L^{-1})	Organic N increase (kg N ha^{-1} yr^{-1})
0	0.133	3	1.4	-7
1	0.169	6	2.8	5
2	0.161	9	4.2	18
3	0.137	11	5.6	31
4	0.109	14	7.0	44
5	0.083	17	8.4	56
6	0.062	20	9.8	69
7	0.045	23	11.3	82
8	0.033	25	12.3	95
9	0.023	28	13.5	108
10	0.016	30	14.8	121
11	0.012	33	16.1	134
12	0.008	35	17.1	147
13	0.006	38	18.2	159
14	0.004	40	19.4	174
Pasture mean		12.5	6.2	36.4

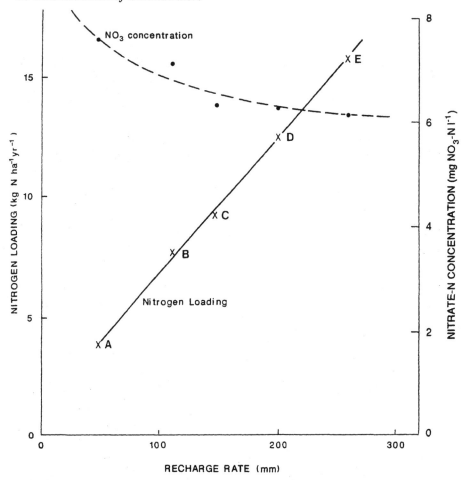

Figure 3 *Nitrogen loading on aquifer and concentration of nitrate-N in recharge related to mean annual recharge rate for different soil groups.*

studied therefore rely on the ability to quantify soil nitrate concentrations during winter.

This depends on the excess of mineralisation and nitrification of pasture residues and animal wastes during summer and autumn over uptake by grasses and young clover plants during late autumn. Generally the excess is a small fraction of the amount nitrified for both perennial and annual pastures.

The NITWIT model was applied using all available data for calibration, for a set of base conditions including 200 mm recharge through an aeolianite soil type with an improved leguminous pasture grazed by beef cattle which consume two-thirds of the pasture each year.Initial tests examined the effect of pasture management on aquifer

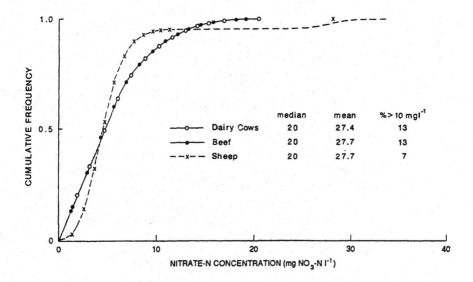

Figure 4 *Cumulative frequency of nitrate-N concentration in recharge for each of three stock types grazing pastures on aeolianite soils.*

nitrogen loads. These were intended to evaluate model output under a wide range of input conditions and check for inconsistencies. Three cases were tested:

(a) No pasture harvested and no livestock.

(b) 67% pasture harvested and not returned, and

(c) 67% pasture eaten by beef cattle - all faeces and urine returned to the paddock.

In (a) grasses would shade clovers so the amount of nitrogen fixation and hence immobilisation is overestimated. One could expect the grass yield to decline in subsequent seasons due to reduced nitrogen input. This illustrates the requirement for the pasture yield and clover content specified in the input to be realistic with respect to pasture management. The high nitrogen load given by the model is due to leaching of the mineralised pasture residue.

(b) is the opposite extreme where no nitrogen is returned from the harvested pasture. This leads to very slow depletion of soil organic nitrogen. For simplicity of modeling subsequent seasons it is assumed that mineralisation from the large organic nitrogen store continues in order to allow the specified pasture yield to be met. In this no return case the nitrogen load on the aquifer 2.8 kg N ha^{-1} yr^{-1} is less than the rainfall nitrogen load on the land surface 4 kg N ha^{-1}yr^{1}. The nitrate-N concentration in recharge, 1.4 mg L^{-1} is at about the 30th percentile of all groundwater nitrate samples outside the influence of identified point sources of contamination. The nitrogen load is apparently in the right range.

(c) This represents normal pastoral practice and includes consumption by livestock and return of animal excreta to the pasture. The results are intermediate between the two previous cases as expected. Urine nitrogen contributes to the recharge concentration as is also evident in the marginal increase in recharge from 200 mm to 202 mm. (Details of the flow of nitrogen are given in Figure 3.)

Urine is distributed over the paddock so that each part of the area receives a discrete number (between zero and fourteen) of urine applications over the 10 years of the simulation. A negative binomial distribution defines the proportion of area subject to each bovine urine loading. Several simulations were performed for each urine loading, varying the month of the first urine application over all possible months, while keeping the interval between applications constant.

This last simplifying assumption (constant intervals between applications) gives a downward bias to the predicted aquifer nitrogen load to the extent that multiple urine applications in the same recharge year are neglected. Loadings in different recharge years are essentially independent. In addition to multiple runs for each urine loading, each run is repeated if necessary until the difference between initial and final inorganic nitrogen storages is less than 5% of the aquifer nitrogen load.

For each urine loading the key outputs are averaged for simulations with different starting months for urine application. The resulting mean key outputs are then weighted by the fraction of area receiving that urine loading combined with the other weighted sub-area means to determine mean figures for the pasture. Table 2 shows the effect of the number of urine applications on sub-area aquifer nitrogen load. This indicates that the 13% of area with more than six urine applications in 10 years recharges water with an average nitrate-N concentration in excess of 10 mg L^{-1}. It will be seen later that the model apparently underestimates aquifer nitrogen loads for areas receiving high urine loadings.

Tests showed that these results were insensitive to the length of each simulation, and ten years was chosen as the standard length. All model testing and sensitivity analyses were performed on one soil type (aeolianite) with 200 mm annual recharge. (More than one third of groundwater recharge in the study area originated in this soil type, Allison and Hughes, 1978). Values for the four parameters affecting flow were physically based or fitted to give the measured annual recharge rate (200 mm) and the sesonal timing of recharge.

The regional nitrogen load and mean nitrate concentration in recharge were determined by fitting hydraulic parameters for six major soil classifications based on recharge rate (Allison and Hughes, 1978) and applying the model for each. The results were then weighted according to the proportion of area occupied by each soil type. The pastoral means were found to be 10 kg N ha^{-1} yr^{-1} and 6.4 mg NO_3N L^{-1}. Figure 3 shows that the aquifer nitrogen load is directly proportional to recharge rate and that the nitrate concentration of recharge is insensitive to recharge rate. This is consistent with lysimeter observations of Kolenbrander (1981).

These results compare favorably with those used by Carey and Lloyd (1985) in modeling the Great Ouse chalk aquifer in eastern England. They reported nitrate concentrations to be typically less than 2 mg NO_3-N L^{-1} beneath unfertilised grass and in the range 4 to 10 mg NO_3-N L^{-1} for fertilized grassland. In modeling grassland and woodland they assumed the nitrate concentration of recharge to be constant and the quantity leached to depend only on the amount of recharge.

Different stock types; sheep, dairy and beef cattle proved to have almost identical effects on the nitrogen budget including the leaching of nitrate. The cumulative frequency of nitrate-N concentrations in recharge beneath a pasture on aeolianite soil is shown for each stock type in Figure 4. It was assumed that the stocking rate was at its maximum sustainable level for each stock type as determined historically for the region.

These results which indicate that the nitrate concentration in recharge is insensitive to stock type and soil type simplifies the approach to land and groundwater management strategies.

Frequency distributions of nitrate concentrations in recharge water (from NITWIT) and in groundwater sampled in areas outside the influence of known point sources of contamination, were compared. Mixing within the aquifer would be expected to reduce the variance of groundwater concentrations with respect to recharge concentrations. This served as a further guide in calibration of the NITWIT model. The procedure is described by Dillon (1988a), along with a sensitivity analysis of the model.

Model outputs proved to be sensitive to only a few parameters. The nitrogen load on the aquifer is most sensitive to the proportion of clover in the pasture and the clover nitrogen content. The product of these two terms is an index of nitrogen fixation - the major source of nitrogen. This affirms conclusions of Burden (1982) and Lawrence (1983) that the amount of nitrogen fixed is an important determinant of the amount of nitrate leached. The clover fraction is particularly important because it also defines the fraction of grass which takes up nitrogen in early winter reducing the stored inorganic nitrogen available for leaching.

The nitrogen load in recharge was found to be sensitive to the fraction of nitrogen consumed by livestock which is returned to the pasture in urine and faeces, and the fraction of pasture dry matter consumed by livestock. Gaseous losses of nitrogen (volatilisation plus denitrification) also had an influence on the nitrogen load of a pasture on groundwater. A number of other variables were examined and found to have little effect on nitrate concentration of recharge or on the nitrogen load in recharge.

Fate of nitrate leached to groundwater

Rsearchers including Gillham (1991) and Schipper *et al.* (1991) have identified natural denitrification zones in groundwater systems. These occur where nitrate enters anaerobic parts of aquifers where bacterial denitrification is facilitated in the presence of labile organic carbon substrates.

In groundwater beneath grazed leguminous pastures in the south east of South Australia the observed ratio of the dissolved organic carbon to nitrate is low. Outside of point source plumes, the saturated zone contains dissolved oxygen (inhibiting denitrification) to depths exceeding 60 metres below water table and nitrate behaves as a conservative solute. From the results of the NITWIT modeling a uniform concentration of nitrate in recharge may be considered for a broad region. The regional spatial scale and a time-span of interest of several hundred years demanded

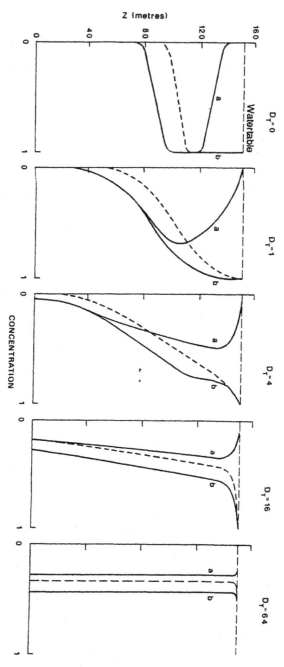

Figure 5 *Concentration profiles in groundwater in 1986 (shown dashed) after 120 years recharge at unit concentration, and 50 years later for cases where unit recharge concentration is (a) eliminated in 1986, and (b) sustained. The effects depend on the coefficient of dispesion in the vertical dimension, D_T (m^2 yr^{-1}).*

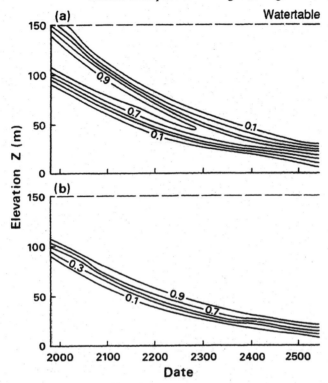

Figure 6 *Predicted concentration profile history for 500 years with unit recharge concentration (a) eliminated in 1986, and (b) sustained. Coefficient of dispersion in the vertical dimension, D_T is 0 ($m^2 yr^{-1}$).*

that careful attention be paid to the discretisation of any numerical model. Kinzelbach *et al.* (1990) describe a range of modeling approaches to predict the fate of nitrate leached to groundwater. The limitations on grid size imposed by the mesh Peclet number and the limitations on the length of time steps imposed by stability conditions suggested that an analytical solution be used instead of a numerical approach.

Consequently a model known as DIVAST was developed (Dillon, 1989a,b, 1988b). This is a Diffuse source Vertical slice Analytical Solute Transport model. It is designed as a management tool for preliminary regional evaluation of potential groundwater contamination due to diffuse solute sources, and to predict the future distribution of solute in the aquifer under alternative land management strategies. Such strategies for long-term control of groundwater contamination in Britain are described by Chilton and Foster (1991).

DIVAST enables recharge concentration histories to be superimposed on surface elements with constant recharge and the consequences to groundwater concentration profiles (concentration versus elevation within the aquifer) to be determined at any location and time. The model allows determination of stratified concentration patterns as this is a common feature of many anthropogenically contaminated

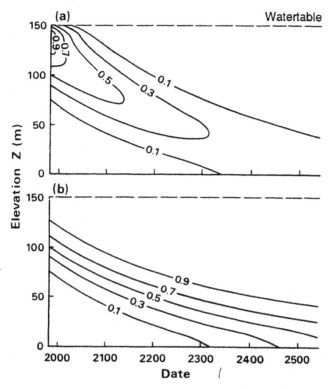

Figure 7 *Predicted concentration profile history for 500 years with unit recharge concentration (a) eliminated in 1986, and (b) sustained. Coefficient of dispersion in the vertical dimension, D_T is 1 $(m^2\ yr^{-1})$.*

aquifers (*e.g.* Spalding, 1984). This capability may disclose management options not previously considered (Foster *et al.*, 1985).

The limestone aquifer underlying the Gambier Plain in southeastern Australia is unconfined. Rainfall averages from 700 to 750 mm year^{-1}. Mean regional recharge to the unconfined limestone aquifer is estimated to be about 150 mm year^{-1} (Allison and Hughes, 1978). The vadose zone consists of sand and clay ranging in thickness to 6 m (Holmes and Colville, 1970). The watertable varies in depth from 0 to 30 m and is generally near the top of the limestone. This is underlain at depths down to 300 m by an aquitard with low leakance which overlies a confined sandstone aquifer. These aquifers are the sole water supply for this region and the unused groundwater discharges into coastal wetlands and the sea.

The model was run by assigning a uniform unit recharge concentration from the year 1866 (when the first cheese factory started) in a previously uncontaminated aquifer. Two options are considered:

(1) To stop immediately all further contaminating activity in 1986, and

(2) To continue contamination unabated.

While the first is an impossible situation, it allows the maximum potential

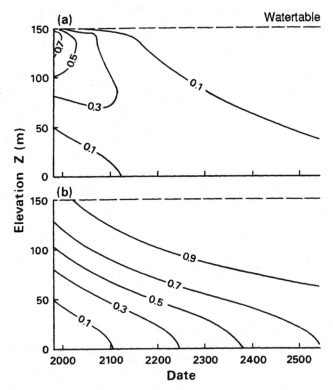

Figure 8 *Predicted concentration profile history for 500 years with unit recharge concentration (a) eliminated in 1986, and (b) sustained. Coefficient of dispersion in the vertical dimension, D_T is 4 (m^2 yr^{-1}).*

improvement in groundwater quality to be estimated if the harshest policies for waste disposal and land use practices policies were imposed immediately. The concentration profile was found to be sensitive to the value of the coefficient of dispersion in the vertical direction.

The forecast nitrate concentration profiles (neglecting unsaturated zone lag) for the year 2036 under the two management policies are shown in Figure 5. For the case of a sudden demise in recharge nitrate concentrations, if vertical mixing is small, bores located deeper than about 60 m are more likely to have higher concentrations than they have at present. Shallow bores (less than 30 m below water table) would experience substantial benefits. For strong mixing, bores at all depths should experience an 18% decline in nitrate concentration. On the other hand if nitrate loadings continue, the affected depth increases for weak mixing but the nitrate concentration increases through the aquifer thickness by 30% for strong mixing. Profile forecasts over 500 years for each 'option' are shown in Figures 6 to 9 for each of four coefficients for vertical mixing.

Matching nitrate concentrations below the watertable to the recharge concentrations determined by the NITWIT model infers that the rate of mixing is

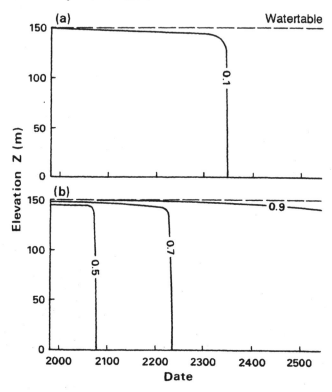

Figure 9 *Predicted concentration profile history for 500 years with unit recharge concentration (a) eliminated in 1986, and (b) sustained. Coefficient of dispersion in the vertical dimension, D_T is 64 (m^2 yr^{-1}).*

very slow in this aquifer. This is confirmed by the contaminant stratification found within the aquifer.

New research directions in diffuse source nitrate transport beneath pastures

Geostatistics has provided valuable tools to investigate the structure of spatial variability of nitrate transport beneath pastures. These provide a means of assessing the correlation length (semivariogram range) for variables describing sources or transport processes. This assists in the design of field sampling programs and often the correlation length itself indicates the scale of the most significant process. For example, correlation lengths less than one metre in surficial nitrate and ammonium concentrations in top soil tend to indicate the dominance of livestock waste application to the pasture as a major determinant of the leaching of nitrate.

Nitrate concentrations in groundwater have also been observed to exhibit spatial correlation (Schouten and Okx, 1987; Dillon, 1988a). Geostatistical methods provide a methodological basis for developing optimal sampling networks for both soil and groundwater.

Empirical models have been successfully used with remote sensing data, data base management systems, and geographic information systems to model nitrate buildups. This approach has had some success in relating changes in management practices to changes in nitrate concentrations (Breeusma and van Duijvenbooden, 1987).

Natural cleanup of nitrate has been observed in riparian zones (adjacent streams) in New Zealand by Schipper *et al.* (1991). This is due to *in-situ* denitrification as the nitrate is conveyed into anoxic soils rich in organic carbon in swampy ground next to streams. Similar features have been observed in duplex soils in pastoral areas of the south east of SA. Lateral interflow above a shallow (0.7m) clay layer results in ponding in swales with the perched water table eventually rising to the ground surface saturating the organically rich top soils. Denitrification is almost instantenous (personal communication. M. Hughes, formerly CSIRO Division of Soils).

Other evidence of denitrification is given by Gillham (1991), where anoxic conditions prevail below a given depth below water table in aquifers which are recharged with oxic water containing dissolved organic carbon. These studies suggest that provided the dissolved organic carbon to nitrogen ratio in recharge water can be maintained, there are opportunities for *in-situ* bioremediation of aquifers. These require further exploration to determine our ability to induce and capitalise on these processes.

Techniques are now available whereby a drilling rig can reem several holes per day to install non-weighing "instant" lysimeters. These intact cores enable small diameter (300mm) lysimeters to be installed with depths of up to 1.5 m. The sampling device attached measures suction at the base of the lysimeter and leachate volume and allows samples to be taken for measurement of solute concentration and calculation of solute flux. Equipment for real time monitoring soil moisture profiles and of nitrate concentrations in water are now becoming commercially available. Such field instrumentation is expected to facilitate model calibration and validation.

An appreciation of the factors affecting microbiological processes in pasture soils has been a weakness of mechanistic models to date. This is a field inviting environmental microbiology research; in particular nitrogen fixation, immobilisation, mineralisation, nitrification and denitrification. Descriptions of these microbially mediated processes under the range of field conditions, normally encountered, would lead to more accurate mechanistic models.

Systems approaches which enable coupling of submodels of appropriate complexity given the modeling objectives and the availablity of data for calibration are currently being developed. It is hoped that these will provide a means of discriminating the effectiveness of submodels to enable integration of the best components of the plethora of leading models currently available.

References

Addiscott, T.M. and Wagenet, R.J. 1985. Concepts of solute leaching in soils: a review of modeling approaches. *J. Soil Sci.,* 36, 411-424.

Allison, G.B. and Hughes, M.W. 1978. The use of environmental chloride and tritium to estimate total recharge to an unconfined aquifer. *Aust. J. Soil Res.*, 16, 181-195.

Aslying, H.C. 1986. Nitrogen balance in crop production and groundwater quality. In: Van, A.L.M. and Wesseling. J. (eds), *Agricultural Water Management*, pp.291-301. Proc. Symp., Arnhem, Netherlands. A.A. Balkema, Rotterdam.

Ball, P.R. 1982. Nitrogen balances in intensively managed pasture systems. In *Nitrogen Balances in N.Z. Ecosystems*, pp.46-66. DSIR Plant Physiology Div, Palmerston North, N.Z.

Bergstrom, L., Jansson, P-E and Johnsson, H. 1990. A model for simulation of nitrogen dynamics in soil and nitrate leaching. In: *Proc. Int. Symp. on Water Quality Modeling of Agricultural Non-Point Sources*, pp.803-807. Logan, Utah, June 1988. USDA ARS-81, June 1990.

Beven, K.J. and Jakeman, A.J. 1990. Complexity and uncertainty in predictive models. In: *Proc. Int. Symp. on Water Quality Modeling of Agricultural Non-Point Sources*, pp.555-576. Logan, Utah, June 1988. USDA ARS-81, June 1990.

Bogardi, I., Fried, J.J., Frind, E., Kelly, W.E. and Rijtema, P.E. 1990. Groundwater quality modeling for agricultural non-point sources. In: *Proc. Int. Symp. on Water Quality Modeling of Agricultural Non-Point Sources*, pp.227-252. Logan, Utah, June 1988. USDA ARS-81, June 1990.

Breeusma, A. and van Duijvenbooden, W. 1987. Mapping of groundwater vulnerability to pollutants in the Netherlands. *Proc. Conf. on Vulnerability of Soils and Groundwater to Pollutants*, pp.309-320. The Hague.

Brockington, N.R. 1972. A mathematical model of pasture contamination by grazing cattle and the effects of herbage intake. *J. Agric. Sci. Camb.*, 79, 249-257.

Burden, R.J. 1982. A model of nitrogen leaching from grazed pastureland to aid groundwater quality management. In: *Nitrogen Balances in N.Z. Ecosystems*, pp.237-244. DSIR Plant Physiology Division, Palmerston North, N.Z.

Carey, M.A. and Lloyd, J.W. 1985. modeling non-point sources of nitrate pollution in groundwater in Great Ouse Chalk, U.K.. *J. Hydrol.*, 78, 83-106.

Chilton, P.J. and Foster, S.S.D. 1991. Control of groundwater nitrate pollution in Britain by land use change. In: *Nitrate Contamination*. Proc. NATO Advanced Research Workshop, Lincoln, Nebraska, Sept 1990.

Close, M.E. and Woods, P.H. 1986. Leaching losses from irrigated pasture: Waiau Irrigation Scheme, North Canterbury. *N.Z. J. Agric. Res.*, 29, 339-349.

Dagan, G., Russo, D. and Bresler, E. 1990. Effect of spatial variability upon subsurface transport of solutes from non-point sources. In: *Proc. Int. Symp. on Water Quality Modeling of Agricultural Non-Point Sources*, pp.523-548. Logan, Utah, June 1988. USDA ARS-81, June 1990.

Dillon, P.J. 1988a. *An evaluation of the sources of nitrate in groundwater near Mount Gambier, South Australia.* CSIRO Div Water Resources Series No 1, Canberra, 68pp.

Dillon, P.J. 1988b. *DIVAST: Diffuse source vertical analytical solute transport model. Users Manual includes CONT contouring program.* CSIRO Division of Water Resources, Tech Memo 88/2, Canberra, 18pp plus appendices.

Dillon, P.J. 1989a. An analytical model of solute transport from diffuse sources in saturated porous media. *Water Resources Research*, 25(6), 1208-1218.

Dillon, P.J. 1989b. Models of nitrate transport at different space and time scales for groundwater quality management. In: Jousma, G. *et al.* (eds), *Groundwater Contamination: Use of Models in Decision Making*, pp.273-284. Kluwer Academic Publishers, Dordrecht.

Dillon, P.J., Close, M.E. and Scott, R.I. 1989. Diffuse-source nitrate contamination of groundwater in New Zealand and Australia. *Hydrology and Water Resources Symp. Christchurch,* Nov 1989, I.E.Aust Nat. Conf. Publ. No. 89/19, pp.351-355.

Duffy, C.J., Kincaid, C. and Huyakorn, P. 1990. A review of groundwater models for assessment and prediction of nonpoint-source pollution. In: *Proc. Int. Symp. on Water Quality Modeling of Agricultural Non-Point Sources,* pp.253-275. Logan, Utah, June 1988. USDA ARS-81, June 1990.

Dwyer, D.D. 1961. *Activities and grazing preferences of cows with calves in Northern Osage County,* Oklahoma. Oklahoma State Univ. Experiment Station Bulletin B-588.

Foster,S.S.D, Geake, A.K., Lawrence, A.R. and Parker, J.M. 1985. Diffuse groundwater pollution: lessons of the British experience. In: *Hydrogeology in the Service of Man,* IAHS Publ. No 154, pp.168-177.

Gillham, R.W. 1991. Nitrate contamination of groundwater in southern Ontario and the evidence for denitrification. in Nitrate Contamination. *Proc. NATO Advanced Research Workshop,* Lincoln, Nebraska, Sept 1990.

Heijde, van der P.K.M. and Prickett, T.A. 1990. Groundwater quality models for planning, management and regulation. In: *Proc. Int. Symp. on Water Quality Modeling of Agricultural Non-Point Sources,* pp.279-306. Logan, Utah, June 1988. USDA ARS-81, June 1990.

Hilder, E.J. 1966. Distribution of excreta by sheep at pasture. *Proc. 10th Int. Grassland Congress,* Helsinki, pp 977-981.

Holmes, J.W. and Colville, J.S. 1970. Grassland hydrology in a karstic region of Southern Australia. *J. Hydrol.,* 10, 38-58.

Jones, R.L. and Hanks, R.J. 1990. Review of unsaturated zone leaching models from a users perspective. In: *Proc. Int. Symp. on Water Quality Modeling of Agricultural Non-Point Sources,* pp.129-139. Logan, Utah, June 1988. USDA ARS-81, June 1990.

Kinzelbach, W.K.H., Dillon, P.J. and Jensen, K.H. 1990. State of the art of existing groundwater quality models of the saturated zone and experience with their application in agricultural problems. In: *Proc. Int. Symp. on Water Quality Modeling of Agricultural Non-Point Sources,* pp.307-325. Logan, Utah, June 1988. USDA ARS-81, June 1990.

Kolenbrander, G.J. 1981. Leaching of nitrogen in agriculture. In: Brogen, J.C. (ed.), *Nitrogen Losses and Surface Runoff,* pp 199-216. Nijhoff/Junk Publ., The Hague.

Lawrence, C.R. 1983. *Nitrate rich groundwaters of Australia.* Aust. Water Resources Council Tech. Paper No 79.

MacLusky, D.S. 1960. Some estimates of the areas of pasture fouled by the excreta of dairy cows. *J. Br. Grassl. Soc.,* 15, 181-88.

Madison, R.J. and Brunett, J.O. 1985. *Overview of the occurrence of nitrate in groundwater of the United States.* USGS Water Supply Paper 2275, pp.93-105.

Nelson, D.W. 1982. Gaseous losses of nitrogen other than through denitrification. p327-363 In: Stevenson, F.J.(ed.), *Nitrogen in Agricultural Soils,* pp.327-363. Agronomy Series No 22, Am. Soc. Agronomy, Madison, Wisconsin.

Petersen, R.G., Lucas, H.L. and Woodhouse, W.W. 1956. The distribution of excreta by freely grazing cattle and its effect on pasture fertility. *Agron. J..* 48, 440-449.

Plate, E. and Duckstein, L. 1990. Stochastic aspects of water quality modeling for non-point sources. In: *Proc. Int. Symp. on Water Quality Modeling of Agricultural Non-Point Sources,* pp.631-654. Logan, Utah, June 1988. USDA ARS-81, June 1990.

Quin, B.F. 1982. The influence of grazing animals on nitrogen balances. In: *Nitrogen Balances in N.Z. Ecosystems,* pp.95-102. DSIR Plant Physiology Div, Palmerston North, N.Z.

Rijtema, P.E. 1986. Nitrate load and management of agricultural land. In: Van Wijk, A.L.M. and Wesseling, J. (eds), *Agricultural Water Management*, pp.303-311. Proc. Symp. Arnhem. A.A. Balkema, Rotterdam.

Rijtema, P.E. and KROES, J.G. 1991. Nitrogen modeling on a regional scale. In: *Nitrate Contamination*. Proc. NATO Advanced Research Workshop, Lincoln, Nebraska, Sept 1990.

Russell, J.S. 1960. Soil fertility changes in the long-term experimental plots at Kybybolite, South Australia, 1. Changes in pH, total nitrogen, organic carbon and bulk density. *Aust. J. Agric. Res.,* 11, 902-26.

Ryden, J.C., Ball, P.R. and Garwood, E.A. 1984. Nitrate leaching from grassland. *Nature*, 311(5981), 50-53.

Schipper, L.A., Cooper, A.B and Dyck, W.J. 1991. Mitigating non-point source nitrate pollution by riparian zone denitrification. In: *Nitrate Contamination*. Proc. NATO Advanced Research Workshop, Lincoln, Nebraska, Sept 1990.

Schouten, C.J. and Okx, J.P. 1987. Identifying regional differences in groundwater quality in a densely populated loess region. Proc. Conf. on Vulnerability of Soils and groundwater to Pollutants, The Hague, pp 459-470.

Scott, R.I. and Dillon, P.J. 1989. *Reconnaissance of soil water nitrate and related microorganisms near Mount Gambier South Australia.* Centre for Groundwater Studies Report No 20, Nov 1989. 29 pp.

Shaw, R.R. and Falco, J.W. 1990. A management perspective on water quality models for agricultural non-point sources of pollution.In: *Proc. Int. Symp. on Water Quality Modeling of Agricultural Non-Point Sources,* pp.697-704. Logan, Utah, June 1988. USDA ARS-81, June 1990.

Simpson, J.R. and Stobbs, T.H. 1981. Nitrogen supply and animal production from pastures. In: Morley, F.H.W. (ed.), *Grazing Animals, World Animal Science, B1*, pp 33-53. Elsevier, Amsterdam

Stevenson, F.J. (ed.). 1982. *Nitrogen in Agricultural Soils.* Agronomy Series No 22, Am. Soc. Agronomy, Madison, Wisconsin.

Snaydon, R.W. 1981. The ecology of grazed pastures. In: Morley, F.H.W. (ed), *Grazing Animals, World Animal Science, B1*, pp.13-31. Elsevier, Amsterdam.

Spalding, E. 1984. *Implication of temporal variations and vertical stratification of groundwater nitrate-nitrogen in the Hall County special use area.* Nebraska Water Resources Center Report, Lincoln, Nebraska. NTIS-PB85-227262/AS. Tanji, K.W. 1982. Modeling of the soil nitrogen cycle. In: Stevenson, F.J. (ed.), *Nitrogen in Agricultural Soils*, pp. 721-72. No. 22 in the Agronomy series, Am. Soc. Agronomy, Madison, Wisconsin.

Till, A.R. 1981. Cycling of plant nutrients in pastures. In: Morley, F.H.W. (ed.), *Grazing Animals, World Animal Science, B1*, pp 33-53. Elsevier, Amsterdam

Vachaud, G., Vauclin, M. and Addiscott, T.M. 1990. Solute transport in the vadose zone: a review of models. In: *Proc. Int. Symp. on Water Quality Modeling of Agricultural Non-Point Sources,* pp.81-104. Logan, Utah, June 1988. USDA ARS-81, June 1990.

Wagenet, R.J., Shaffer, M.J. and Green, R.E. 1990. Predictive approaches for leaching in the unsaturated zone. In: *Proc. Int. Symp. on Water Quality Modeling of Agricultural Non-Point Sources,* pp.63-79. Logan, Utah, June 1988. USDA ARS-81, June 1990.

Willigen de P., Bergstrom, L. and Gerritse, R.G. 1990. Leaching models of the unsaturated zone: their potential use for management and planning. In: *Proc. Int. Symp. on Water Quality Modeling of Agricultural Non-Point Sources,* pp.105-128. Logan, Utah, June 1988. USDA ARS-81, June 1990.

Woolhiser, D.A., Jury D.A. and Nielsen, D.R. 1990. Effects of spatial and temporal variability on water quality model development. In: *Proc. Int. Symp. on Water Quality Modeling of Agricultural Non-Point Sources*, pp.505-522. Logan, Utah, June 1988. USDA ARS-81, June 1990.

4

Modeling point sources of groundwater contamination: a review and case study

G. B. Davis and R. B. Salama

CSIRO Division of Water Resources, Private Bag PO Wembley, Western Australia 6014

Abstract

Modeling is a tool which aids in assessing and predicting the likely impact of point sources of contamination on groundwater quality. In this chapter the results of recent modeling work on characterising the scales of hydrodynamic dispersion are reviewed. We also show in the chapter the usefulness of analytical models for initial assessment of the impact of a point source, and compare analytical models extensively to two and three dimensional numerical models for a test-case of a landfill-leachate plume in groundwater.

The review highlights the need to consider field-scale aquifer heterogeneity, since it complicates groundwater flow paths and leads to scale dependency of the hydrodynamic dispersion coefficient. Both deterministic and stochastic approaches to dispersion modeling suggest that the dispersion coefficient, although dependent on the scale of measurement, asymptotes at large measurement scales (in the far-field) to a constant value. For predictions close to the point source (in the near-field) fine detail of the variability of the hydraulic conductivity are required to adequately define the dispersion coefficient. It is clear that further research is needed to develop cost-effective means of identifying the appropriate scale of measurement of the hydraulic conductivity in heterogeneous aquifers.

Analytical modeling is shown to allow reliable testing of modeling assumptions. Analytical models are used to aid definition of the finite dimensions of a "point" source in relation to plume dimensions, and to show the effect on plume development of the occurrence of boundaries within the flow domain. Also the chapter highlights the advantages of three-dimensional modeling over two-dimensional modeling.

Finally three point-source modeling approaches (three-dimensional semi-analytical: SAM3D, and two- and three-dimensional numerical: MOC and HST3D) are compared for a test case of landfill-leachate contamination of groundwater. Uniform groundwater velocities are assumed but the irregular landfill shape is retained. Significant numerical error is observed for MOC for larger grid sizes (Peclet numbers ≥ 3.5). HST3D displays less numerical error than MOC and gives results comparable to those from the semi-analytical model SAM3D.

Introduction

Instances of point source pollution of groundwater are legion (Keeley, 1977; Cheremisinoff *et al.*, 1984). Chemical spills and leakages, industrial and domestic waste disposal, leachate from landfilling, septic tank effluent and industrial waste water outfalls are some of the many potential sources of point source contamination. Growing awareness of the occurrence of groundwater contamination from such sources has led to the need to be better equipped to monitor and predict the extent of pollution in groundwater and to assess the migration rates of contaminant species in groundwater and their dispersion potential.

Also water and environmental regulatory authorities need to be able to assess and predict the impact of such point sources on groundwater and its likely effect on community water supplies and the environment into the future. Such demands can be met partly by appropriate emplacement and monitoring of observation boreholes (Barber *et al.*, 1991). But where predictions of impact are required, modeling becomes an essential element of any assessment.

Models of contaminant transport in groundwater can take many forms (Javandel *et al.*, 1984; Bear and Verruijt, 1987; Kinzelbach, 1988) depending on contaminant species present in groundwater, knowledge of the region of interest and the objectives of any predictive assessment. Whereas fully three-dimensional numerical modeling of transient groundwater flow and contaminant transport may be feasible where an abundance of data is available, it is hardly appropriate if little or no borehole data is available and the site contains low priority pollutants, in a relatively isolated area. Simple models are useful for providing broad indications of likely contaminant motion for planning of monitoring networks (Davis *et al.*, 1988). However simple models would not make full benefit of an abundance of borehole data, and would not provide the better definition of flow paths and likely impact. High priority pollutants in heterogeneous aquifers, especially warrant more detailed modeling.

A variety of physical and chemical processes affect the behaviour of contaminants in groundwater systems. These include advective transport by groundwater motion, dispersion, diffusion, density driven motion, adsorption/desorption and other chemical reaction processes, and radioactive decay. Each of these can play a major role in the migration and development of contamination within groundwater. Generally groundwater motion is of principal importance since it has the greatest capacity for transport of pollution. Dispersion, adsorption and other chemical reaction processes would also be important, depending on the hydrogeology of the aquifer, its flow characteristics, the physicochemical interactions between the contaminant and the aquifer matrix, amongst other factors. These latter processes are increasingly being integrated in transport models (*e.g.* Schäfer and Kinzelbach, 1992). Where groundwater is slow moving (or stagnant), however, diffusional and possibly density driven transport may predominate. Of all the transport processes affecting contaminant migration in groundwater, hydrodynamic dispersion appears to be the one most difficult to describe mathematically and estimate in the field. Solute-matrix interaction (*e.g.* sorption/desorption) may not be as significant as dispersion in affecting contaminant migration, but it is also highly complex and difficult to quantify.

The groundwater scientist therefore needs to have available a range of models which can address issues specific to a site. Some models will be reviewed later in this paper, ranging from simple analytical and semi-analytical models to one, two or three dimensional numerical models. The advantages, limitations, and difficulties of application of each type of model will be illustrated by reference to a case study.

Dispersion

Dispersion refers to the enhanced spreading of a mass of a chemical species in groundwater due to groundwater motion. The term is used similarly to describe gaseous or particulate dispersion in the atmosphere, and dispersion due to turbulent energy fluxes in surface water bodies, such as rivers and streams (Fischer *et al.*, 1979). In groundwater, dispersion leads to spreading of a contaminant over a much larger volume of the aquifer than would occur by advective transport only. The enhanced spreading, under laminar flow conditions, is a consequence of the transport of a portion of the contaminant mass at a range of velocities, due to micro-scale boundary layer effects and soil heterogeneities.

One of the first to describe the principal features of the effects of dispersion on solutes in groundwater was Slichter (1905). He conducted a field tracer test and laboratory tests to estimate groundwater velocities. Instead of the expected step breakthrough of the tracer due to advective transport he observed the now typical s-shaped breakthrough curve. Since Slichter, many researchers have studied breakthrough characteristics of tracers in laboratory columns and field tests to provide a better quantitative understanding of dispersion.

Scheidegger (1961) demonstrated, assuming negligible diffusion, that the hydrodynamic dispersion coefficient was proportional to groundwater velocity, such that under an isotropic uniform groundwater flow velocity (v)

$$D_L = \alpha_L v \qquad\qquad (2.1)$$

$$D_T = \alpha_T v \qquad\qquad (2.2)$$

where D_L and D_T are the longitudinal and transverse dispersion coefficients ($m^2 s^{-1}$) respectively, and α_L and α_T are the longitudinal and transverse dispersivities (m). More complex formulations of the relationship between the dispersion coefficient and groundwater velocities can be found in Bear (1979) and others.

It is reasonably well accepted now that the dispersivity varies depending on the scale at which it is measured (Anderson, 1984; Gelhar *et al.*, 1985), although there is continuing argument with this view (Robbins and Domenico, 1984; Taylor and Howard, 1987). Evidence suggests that dispersivities estimated at the column scale represent microscale dispersion processes. Column scale dispersivities are often 10 to more than 100 times smaller than macroscale dispersivities estimated from field tracer trials or from contaminant plume monitoring (see *e.g.* Gelhar *et al.*, 1985). The large discrepancy between laboratory and field scale estimates of the dispersivity are thought to be due to the inherent heterogeneity of field soils and aquifers. A tracer experiment at the column scale (say, of the order of 1 m) "sees" only microscale variability - a small amount of the likely variability in an aquifer. On the other hand, a tracer or pollutant monitored at the field scale (say, 10 m up to kilometres) encounters a larger volume of the aquifer containing both microstructural (*e.g.* pore size or intergranular) and macrostructural (*e.g.* clay interbedded with sand) features, leading to greater dispersion.

Some authors (Lallemand-Barres and Peaudecerf, 1978) have indicated a linear increasing relationship between the magnitude of the dispersivity and the scale of measurement. Others (Gelhar and Axness, 1983) have suggested that the dispersivity varies with scale in the near-field (*i.e.* close to the source of contamination) but asymptotes to a constant value in the far-field (*i.e.* at large distances from the source). Mishra and Parker (1990) used an empirically fitted hyperbolic curve in their modeling to depict this trend. Field studies at the Borden landfill site (Sudicky, 1986; Freyberg, 1986), and theoretical studies by Matheron and de Marsily (1980) and Dagan (1983), also indicate that dispersion coefficients are time dependent, reaching a constant asymptotic value after large travel times. Dagan (1982) developed a functional relationship for an increasing dispersivity.

The classical equations governing field-scale solute dispersion in groundwater have therefore been under question (Gelhar *et al.*, 1979; Matheron and de Marsily, 1980). Classically, dispersion coefficients are assumed to be independent of scale, although scale-dependent behaviour, as described above, has been observed (Fried, 1975), and alternate formulations of the underlying process suggested (Tompson, 1988). Recently Gelhar *et al.* (1979), Dagan (1983; 1986; 1987), Gelhar and Axness (1983), Philip (1986), Wheatcraft and Tyler (1988), Tyler and Wheatcraft (1992) and others have applied stochastic and fractal methodologies to relate the variability of the hydraulic conductivity of the aquifer to the dispersion coefficient. The stochastic hypothesis is that if the hydraulic conductivity can be characterised in enough detail (*i.e.* in terms of the mean, variance and correlation structure) then the macro-dispersivity is calculable, with the addition only of laboratory-scale dispersivity estimates. Frind *et al.* (1988) has shown this to be true for the Borden aquifer. Microscale modeling in two dimensions showed convergence in the far-field (after 50 correlation lengths) of the effective dispersivity to the theoretical macrodispersivity value. Therefore, given detailed information on the variability of the aquifer hydraulic conductivity, fitting of the solute transport model to the far-field plume to estimate the dispersion coefficient would then not be necessary. A challenge with this approach is the need to develop cost-effective techniques for obtaining large numbers of estimates of the hydraulic conductivity at small spatial intervals over the aquifer (Sudicky, 1986; White, 1988; Sudicky, 1988).

In the case of point sources the scale dependency of the dispersivity needs further resolution. This is especially so if model prediction is required of early plume development in the near-field. Far-field predictions are less problematic, if one accepts that the dispersivity asymptotes to a constant value at the far-field scale. In this case model results can then be fitted to observed concentrations to determine the dispersivity for subsequent prediction. Limited resources for site investigation often dictate the density and frequency of data collection at a field site. So for prediction, research is needed to identify the scales of measurement which adequately characterise the variability of the hydraulic conductivity, so as to better define the dispersivity.

Flow and transport

To model contaminant transport in groundwater it is first important to model adequately groundwater flow processes to estimate groundwater velocities. This in itself can be difficult and should be undertaken with care, since if the magnitude or direction of groundwater flow velocities are significantly in error then so will be predicted solute concentrations and the direction of plume motion. This is of added concern for point source modeling since local groundwater velocities or directions of flow may differ markedly from those estimated from regional groundwater monitoring or modeling (*e.g.* Curtis *et al.*, 1986).

Groundwater flow equations
The hydrodynamic laws governing groundwater flow are well described by Freeze and Cherry (1979), Bear (1979) and others. However since groundwater flow modeling is fundamental to solute transport simulation, we briefly review the flow equations below.

Balancing liquid fluxes into and out of an elemental volume (or REV of Bear, 1979) of a saturated aquifer of porosity n, and assigning the differences in fluxes to the accumulation or depletion of liquid mass within the element, leads to the classical conservation of mass equation;

$$\frac{\partial(n\rho)}{\partial t} = -div(\rho q) + r\rho* \tag{3.1}$$

where ρ is the density of water (kg m^{-3}), q is the flux of water (m s^{-1}), r is the volumetric fluid source/sink flow rate per unit volume of the aquifer (s^{-1}), $\rho*$ is the density of a fluid source (kg m^{-3}) and t is time (s). Note that if r < 0 (*i.e.* a fluid sink) then $\rho* = \rho$.

If the density ρ is assumed to only vary with pressure p (Pa) then the left hand side of equation (3.1) can be expanded in terms of the compressibility of the aquifer, α (Pa^{-1}) and of the liquid, ß (Pa^{-1}) to give,

$$\frac{\partial(n\rho)}{\partial t} = \rho S_o \frac{\partial\phi}{\partial t} \tag{3.2}$$

where

$$\frac{\partial\phi}{\partial t} = \frac{1}{\rho g} \frac{\partial p}{\partial t} \tag{3.3}$$

$$S_o = \rho g \, [(1-n) \, \alpha + n\beta] \qquad\qquad (3.4)$$

$$\alpha = \frac{1}{(1-n)} \frac{\partial n}{\partial p} \qquad\qquad (3.5)$$

$$\beta = \frac{1}{\rho} \frac{\partial \rho}{\partial p} \qquad\qquad (3.6)$$

In equations(3.2) to (3.4) S_o is the specific storativity of the aquifer (m^{-1}), g is acceleration due to gravity (m s^{-2}) and ϕ is the piezometric water head (m). Combining equations (3.1) and (3.2) and assuming that changes in the density of water are small spatially in comparison to transient density changes gives,

$$S_o \frac{\partial \phi}{\partial t} = -div(q) + r \qquad\qquad (3.7)$$

If, however significant liquid density changes occur due to the addition of a solute to the aquifer (*i.e.* density driven flow occurs) then equation (3.7) is no longer valid. ϕ cannot be sensibly defined and the governing equations need to be altered (*i.e.* equation (3.2)) to incorporate the influence of solute concentration (c) changes on liquid densities. Pressure and/or density remain then as the dependent variables giving,

$$(S_o/g) \frac{\partial p}{\partial t} + n. \frac{\partial \rho}{\partial c} . \frac{\partial c}{\partial t} = -div(\rho q) + r \, \rho* \qquad\qquad (3.8)$$

Note that equation (3.8) requires an extra identity, similar in form to equations (3.5) and (3.6), relating solute concentration (or solute mass fraction) to the liquid density. The influence of liquid density changes due to added solute is of increasing interest in solute transport studies but will not be considered further here.

Equations (3.7) and (3.8) require an equation of motion to define the liquid flux q in terms of the dependent variables, pressure or piezometric head. Darcy's empirical law and some theoretical studies, for "moderate" groundwater flows, relate the flux of water flowing through a porous medium, linearly to the driving head gradient. Darcy's law can be written as

$$q = \frac{-k\rho g}{\mu} \, grad(\phi) \qquad\qquad (3.9)$$

where k is the permeability of the porous matrix (m^2) and is the viscosity ($kg\ m^{-1}s^{-1}$) Alternate non-linear formulations have been proposed, for situations of extremely slow moving groundwater or rapid flow, but equation (3.9) is widely accepted for most purposes. Substitution of (3.9) into (3.7) gives;

$$S_o\ \frac{\partial \phi}{\partial t} = div\ [K.grad\ (\phi)] + r \qquad (3.10)$$

where K is the hydraulic conductivity tensor ($m\ s^{-1}$). Equation (3.10) and simplifications of it (assuming homogeneity of aquifer materials, isotropy *etc.*) are generally sufficient for groundwater flow problems of practical interest. However more complex flow equations may be required in particular circumstances (*e.g.* equation (3.8) for density driven flows). Note that although equation (3.4) establishes a method to calculate S_0, most often it is estimated from inverse methods (such as a pump test) because the parameter α is difficult to quantify.

To complete the groundwater flow model, boundary and initial conditions must be specified. Initially (or at some set time) the pressure or piezometric head needs to be specified throughout the flow domain. The boundary conditions may take a number of forms and should be chosen to correspond to the physical barriers and features of the aquifer system under study. On the boundary between an aquifer and a surface water body, for example, a specified head (or potential) boundary condition is appropriate. A specified flux boundary condition will best simulate recharge to an aquifer or where a geological barrier is present. Mixed (head and flux) boundary conditions are applicable where a thin layer separates two regions of an aquifer. This concept is often used to define vertical leakage between aquifers and also for surface water bodies where accumulated sediments impede water exchange with an underlying aquifer. To account fully for phreatic conditions a complex non-linear boundary condition needs to be satisfied at the water table (Bear, 1979). Usually, however some approximation to this condition is assumed, especially if there are no large head changes (drawdowns) within the domain of interest.

Calibration and validation of the groundwater flow equations needs also to be undertaken prior to estimation of groundwater velocities for solute simulation. This may require carrying out of pump, slug or tracer tests to estimate aquifer parameter variability, and installation and monitoring of wells to determine local groundwater flow changes and short term fluctuations.

Solute transport equations
Detailed accounts of the derivation of the equations governing solute transport in porous media are given in various texts (Bear, 1979; Bear and Verruijt, 1987). As for the groundwater flow equations, mass flux balances can be considered within an elemental volume of the aquifer. For solutes in groundwater, source/sink terms such as adsorption, reaction kinetics and radioactive decay may influence the mass balance. To conserve mass within an elemental volume of the aquifer, mass flux differences (or divergence - div) plus solute mass changes due to sources or sinks in

the aquifer are balanced by the solute mass accumulated (or reduced) per unit time and can be written as,

$$\frac{\partial(nc)}{\partial t} = -div(Q) - f + rc* \tag{3.11}$$

where c is solute concentration (kg m^{-3}), Q is the total mass flux (kg m^{-2} s^{-1}), f is the loss term for water-solid interactions (kg m^{-3} s^{-1}) and c* is the solute concentration in recharged water (kg m^{-3}). Note that if r < 0 then c* = c.

The total flux of pollutant passing through a unit area of the porous medium (Q) can be derived from averaging of the microscopic diffusional and advective fluxes to give,

$$Q = -nDgrad(c) + ncv \tag{3.12}$$

where v is groundwater velocity = q/n (m s^{-1}) and D is the coefficient of hydrodynamic dispersion (m^2 s^{-1}). Substituting equation (3.12) into equation (3.11) and assuming porosity is constant gives the general partial differential equation governing solute transport as,

$$\frac{\partial c}{\partial t} = div[D.grad(c)] - div(vc) - \frac{f}{n} + \frac{rc*}{n} \tag{3.13}$$

Assuming homogeneity of aquifer properties, uniform groundwater flow in the x direction, and a linear equilibrium isotherm for liquid-solid adsorption, allows (3.13) to be reduced to the equation,

$$R\frac{\partial c}{\partial t} = D_x\frac{\partial^2 c}{\partial x^2} + D_y.\frac{\partial^2 c}{\partial y^2} + D_z\frac{\partial^2 c}{\partial z^2} - v.\frac{\partial c}{\partial x} \tag{3.14}$$

where D_x, D_y and D_z are the dispersion coefficients in the x, y and z directions respectively and R is the retardation coefficient defined as

$$R = 1 + \frac{\rho_s K_d}{n} \tag{3.15}$$

In equation (3.15) ρ_s is the aquifer bulk density (kg m^{-3}) and K_d is an empirical distribution constant (m^3 kg^{-1}).

As with the flow equations, boundary and initial conditions are required to define solute transport adequately. A constant concentration boundary condition may be appropriate where a boundary abuts a well mixed reservoir of solute, such as a solute tracer experiment in a laboratory soil column. More often mass flux boundary conditions are applied, for example, where leachate from a landfill infiltrates to groundwater. Mixed type boundary conditions are also assumed where the total dispersive and advective flux of mass at a boundary is known.

Modeling appproaches

Analytical and semi-analytical modeling

Equations governing solute transport in groundwater have been simplified and solved in many ways. Simple analytical modeling is feasible in homogeneous, isotropic aquifers with uniform groundwater velocities. Analytical solutions may be possible for a portion of the domain of interest even when some of these assumptions do not hold. Results from analytical models provide simplified predictions which often are useful in further detailed modeling.

There are a vast number and variety of analytical models and solutions to the advection-dispersion equations for point sources. Van Genuchten and Alves (1982) published a compendium of one dimensional solutions, in some cases accounting for linear equilibrium adsorption, zero-order production and first-order decay terms. More recently Lindstrom and Boersma (1989) derived analytical one dimensional solutions for a variety of boundary (some time varying) and initial conditions, modeling landfill leachate contamination of groundwater. Two dimensional point source solutions have been developed and applied at field sites (Wilson and Miller, 1978; Kent *et al.*, 1985). One, two and three dimensional instantaneous (slug) and continuous point source solutions to the advection-dispersion equations were also considered by Hunt (1978). Numerous three dimensional strip source analytical models have also been developed (Yeh, 1981; Domenico and Robbins, 1985; Huyakorn *et al.*, 1987).

Analytical expressions describing solute transport can be complex and sometimes difficult to evaluate and apply (Davis and Johnston, 1984). Often it is desirable to reduce the level of complexity for simple application and ease of understanding. However, assumptions required for simplification need then to be clearly stated and tested. Below we illustrate the usefulness of analytical models in testing common modeling assumptions. Three questions will be considered:

(i) under what circumstances is the two dimensional solution representative of the three dimensional solution?

(ii) considering aquifers are of finite depth, and often shallow, under what conditions can a solution be applied assuming an infinitely deep aquifer?

(iii) since all point sources are of finite size (not a single point in the domain) when does the simpler solution for a single point source apply?

(i) *Two versus three dimensional solutions*. The two dimensional solution for plume development from a continuous point source has had some prominence in the literature since its introduction by Wilson and Miller (1978) and Hunt (1978). It has also been recommended to the United States Environmental Protection Authority as the basis for initial prediction of leachate movement when assessing permit applications associated with hazardous waste facilities (Kent *et al.*, 1985). The two dimensional solution can be written as

$$c_c(x,y,t) = \frac{M_2 \exp\left(\dfrac{xv}{2D_x}\right)}{4\pi n\sqrt{D_xD_y}} W\left(\frac{B_2^2}{4D_xt}, \frac{VB_2}{2D_x}\right) \tag{4.1}$$

where M_2 is the continuous rate of mass injection per unit depth of aquifer (kg $s^{-1}m^{-1}$), W is the leaky aquifer function of Hantush and Jacob (1955) and B_2 is given by

$$B_2 = \left(x^2 + y^2\, D_x\,/\,D_y\right)^{1/2} \tag{4.2}$$

The subscript c in (4.1) refers to the continuous point source solution. At steady-state $t \rightarrow \infty$ equation (4.1) reduces to

$$c_c(x,y,\infty) = \frac{M_2 \exp\left(\dfrac{xv}{2D_x}\right)}{2\pi n\sqrt{D_xD_y}} K_o\left(\frac{vB_2}{2D_x}\right) \tag{4.3}$$

where K_o is the modified Bessel function of the second kind of order zero.

The three dimensional continuous point source solution in a semi-infinite (z>0) region can be written as

$$c_c(x,y,z,t) = \frac{M_3 \exp\left(\dfrac{xv}{2D_x}\right)}{4\pi nB_3\sqrt{D_yD_z}} \left\{ \exp\left(\frac{-B_3v}{2D_x}\right) erfc\left(\frac{B_3-vt}{2\sqrt{D_xt}}\right) \right.$$

$$\left. + \exp\left(\frac{B_3v}{2D_x}\right) erfc\left(\frac{B_3+vt}{2\sqrt{D_xt}}\right) \right\} \tag{4.4}$$

where M_3 is the rate of mass "injected" at the source (0,0,0) (kg s^{-1}), erfc (z) is the complementary error function of argument z and B_3 is defined as

$$B_3 = \left(x^2 + y^2\, D_x\,/\,D_y + z^2\, D_x\,/\,D_z\right)^{1/2} \tag{4.5}$$

Note that equation (4.4) is the solution for a semi-infinite domain (z ≥ 0) to represent solute concentrations with depth below a groundwater water table; the solution given by Hunt (1978) is for an infinite domain and is half that given in equation (4.4). The steady-state equivalent to equation (4.4) is

$$c_c(x,y,z,\infty) = \frac{M_3 \exp\left[\left(\dfrac{x-B_3}{2D_x}\right)v\right]}{2\pi nB_3\sqrt{D_yD_z}} \tag{4.6}$$

Equations (4.1) and (4.3) assume complete mixing of the solute mass over some specified depth of the aquifer or the total aquifer depth. Equations (4.4) and (4.6), however, assume the solute mass is confined to a shallow region close to the water table and as the solute migrates downgradient the solute mixes via dispersion over a greater depth of the aquifer. When, then, is it valid to use the two-dimensional equations in preference to the three dimensional ones?

Taking the ratio of the two-dimensional steady-state solution (4.3) to that for three dimensions (4.6) gives,

$$\frac{c_c\,(x,y,\infty)}{c_c\,(x,y,z,\infty)} = \frac{M_2}{M_3}\,B_3\,(D_z\,/\,D_x)^{1/2}\,K_o\left(\frac{vB_2}{2D_x}\right)\exp\left(\frac{B_3 v}{2D_x}\right) \qquad (4.7)$$

Consider the concentration ratio along the centre line of the plume by setting y=z=0. Now since M_2 is the mass "injection" rate per unit depth of aquifer (or mixing depth L) then

$$M_2\,/\,M_3\ =\ 1/L \qquad (4.8)$$

and if vx/2Dₓ >> 1, equation (4.7) reduces to

$$\frac{c_c\,(x,0,\infty)}{c_c\,(x,0,0,\infty)} \approx \frac{1}{L}\left(\frac{D_z \pi x}{v}\right)^{1/2} \qquad (4.9)$$

Therefore the two solutions agree, *i.e.* $c_c(x,0,) = c_c(x,0,\infty)$, only at the point

$$x_d\ =\ \frac{L^2}{\pi\,\alpha_z} \qquad (4.10)$$

If L = 20 m and $\alpha_z = 1$ m, then the solutions agree at a distance of 127 m. For α_z reduced to 0.1 m the distance x_d increases to 1,273 m. However, for x < x_d, $c_c(x,0,\infty)$ < $c_c(x,0,0,\infty$) and for x > x_d, $c_c(x,0,\ \infty)$ > $c_c(x,0,0,\infty$). Perhaps this is expected since the two dimensional result should be more correctly compared to the three dimensional concentration averaged over the depth of the aquifer. Even then, however, significant differences in calculated concentrations are likely to be evident.

The two dimensional solution should therefore be used with caution especially if the point source of interest is not well mixed over some depth L. This is often the case, since contaminants in leachate or from spills enter groundwater from the top surface of the aquifer (usually the water table), unless artificially injected. A disadvantage of the three dimensional solution is that it requires an estimate of the vertical dispersion coefficient. The two dimensional solution, however, also requires an estimate of the mixing depth L, unless M_2 can be independently estimated. Therefore the advantages of the 'simpler' two dimensional approach over the three dimensional approach are not significant.

(ii) *Finite versus infinite depth solutions.* Here we follow the analysis of Hunt (1978). Equation (4.6) represents the steady-state three dimensional solution to a point source of solute in an infinitely deep aquifer. To model an aquifer of finite depth 2d we can apply the method of images to the linear equation (4.6) to ensure no mass is transported below the base of the aquifer. The resultant concentration c' is

$$\dot{c}\,(x,y,z,\infty) = c_c\,(x,y,z,\infty) + \sum_{m=1}^{\bullet} \left[c_c\,(x,y,z-2md,\infty) + c_c\,(x,y,z+2md,\infty) \right] (4.11)$$

From the ratio of (4.6) and (4.11) it is possible to define the region B_o, at the water table (z=0) where the simpler infinite depth solution is within 1% of the more complex (but more realistic) finite depth aquifer solution. A Figure relating the region B_o to transport parameters is given in Hunt (1978), and can be graphically approximated by the equation

$$B_o = \frac{0.167 \; vd^2}{D_z} \tag{4.12}$$

or in terms of the vertical dispersivity

$$B_o = \frac{0.167 \; d^2}{\alpha_z} \tag{4.13}$$

So for an aquifer 20 m deep, with a vertical dispersivity of 1 m, B_o = 16.7 m. Therefore the region where the two solutions for the maximum concentration at the water table differ by less than 1% is bounded by the ellipse centred on the origin

$$x^2 + y^2 \; D_x \,/\, D_y \; = 280 \tag{4.14}$$

where we have used the definition of B_o given by equation (4.2). A much deeper aquifer, say 50 m deep would give a much larger elliptical region with B_o = 105 m.

Clearly the elliptical region of agreement between the two solutions is not large for shallow aquifers. In the near-field, close to the source of solute (and at short times), however, the infinite depth solution does mimic the finite depth solution. This is because the solute has not migrated to a depth in the aquifer where the base of the aquifer influences the solute concentration significantly. This points to the need in modeling to be wary of boundaries within the flow field that may affect solute migration, either laterally or vertically.

(iii) *Point source versus finite dimensional source solutions.* Often, so-called point sources have large dimensions. Intuitively we might suggest that the size of the source would only have an impact on the predicted plume dimension in the initial stages of plume development or at locations near the source. But at what stage are the dimensions of the source no longer important, and when is the simpler point source solution valid? Below we investigate these questions for an instantaneous slug of solute.

The solution for an instantaneous slug of solute of mass M_4 at the point $(0,0,0)$ at time zero in an infinite domain can be written as

$$c_i(x,y,z,t) = \frac{M_4 \exp\left[\dfrac{-(x-vt)^2}{4D_x t} - \dfrac{y^2}{4D_y t} - \dfrac{z^2}{4D_z t}\right]}{8n\left(\pi^3 t^3 D_x D_y D_z\right)^{1/2}} \qquad (4.15)$$

Using equation (4.15) and the solution for an instantaneous slug input of finite dimensions $|x|$, $|y|$, $|z| < H/2$, Hunt (1978) found a characteristic time defined by

$$\left(H^2/t\right) \cdot \left(1/D_x + 1/D_y + 1/D_z\right) < 0.48 \qquad (4.16)$$

beyond which the simpler point source solution peak concentrations ($x = vt$, $y = z = 0$) were within 1% of the finite dimensional source solution. Clearly the characteristic time given by inequality (4.16) is highly sensitive to the smallest of the dispersion parameters and if, as is often the case, $D_z < D_y \ll D_x$ then (4.16) approximates to

$$H^2 / \left(D_z t\right) < 0.48 \qquad (4.17)$$

For a small scale source $H=10$ m with D_z in the range 10 to 100 m^2 yr^{-1} then the lapsed time before the point source adequately represents the finite dimensional source peak concentration, is greater than 20 and 2 years respectively. This may be an impractical delay time for field application. It also points to the need to consider carefully the effects of applying point source solution estimates where the dimensions of the source are significant or dispersion coefficients are small. Criteria, such as those given in equations (4.16) and (4.17), have not been developed for the case of a continuous point source. Pereboom *et al.* (1992) have recently considered this further using both approximate analytical and three-dimensional numerical models.

Numerical modeling
Commonly aquifers are heterogeneous, have irregular shaped boundaries (undulating bedrock *etc.*) and often groundwater velocities vary spatially and temporally depending on rainfall and subsequent recharge. Such variability can rarely be tackled with analytical and semi-analytical approximations.

Numerical modeling can, however, deal with variability, and can allow coupling of quite complex transport processes such as density driven flows and chemical reaction kinetics (Kincaid *et al.*, 1984; Schäfer and Kinzelbach, 1992). These capabilities are of increasing importance when assessing the siting of sources of potential groundwater contaminants. Research on the intercomparison and validation of numerical models for prediction of likely plume development from such sources is on-going (Huyakorn *et al.*, 1984; Hamilton *et al.*, 1985; Hayden, 1985; Intracoin,

1986a,b). A number of reviews of the state of solute transport modeling have been published (*e.g.* Anderson, 1979; Javandel *et al.*, 1984; van der Heijde *et al.*, 1985).

Of concern with numerical modeling are numerical errors associated with the approximation of the spatial and temporal derivatives of the advection-dispersion equation. This sometimes leads to accuracy and stability problems and numerical dispersion. Numerical dispersion is the additional dispersion of mass due to discretisation of the modeled region and approximation of the model equations (not due to physical processes).

Criteria to achieve stable and accurate numerical solutions have been established for the simple one dimensional case. Analysis of numerical schemes to establish criteria for two and three dimensional and more complex models is more difficult. The two criteria for one dimension are often called the Peclet number (or cell Reynolds number) and Courant number criteria and are defined as (Daus *et al.*, 1985)

$$Pe = \frac{v.\Delta x}{D_x} = \frac{\Delta x}{\alpha_L} \leq 2 \tag{4.18}$$

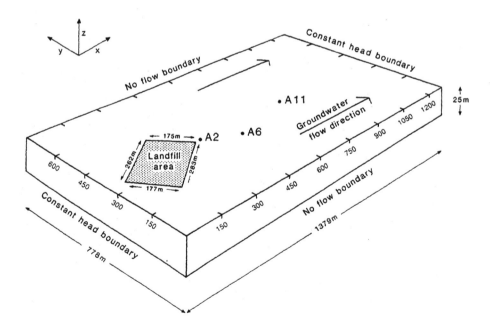

Figure 1 *modeled region showing area of landfill and borehole locations A2, A6 and A11.*

$$Co = \frac{v.\Delta t}{\Delta x} \leq \frac{Pe}{2} \qquad (4.19)$$

Often equation(4.19) is stated as,

$$\frac{D_x \Delta t}{(\Delta x)^2} \leq 0.5 \qquad (4.20)$$

Significant numerical instability and numerical dispersion are generally thought to occur when the Peclet number is 5 or greater (Dillon, 1987). However, results for much larger Peclet numbers are often quoted. For example, Hamilton *et al.*, (1985), when comparing three mass transport models presented results for Peclet numbers of 26 and 50. Large Peclet numbers often result from field application of models (especially for regional modeling) and limited computer memory and speed requiring coarse grid sizes. Increased computer storages and faster computers will soon allow use of much finer grids, thus reducing the Peclet number.

Analogous criteria to equations (4.18) and (4.19) are often quoted for two and three dimensional systems but rigorous mathematical derivations are mostly lacking. Kinzelbach and Frind (1986), for example, showed through numerical experimentation using a two dimensional finite element code that the Peclet and Courant number criteria are not sufficient to maintain the accuracy of the solution

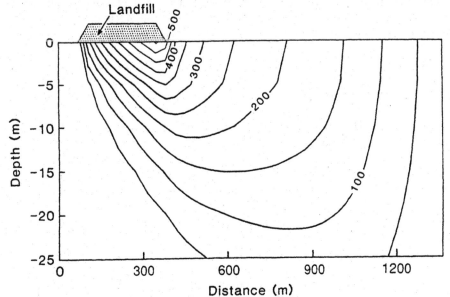

Figure 2 *Vertical cross-section along the mid-section of the plume of contamination after 10 years, predicted by SAM3D. Contours are shown at 50 mg L^{-1} intervals.*

when anisotropy occurs. Where stability and accuracy criteria are difficult to derive, say in complex flow systems, then numerical algorithms can be tested for stability and accuracy using parameter sensitivity studies, or tested for accuracy against analytical solutions of less complex test cases.

Much effort has been expended on developing numerical algorithms which solve the advection-dispersion equation accurately *i.e.* maintain numerical stability and limit numerical dispersion. The proceedings of groundwater modeling conferences (*e.g.* Jousma *et al.*, 1987; Custodio *et al.*, 1988; Kobus and Kinzelbach, 1989) contain many such numerical algorithms. The most popular schemes have been finite difference, finite element, method of characteristics (or random walk), and more recently a number of hybrid schemes have been developed. An overview was given by Pinder (1988).

The finite difference method was one of the first discretization techniques developed for flow and transport equations (Oster *et al.*, 1970; Bredehoeft, 1971). The method consists of subdividing a modeled region into rectangular cells and labelling the central or peripheral nodes. The governing partial differential equation is then expanded in terms of a Taylors series about each node, and reduced to a set of algebraic equations. The resulting set of equations (matrix equation) is usually then amenable to solution. The finite difference method is straightforward to apply and is still prominently employed (Kipp, 1987).

The finite element method provides flexibility in dealing with complex geometries (*e.g.* Voss, 1984; Sa da Costa *et al.*, 1986). As with the finite difference method it also involves discretisation of the modeled region, but uses a method of weighted residuals (*e.g.* Galerkin or collocation) over cell elements of different geometries (*e.g.* triangular or iso-parametric). The resultant algebraic equations are more difficult to evaluate than those derived for the finite difference method, but an appropriate choice of weightings can give good accuracy.

The method of characteristics is also often employed to solve the advection-dispersion equations (Pinder and Cooper, 1970; Konikow and Bredehoeft, 1978), and has the reputation of limiting numerical error in advection dominant systems. The method consists of first estimating groundwater flow velocities. Characteristic curves are then established along which solute mass is advectively transported with the groundwater velocity, 'attached' mathematically to a cluster of migrating particles. The resulting concentration, averaged over the grid cell dimensions, is then corrected for dispersion by a finite difference approximation to the remaining terms of the characteristic equation. MOC, a method of characteristics code, developed in 1978 by Konikow and Bredehoeft has found wide application, and further modification and development of the code has occurred. Although the method of characteristics is popular and simply applied, care should be taken in validation of model results, as large errors can occur.

To reduce numerical dispersion and maintain stability a number of hybrid schemes have also been developed (Babu and Pinder, 1982; Taigbenu and Liggett, 1986; Sudicky, 1989). Sudicky (1989), for example, coupled Laplace transforms of the time domain with finite element (Galerkin) discretisation of the spatial domain (LTG). LTG achieved negligible numerical dispersion in predicted solute concentrations for Peclet numbers in excess of 30 (cf. equation (4.18)). Application of LTG is, however, restricted to situations where the model parameters are not time dependent. It is often the case with hybrid schemes that added accuracy and stability are achieved, but general application of the scheme is restricted.

Three models: a case study

In this section we compare two numerical approaches and one semi-analytical approach. The three approaches are applied to an idealised test case of contaminant plume development in groundwater due to leachate infiltration from a domestic landfill. Since modeling reactive species transport poses extra difficulties and to keep the test case simple, only conservative species transport is considered. The three approaches consist of:
(i) three-dimensional semi-analytic modeling using SAM3D (Davis *et al.*, 1988),
(ii) two-dimensional method of characteristics modeling using MOC (Konikow and Bredehoeft, 1978), and
(iii) three-dimensional finite difference modeling using HST3D (Kipp, 1987).

The test case
The test case is based on results of a three year investigation to develop procedures for more efficient monitoring and assessment of groundwater contamination from a landfill site (Barber *et al.*, 1991). The landfill of interest is a currently operating sand pit in Morley, Western Australia, partially backfilled with solid domestic wastes. The site has been operational since 1980 and the landfill covers an area of approximately 5 ha.

The saturated aquifer beneath the site consists of fine to coarse sands 20 to 35 m thick. Beneath the sands are intercalated clays and sands of the Osborne formation forming a leaky/confining basement (Salama *et al.*, 1989). The unsaturated zone in the vicinity of the landfill varies in thickness from exposed groundwater within the sand pit to 40 m on the nearby dune ridge. The thickness of unsaturated zone beneath the base of the landfill ranges from 0 to 5 m, and is mostly less than 5 m.

The aeolian sand aquifer, although generally considered to be fairly uniform (homogeneous), is locally quite variable (distinctly layered, current bedded) but contains only small quantities of clay. Hydraulic conductivities, determined from a tracer test and grain size analyses, lie in the range 19 to 70 m day^{-1} with a mean value of 30 m day^{-1}. The local hydraulic gradient across the site was estimated at 0.003, and showed only small changes seasonally. With a porosity of 30%, groundwater velocities have been estimated to be of the order of 100 m yr^{-1}.

The test case is taken to represent a simplified model of plume development, consistent with data and observations from the Morley site (see Figure 1). The aquifer is assumed homogeneous and isotropic, to be 25 m thick and to have a uniform groundwater velocity of 100 m yr^{-1} over the modeled region. The irregular shaped landfill area appropriate to the Morley landfill is retained, to assess the impact of the dispersion coefficients on the plume shape (see Pereboom *et al.*, 1992). For the test case we model chloride as a conservative tracer of pollution.

To compare the three approaches it is important that each approach is modeling the same test case. One aspect that is difficult to incorporate into each of the three approaches is the condition at the source (*i.e.* leachate concentration and recharge rate to groundwater beneath the landfill). In the semi-analytical model we simply multiply the leachate concentration (2500 mg L^{-1} as Cl) and the recharge rate (200 mm yr^{-1}) to give a mass flux to groundwater per unit area of the landfill. HST3D allows a comparable approach giving the option to set a diffusive mass flux over a portion of a boundary, without water injection. MOC, however, does not allow a solute mass flux boundary condition independently of the water flux, *i.e.* a concentration boundary condition is only allowed. The difficulty with assigning a water flux is that it disturbs the flow regime leading to a spatially variable groundwater gradient and velocity. The semi-analytical approach, however, is founded on the groundwater velocity being uniform.

To resolve this difficulty, results are presented for cases which assume two different source conditions. HST3D and the semi-analytical model results are compared assuming a constant mass flux condition over the landfill area and, MOC and HST3D results are compared assuming constant concentration and liquid flux boundary conditions over the landfill area. In the latter simulation therefore the flow regime is non-uniform, but facilitates comparison of MOC and HST3D.

Three-dimensional semi-analytical model (SAM3D)
The model accounts for advective transport in a uni-directional flow field and dispersion in the three principal directions. Aligning the x axis (direction of the constant groundwater flow) with the principal flow direction in the field we can write the continuous mass injection solution at a point as that given in equation (4.4). The solution for a semi-infinite domain integrated over the finite dimensions of the landfill area is given by,

$$c\ (x,y,z,t) = \frac{1}{A} \int_{x'} \int_{y'} c_c\ (x-x',\ y-y',\ z,t)\ dx'\ dy' \qquad (5.1)$$

where A is the surface area of the landfill (m^2), and x and y are the planar dimensions of the landfill (m). Equation (5.1) is evaluated using a nested integration scheme based on rhomberg extrapolation (de Boor, 1971).

Figure 2 shows a typical vertical cross-section through the mid-section of the plume after 10 years simulation for a longitudinal dispersivity of 10 m and transverse dispersivities of 0.3 m. After 10 years the 100 mg L^{-1} contour extends some 780 m downgradient at the water table and has a lateral spread of 320 m which is similar to the width of the landfill. Also of interest is the penetration of the 50 mg L^{-1} solute contour to the base of the aquifer. SAM3D does not account for the base of the aquifer but assumes a semi-infinite domain in the z direction. This may lead to differences in predicted concentrations when compared to numerical model results, especially at large times or in the far-field. The basement of the aquifer can be accounted for by modification of SAM3D to sum over an infinite number of image sources in a similar manner to that described by Hunt (1978).

For the same parameter values, Figure 3 shows breakthroughs for solute concentrations predicted at locations 35 and 460 m downgradient of the landfill and

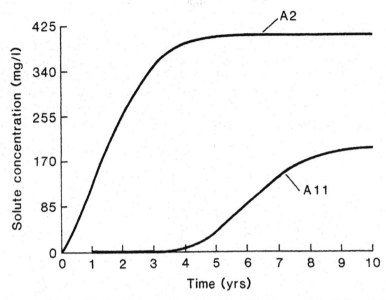

Figure 3 *Breakthrough curves predicted by SAM3D for two sites (A2 and A11) located mid-plume and down gradient of the landfill 35 m and 460 m, respectively.*

along the mid-line of the plume. Near-constant (steady-state) concentrations are attained nearest the landfill after only 4 to 5 years whereas further downgradient, solute concentrations are continuing to increase after 9 years. Clearly, it is important for proper aquifer management to be able to determine the extent of a steady-state plume. This will also indicate the area which needs to be monitored for validating and adjusting predictions.

MOC
The method of characteristics code MOC is a two dimensional flow and solute transport code developed by Konikow and Bredehoeft (1978). It is simple to use and was one of the first flow and solute codes to be widely distributed for use on personal computers. Recent modifications of MOC have incorporated density driven flows, and pre- and post-processors are available to aid data input and interpretation of results.

To run MOC the modeled region was divided into a 20 x 12 evenly spaced grid. The grid spacing in the y direction was taken as $\Delta y = 70.71$ m. Two different spacings were taken in the x direction, $\Delta x = 35.355$ m and $\Delta x = 70.71$ m. This was done to assess error in the calculated results due to numerical dispersion.

To simulate the test case, uniform groundwater velocities are required. MOC does not allow constant head boundary conditions *per se*. But by assigning a large leakance factor at the boundaries at each end of the x domain a constant head condition can be approximated. With no sources or sinks this then gives a uniform groundwater velocity in the x direction.

As indicated previously MOC only allows solute input to the modeled region through water inflows, through the boundaries or through injection wells within the domain. This disturbs the uniform groundwater velocity field near the landfill. For the simulations described here we allocated water injection rates of 200 mm yr^{-1} (as in SAM3D) to wells centred on each cell. The volumetric injection rate was obtained by multiplication of 200 mm y^{-1}r by the cell area overlain by the landfill surface area. The injection concentration was taken as uniform throughout the simulations.

Figure 4 shows solute breakthrough curves calculated using MOC, at two

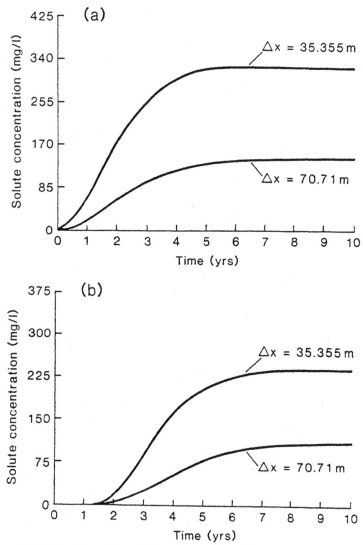

Figure 4 *Comparison of breakthrough curves predicted by MOC for two different grid spacings Δx = 35.355 m and Δx = 70.71 m, and for two locations down gradient of the landfill. (a) A2-35 m downgradient (b) A6-150 m downgradient.*

locations downgradient of the landfill (Figure 1). One location is mid-plume and 35 m from the leading edge of the plume (A2). The other is approximately 150 m downgradient of the leading edge and 70 m off mid-plume (A6). Breakthrough for A2 occurs at 5 to 6 years and for A6 after 7 to 8 years.

Breakthroughs are shown for two Δx values (*i.e.* two Peclet values) at each location. The breakthroughs are clearly different, and illustrate the need for care in selection of grid sizes and model parameters. The breakthrough concentrations for the coarser grid size display appreciable numerical error. At both locations, peak concentrations are lower for $\Delta x = 70.71$ m (Pe = 7.1) than for $\Delta x = 35.355$ (Pe = 3.5). These differences (errors) are difficult to explain since the method of characteristics seeks to minimise numerical dispersion error, however, the coarser grid size gives a Peclet number much greater than the critical value quoted in equation (4.18). A portion of the numerical error may also be due to the less accurate representation of the landfill area with the coarser grid ($\Delta x = 70.71$ m).

HST3D

HST3D is a three dimensional finite difference code developed by the United States Geological Survey (Kipp, 1987). It is a large code allowing coupled heat and solute transport, accounting for density driven flows. The coded equations are cast in terms of pressure, solute mass fraction and temperature. Coupling of the equations occurs through the dependence of the fluid viscosity on temperature and solute concentration, and the dependence of fluid density on pressure, temperature and solute concentration.

To simulate a uniform groundwater velocity in the test case, and to compare with SAM3D results, confined aquifer conditions were modeled. Constant head boundary conditions were set at either end of the x domain, and no flow boundaries were set at the base and the top of the aquifer, and at the lateral extremities (y direction). The modeled region was divided into 39 x 11 x 5 evenly spaced cells with dimensions (in metres) of 35.355 (x) x 70.71 (y) x 5 (z). Solute transport was simulated after steady flow was achieved - this took only a few time steps.

A number of differencing and solution method options are incorporated in HST3D. Each of the options has the potential to alter the calculated solute concentration. Two options are given for spatial differencing of the advective transport terms; centred in space differencing (CS) and backwards in space (or upstream) differencing (BS). Also two temporal difference options are possible; centred in time (Crank-Nicholson) differencing (CT) and backward in time (fully implicit) differencing (BT). After discretisation and development of the difference equations, the resultant matrix equations can be solved either using a direct equation solver, or by the successive over-relaxation equation solver (the direct equation solver is used in all results presented here).

To assess the influence of the type of differencing on the calculated solute concentration, four calculations were carried out keeping the model parameter values constant but pairing each of the differencing options available (*e.g.* CS/CT). The aim was to assess the accuracy and stability of each of the pairings. For a time step of 0.01 years, both the BS/CT and CS/CT pairings showed oscillatory but similar behaviour in predicted groundwater velocities, in the x and z directions. For the BS/BT and CS/BT pairings, similar groundwater velocities were again obtained but no oscillatory behaviour was observed. It appears then, as is noted by Kipp (1987), that centred in time differencing leads to non-physical velocity fluctuations.

Unlike the groundwater velocities, regardless of the time differencing, solute concentrations predicted by HST3D with the same spatial differencing were nearly identical. This suggests that for the test case under consideration oscillations in the velocity due to time differencing are inconsequential, and are largely overridden by spatial differencing options. Even increasing the time step to 0.5 yr (Co = 1.4, Pe = 3.5) from 0.01 yr (Co = 0.03) did not significantly alter the calculated solute concentrations, for either time difference option. The two spatial differencing options did however, yield different results. The CS calculation gave slightly elevated solute concentrations downgradient and within the highly contaminated region of the plume ($\geq$ 100 mg L $^{-1}$) when compared with BS results. Upgradient and further downgradient CS concentrations were lower than BS concentrations.

Intercomparison of models

Two intercomparisons are given. Results from SAM3D and HST3D are compared to establish the accuracy and reliability of HST3D. MOC results are then compared to HST3D results to assess the relative usefulness of the two or three dimensional numerical approaches. As indicated previously it is not possible to directly compare SAM3D with MOC.

SAM3D versus HST3D
Analytical models if evaluated accurately display no artificial dispersion. HST3D is expected to display some of the effects of numerical dispersion since the grid (cell) Peclet number is held at 3.5, which is above the critical number (2) suggested in equation (4.18).

Figure 5 shows a comparison of breakthrough curves calculated by SAM3D and HST3D at locations A2 and A6 (see Figure 1). The effect of numerical dispersion is evident at A2 where the breakthrough curve calculated by HST3D largely underlies that calculated by SAM3D. This is a typical effect of numerical dispersion - causing artificial decreases in the peak solute concentration. At A6, however, further from the landfill than A2 and off-centre (see Figure 1) the breakthrough curves predicted by HST3D and SAM3D closely correspond. Prior to 2 years the breakthrough curve predicted by HST3D at A6 marginally overlies the SAM3D curve and this is likely to be caused by the accelerated transport of solute mass downgradient due to added numerical dispersion.

Despite these effects, prediction errors for HST3D results, for this test case, appear small. The largest differences in the breakthrough curves are less than 50 mg L $^{-1}$, and these occur close to the wastes. A vertical section through the mid-line of the plume (as in Figure 2) calculated for HST3D mimics that of SAM3D, except close to the base of the aquifer (see previous discussion). Aquifer variability and errors associated with sampling of groundwater quality are therefore likely to be

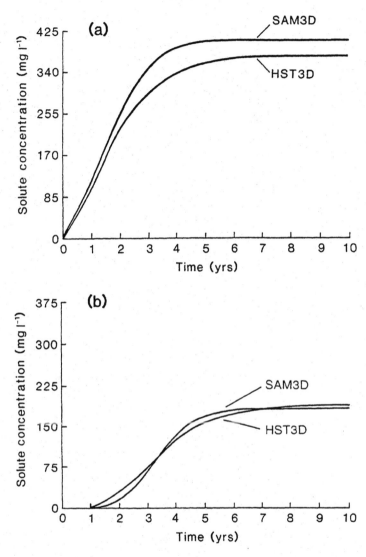

Figure 5 *Comparison of solute breakthrough curves predicted by SAM3D and HST3D for locations (a) A2 and (b) A6.*

greater than numerical dispersion errors evident in HST3D calculations for this test case.

MOC versus HST3D
MOC and HST3D do not give equivalent results. Figure 6 gives breakthrough curves for HST3D at $z=0$ and MOC for equivalent model parameters and grid spacings. The same two locations are shown; A2 and A6 as described previously (see Figure 1). At

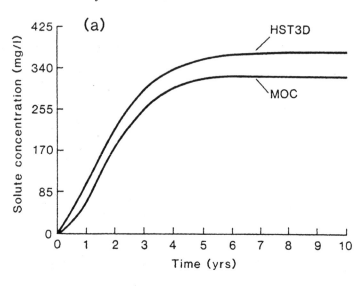

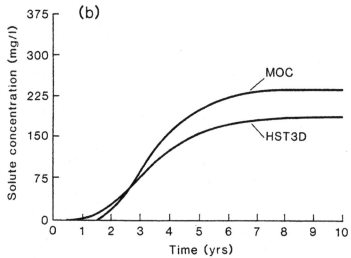

Figure 6 *Comparison of solute breakthrough curves predicted by MOC and HST3D (z = 0 m) for locations (a) A2 and (b) A6.*

A2 close to the landfill solute concentrations calculated by MOC are lower than those calculated by HST3D by approximately 50 mg L^{-1} , for all times to 10 years. At A6, further downstream, MOC results underlie those of HST3D for 2 years, and then increase to more than 50 mg L^{-1} above those of HST3D.

If we are modeling the same physical processes with both MOC and HST3D then they should produce comparable results or alternatively should be closely scrutinised for modeling inconsistencies. MOC, because it is a two dimensional code assumes complete mixing of the input concentration over some vertical aquifer thickness (often the entire aquifer thickness). HST3D, in this simulation, assumes a gradual penetration of the input solute from the water table to deeper depths in the aquifer. At no stage in the HST3D simulation is it possible for the source concentration (at the source) to be evenly mixed throughout the aquifer depth. A high concentration is maintained at the source near the water table throughout the simulation, and lower concentrations at deeper depths and downgradient result from advective and dispersive transport. For longer simulation times (> 5 years) and downgradient of the simulated landfill, HST3D does show uniform concentrations with depth. These do not correspond, however, to uniform mixing of the input concentration as assumed by MOC.

It is, therefore, difficult to sensibly compare MOC, a two dimensional numerical code with HST3D a three dimensional code. It is more appropriate to compare vertically integrated profiles calculated from HST3D with MOC results. Even so, by averaging HST3D solute concentrations over the aquifer depth at A6 leads to lower integrated solute breakthrough concentrations than those calculated for HST3D at $z = 0$ (*i.e.* at the water table), since the solute concentration at $z = 0$ (Figure 6) is a maximum over the thickness of the aquifer. So we deduce that differences between breakthrough curves predicted by MOC and the vertically integrated HST3D profiles would be even greater at A6 than the difference shown in Figure 6. Results from MOC should therefore be cautiously applied.

Summary and conclusions

The equations describing solute transport in groundwater are well established (although still debated), but quantifying parameters that appear in these equations remains somewhat of an art. Quantifying the dispersion coefficient (dispersivity) is of intense research interest currently. Much research is still to be done to provide adequate field-scale estimates of the dispersivity.

Analytical and numerical models have been successful in estimating contaminant plume migration under a variety of aquifer conditions (Sudicky, 1986; Barber *et al.*, 1991; Pereboom *et al.*, 1992). Some of the advantages and limitations of the different modeling approaches have been outlined in this chapter. Analytical models (solutions) have been shown to be particularly useful at the initial stages of an investigation, giving order-of-magnitude estimates of likely pollution plume concentrations and pollution distribution. They have also been shown to be useful in assessment of modeling assumptions. Indeed where coarse grids must be used, and numerical dispersion errors are potentially large, then analytical modeling is preferable (McLaughlin, 1984).

Comparison of three model types was carried out for a single test case. HST3D was shown to display limited numerical dispersion for Pe = 3.5 and gave good correspondence to results obtained for the semi-analytical model (SAM3D). MOC displayed large numerical error for large Peclet number. Significant discrepancies were found between results from HST3D and MOC at a Peclet number of 3.5. Since

HST3D results were in reasonable agreement with those of SAM3D it is concluded that for this test case and for Pe > 3.5, MOC exhibits significant numerical error.

modeling and assessment of point source contamination of groundwaters is a 'growth industry'. Many centres of research have developed throughout the world. To effectively deal with the variability of natural aquifer systems a multi-skilled approach is required. Good quality field data (hydrogeological) is essential for model conception and later validation. Better numerical algorithms are required to reduce numerical error, to increase computational speed, and thus allow 'larger' problems to be solved, or finer detail to be obtained. Further development of theory with experimental quantification is also essential to determine appropriate, and cost-effective scales of measurement of the model parameters, especially the dispersion coefficient.

Acknowledgments

The efforts of Dean Laslett in the initial stages of modeling with HST3D are gratefully acknowledged. Constructive and thoughtful comment on an earlier draft of the manuscript by Dr Chris Barber and Mr Colin Johnston of CSIRO Division of Water Resources are appreciated. Miss Tracy Keevers is also acknowledged for her patient retyping of the manuscript.

References

Anderson, M.P. 1979. Using models to simulate the movement of contaminants through groundwater flow systems. *Crit. Rev. in Environmental Control,* 9(2), 97-156.

Anderson, M.P. 1984. Movement of contaminants in groundwater, groundwater transport - advection and dispersion. In: *Studies in Geophysics, Groundwater Contamination, pp.37-45.*. National Academy Press, Washington D.C.

Babu, D.K. and Pinder, G.F. 1982. A three-dimensional hybrid finite element-finite difference scheme for groundwater simulation. *10th IMACS World Congress on System Simulation and Scientific Computation,* Montreal, pp.292-294.

Barber, C., Davis, G.B., Buselli, G. and Height, M.I. 1991. *Monitoring and predicting the development of leachate plumes in groundwater , a case study on alternative strategies. CSIRO Division of Water Resources.* Water Resources Series No. 5.

Bear, J. 1979. *Hydraulics of Groundwater.* McGraw-Hill Book Co., New York, 569 pp.

Bear, J. and Verruijt, A. 1987. *modeling Groundwater Flow and Pollution.* Reidel Publ. Co., 414 pp.

Bredehoeft, J.D. 1970. Comment on "Numerical solution to the convective diffusion equation" by Oster *et al. Water Resources Research,* 7, 755

Cheremisinoff, P.N., Gigliello, K.A. and O'Neill, T.K. 1984. *Groundwater-Leachate, modeling/Monitoring/Sampling.* Technomic Publ. Co. Inc., Lancaster, Pennsylvania.

Curtis, G.P., Freyberg, D.L., Goltz, M.N., Hopkins,G.D., Mackay, D.M., McCarty, P.L., Reinhard, M. and Roberts, P.V. 1986. *A Natural Gradient Experiment on Solute Transport in a Sand Aquifer.* Stanford Univ. Techn. Report No. 292.

Custodio, E., Gurgui, A. and Lobo Ferreira, J.P. (eds). 1987. *Groundwater Flow and Quality modeling.* NATO ASI, Series C, Mathematical and Physical Sciences Vol. 224, 843 pp.

Dagan, G. 1982. Stochastic modeling of groundwater flow by unconditional and conditional probabilities. 2. The solute transport. *Water Resources Research*, 18, 835-845.

Dagan, G. 1983. Stochastic modeling of solute transport by groundwater flow, state of the art. In: *Relation of Groundwater Quantity and Quality*. IAHS Public. No. 146, pp. 91-101.

Dagan, G. 1986. Statistical theory of groundwater flow and transport, pore- to laboratory-, laboratory- to formation- and formation- to regional-scale. *Water Resources Research*, 22, 120S-135S.

Dagan, G. 1987. Review of stochastic theory of transport in groundwater flow. In: Custodio, E. *et al.* (eds), *Groundwater Flow and Quality modeling*, pp.1-32.

Daus, A.D., Frind, E.O. and Sudicky, E.A. 1985. Comparative error analysis in finite element formulations of the advection-dispersion equation. *Advances in Water Resources*, 8, 86-95.

Davis, G.B., Barber, C. and Buselli, G. 1988. Borehole and surface geophysical monitoring, and simple modeling of groundwater polluted by waste leachates. *The 3rd Int. Mine Water Congress*, Melbourne Australia, pp.261-270.

Davis, G.B. and Johnston, C.D. 1984. Comment on "Contaminant transport in fractured porous media, analytical solutions for a system of parallel fractures" by Sudicky, E.A. and Frind, E.O. *Water Resources Research*, 20(9), 1321-1322.

de Boor, C. 1971. An algorithm for numerical quadrature. In: Rice, J. R. (ed), *Mathematical Software*, Chap. 7, pp.417-449. Academic Press, San Diego,

Dillon, P. 1987. Discussion of "Field comparison of three mass transport models". *ASCE J. Hydraulic Engineering*, 113(5), 683-684.

Domenico, P.A. and Robbins, G.A. 1985. A new method of contaminant plume analysis. *Ground Water*, 23(4), 476-485.

Fischer, H.B., List, E.J., Koh, R.C.Y., Imberger, J. and Brooks, N.H. 1979. *Mixing in Inland and Coastal Waters*. Academic Press, New York, 483 pp.

Freeze, R.A. and Cherry, J.A. 1979. *Groundwater*. Prentice-Hall Inc., Englewood Cliffs, New Jersey.

Freyberg, D. 1986. A natural gradient experiment on solute transport in a sand aquifer. 2 Spatial moments and advection and dispersion of nonreactive tracers. *Water Resources Research*, 13, 2031-2046.

Fried, J.J. 1975. *Groundwater Pollution, Theory, Methodology, modeling and Practical Rules*. Elsevier Sci. Publ. Co., Amsterdam, 330 pp.

Frind, E.O., Sudicky, E.A. and Schellenberg, S.L. 1988. Micro-scale modeling in the study of plume evolution in heterogeneous media. In E. Custodio *et al.* (eds), *Groundwater Flow and Quality modeling*, pp.439-461.

Gelhar, L.W. and Axness, C.L. 1983. Three-dimensional stochastic analysis of macrodispersion in aquifers. *Water Resources Research*, 19(1), 161-170.

Gelhar, L.W., Gutjahr, A.L. and Naff, R.L. 1979. Stochastic analysis of macrodispersion in a stratified aquifer. *Water Resources Research*, 15(6), 13871396.

Gelhar, L.W., Mantoglou, A., Welty, C. and Rehfeldt, K.R. 1985. *A Review of Field-Scale Physical Solute Transport Processes in Saturated and Unsaturated Porous Media*. EPRI EA-4190, Project 2485-5.

Hamilton, D.A., Wiggert D.C. and Wright, S.J. 1985. Field comparison of three mass transport models. *ASCE J. Hydraulic Engineering*, 111(1), 1-11.

Hantush, M.S. and Jacob, C.E. 1955. Non-steady radial flow in an infinite leaky aquifer. *Trans. Amer. Geophys. Union*, 36(1), 95-100.

Hayden, N.K. 1985. Nevada Nuclear Waste Storage Investigations Project. *Benchmarking NNWSI Flow and Transport Codes, Cove 1 Results.* Sandia Report. SAND-84-0996.

Hunt, B. 1978. Dispersive sources in uniform ground-water flow. *J. Hydraulics Div. ASCE,* 104(HY1), 75-85.

Huyakorn, P.S., Kretschek, A.G., Broome, R.W., Mercer, J.W. and Lester, B.H. 1984. *Testing and Validation of Models for Simulating Solute Transport in Groundwater, Development, Evaluation, and Comparison of Benchmark Techniques.* Int. Ground Water modeling Centre Report, HRI Report No. 35, 419 pp.

Huyakorn, P.S., Ungs, M.J., Mulkey, L.A. and Sudicky, E.A. 1987. A three-dimensional analytical method for predicting leachate migration. *Ground Water,* 25(5), 588-598.

INTRACOIN. 1986a. *International Nuclide Transport Code Intercomparison Study. Final Report Level 1, Code Verification.* Swedish Nuclear Power Inspectorate, Report No. SKI 84,3.

INTRACOIN. 1986b. *International Nuclide Transport Code Intercomparison Study. Final Report Levels 2 and 3, Model Validation and Uncertainty Analysis.* Swedish Nuclear Power Inspectorate, Report No. SKI 86,2.

Javandel, I., Doughty, C. and Tsang, C.F. 1984. *Groundwater Transport, Handbook of Mathematical Models.* AGU Water Resources Monograph 10, 228 pp.

Jousma, G., Bear, J., Haimes, Y.Y. and Walter, F. (eds). 1987. *Groundwater Contamination, Use of Models in Decision-Making.* Kluwer Academic Publ., Dordrecht, 656 pp.

Keeley, J.W. 1977. Magnitude of the ground water contamination problem. In: Kearns, W.R. (ed.), *Proc. Nat. Conf. Public Policy on Ground Water Protection.* Blacksburg, Virginia, pp.2-10.

Kent, D.C., Pettyjohn, W.A. and Prickett, T.A. 1985. Analytical methods for the prediction of leachate plume migration. *Ground Water Monitoring Review,* 5(2), 46-59.

Kincaid, C.T., Morrey, J.R. and Rogers, J.E. 1984. *Geohydrochemical Models for Solute Migration Vol. 1, Process Description and Computer Code Selection.* EPRI EA-3417, Project 1619.

Kinzelbach, W.K.H. 1986. *Groundwater modeling, An Introduction with Sample Programs in Basic.* Elsevier, Amsterdam, 333 pp.

Kinzelbach, W.K.H. and Frind, E.O. 1986. Accuracy criteria for advection-dispersion models. In: da Costa, Sa *et al.* (eds), *Proc. 6th Int. Conf. on Finite Elements in Water Resources,* pp. Springer-Verlag.

Kipp, K.L. 1987. *HST3D, A Computer Code for Simulation of Heat and Solute Transport in Three-Dimensional Ground-Water Flow Systems.* US Geological Survey Water-Resources Investigations Report 86-4095, 342 pp.

Kobus, H. and Kinzelbach, W.K.H. (eds). 1989. *Contaminant Transport in Groundwater,* Balkema, Rotterdam.

Konikow, L.F. and Bredehoeft, J.D. 1978. *Computer Model of Two-Dimensional Solute Transport and Dispersion in Ground Water.* US Geological Survey, Techniques of Water-Resource Investigations, Book 7 Chapter C2, 90 pp.

Lallemand-Barres P. and Peaudecerf, P. 1978. Recherche des relations entre la valeur de la dispersivite macroscopique d'un milieu aquifere, ses autres caracteristiques et les conditions de mesure. *Etude Bibliographique, Bull. BRGM,* 3(4), 277-284.

Lindstrom, F.T. and Boersma, L. 1989. Analytical solutions for convective-dispersive transport in confined aquifers with different initial and boundary conditions. *Water Resources Research,* 25(2), 241-256.

Matheron, G. and de Marsily, G. 1980. Is transport in porous media always diffusive? A counter example. *Water Resources Research,* 16(5), 901-917.

McLaughlin, D.B. 1984. A Comparative Analysis of Groundwater Model Formulation -The San Andreas-Glorieta Case Study. HEC Report, Davis California, 75 pp.

Mishra, S. and Parker, J.C. 1990. Analysis of solute transport with a hyperbolic scale-dependent dispersion model. *Hydrological Processes* 4, 45-57.

Oster, C.A., Sonnichsen, J.C. and Jaske, P.T. 1970. Numerical simulation of the convection diffusion equation. *Water Resources Research,* 6, 1746-1752.

Pereboom, D., Davis, G.B. and Laslett, D. 1992. modeling assessment of long-term impact of landfill leachate on groundwater quality. *1st National Hazardous and Solid Waste Convention,* Sydney, AWWA, 80.1-80.8.

Philip, J.R. 1986. Issues in flow and transport in heterogeneous porous media. *Transport in Porous Media,* 1, 319-338.

Pinder, G.F. 1988. An overview of groundwater modeling. In: Custodio, E. *et al.* (eds) *Groundwater Flow and Quality modeling,* pp.119-134.

Pinder, G.F. and Cooper, H.H. Jr. 1970. A numerical technique for calculating the transient position of the saltwater front. *Water Resources Research,* 6, 875-882.

Robbins, G.A. and Domenico, P.A. 1984. Determining dispersion coefficients and sources of modeldependent scaling. *J. Hydrology,* 75, 195-211.

Sa da Costa, A., Baptista, A.M., Gray, W.G., Brebbia, C.A. and Pinder, G.F. (eds). 1986. *Finite Elements in Water Resources.* Proc. 6th Int. Conf. Portugal, 811 pp.

Salama, R.B., Barber, C. and Davis, G.B. 1989. Characterising the hydrogeological variability of a sand aquifer in the region of a domestic waste disposal site. In: *Int. Symp. on Groundwater Management, Quantity and Quality.* IAHS Publ. No. 188, pp.215-226.

Schäfer, W. and Kinzelbach, W. 1992. Stochastic modeling of *in-situ* bioremediation in heterogeneous aquifers. *Journal of Contaminant Hydrology,* 10(1), 47-73.

Scheidegger, A.E. 1961. General theory of dispersion in porous media. *J. Geophysical Research,* 66, 3273.

Slichter, C.S. 1905. *Field Measurements of the Rate of Underground Water.* US Geological Survey Water Supply Paper No. 140, 9-85.

Sudicky, E.A. 1986. A natural gradient experiment on solute transport in a sand aquifer, spatial variability of hydraulic conductivity and its role in the dispersion process. *Water Resources Research,* 22(13), 2069-2082.

Sudicky, E.A. 1988. Reply to comment of White (1988). *Water Resources Research,* 24(6), 895-896.

Sudicky, E.A. 1989. The Laplace transform Galerkin technique, A time-continuous finite element theory and application to mass transport in groundwater. In: Kobus, H.E. and Kinzelbach, W.K.H. (eds) *Contaminant Transport in Groundwater,* pp.317-325.

Taigbenu, A. and Liggett, J.A. 1986. An integral solution for the diffusion-advection equation. *Water Resources Research,* 22(8), 1237-1246.

Taylor, S.R. and Howard, K.W.F. 1987. A field study of scale-dependent dispersion in a sandy aquifer. *J. Hydrology,* 90, 11-17.

Tompson, A.F.B. 1988. On a new functional form for the dispersive flux in porous media. *Water Resources Research,* 24(11), 1939-1947.

Tyler, S.W. and Wheatcraft, S.W. 1992. Reply to Philip (1992). *Water Resources Research* 29(5), 1487-1490.

van der Heijde, P., Bachmat, Y., Bredehoeft, J., Andrews, B., Holtz, D. and S. Sebastian. 1985. *Groundwater Management, The Use of Numerical Models.* American Geophysical Union, Washington D.C. 180 pp.

van Genuchten, M.Th. and Alves, W.J. 1982. *Analytical Solutions of the One-Dimensional*

Convective-Dispersive Solute Transport Equation. US Dept of Agric. Technical Bulletin No. 1661, 149 pp.

Voss, C.I. 1984. *A Finite-Element Simulation Model for Saturated-Unsaturated, Fluid-Density-Dependent Ground-Water Flow with Energy Transport or Chemically-Reactive Single-Species Solute Transport*. US Geological Survey, Reston Virginia, 409 pp.

Wheatcraft, S.W. and Tyler, S.W. 1988. An explanation of scale-dependent dispersivity in heterogeneous aquifers using concepts of fractal geometry. *Water Resources Research,* 24(4), 566-578.

White, I. 1988. Comment on "A natural gradient experiment on solute transport in a sand aquifer, spatial variability of hydraulic conductivity and its role in the dispersion process" by Sudicky, E.A. *Water Resources Research,* 24(6), 892-894.

Wilson, J.L. and Miller, P.J. 1978. Two-dimensional plume in uniform ground-water flow. *J. Hydraulics Div. ASCE,* 104(HY4), 503-514.

Yeh, G.T. 1981. *AT123D, Analytical Transient One-, Two- and Three-Dimensional Simulation of Waste Transport in the Aquifer System*. Oak Ridge National Lab., Report No. ORNL-5602, 83 pp.

5 Representative sampling of groundwater from boreholes

C. Barber and G. B. Davis

CSIRO Division of Water Resources, Private Bag, P O Wembley, Western Australia

Abstract

It is generally recognised that proper management of groundwater resources requires good quality monitoring data. There is consequently a need to obtain samples which are 'representative' of groundwater present within specified depth intervals in aquifers.

To obtain a sample which is representative of groundwater, requires flushing of either stagnant casing storage (in well or borehole) or water present in a sampling device (e.g. lysimeter). With boreholes, at least three well volumes need to be pumped before taking a sample from pump discharge, based on theory and experimental studies. Pumping rate can be optimised depending on requirements using a nomographic technique, given basic aquifer properties.

Concentrations of inorganic substances in pumped samples can be changed by reactions taking place as a result of pumping. Gas-lift pumping devices are particularly prone to this problem due to oxidation as a result of introduction of molecular oxygen, or changes in pH due to removal of carbon dioxide. Displacement pumps (bladder or diaphragm pumps) generally give the least problems because with these there is no direct contact with a gas-phase and no applied suction.

Volatile constituents of groundwater (organics (VOC's) and dissolved gases) are also affected by gas lift devices, and particularly by sorption. Absorption of VOC's and gases by flexible tubing used for pump discharge lines is perhaps the major difficulty. To avoid this, several schemes for obtaining in-situ samples or measurements have been devised. Further development of these devices is needed.

Introduction

Increased environmental awareness has led to greatly increased interest in water quality and its degradation. This has led to the development of a wide range of analytical techniques for the determination of low levels of contaminants which are of concern because of effects on humans or on the environment in general. Groundwater quality in particular has become a major issue because of the reliance on groundwater for domestic supply in the USA and parts of Western Europe.

Barcelona *et al.* (1984) indicated that one of the principal obstacles to reasoned management of the groundwater regulatory process [in the USA] was the " assurance of high quality, reliable [groundwater quality] monitoring results". Thus given the availability of high precision analytical instrumentation, there is an overriding need for obtaining a sample which is representative of the groundwater both physically and chemically.

The difficulties in obtaining representative samples have been highlighted in a number of studies of groundwater sampling techniques (Gibb *et al.*, 1981; Scalf *et al.*, 1981; Unwin and Huis, 1983; Barber and Davis, 1987; Reynolds *et al.*, 1990). Some recent research has been summarised by Davis *et al.* (1992). It is clear from these studies that the term 'representative sample' can be defined differently in different situations. For example, samples obtained by pumping from open boreholes or fully screened wells are often used to provide a bulked sample giving overall, indications of changes in groundwater quality. These may be sufficient to provide "early warning of contamination at relatively low cost" (McNicholls and Davis, 1988), or for monitoring broad changes in quality of groundwater in an aquifer used for supply. This approach would be inadequate for determination of the 3-dimensional variability in groundwater quality, as in field validation of model simulations, monitoring of efficiency of contaminant remediation and other applications requiring detailed information. In these cases, taking of discrete samples with depth (representative of groundwater in short lithostratigraphic intervals in aquifers) would be needed from appropriate networks of boreholes.

A working definition of the term representative sample is needed, which applies to both the above (and other) situations which are encountered by hydrogeologists involved with groundwater quality assessment. In this paper, for convenience, the following definition is used.

A representative sample is a volume of water recovered from a well, borehole or *in situ* sampling device, which has the same average chemical composition as groundwater flowing radially towards the well or device intake (screen) under natural or induced hydraulic head, and from a specified depth-interval.

Two aspects of representivity are discussed. First, the hydraulic basis for recovery of representative samples which allows an assessment to be made of the proportion of fresh groundwater, as opposed to casing storage water in a discharge at the surface. Second, physicochemical aspects, which are important considerations to ensure that the chemical integrity of water being pumped from an aquifer to a wellhead is maintained. Precautions needed to maintain sample integrity following recovery are not covered in the paper. Neither are aspects involving biological sampling from boreholes.

Hydraulic basis for representative sampling of groundwater

A wide range of techniques are available for recovery of samples of groundwater for monitoring of groundwater quality. These range from conventional borehole constructions, to partially penetrating boreholes and piezometer installations and other multilevel sampling devices. The main types of these are discussed below.

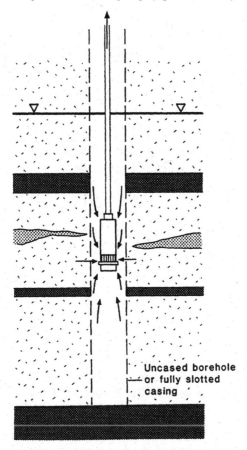

Figure 1 *Sample collection by pumping from an open borehole or fully-screen well.*

Fully screened and open (uncased) boreholes
Fully screened boreholes (used in unconsolidated materials) and uncased boreholes (in stable, competent strata) have been used extensively in the past for monitoring of groundwater quality. Techniques used for sampling of these have varied considerably.

Early attempts at obtaining representative samples of groundwater from fully screened/penetrating boreholes focussed on obtaining these by pumping to obtain a single depth-averaged sample, or by recovering a number of samples from borehole standing water using surface activated "grab" samplers (*e.g.* see Figure 1). Water quality profiles with depth which have been obtained using the latter have been shown to be a poor reflection of natural stratification of groundwater composition (Barber and Baxter, 1983).

Gibb *et al.* (1981), Scalf *et al.* (1981) and others have stressed the need to pump boreholes to remove stagnant water present within the screen and casing and replace

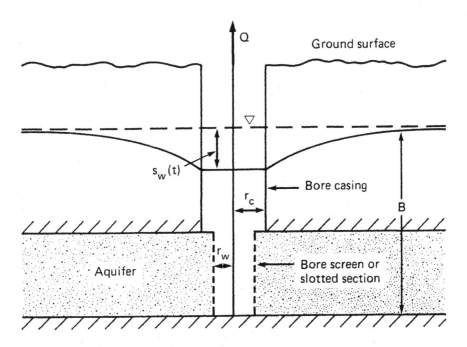

Figure 2 *Schematic diagram of a finite diameter well in a confined aquifer showing drawdown s_w, casing radius r_c screen radius r_w, pumping rate Q and initial depth of water in the well B after Barber and Davis, 1987.*

this with fresh groundwater from the surrounding aquifer. Pumping of boreholes has thus largely replaced sampling of standing water, despite increased equipment and labour costs involved with borehole pumping.

Recovery of samples from boreholes penetrating layered aquifers has been investigated in laboratory and field tests by Barczewski and Marschall (1989). They showed that the composition of pump discharge quickly stabilised to reflect groundwater quality in more transmissive layers. Pumped samples were termed depth-averaged or "transmissivity-averaged".

The extent of pumping required to flush a well or borehole and achieve a sample which contains a negligible proportion of standing water or casing storage, has been the focus of considerable debate. A variety of investigations have attempted to determine this experimentally. Usually, this has involved recording of the time taken for groundwater to stabilise in composition (*e.g.* by measuring pH, EC or some other indicator) during pumping. Estimates have varied considerably (Bryden *et al.*, 1986) and a need for a viable theoretical basis for sampling is clearly required.

Casing storage effects in pumping boreholes have been considered by Schafer (1978) and Gibb *et al.* (1981). Barber and Davis (1987) developed this approach and presented a more detailed theoretical approach for recovery of representative samples, by consideration of well hydraulics during pumping. They considered the case of a fully penetrating, finite diameter well in a confined aquifer (Figure 2) using the approach of Papadopulos and Cooper (1967). This was a close approximation to the more usual case of partial penetrating well in unconfined conditions. Barber and Davis (1987) showed that if fresh groundwater and casing storage water did not mix during pumping the proportion of fresh groundwaterwater in pump discharge (*faq*) could be calculated from:

$$faq = 1 \left[r_e^2 \mid 4T' \; dF \mid dt \left(u_w', \; \alpha \right) \right] \qquad (1)$$

Here T' is the effective transmissivity of the screened or open section of borehole (Davis, 1987), r_e is effective borehole casing radius allowing for radius of pump discharge line, F is the well function given by Papadopulos and Cooper (1967), t is time, S is storativity and:

$$u_w' = r_e^2 \mid 4T' \; t \qquad (2)$$

$$\alpha = S \; r_w^2 \mid r_e^2 \qquad (3)$$

The term in square brackets in equation (1) is the tabulated function H/H_0 given in Cooper *et al.* (1967) and Papadopulos *et al.* (1973), which allows easy estimation of *faq*. Generally, *faq* > 0.98 when $u_w/\alpha < 10^{-2}$, giving

$$t > 25 \; r_e^2 \mid T' \qquad (4)$$

In equation (4) t is the minimum pumping time to achieve better than 98% of fresh groundwater in pump discharge, assuming no mixing of casing storage water and fresh groundwater during pumping.

Barber and Davis (1987) used a mass balance approach for cases where mixing of fresh groundwater and casing storage water takes place. They showed that minimum pumping time t_a could be estimated as:

$$t_a = -V \mid Q \; \ln \; (m) \qquad (5)$$

Here, V is borehole storage volume and Q is pumping rate. The term m is derived from the concentration of some conservative chemical constituent in fresh groundwater (Cg), initially in casing storage (Cc) and that in pumped discharge (Ct), as

$$m = [Ct - Cg] \mid [Cc - Cg] \qquad (6)$$

Equation (5) can be used to assess the (commonly used) "number of borehole volumes" to achieve a representative sample of groundwater. Thus, setting the value of m to 0.1, 0.04 and 0.01 (equivalent to coefficients of variation for sampling error of 2.5%, 1% and 0.25%), requires the pumping of 2.3, 3.2 and 4.5 well volumes.

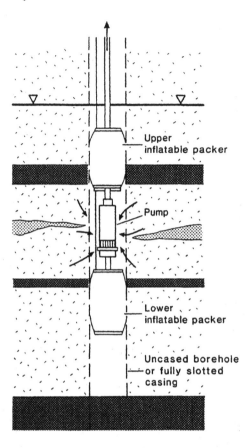

Figure 3 *Sample recovery by pumping from between inflatable straddle packers.*

These compare favourably with results of experimentally determined values (Barber and Davis, 1987; Bryden *et al.*. 1986 and others).

Pumped discharges from open or fully slotted boreholes, as noted above, would provide depth- or transmissivity-averaged samples of groundwater. Although, such broadly representative samples would be of little value in detailed studies of aquifer behaviour, they would be representative of water abstracted for supply or for some other beneficial use. They would also be appropriate for providing initial, broad indications of contamination.

Straddle packers and multipackers
More detailed information on groundwater quality can be obtained by isolation of sections within open or fully screened boreholes using inflatable packers or other devices which restrict vertical movement of water within borehole casing (Figure 3). Pumping from between isolated intervals thus induces fresh groundwater to replace casing storage water within the isolated section of borehole. Under these conditions,

groundwater could be largely derived from that part of the aquifer adjacent to the isolated section.

Relatively small volumes of water need to be pumped from the isolated sections because the volume of casing storage within the section is generally small. Equation (5) indicates that to achieve suitably representative samples requires pumping of only between 2 and 3 times the volume of isolated casing storage for the case of zero drawdown. In general, this could be achieved within a few minutes assuming that fresh groundwater was present immediately outside the borehole.

Barber and Baxter (1983) and Suffet *et al.* (1983) present evidence that straddle packers can be used successfully for recovery of representative samples of groundwater from discrete intervals in open boreholes. However, both these studies indicate difficulties with borehole wall irregularities which may give rise to operational problems. These problems can be overcome by identifying irregularities using downhole caliper logs.

There has been less success in sampling fully screened boreholes in unconsolidated aquifers. In laboratory tests, Barczewski and Marschall (1989) found that straddle packers gave depth-averaged samples, similar to those obtained by pumping without packers. It was clear that permeable sand-pack grout emplaced around casing negates the benefits of using packers in their case .

Several studies have attempted to overcome problems of pumping from packer-isolated sections by developing multipacker assemblages to give vertical profiles of water quality within fully screened boreholes. Ronen *et al.* (1986) used an *in situ* system. They incorporated dialysis cells separated by low permeability flexible rubber seals which acted as packers. Each cell was initially filled with distilled water but after several weeks the cells equilibrated with groundwater. The composition of water present in cells recovered at the surface thus reflected natural stratification in groundwater.

Barczewski and Marschall (1989) used a multipacker assembly with variable rate mini-pumps. The pumping rate was proportional to hydraulic conductivity for each isolated section, which produced a set of representative samples which related to water quality in their layered simulated aquifer. Use of this equipment in the field would require prior knowledge of variations in hydraulic conductivity of aquifer units penetrated by a borehole.

More commonly, groundwater investigations do not use fully screened boreholes to obtain vertical water quality profiles in groundwater, largely because of excessive equipment and manpower requirements (for example with use of large pumps and packers). Instead, other borehole constructions have been developed to simplify and streamline operational monitoring, whilst still recovering representative samples from different depths in an aquifer. Some of these constructions are described below.

Short-screened boreholes and piezometers

The simplest, and probably the most widely used multilevel construction is achieved by emplacement of nested piezometers, or piezometer bundles within large diameter boreholes (Figure 4) or drilling of multiple boreholes in one location. Generally, these consist of a series of short-screened boreholes, with 1-2 m long screens, each

Table 1 *Worked examples illustrating the use of Figure 5 in optimising pumping rates to remove casing storage water prior to sampling groundwater.*

Borehole
Effective radius r_e	0.05 m
Depth of standing water in borehole B	5 m
Casing storage volume V	0.039 m^3
Screen length L	1 m

Aquifer

	Sand aquifer			Low permeability clay soil			
Hydraulic conductivity of screened interval K	10^{-4} m s^{-1}			4×10^{-6} m s^{-1}			
Effective transmissivity T	10^{-4} m^{2}s^{-1}			4×10^{-6} m^{2}s^{-1}			
Storativity	10^{-2}			10^{-2}			

Pumping Scheme

Pumping rate (Q) (L min^{-1})	4	4	10	4	4	1.2	3
Mixing/No Mixing Fig 4	Mixing	No Mixing	Mixing	Mixing	No Mixing	Mixing	Mixing
δ Fig 5	188	-	75.4	7.54	-	25	10
m Fig 5	0.04	0.04	0.04	0.04	0.04	0.04	0.4
∈ sampling error as in Fig 5	1%	1%	1%	1%	1%	1%	10%
Pumping time required before sampling (-min)	33	4	13	14	100	100	14
Fractional drawdown (Sw/B)	0.1	0.08	0.3	0.82	>1	0.56	0.6

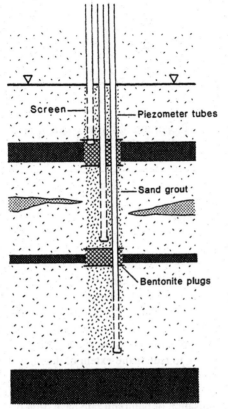

Figure 4 *Multilevel sampling by pumping from nested piezometers. Alternatively, individual short-screened boreholes can be emplaced at different levels to avoid vertical leakage through bentonite plugs.*

set at different depths and separated by low permeability grout/bentonite seals. These thus provide depth-averaged samples of groundwater over short intervals. Appropriate location of screens, for example taking account of natural layering of aquifers, can provide an effective mechanism for characterising variability in water quality with depth at low cost. Inevitably, purging of casing storage from short-screened piezometer tubes would be required, as stressed by Gibb *et al.* (1981). However, the availability of a wide variety of small diameter pumps, and the requirement for removal of relatively small volumes of casing storage water (within small diameter casing) makes sampling using these constructions effective and efficient in manpower and time.

The extent of flushing of casing storage which is required to obtain a representative sample can be estimated using equation (5) for short screened boreholes, as shown by Barber and Davis (1987). In order to achieve an acceptable degree of representivity, Barber and Davis (1987) recommended pumping for at least 2.3 borehole volumes. Experimental work by Barker *et al.* (1987), similarly

Table 2 *Effect of gas/liquid exchange in reducing concentrations of dissolved VOC's, assuming equilibrium conditions given by Henry's Law.*

	Methane	Perchloroethylene	Benzene
Henry's coefficient	29	0.94	0.25
Gas/liquid volume ratios	Proportion of initial concentration remaining in liquid phase		
0.1	0.26	0.91	0.98
1.0	0.03	0.52	0.8

recommended flushing of three times the volume of stagnant casing storage to o a representative sample from small diameter piezometer bundles.

In some cases there may be a need to optimise the rate of pumping before t a representative sample. For example, higher rates of pumping in transmi: aquifers would decrease the time needed for sampling giving potential cost sav Alternatively, pumping of groundwater from wells in low conductivity soils require low rates of pumping to reduce the extent of drawdown. This may be ne to avoid damage to submersible pumps which would occur if boreholes are em of water during pumping.

Mechanisms for optimisation of pumping rate have been described by Barbe: Davis (1987), who presented a nomograph for this, based on theore considerations of well hydraulics. The nomograph is shown in Figure 5 dimensionless times plotted against 'sampling error' (ε) expressed as a coefficie variation, or alternatively the value m as given in equation (6). Different curve the reciprocal of dimensionless pumping rate (δ) are shown for cases where m of stagnant casing storage and groundwater takes place during pumping. Also, st is a single curve for the no-mixing case.

Worked examples are shown in Table 1 to illustrate the use of the nomograp optimising pumping rate and drawdown. Two cases are considered, a sand aq and a low permeability clay soil.

In order to achieve a concentration of a conservative constituent in p discharge to within 1% of its true concentration in groundwater in the sand aq ($m = 0.04$), would require pumping for over 30 minutes at 4 L min^{-1}, assu: mixing occurs. Only five minutes pumping would be needed if no mixing took p Thus the time taken for the discharge to stabilise in composition would be bet' four and thirty minutes at this pumping rate depending on the extent of mi Increasing the rate to 10 L min^{-1} reduces the time to between four and fifteen mi: pumping (schemes 2 and 3 in Table 1). One could anticipate that the actual pum time would be closer to the higher figure, *i.e.* mixing of fresh groundwater and c: storage water would inevitably take place.

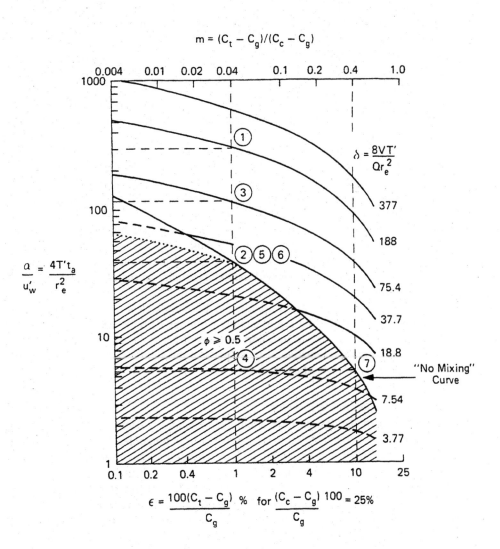

Figure 5 *Nomograph for estimating pumping time and optimising pumping rate and time, for flushing of stagnant casing storage water before taking a representative sample. C_c, C_g and C_t are concentrations of a conservative constituent in casing storage water, fresh groundwater and pump discharge respectively. Other terms are defined in Table 1. Circled numbers are pumping schemes used in Table 1. After Barber and Davis, 1987.*

The same borehole configuration is used in the less transmissive soil in schemes 4 to 7 in Table 1. In these cases, time required for flushing of casing storage assuming no mixing is larger than for the mixing case, as a result of the large drawdown produced during pumping. It is clear in cases 4 and 5 that using pumping rates of 4 L min^{-1}, produces excessive drawdown, with the bore being completely emptied after approximately 15 minutes. To avoid this, and still achieve the same level of representivity (m,ε) and pumping time, an 'optimum' pumping rate of just over 1 L min^{-1} is estimated (case 6), from the appropriate value of δ for the 'mixing' case. Fractional drawdown in this case is significantly lower, although pumping time is long. Accepting a lower level of representivity (scheme 7) significantly decreases the required pumping time, optimising sampling time at the expense of sampling accuracy.

The results given in Table 1 indicate the potential usefulness of the nomograph in planning of sampling programs. It is more than likely, however, that in practice in the field, most investigations will still rely on stabilisation of indicator parameters (pH, Eh, DO, EC *etc.*) in pump discharge. The nomograph provides an understanding of why indicator parameters stabilises in a given time.

In all these cases, it is assumed that the pump is placed immediately above or within the screened interval in the borehole. Additionally, it is assumed that chemical constituents in the groundwater behave conservatively during pumping. Sorption, for example on pump discharge lines will inevitably require an increased pumping time if sampling from the pump discharge.

Samples recovered from shortscreened boreholes in both high and low permeability media would be representative (in terms of bulk volume) from those parts of the aquifer adjacent to the short sections of borehole screen. These samples are not necessarily representative of those parts of the aquifer between the screened sections especially in layered aquifers.

Multiport boreholes and lysimeters

A number of borehole designs have attempted to simplify sampling, and particularly the need to remove stagnant casing storage water from boreholes, by reducing screen length to essentially a point inlet, and by reducing access tube diameter and consequently reducing the volume of casing storage. Multiport boreholes (*e.g.* Figure 6) and lysimeters (*e.g.* Barber and Baxter, 1983) thus provide numerous point measurements, as opposed to depth averaged measurements. Often these devices use gaslift or suction to remove water present in access lines and sampling devices, and for recovery of samples. Some problems with the use of gas-lift and suction for recovery of groundwater are discussed further in the next section .

Chemical considerations of representivity.

A number of studies have indicated that groundwater is often altered chemically during sampling. This has been attributed partly to mixing effects between fresh groundwater and stagnant casing storage as discussed above.

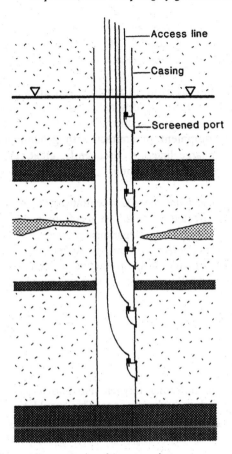

Figure 6 *Multilevel sampling using multiport casing.*

Composition of casing storage water

Casing storage water may undergo significant chemical changes on standing under near atmospheric conditions in a borehole. In particular, microbiological reactions can take place. Reactions involving nitrification/denitrification, sulphate reduction and sulphide oxidation, oxidation of iron and manganese and methane production and utilisation are all microbiologically mediated (Stumm and Morgan 1981). Gibb *et al.* (1981) indicated that constituents which were most sensitive to change induced by well flushing, which by inference would be most likely to be affected by standing in boreholes which usually are open to atmosphere, were pH, Fe, Mn, Mg, Zn, Cd, Cu, As, Se and B. Chloride, Na, K, Mg and Ca were relatively insensitive to change. Changes which do occur are clearly dependent on local aquifer conditions. The most extensive changes would be expected where anoxic groundwater surrounds large diameter boreholes open to atmospheric conditions.

Chemical and physicochemical changes during sampling
Flow of groundwater from an aquifer into a borehole changes physicochemical conditions such as temperature, pressure and concentrations of dissolved gases and other volatiles. These can give rise to acid-base reactions, redox reactions, precipitation/dissolution reactions, sorption and in some cases microbial transformations (Gibb *et al.*, 1981). Additionally, dissolved concentrations of specific substances can be changed by physicochemical reactions taking place on exposed surfaces of sampling equipment in contact with groundwater (principally the pumping device and discharge lines). Consequently, great care has to be taken to minimise any changes induced by sampling.

Inorganic constituents of groundwater: Concentrations of inorganic constituents of groundwater are mainly affected by acid/base reactions and redox reactions during pumping. These are brought about by degassing (particularly of carbon dioxide) and by addition of oxygen from contact with the atmosphere.
   Degassing of carbon dioxide from groundwater increases pH value, moving equilibria to the right in the following reactions in equations (8) and (9).

$$H^+ + HCO_3^- = H_2O + CO_2 \qquad\qquad (8)$$

$$H^+ + CO_3^{2-} = HCO_3^- \qquad\qquad (9)$$

   Likewise, injection of molecular oxygen into anoxic systems induces several redox reactions which produce hydrogen ions. This would give rise to decreases in pH value of pumped groundwater. Sulphide oxidation (equation 10) and iron oxidation and hydrolysis (equation 11) are common reactions which can occur.

$$2O_2 + H S^- = SO_4^{2-} + H^+ \qquad\qquad (10)$$

and

$$O_2 + 4Fe^{2+} + 10H_2O = 4Fe(OH)_3 + 8H^+ \qquad\qquad (11)$$

   Significant changes in water chemistry can be induced by particular types of pumping mechanisms. Gibb *et al.* (1981) showed that gas-lift devices induced significant changes in concentrations of a range of metals and pH where gas was in direct contact with groundwater. This was more significant where air-lift was used, but changes also occurred where nitrogen gas was used to recover samples. Use of a diaphragm (bladder) pump was considered preferable to other recovery methods in terms of achieving reproducible, representative samples of groundwater from test boreholes at six sites.
   Barcelona *et al.* (1984) conducted laboratory tests on a wide range of pumping mechanisms to determine the effect of pumping on water quality parameters. Again, bladder pumps were ranked above average for recovery of samples containing

dissolved gases (and volatile halocarbons). Gas lift devices, positive (mechanical) displacement pumps and suction devices gave significant losses of dissolved volatile and gaseous constituents.

Exposure of samples to atmospheric oxygen seems to be a major difficulty in maintaining the inorganic, chemical integrity of a pumped sample of groundwater. Exposure to oxygen could occur either directly at the point of discharge, or by diffusion through pump discharge lines. Holm *et al.* (1988) found that diffusion of atmospheric oxygen through polymeric tubing used for pump discharge lines could significantly affect redox conditions when pumping anoxic groundwater. Subsequent reactions could occur such as oxidation of iron, precipitation of iron oxyhydroxides, and scavenging of other dissolved constituents particularly trace metals.

There is little information available on sorption of inorganic compounds on borehole materials/sampling equipment, or on leaching (*e.g.* of trace metals) from these materials. Clearly the latter needs careful consideration.

Organic constituents: Interest in organic pollutants in groundwater, particularly volatile organic compounds (VOC's), has required the development of new sampling procedures for recovery of representative samples of groundwater containing trace (ppb) levels of VOC's (Scalf *et al.*, 1981). As with inorganic constituents, degassing and contact of the sample with a gas phase is important. The extent of any effect on concentration of VOC's in groundwater depends on Henry's coefficients for individual compounds (which determine partition between gas and water phases), the extent to which gas and water phases have equilibrated and the relative volumes of these phases.

Assuming that equilibration is rapid, then a given volume of groundwater (V_l) initially containing a concentration C_i of a VOC, will partition between liquid and gas phases (the latter of volume V_g) according to the following equation, by mass balance.

$$V_l \cdot C_i = C* (V_l + V_{gH}) \tag{12}$$

Here $C*$ is the new equilibrium concentration in liquid phase, H is Henry's coefficient. The effect of two different ratios of liquid and gaseous phase volumes on three VOC's of different volatility are shown in Table 1. A relatively low gas to liquid volume (for example which might be achieved in a gas displacement pump) does not greatly decrease concentrations, except for the gas methane. However, where the gas and liquid phases are equal, as in devices which rely on entrainment of liquid in pressurised gas streams, significant changes are induced, particularly for methane and PCE. This simple analysis supports the conclusions of Barker *et al.* (1987) that loss of VOC's can be minimised by reducing direct contact between gas and liquid phases, by using small diameter discharge lines and gas displacement devices.

In contrast to inorganic constituents, sorption effects are probably more important in affecting concentrations of dissolved organic constituents in groundwater. Barcelona *et al.* (1985) found sorption of dissolved VOC's on polymeric tubing following short term exposure in laboratory tests. Teflon, polyethylene, polypropylene, PVC and silicone rubber all sorbed chloroform, TCE, TCA and PCE from aqueous solution. Teflon showed the least sorption for contact periods of as little as 10 minutes. Flexible PVC tubing and silicone rubber showed the greatest sorptive effects, removing 80% of VOC's from solution. Poor desorption of VOC's led to the conclusion that these compounds were absorbed into the polymeric matrix.

Later studies by Barker *et al.* (1987) on multilevel piezometer bundles noted similar problems. They found that VOC's were able to diffuse from groundwater through flexible access tubing, contaminating groundwater samples inside the tube.

Reynolds *et al.* (1990) further evaluated sorption effects and concluded that sorption was more prevalent for the more flexible polymeric materials. Thus uPVC and PTFE materials used in borehole construction and sampling devices sorbed least. Flexible PVC and rubber materials used in pump discharge lines, sorbed the most. Stainless steel, galvanised steel and aluminium were also found to reduce concentrations of selected VOC's from solution to some extent in the latter evaluation. The only material not giving rise to sorptive effects was borosilicate glass!

Sorption clearly is a major difficulty in obtaining a representative sample for VOC analysis, even more so than volatilisation (Barker *et al.*, 1987). To overcome this problem Pankow *et al.* (1984) developed glass syringe samplers for sampling groundwater for VOC analysis. This system concentrated organics on a concentrator tube attached to the syringe *in situ* within the borehole, following conventional flushing of casing storage water. A similar scheme using glass syringe samplers was adopted by Barber *et al.* (1992). In the latter case, VOC's in water samples were concentrated using purge-and-trap technique at the surface. Purging of VOC's took place within the syringe sampler which avoided losses brought about by sample transfer. Also, more recently, stainless steel multiports sampled via suction into a syringe were used successfully to characterise at fine scale (25-50 cm vertical intervals) VOC's in groundwater due to a gasoline leakage (Thierrin *et al.*, 1992; Davis *et al.*, 1992).

These techniques offer a relatively simple method which can be used in shortscreened piezometers and boreholes, although they have not been evaluated for monitoring on a routine basis. Monitoring for VOC's on a routine basis still poses significant problems which require further method development (see Patterson *et al.*, 1993).

Recent developments

The difficulties in obtaining samples for determination of concentrations of VOC's and gases have prompted development of several new techniques. These essentially are *in situ* devices which preclude the need for pumping groundwater. Two types have been developed involving *in situ* monitoring and *in situ* measurement.

In situ monitoring
In situ monitoring devices are a relatively recent development, which avoid the need for pumping of groundwater samples to the surface for later laboratory analysis.

Essentially, these techniques provide minimal disturbance of the groundwater system. A system for obtaining point measurements within a cased borehole have been described by Ronen *et al.* (1986, 1989). They used dialysis cells for sampling groundwater profiles for analysis of inorganics and for determination of concentrations of dissolved oxygen.

Similar equipment, consisting of diffusion cells which were installed within backfilled boreholes, was developed by Barber and Briegel (1989) for the determination of dissolved methane in groundwater. In this system gas within each cell equilibrated with dissolved gas in groundwater and was flushed to the surface by a gas stream via small diameter access lines. Gas was collected at the surface for later analysis. Alternatively, analysis of the gas stream was carried out directly in the field. The method was also successfully used for determination of dissolved oxygen in groundwater (Barber *et al.*, 1990).

In situ measurement

In situ measurement of organic compounds dissolved in groundwater has been accomplished using fibre optic chemical sensors or FOC's (Milanovich, 1986; Klainer *et al.*, 1988). These sensors pass light of an appropriate wavelength to the measurement point by optical fibre. The fibre is terminated with a chemical sensor; interaction of the sensor with a target organic molecule changes fluorescence, absorption or reflectance. Optical couplers at the surface are able to separate reflected light from excitation light, and this is analysed using a spectrometer at the surface. These devices are reported to be able to determine single organic compounds in complex mixtures. However, individual FOC's are required for each compound of interest which detracts from the usefulness of these for *in situ* measurement in groundwater where several organic compounds are often present together.

Summary

The term 'representative sample' is ill-defined in the literature. For convenience in this paper we have defined that a sample is representative of groundwater contained within a specified depth interval, if the composition of the sample is the same as the average composition of groundwater obtained from that interval under flowing conditions. These samples would thus be depth-averaged, or perhaps transmissivity-averaged.

Samples which are representative of gross changes in groundwater quality can be obtained by pumping from an open borehole or fully screen observation well penetrating a significant proportion of aquifer or aquifer unit. The pumping time required to remove stagnant casing storage under these conditions can be estimated using equation (5). Normally, this would involve removal of at least 2 to 3 well volumes of water.

More detailed sampling, for example to investigate three dimensional variability in groundwater quality in an aquifer, requires a different approach. A wide range of well constructions and multiport devices are available for collection of samples which are representative (at least in hydraulic terms) of groundwater present within short sections, or at points within a vertical profile. As with more conventional observation wells, stagnant water present as casing storage needs to be removed by pumping from short-screened boreholes, piezometers, and multiport access lines.

The chemical integrity of samples of groundwater can be affected by sampling, for example during pumping. Inorganic constituents are affected by acid-base reactions and redox reactions, involving precipitation and dissolution of solid phases. These reactions are promoted by degassing, particularly of carbon dioxide, and by addition of oxygen to anoxic samples. Devices which use gas-lift for sample recovery are particularly prone to these difficulties.

Volatile organic compounds (VOC's) and dissolved gases can also be affected by degassing and by the use of gas-lift devices. Losses of volatiles can be minimised by reducing the contact between gas and liquid phases.

Absorption of polymeric materials used in construction of sampling devices and borehole casing is a major problem for organic compounds, particularly VOC's. Transfer of absorbates to water or gas within polymeric access lines (oxygen and VOC's) has been observed. Use of *in situ* syringe sampling devices which recover samples within a borehole following pumping offer a possible alternative to overcome these difficulties. Other *in situ* devices such as diffusion cells and fibre optic chemical sensors also show promise. Further development is required before these devices can be used for routine monitoring programs.

References

Barber, C. and Baxter, K. 1983. Sampling of groundwater contaminated by organic substances. *Proc. 1st Atlantic Workshop "Organic Chemical Contamination of Groundwater", Dec. 1982,* pp.59-70. Nashville, Ts. Amer. Water Works Assoc.

Barber, C. and Davis, G.B. 1987. Representative sampling of groundwater from short-screened boreholes. *Ground Water,* 25 (5), 581-587.

Barber, C. and Briegel, D. 1989. A method for the *insitu* determination of dissolved methane in groundwater in shallow aquifers. J. Contam. Hydrol., 2, 5160.

Barber, C., Davis, G.B. and Farrington, P. 1990. Sources and sinks for dissolved oxygen in groundwater in an unconfined sand aquifer, Western Australia. In: Durrance *et al.* (eds), *Geochemistry of gaseous elements and compounds,* pp. 353-368. Theophrastus Publications SA, Athens.

Barber, C., Briegel, D., Power, T. and Hosking, J.K. 1992. Pollution of groundwater by organic compounds leached from domestic solid wastes; a case study from Morley, Western Australia. In: Lesage S and Jackson R (eds), *Groundwater Contamination and Analysis at Hazardous Waste Sites,* pp.357-380. Marcel Decker Inc. NY.

Barcelona, M.J., Helfrich, J.A., Garske, E.E. and Gibb, J.P. 1984. A laboratory evaluation of ground water sampling mechanisms. *Ground Water Monitoring Review,* 4 (2), 3241.

Barcelona, M.J., Helfrich, J.A.and Garske,E.E. 1985. Sample tubing effects on groundwater samples. *Anal. Chem.,* 57, 460464.

Barczewski, B. and Marschall, P.1989. The influence of sampling methods on the results of groundwater quality measurements. In: H Kobus and W Kinzelbach (eds), *Contaminant Transport in Groundwater.* Balkema, Rotterdam.

Barker, J.F., Patrick, G.C., Lemon, L. and Travis, G.M. 1987. Some biases in sampling multilevel piezometers for volatile organics. *Ground Water Monitoring Review,* 7 (2), 48-54.

Bryden, G.W., Mabey, W.R. and Robine, K.M. 1986. Sampling for toxic contaminants in groundwater. *Ground Water Monitoring Review,* 6 (2), 67-72.

Cooper, H.H., Bredehoeft, J.D. and Papadopulos, I.S. 1967. Response of a finitediameter well to an instantaneous discharge of water. *Water Resources Research,* 3 (1), 263-269.

Davis, G.B. 1987. A theoretical assessment of casing storage effects when pump sampling a partially penetrating borehole. *Hydrological Sciences Journal,* 32 (2), 133-141.

Davis, G.B., Barber, C., Patterson, B.M., Briegel, D. and Lambert, M. 1992. Sampling groundwater quality for inorganics and organics: some old and new ideas. *Drill '92. Australian Drilling Industry Association Conference, Perth Oct. 1992.* 24.1-24.9.

Gibb, J.P., Schuller, R.M. and Griffin, R.A. 1981. Procedures for the collection of representative water quality data from monitoring wells. *Cooperative Resources Report* No. 7, p.61. Illinois State Water Survey and State Geological Survey.

Holm, T.R., George, G.K. and Barcelona, M.J. 1988. Oxygen transfer through flexible tubing and its effect on groundwater sampling results. *Ground Water Monitoring Review,* 8(3), 83-89.

Klainer, S.M., Koutsandreas, J.D. and Lawrence, A. 1988. *Monitoring groundwater and soil contamination by remote fibre spectroscopy,* Publ.963, pp. 370-380. ASTM Special Technical Publication.

McNicholls, R.J and Davis, G.B. 1988. Statistical issues and problems in groundwater detection monitoring at hazardous waste facilities. *Ground Water Monitoring Review,* Fall, 135-150.

Milanovich, F.P. 1986. Detecting chloroorganics in groundwater. *Environ. Sci. Technol.,* 20(5), 441-442.

Pankow, J.F., Isabelle, L.M., Hewetson, J.P. and Cherry, J.A. 1984. A syringe and cartridge method for downhole sampling for trace organics in groundwater. *Ground Water,* 22(3), 330-339.

Papadopulos, I.S and Cooper, H.H. 1967. Drawdown in a well of large diameter. *Water Resources Research,* 3(1), 241-244.

Papadopulos, I.S., Bredehoeft, J.D. and Cooper, H.H. 1973. On the analysis of slug test data. *Water Resources Research,* 9 (4), 1087-1089.

Patterson, B.M., Power, T.R. and Barber, C. 1993. Comparison of two integrated methods for collecting and analysis of volatile organic compounds in groundwater. *Ground Water Monitoring and Remediation, 13(5), 118-123.*

Reynolds, G.W., Hoff, J.T. and Gillham, R. W. 1990. Sampling bias caused by materials used to monitor halocarbons in groundwater. *Environ. Sci. Technol.,* 24 (1), 135-142.

Ronen, D., Magaritz, M., Alman, E. and Amiel, A.J. 1987. Anthropogenic anoxification (eutrophication) of the water table region of a deep phreatic aquifer. *Water Resources Research,* 23 (8), 1554-1560.

Ronen, D., Magaritz, M. and Itzhak, L. 1986. A multilayer sampler for the study of detailed hydrochemical profiles in groundwater. *Water Res.,* 20 (3), 311-315.

Scalf, M.R., McNabb, J.F., Dunlap, W.J., Cosby, R.L. and Fryberger, J. 1981. *Manual of ground water sampling procedures,* p.93. NWWA/EPA Series. National Water Well Assoc., Wothington, Ohio.

Shafer, D.C.1978. Casing storage can effect pumping test data. *The Johnson Drillers Journal.* Jan-Feb. 1978, 1-5 and 11-11.

Stumm, W. and Morgan, J.J. 1981. *Aquatic Chemistry; An introduction emphasising chemical equilibria in natural waters,* 2nd Edn. John Wiley & Sons, New York.

Suffet, I.H. 1983. Organic chemical analysis of groundwater contamination: innovations and applications in. *Proc. 1st. Atlantic Workshop, Dec 1982,* Nashville Ts. Am. Water Works Association. 95-113.

Thirerrin, J., Davis, G.B., Barber, C., Patterson, B.M., Pribac, F., Power, T.R. and Lambert, M. 1992. Natural degradation rates of BTEX compounds and naphthalene in a sulphate reducing groundwater environment. In: Lesage, S. (ed.), In-situ *Bioremediation Symposium '92* pp.108-123. Ontario.

Unwin, J.P. and Huis, D. 1983. A laboratory investigation of the purging behaviour of small diameter monitoring wells. *Proc. 3rd Nat. Symp. on Aquifer Restoration and Groundwater Monitoring, Columbus,* Ohio, May 1983. Nat. Water Well Assoc.

6   Modeling N and P loads on surface and groundwater due to land-use

C.F. Hopstaken [1] **and E.F.W. Ruijgh** [2]

[1]*Public Works Rotterdam, Environmental Department, P.O. Box 6633, 3002 AP Rotterdam, The Netherlands*
[2]*Delft Hydraulics, Water Resources and Environment Division, P.O. Box 177, 2600 MH Delft, The Netherlands*

Abstract

The central theme of the chapter is the development of the non-point source NITrogen and PHOsphorus SOiL model (NITSOL/PHOSOL). The models describe the behaviour of N and P in soils. Currently the models operate within an extended computational framework for integral water management and have proved to be a solid instrument of support in decision making. We used the models during the past 12 years for studies on a local, regional and national scale. From these experiences we subtracted our ideal set of data for calibration and verification. The practical potential of the models is illustrated using examples from two catchment areas, Horst and Hupsel.

We conclude that the models are especially useful to test the effects of legislation and to select the Best Management Practices (BMP). In this context, we would advise the use of models as a sensitivity instrument, especially with respect to critical input data. More generally, we conclude that legislation in the Netherlands should specify, apart from P quantities, the maximum admissible supply in terms of N-quantities. We finally strongly recommend the use of models in an early stage of environmental research, stretching the mind of the manager.

Introduction

In the past decade, non-point source models coupled to water quality models have become a very important tool in developing public policy. Recently, a compilation of research results and recommendations was published (US-EPA, 1992). In the same period, in the densely populated low land countries of Europe, the agricultural practices changed to more and more intensive forms of land use (Goeller *et al.*, 1983, Grashoff *et al.*, 1989).

In this context we developed mathematical models for the description of N and P behaviour in soils: the NITrogen SOiL model (NITSOL) and the PHOsphorus SoiL model (PHOSOL). The models were developed to estimate the impact of processes in the unsaturated zone of the soil on the ultimate loads of N and P on the surface and groundwater (Hopstaken *et al.*, 1986, 1988).

The model results are helpful to address the environmental problems of surface water eutrophication and high nitrate concentrations in groundwater. Therefore, both models were adopted as analytical tools in decision making (Hopstaken *et al.*, 1989, Ruijgh *et al.*, 1990, Gerrits *et al.*, 1993). Currently the models are incorporated in an extended computational framework for integral water management. This framework consists, in addition to NITSOL/PHOSOL, of the agro-hydrological model DEMGEN, the groundwater quantity model MODFLOW, the groundwater quality model STYXZ and the surfacewater quantity and -quality models HYDSIM, DELWAQ and BLOOM (Hopstaken *et al.*, 1991; USGS, 1988).

Since the analysis and description of ecosystems with the use of computational methods has become unavoidable in public decision making, we will discuss the potential of the models in the analysis of water quality problems related to agricultural practices. Although the advantages of mathematical models are obvious with regard to the preservation and combination of scientific knowledge, its application may also introduce difficulties. Because of the enormous potential of hardware, software and especially presentation techniques, the mathematical modeler sometime confuses calculation results with reality. We state that, to avoid misinterpretations and to enhance the link between computer simulation and reality, the verification and validation of models deserve intensified attention. This topic will be under discussion for the next decade (Konikow, 1992).

NITSOL/PHOSOL are proposed to serve as instruments on a regional and national scale. However, it is almost impossible to verify models at this level.

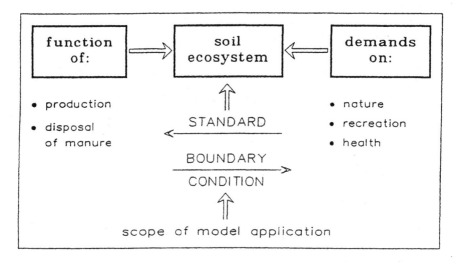

Figure 1 *Interaction between the functions of (human impacts on) and the (human) demands on the soil ecosystem.*

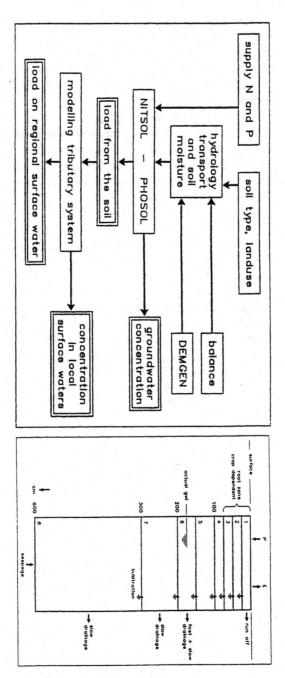

Figure 2 *Set up of the DEMGEN-NITSOL/PHOSOL computational framework including the applied vertical schematization.*

Therefore, we choose two local case studies on catchment areas of 360 and 650 hectares in the Netherlands to demonstrate the potential of the present models and the difficulties of verification on each of the data sets.

Both studies were performed with the same setting of coefficients for NITSOL and PHOSOL. Before discussing the case studies,we will briefly discuss the set up of the models and the applicability of field data for the verification of models .

In the use of models in the decision making process one has to consider the following interactions. Describing the behaviour of an ecosystem we distinguish the *functions of* the ecosystem on one side and the *demands on* the ecosystem on the other (Figure 1). NITSOL/PHOSOL aim at the deterministic description of the agricultural function of the soil system in interaction with human demands for acceptable loads of N and P to the environment.

Figure 1 schematically presents this strained interaction. The demand for a productive function of the soil (and especially in the Netherlands the still increasing demand for disposal sites for manure) cause human impacts on the ecosystem. Therefore, standards have been set for the protection of the ecosystem and its functions. Human interests relevant to these standards are nature preservation, recreational facilities and high-quality surface and drinking water. The functions of the ecosystem form the boundary conditions for the realization of human demands.

In the description of the case studies we will highlight some of the most effective measures, or Best Management Practices (BMP). Most BMP's are directed to the boundary conditions, *e.g.* reduction of fertilizer or manure application. Moreover, we will discuss the potential of models in decision making and the dilemma of the modeler choosing between model development or application.

Methodology

Scale of application and set up of the models
Primarily NITSOL and PHOSOL were applied on a regional scale (NW Veluwe, 61,000 ha; De Bruyn *et al.*, 1986). In this stage of development, only the unsaturated zone of the soil was described because of the dominant role of this part of the soil in nutrient balances. Under humid climatic conditions the major changes in N and P fluxes from agricultural activities towards ground- and surface water take place in this part of the soil. The first regional application of the models indicated the importance of chemical and microbial processes as well as the importance of crop uptake (Hopstaken *et al.*, 1986, 1988). We concluded that rather low percentages of N and P supply are transported into the environment; approximately 2-3% of N and P to the surface water and 13% of N (<1% for P) to the groundwater. Although the percentages are low, the impact of the resulting fluxes could be enormous. Therefore, managers of water systems are especially interested in the small part of the total supply that surpasses the unsaturated zone. Recent development of the models and the integration of models into several computational frameworks as described in the introduction section are consistent with this management objective.

For the verification of the models in the small catchment areas of Horst and Hupsel, the linkage of NITSOL/PHOSOL to the agro-hydrological model DEMGEN

Table 1 *Process formulation of relevant processes in NITSOL/PHOSOL on various substances (C) and used coefficient values (K) for the Horst and Hupsel area.*

General formulation: $dC = K * 1.1^{(T20°C)} * [C]$

Process	Substance C	Coefficient K	Unit
Mineralization	N-refractair	k = 0.002	day^{-1}
	N-organic	k = 0.025	day^{-1}
Nitrification	NH_4-N	k = 0.030	day^{-1}
Denitrification	NO_3-N	k = 0.011	day^{-1}
Adsorption	PO_4-P	k = 0.100	day^{-1}
Desorption	P-adsorbed	k = 0.,0001	day^{-1}
		max.ads = 8350	$kg\ P_2O_5\ ha^{-1}$
Dissolution	CaMg-P	k = 0.050	day^{1}
Precipitation	PO_4-P	k_sat. = 90	$mg\ PO_4$-PL^{-1}

(DEMand GENerator; Abrahamse *et al.*, 1982, Grashoff *et al.*, 1990) is of great importance. For the prediction of effects in the surface water of the small tributaries a computational module of NITSOL/PHOSOL, the DIstrict WAter Model (DIWAMO; Ruijgh and Hopstaken, 1990) is used. The set up of the DEMGEN-NITSOL/PHOSOL computational framework, referred to as DEMNIP, is presented in Figure 2. DEMGEN, as well as NITSOL/PHOSOL and DIWAMO are numerical models (FORTRAN77) developed by DELFT HYDRAULICS for the Ministry of Transport and Public Works in the Netherlands.

DEMGEN
DEMGEN is the central hydrologic model in the Policy Analysis studies for the Water Management in the Netherlands (PAWN) and was originally developed for quantitative purposes (Abrahamse *et al.*, 1982). A recent overview of the DEMGEN model is given by Van Vuuren (1990). DEMGEN has been fully adapted for the purpose of water quality modeling (Grashoff *et al.*, 1990). It describes the vertical and horizontal fluxes and the soil moisture content in a soil column. In general, a 10 day period (decade) is used as the basic time interval.

The calculation of the surface runoff is divided into a summer and a winter algorithm (De Bruyn *et al.*, 1986). In summer, rain intensity and infiltration capacity determine the surface runoff; in winter, groundwater table and soil moisture. In both calculations, a variable on surface storage is taken into account. For the case studies the surface storage was calibrated on 10 mm.

Subsurface runoff or fast drainage is calculated using the Kraayenhofv/d Leur relation (Grashoff *et al.*, 1990). It represents the amount of drainage water entering the tributary system with a residence time in the saturated zone of less than 10 days.

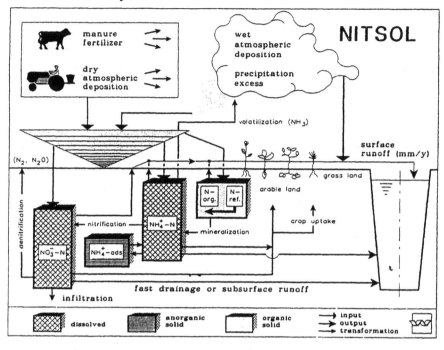

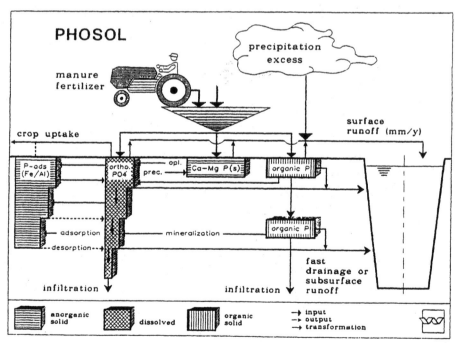

Figure 3 *Simulated processes by NITSOL and PHOSOL.*

Fast drainage occurs near the drainage system in case of sudden rises in groundwater table (Figure 2, right).

The residual or slow drainage forms the largest contribution to the discharge of the tributary and/or the district. It is assumed to be distributed over the saturated zone of the soil proportionally to the thickness of the soil layer.

The number of segments in soil column is determined by the accuracy needed. For the rather thin unsaturated zones in the Horst and Hupsel areas we used 8 segments (Figure 2). In arid areas, 15 or even more segments are used (Facchino *et al.*, 1993). An indication of the interaction of the soil with the first aquifer is achieved by the calculation results in the lowest segment. The vertical size of this segment has been calibrated using chloride measurements (Abrahamse *et al.*, 1982). Between the unsaturated zone and the groundwater segment, always one or more segments are saturated and have a buffering function. At present DEMGEN is directly linked to the groundwater model MODFLOW (USGS, 1988).

NITSOL

The Nitrogen Soil Model NITSOL describes the N behaviour in the soil and distinguishes among the adsorbed and dissolved ammonium, dissolved nitrate and dissolved and solid organic nitrogen concentrations in the soil. Linked to DEMGEN, NITSOL computes the transport of N components to and from the soil.

The processes simulated by NITSOL are shown schematically in Figure 3. The microbial processes (mineralization or ammonification, nitrification and denitrification) dominate N behaviour in soils. These processes are simulated using first order kinetics (Table 1). Further important processes are volatilization and equilibrium adsorption of ammonium, using a Langmuir formulation with pH and Ca^{2+} corrections. The cation exchange capacity (CEC) for ammonium of the soil is calculated on basis of specific CEC of organic matter and clay and the organic matter content and clay content of the soil.

The coefficient values for the first order processes are derived from literature (Bhat *et al.*, 1980; Beasley and Huggins, 1980; Donigian *et al.*, 1983; Reddy *et al.*, 1979; Rijtema, 1980; Van Veen, 1977) or calibrated on both regional and local data sets. The value of the denitrification coefficient shows a great resemblance to values obtained for marine ecosystems with the ECOLUMN model (0.011; De Vries and Hopstaken, 1984). Besides the dependency on temperature (Table 1), the given first order coefficients of the microbial processes are dependent on pH, soil moisture and organic matter concentration (Ruijgh, 1992).

PHOSOL

The Phosphorus Soil Model PHOSOL describes orthophosphate, adsorbed phosphate, mobile and solid organic P and solid calciummagnesium phosphate in the soil on the same basis as NITSOL (Figure 3). However, the adsorption process of orthophosphate on Fe and Al components in the soil is dominant compared to the microbial processes. Adsorption and desorption are simulated by means of kinetic formulations with an adsorption maximum, partly similar to Langmuir adsorption.

Desorption is counterbalancing the adsorption process when the adsorption maximum is reached. In practice net adsorption decreases with increasing quantities of adsorbed P. This phenomenon is empirically described by Van Riemsdijk *et al.* (1984). Table 1 gives the coefficient values derived from literature and calibrated on field data.

DIWAMO

The modeling of tributary systems or water districts is of great importance for the calculation of the resulting load of N and P on regional surface waters (Figure 2; Ruijgh and Hopstaken, 1990). During the presence of N and P components in a district a large change in load may occur due to chemical and physical processes. The magnitude of this change is in most cases dominated by the residence time in the district.

Apart from the calculation of change in load, the simulation of concentrations in a tributary or water district provides an important possibility for model calibration and verification. Basically, for both presented case studies calculated concentrations in the main tributary were extremely useful for model verification. On the other hand, the residence time in the main tributary of the studied catchment areas lies in the order of 4 hours to 2 days, so the impact of microbial processes is negligible. Moreover, the chemical changes, sedimentation and erosion only cause a shift in time of the P (not the N) load to the regional surface water. On an annual basis, the retention of the tributaries turned out to be negligible. The formulation of processes is discussed in Ruijgh and Hopstaken (1990).

The applicability of field data for model verification

Easily made mistakes

A brief introduction might help to clearify this subject. The development of a mathematical model in general ends with the calibration on a specific set of field data. Usually, up to then most attention has been paid to the selection of the type of formulations, the programming and the presentation of results. One is satisfied to have made a mathematical - sometimes even object oriented - abstraction of observed phenomena and processes. Most often coefficient values are derived from the literature, experiments and all kinds of field data. Suddenly the model is ready for a first application, a test or calibration. The model produces output, but is it calibrated? As a modeler one is easily inclined to state that the model is calibrated. But of course, this depends on the extension and quality of the data that are used. A sensitivity analysis on the most critical assumptions and hypotheses might help to quantify the result. In most cases it turns out that not one set of field data is ideal for verifying your hypotheses.

What is the implication of this conclusion? At this stage of development we advise to apply the model with reservation and not for the purpose of decision-making. Further study and testing is needed to verify or falsify your set of mathematical equations. This process will bring the model at a higher level of verification, if it takes place in the context of *the purpose of the model.*

Let's return to NITSOL/PHOSOL. Since the ultimate stimulus for the development of NITSOL/PHOSOL were problems related to the quality of the surface water, we searched for load measurements of catchment areas (*e.g.* a river, canal or lake) for calibration and verification. A satisfactory reproduction of the load at the end of a tributary or at a pumping station gives you confidence in the validity of your hypotheses. Especially when the data set covers a long period with a variety of meteorological and hydrological conditions.

Three questions remain. Which data are relevant? Can the model be applied for all purposes? And what is the ideal set of data?

The relevant data

Besides the data on the loads to the regional surface water, other data can be useful. In principle, all other data that can be used to support or disprove an assumptions incorporated in the model should be used.

(a) The *concentration measurements* in the soil or aquifer are most relevant. However, these data are difficult to interpret. The vertical position of the sample is very important. Sometimes this position is known but in many cases the concentration measurement is based upon a withdrawal from different depths. In this case, the concentration more or less represents an average over a certain depth. But even if the vertical position is known the lateral or geographical heterogeneity can negate the use for calibration of verification. The heterogeneity in the unsaturated zone is enormous, for example due to the patchwise disposal of manure by animals. One has to collect a lot of samples for the computation of a reliable statistical average.

Is the sampling and analysis of concentrations in the soil or groundwater useless? No, certainly not. Every concentration measurement that lies within a limited range of the calculation results gives more confidence to the performance of the model and the underlying assumptions. In general, concentration data in the soil are not reliable enough for a complete verification, unless you have an extremely large budget for the collection and analysis of a statistically sufficient amount of samples.

These arguments are also applicable to unsaturated zone hydrology. Soil moisture content data are difficult to interpret. However, groundwater levels might give reliable information, but always show some local variability.

(b) Data concerning crop uptake, volatilization or other *balance components* might be used to test modelresults from another perspective. Figure 3 represents the most important balance components for the models, some of which can be studied by flux analysis. For example crop uptake, is responsible for a relatively large flux of water and nutrients. It can be studied by measuring the net yearly recovery. Crop uptake affects both the hydrology as well as the nutrient balances of the soil. However, is it really necessary to calibrate your model on this flux or does it serve as a boundary condition? In alignment with some of the hydrological balance components (precipitation, evaporation, seepage) we chose the last option. The net crop uptake is described in NITSOL and PHOSOL by a fixed yearly flux and a fixed distribution of this flux in time. The extent of the flux is based on agricultural data and we assume that the yearly production is achieved unless water or nutrient limitation occurs. As the process is only to a minor extent interacting with the proposed outputs of the model, *i.e.* the nutrient fluxes to the environment, this concept restricts the number of coefficients and avoids unnecessary noise. Still, you have to verify the reliability of your input data.

Also for processes like volatilization, which are only studied at a laboratory scale, we use equations reflecting the status of a boundary condition or input data. One coefficient determines which amount of mineral nitrogen volatilizes. The value of this coefficient depends on the technique of manure application.

(c) *Experimental information* is most essential to the modeler. A lot of balance components can only be studied on a laboratory scale. One has to prove that the experimental conditions are equal or at least comparable to the field conditions one is studying. In most cases this is extremely difficult. For example, experiments with lysimeters (*e.g.* Bergström and Johansson, 1991) or conditioned parcels may not reflect field conditions. However, it is often necessary to use all available information. Keep in mind, never verify your model only on this kind of information.

Applicability of a verified model for various purposes

Once your model is calibrated on a specific area with the help of all available and relevant data for one or more years, you are eager to verify the model. A valuable verification is performed using data from a different area. However, you should be very careful. First determine the sensitivity of the system before testing the model for different conditions and then always check if these conditions are really different in this area.

This might be illustrated by the erroneous verification of a simple statistical model for the day-length. The model was based on measurement data from Amsterdam. Day-length was calculated as a function of day-number, but the model did not include the dependency of day-lenght to latitude.

The model was verified using measurement data from Vancouver. Unlucky enough the information matched, because both cities are located on exactly the same latitude. So, the model in question was only valid for the latitude of Vancouver and Amsterdam. You can imagine what happened when it was applied in Indonesia.

Every application will intensify the verification of your model concept, especially when the right data are available. If your model has been applied several times, but has only been checked on groundwater level or the organic carbon content, it is not verified at all. In general, the applicability for specific purposes depends on: (a) the number of applications, (b) the variability of the most significant conditions between the various applications, and (c) the extent of useful and reliable data used for the verification. Always keep in mind that the degree of fulfillment of these criteria determines the reliability of your output. Never forget to remark and if possible to illustrate this along your conclusions and recommendations. Never stop looking for opportunities to intensify the verification of your model on new data. In other words, the curriculum vitae of your model determines the level of verification.

The ideal set of data

Does it exist? Probably not. One way to obtain knowledge is to define and specify what questions remain and how they might be answered. Computational frameworks are effective instruments to demonstrate the gaps in knowledge and the lack of calibration or verification data.

The ideal set of data for a specific application is related to: a) the level of verification for a comparable area and b) the purpose of the application. In case NITSOL/PHOSOL are applied on an area with a lot of new phenomena and with the purpose to support the decision making process with respect to the pollution of the groundwater (first aquifer) and the regional surface water, the ideal set for calibration consists of the following data.

(1) To start your model you need reliable data for the boundary conditions and inputs. For example precipitation data per day with an indication of rainfall intensity (for the purpose of runoff calculations); manure and fertilizer supply data with an adequate description of the distribution over crops, soils and time.

(2) A reliable network of groundwater monitoring stations with separately useable filters on different depths will provide you with data on the horizontal and vertical variation within the study area. An adequate description of the local conditions (soiltype, groundwater level, pumping rate effects, agricultural practice) should be available and the monitoring interval should be tuned in relation to the residence time and the seasonal fluctuations.

(3) To verify the calculated N- and P-loads to surface water you will need flow proportional surface water concentration measurements in the main tributary of the area; collecting a sample of every millimeter precipitation excess will be appropriate.

(4) A balanced set of various measurements concerning fluxes and concentrations in the unsaturated zone is quite useful as well. For example one analysis of P-saturation for every combination of soil-type and soil-use (plot) with some measurements in duplicate; nitrate concentrations at different places and for different seasons in the unsaturated zone. The balance should be found in the required reliability and the available budget, but these data are only additional to the data mentioned above.

The mentioned data should be available for a certain time-interval. In general, one

Table 2 *Supply of N and P to the Horst catchment area.*

	Nitrogen kg N ha^{-1} yr^{-1}		Phosphorus kg P$_2$O$_5$ ha^{-1}yr^{-1}	
	grassland	arable land	grassland	arable land
Fertilizer	194	74	11	15
Atm.deposition	48	48	1	1
Manure	360	1,052	164	888
Total	602	1,174	176	904

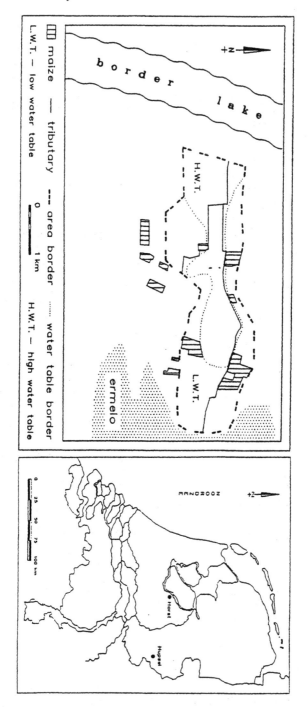

Figure 4 *The Horst catchment area, divided into agro-hydrological subsystems; including geographical location of the case study area in the Netherlands.*

year is not sufficient as hydrological conditions may vary to great extent from one year to another. A time interval of three to five years is required for a serious test.

Another thing to keep in mind: Always ask for the reliability of the data or a statistical analysis on the data. In many cases one is satisfied when the required data exist. But, a quality control pays off.

The Horst catchment area

Problem description and schematization
Before we used the Horst catchment area as a model calibration area, we studied the NW Veluwe district (61,000 ha; De Bruyn *et al.*, 1986). The Horst catchment area is situated within the NW Veluwe district (Hopstaken *et al.*, 1986, 1988, 1989). Although we calibrated NITSOL/PHOSOL on overall load estimates for 1976 to 1983 for the district, we did not consider this calibration to be a profound basis for predictions. The large district contains a considerable variety in hydrology, soil types and land-use. Therefore, the Horst tributary was chosen for a more detailed application and verification of the model concept.

The Horst catchment area (340 ha) is almost entirely used for agricultural purposes. The lower part of the catchment area near the border lakes of Lake IJssel consists of grassland, whereas the higher part near the slopes of the glacial sand massive of the Veluwe is partly used for maize cultivation (Figure 4). The total area consists of sandy soils with an organic top layer. On basis of hydrology, land-use and soil type we delineated three different plots:
(1) Grassland with high groundwater table (150 ha).
(2) Grassland with low groundwater table (175 ha).
(3) Arable land with low groundwater table and maize cultivation (15 ha).

The objectives of the Horst catchment area study were to quantify both N and P loads on the border lakes as well as the N load on the groundwater. Additionally, we tried to define the critical boundary conditions determining the extent of the loads.

Model set up
Primarily the period 1976-1979 was used for calibration because of the availability of monitoring data for the tributary system. Sufficient data were available to simulate hydrology with DEMGEN. The seepage in the lower part was calibrated on tributary flow and amounts to 0.3 mm yr^{-1} No upward or downward seepage is assumed for the higher part of the catchment. Figure 5 shows the results and also reflects the differences in meteorology between the extremely dry 1976 and the moderately wet period 19771979. So far the chosen period matched with the criteria for calibration.

Input data on N and P, concerning manure supply were available from the Central Bureau for Statistics (CBS, 1983). Additional information on fertilizer application (Wijnands *et al.*, 1983), sewage sludge application and atmospheric deposition (Van Aalst and Diederen, 1983) have been used to determine the total supply.

It appeared, there exists a surplus of manure in the area. Only a limited amount of the surplus can be used on grassland due to the maximum admissible potassium

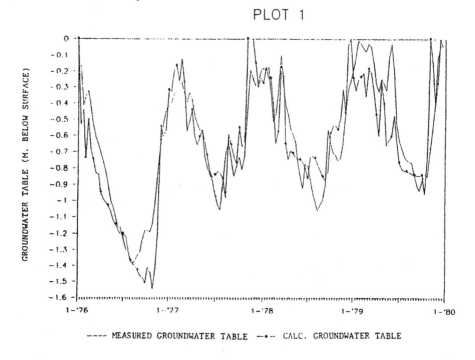

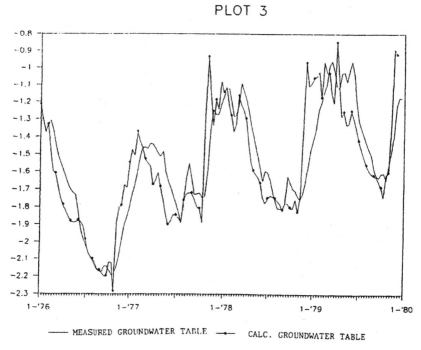

Figure 5 *Hydrological calibration o DEMGEN on groundwater table*

application on grassland (Van Dijk, 1984). Most of the manure surplus is therefore dumped on arable land (Table 2).

Results for 1976–1979

Phosphorus

According to current P supply within the Horst catchment area, groundwater pollution and subsequently surface water pollution may occur at arable land. Large loads on the groundwater will occur when the soil becomes saturated with P down to the shallowest occurring groundwater table. For moderately wet years, this means 120 cm below the surface (Figure 5). On basis of our calculations the soil under grassland will only be saturated with P in the very long term. For the arable land we concluded that saturation of the total unsaturated zone of the soil will occur at approximately year 2020, if present P supply is continued (Hopstaken et al., 1988).

We tried to calibrate the calculated P-concentration on monitoring data of the tributary system (surface water) from 1976-1979, but this turned out to be impossible. Our model predicted 5 to 10 fold lower tributary loads and concentrations than observed.

However, we never got the information of a small feather treatment plant that discharged its treatment water with a lot of phosphorus directly into the tributary. When we visited the area, in 1987, the plant was gone.

We could not calibrate our calculations, nor did we have information to falsify the assumptions we made. We asked to collect more recent data and received 1987 monitoring data. Additionally we performed a small field campaign of one day. It was directed to the analysis of P-saturation and N concentrations in the unsaturated zone at 6 sampling locations distributed over the three plots .

Nitrogen

The calculation for N gave better results. Figure 6 shows the calculated total N load of the tributary compared to the available data. Although the data show a certain scatter, probably introduced by point loads, we put some confidence in the predicted level, especially on an annual basis. On basis of the annual concentration variations, we concluded that the N behaviour was almost in steady-state according to the present N supply. In general the observed annual variations in soil concentrations and tributary load were induced by hydrology. Dry conditions as in 1976 caused higher N-concentrations in the top soil but not in the groundwater and certainly not induced an increase in tributary N-load.

We started to gain some confidence in the simulation of the systems behaviour for N and conducted several sensitivity calculations. We learned that even this small area showed large variations. Since the lower area is influenced by seepage of deep groundwater with a negligible nitrate concentration (for the moment), the resulting concentration in the groundwater (2-3 m below surface) is almost zero. For the elevated parts of the area, where no seepage occurs a higher nitrate concentration is calculated. Figure 6 (right) shows the simulated nitrate concentration at 3 to 6 m below surface. Typical are the annual fluctuations. These kind of fluctuations are often found for nitrate (Tuckwell and Knight, 1988) and are smoothed with increasing depth. Figure 6 also shows what we predict if the supply is reduced to 50%, starting 1976. The effect is drastic: 50% reduction of groundwater concentration and tributary load (not presented) within 4 years.

Again a sensitivity analysis gave us another lead. The simulation results for arable land, assuming a 0.3 mm day^{-1} downward seepage, showed an increase in groundwater concentration to approximately 7 mg NO_3-N L^{-1} (Hopstaken *et al.*,

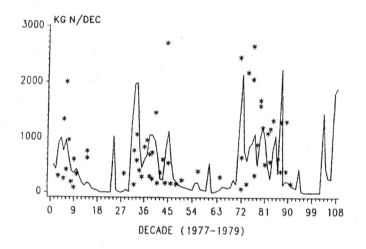

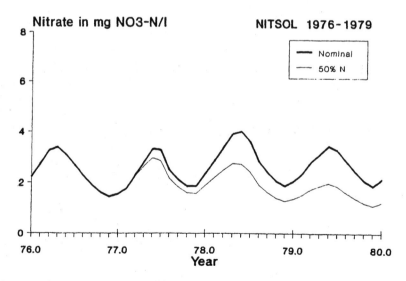

Figure 6 *Calculated and measured N-load of the tributary (top) and calculated NO_3-N concentration in the soil at 3 to 6 m below surface for the higher parts of the Horst area (bottom) at a nominal situation and with 50% N supply reduction.*

1988). On the contrary, the tributary load slightly decreased. We concluded that dry soils, due to short residence time and fast vertical transport, induce high nitrate concentrations in the soil and high loads on the groundwater. A verification is found in almost all data on dry soils and especially data on deserts (Dillon, 1989). In these soils denitrification is almost zero, because of the moisture dependent (de)nitrification. The calculated effects for the Horst area, still for relatively humid conditions, purely evolved from residence time and transport. In our latest (mediterranean) model versions moisture dependent (de)nitrification is incorporated (Facchino, 1993).

The most important conclusion was that N behaviour and resulting nitrate concentrations react rapidly upon changes in N supply. This indicates that both the N load on the tributary and the nitrate load on the groundwater can be decreased in due time, opening possibilities for a nitrogen directed eutrophication control. After 20 years of management focused on phosphorus load reduction, in 1989 for the first time two out of five eutrophication strategies in the policy analysis document of the ministry focus on reducing nitrogen availability.

Results for 1987

Phosporus

As the P monitoring data for 1976-1979 were strongly influenced by the discharge of a small duck-feather treatment plant, we could not calibrate our model on these data. Moreover, we learned from previous excercises with PHOSOL that P loads to surface water are governed by the actual hydrological conditions and P content of the unsaturated zone. Hydrological conditions (groundwater table, moisture content, drainage, surface runoff) show considerable variation in time (Figure 5) and were calculated using DEMGEN. However, total P content of agricultural soils only slightly change within a year. When we had the opportunity to verify PHOSOL for 1987 we decided to sample the actual P content of the soil and use these values as a initial boundary condition in the calculations.

In small catchment areas such as the Horst tributary (340 ha) residence times are extremely short. After a heavy shower, discharge reacts almost immediately. Consequently, P concentrations in surface water will show considerable variation in time. On the contrary, the total load to regional surface water only shows a clear seasonal variation. As the aim of PHOSOL is to describe the P load to regional surface water, we decided to calibrate the model on the loads rather than concentrations.

Figure 7 shows the results for 1987. The calculated and measured cumulative loads both indicate that total yearly loads are governed by the loads in winter. Although the model slightly underestimates P loads in Januar and overestimates in November and December we were quite satisfied by these results. Especially at the end of 1987 some loads might be missed in the sampling routine. This underlines the need of discharge proportional measurements.

Nitrogen

The results for N were even better and support our confidence in the model after the calculations for 1976-1979. Similar to P, total yearly N loads were dominated by discharge in winter. Even calculated N loads for November and December are confirmed by the measurement data.

Conclusions

We have stated that every test of the model is a step forwards to a higher level of verification, if the data are appropriate for testing the purpose of the model. This has been illustrated by the application of the modelsa for the Horst area. Annual loads turned out to serve as the proper test data, concerning system dynamics in the Horst area. Moreover, we proved that the linkage with hydrology is extremely important for the description of phenomena.

We also proved that an omission in data might give confusing test results. We advise to never trust your model nor your data, but to use sensitivity analysis as a cheap instrument to improve your data. For example, the use of NITSOL/PHOSOL guided the additional field investigation and resulted in a cut on laboratory costs. However, always conduct a sensitivity analysis or calculation with understanding and never because computation time is endless.

Although the application for the Horst area served as a verification, the simulation results and especially the sensitivity analysis gave indications for Best Management Practices (BMP's). Some BMP's were tested and elaborated in later studies (Ruijgh *et al.*, 1990; Gerritse *et al.*, 1993).

The Horst area is part of the NW Veluwe region. The Veluwe is a sand massive with a highest altitude of 100 m above sea level. The Horst area forms the lowest part, with altitudes of 10 m above to 0,1 m below sea level (near the border lakes). We found that hydrology in this area is of great importance with respect to the water quality problems. The higher Veluwe, with relatively low groundwater levels, is vulnerable with respect to high nitrate loads to the groundwater. Vasak *et al.* (1981) already described the effects, but not the causes. On the other hand the lowlands near the border lakes are vulnerable with respect to phosphorus saturation. As a BMP we advised to reduce P supply near the border lakes to the level of crop uptake. This BMP became known as equilibrium supply. For the higher Veluwe a N supply equal or close to equilibrium was advised. The least vulnerable areas to N and P supply are the higher parts of the Horst area (plots 2 and 3). Manure excesses should preferably be applied in these parts.

The Hupsel catchment area

Problem description and schematization

From the study in the Horst area we learned to appreciate a good hydrological model. Just after this study we were asked to run our models on the Hupsel research area. We liked the idea because the Hupsel area had always been the verification area for the hydrological models of the ministry and the universities and thus also for DEMGEN.

The Hupsel area (665 ha) is situated in the eastern part of the Netherlands, near Eibergen (Figure 4). The presence of a shallow impermeable layer of boulder clay (till) excludes interactions to the deeper aquifers and surrounding areas and makes the area very suitable to verify models. The Hupsel, a rather small stream of about 4 km length with some tributaries, is influenced by local agricultural activities. Data on precipitation, evaporation, discharge and water quality are collected by the Ministry of Transport and Public Works for over 20 years. The application was driven from a research point of view. We never had the idea or intention to find shocking theories about agricultural management.

Similar to the Horst case DEMGEN, NITSOL and PHOSOL calculations are

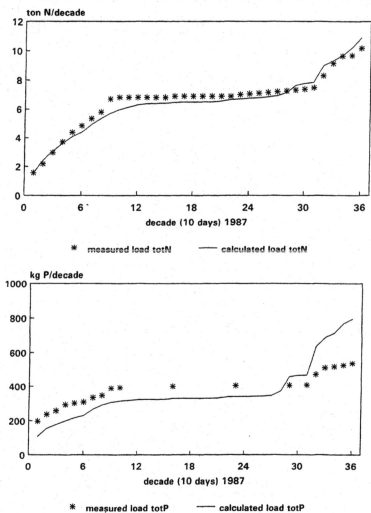

Figure 7 *Calculated and measured cumulative load of total N (top) and total P (bottom) at the mouth of the Horst tributary for 1987.*

Figure 8 *Schematization of the Hupsel catchment area.*

Plot	land-use 1941-1982	1983-1987	till at cm below surface	ha	Plot	land-use 1941-1982	1983-1987	till at cm below surface	ha
1	corn	corn	160	22	7	nature	nature	40-120	18
2	corn	corn	40-120	13	8	grass	nature	40-120	1
3	nature	corn	160	1	9	corn	grass	160	20
4	grass	corn	160	35	10	corn	grass	40-120	22
5	grass	corn	40-120	38	11	nature	grass	40-120	1
6	nature	nature	160	11	12	grass	grass	160	226
					13	grass	grass	40-120	256

performed for plots with a specific soil characteristic and land use type. The parcels belonging to a plot may be situated anywhere within the study area; they do not have to be adjacent.

We used the ArcInfo Geographic Information System (GIS) to schematize the research area. The two most widespread soil types in the analysis are both sandy soils. In one type, boulder clay (till) starts between 0.4 and 1.20 m. below the surface, in the other one at 1.6 m.

Land use inventories of the Ministry of Transport and Public Works in 1979 and 1986 indicate crop rotations in the study area. Crop rotation is taken into account in the plot definition. Until 1982 we used the 1979 data on land use; starting from 1983 we used the 1986 data.

By means of overlay techniques thirteen different plots are distinguished, as shown in Figure 8. These plots are the basic geographic units used for hydrological calculations with DEMGEN and for chemical calculations with NITSOL and PHOSOL.

Model set up

An enormous amount of hydrological data was available, but data on chemical soil conditions were insufficient to start the calculations in 1987. Initial chemical soil conditions were computed with the models by repeating meteorological data of 1976-1982 six times, simulating a period of 42 years (1941-1982).

The use of manure and fertilizer in the study area was estimated by Witte (1987) for 1982 and by Van Engelenburg *et al.*, (1988) for 1987 (Table 3). For the initializing period (1941-1982) we assumed that the use of manure and fertilizer increased linearly to the 1982 level. The 1987 level was used in the calculations from 1983 to 1987.

According to Aalst and Van Diederen (1983) the average atmospheric deposition amounts to 46 kg N ha^{-1}yr^{-1}. In nature areas (woods) leaves fall contributes to 50 kg ha^{-1} N and 10 kg ha^{-1} P$_2$O$_5$ per annum (Ruijgh *et al.*, 1991). The contribution to surface water pollution from scattered buildings in the research area is estimated at 3 kg N day^{-1} and 1 kg P$_2$O$_5$ day^{-1}. Dramatic point loads as in the Horst area before 1987 are absent.

Results

The application of NITSOL/PHOSOL on the Hupsel area can be considered as a serious attept to verify the models. After the request to run the models we only received input data. Measurement data on the concentration N and P in the tributary became available after the calculations had been finished. Therefore, we started from the model set up and coefficient values of the Horst application and never changed them. When we first ran the model we calculated extremely high nitrate concentrations in the tributary and expected the model had ran out of control. However, the 1987 data on surface water discharge and concentrations, collected by the Ministry of Transportation and Public Works and the local surface water controlling board (Zuiveringschap Oostelijk Gelderland, ZOG) proved the opposite (Figure 9).

Calculated nitrogen concentrations in the tributary appear to be very high and dominated by nitrate (between 30-34 mg N L^{-1}, compare to Figure 9). The very high nitrate concentration in the surface water is presumably caused by the contribution from the subsoil of grassland, where nitrate concentrations amount to 50 mg NO$_3$-N L^{-1}(Figure 10).

Calculated and measured nitrogen concentration levels show a good resemblance. However, measurements indicate more variations than calculations with a timestep of 10 days can reproduce.

Unpublished data of the local surface water control board (ZOG) on arable land drainwater composition, indicate nitrate concentrations of 50-80 mg NO$_3$-N L^{-1}. These data confirm calculated nitrate concentrations at 75-100 cm below surface under arable land.

What amazed us when we considered the calculated concentrations at the depth from which the bulk of the drainage originates (2-3 m below surface) is that the highest nitrate concentration (45-50 mg NO$_3$-N L^{-1}) occured underneath grassland. Lower nitrate concentrations (20-30 mg NO$_3$-N L^{-1}) are calculated in the subsoil underneath fields recently converted from grassland to maize and the lowest concentrations (10-15 mg NO$_3$-N L^{-1}) in the subsoil under arable land (Figure 10).

Table 3 *Use of manure and fertilizer on grassland and arable land in the Hupsel research basin for 1982 (Witte, 1987) and for 1986 (Van Engelenburg et al., 1988)*

| | 1982 | | | | 1986 | | | |
| | Manure | | Fertilizer | | Manure | | Fertilizer | |
	N	P$_2$O$_5$	N	P$_2$O$_5$	N	P$_2$O$_5$	N	P$_2$O$_5$
Grassland	260	150	300	10	303	121	400	170
Arable land	1,005	735	100	10	525	340	200	100

On arable land the application of manure provides the input of organic matter to the soil. On grassland less manure and at lot of fertilizer is applied. The amount of denitrification is not only residence time and moisture-dependent as we saw in chapter 4, but also dependent on the availability and thus the supply of readily decomposable organic matter (Kroeze *et al.*, 1989). In the subsoil of grassland the low organic matter content limits denitrification and consequently higher nitrate concentrations appear. This hypothesis is discussed later.

Calculated ortho-phosphorus concentrations in the topsoil of arable land are extremely high (20-80 mg L^{-1} PO$_4$-P). At a depth of 75-100 cm concentrations are

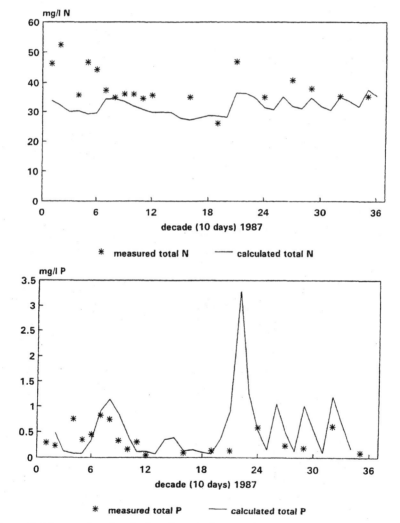

Figure 9 *Measured and calculated total nitrogen (top) and total phosphorus (bottom) concentrations in the Hupsel tributary at the departure of the area for 1987.*

lower (on average 2-5 mg L^{-1} PO_4-P) but still above the measured arable land drainwater concentrations (0.02 mg L^{-1} PO_4-P, unpublished data ZOG).

For grassland, used as arable land (plot 9-10) until 1983, high ortho-phosphorus concentrations are calculated because of the desorption of formerly adsorbed phosphorus. For arable land, before 1983 used as grassland (plot 4-5), calculated concentrations at 50-75 cm depth are still comparable with those of grassland (Figure 10). Probably, drainwater has been collected at a parcel only recently in use as arable land and should be compared to the concentrations calculated for grassland.

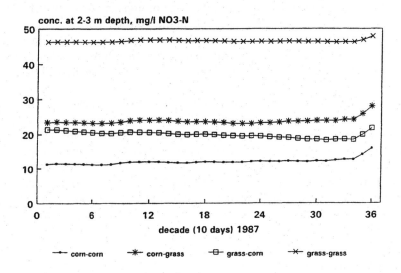

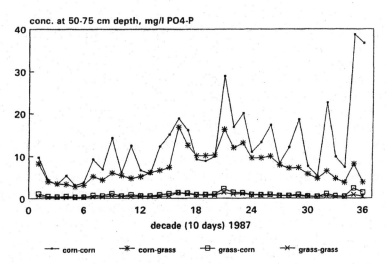

Figure 10 *Calculated nitrate (top) and ortho phosphorus (bottom) concentrations in the subsoil (2-3 m for N and 50-75 cm for P) of grassland and arable land.*

Conclusions

The Hupsel area is extremely suitable for a verification because of the extent of available data. We certainly reached a higher level of verification and increased the confidence in the models. On the other hand we concluded that measured discharge, and measured N and P concentrations show more temporal variability than can be reproduced by the model (Figure 9). However, we consider the calculated decade averaged discharges and concentrations to give an indication for the in situ situation. The models were also used to indicated BMP's for the area. We conclude that the application of fertilizers instead of manure might lead to higher nitrate concentrations in both surface and groundwater, caused by shortage of organic matter available for denitrification. In the subsoil of grassland, higher nitrate concentrations are calculated than underneath arable land. Under these conditions, replacement of fertilizer on grassland by manure to decrease nitrate concentrations in surface water, is the best management practice. Measurements of groundwater quality will have to confirm this.

Land use history is of great importance for the phosphorus concentrations in soil. Desorption of phosphorus accumulated in the past leads to increased ortho-phosphorus concentrations, whereas recent intensive manuring only influences topsoil conditions. Even after the change of all arable land to grassland, phosphorus concentrations and loads to surface water will remain high because of desorption.

All calculations should be preceded by an inventory of chemical soil conditions or land use history in order to estimate the initial phosphorus concentrations. At the moment we are working on this in the Hupsel area, in order to verify the model-set in more detail.

Discussion and conclusions

Applicability of the developed models

Twelve years after we started the development, NITSOL and PHOSOL have proven to serve as solid instruments for the deterministic description of nutrient behaviour in the unsaturated zone, the groundwater and the tributary system. Linked to hydrological and groundwater models like DEMGEN and MODFLOW simulations on local, regional or even national scale are carried out.

The model setup developed in the same manner as other agricultural nonpoint source pollution models like CREAMS (Knisel, 1980), ANSWERS (Beasley and Huggins, 1980) and HSPF (Donigian *et al.*, 1983). Also these models are still applied for several purposes (Wright *et al.*, 1992; Moore *et al.*, 1992; Cooper *et al.*, 1992; Shiromohammadi *et al.*, 1992). Different to these models, NITSOL and PHOSOL are primarily directed to the excessive manuring problems in humid and mediterranean climates. Especially polderlands and areas with gentle slopes were studied. Sediment run off was always of minor importance, in contrast to other NPS models (Binger *et al.*, 1993). NITSOL and PHOSOL were tested several times. NITSOL already in 1985, when it performed equal to the best of all tested models (Willigen and Neeteson, 1985). The models are used on all kinds of soils, among which clay and peat soils (Ruijgh *et al.*, 1993), and for all kind of nutrient supply levels.

Use of NITSOL/PHOSOL in decision making

Although model improvements are necessary, the extent of the excessive manuring in the Netherlands have forced the water management to take decisions. In our opinion, model results can support these decisions. At present, an overall reduction of manure application is imposed by legislation. The dutch soil protection law restricts phosphorus supply to 175, 250 and 350 kg P_2O_5 ha^{-1}yr^{-1} for arable land, grassland and maize cultivation, respectivey. A further decrease towards equilibrium supply (supply equal tot crop uptake) is forseen for 2,000. For arable land this results in a large reduction of manure application. However, due to the limited area of arable land the surplus is likely to be applied on grassland, resulting in an increased supply of at least 25 kg P_2O_5 ha^{-1}yr^{-1}. This implies an increase of manure application in the lower parts of *e.g.* the Horst area. We recommend that the supply of nutrients tot these vulnerable areas is separately regulated.

As the protection of the groundwater against nitrate contamination is an important objective of the environmental legislation, we recommend to specify manure application also in terms of N-quantities. N supply is not necessarily comparable to P supply on a stoichiometric basis, especially when N-specific fertilizer is used (Table 3). Thus the supply of nitrogen is not effectively restricted by a P-based legislation. NITSOL provides an instrument to evaluate N supply restrictions with respect to both the prevention of load on the surface water as with respect to concentrations in the groundwater.

For the future we recommend to specify management decisions or legislation if necessary in terms of hydrology and soil type. For the Netherlands this information is available on a national scale. In areas with severe problems a local or regional differentiation could be effective. Models should be used to test legislation and to select BMP's. Again sensitivity analysis can be used on:
-reduction of supply,
-hydrological changes (especially areas with hydrological control),
-land use changes,
-alternative supply forms (*e.g.* manure injection, organic fertilizers),
-distribution of supply on local, regional, national scale,
-relative contribution of other (non point) sources,
-time horizon of the effect of changes (N compared to P behaviour).

The dilemma of the mathematical modeler

In general the mathematical modeler is positioned between the reality he has created by his model and the questions from managers to support their decisions using the model. From the one hand the modeler is inclined to choose for his model and to ask more money and time for further verification. On the other hand he is convinced that the model is the best decision support system. How to deal with this dilemma?

From our experiences we would advise as a first step to use the model to scope the environmental and economic objectives of the decision maker into optional best management practices. The optional BMP's can be described by some kind of sensitivity analysis. The second step is to present draft results so that a provisional choice for one or more BMP's can be made. However, simultaneously the most critical assumptions and the quality of input and/or calibration data should be presented . On basis of these provisional results the decision maker is able to define the quality of the final recommendations and can trade off between time, costs, number of alternatives and quality, in terms of input, calibration or verification.

In this manner the modeler proposes the alternatives and is able to stay in control of his natural dilemma. The manager chooses. However, it is advisable to keep contact with the model and he cariied out simulations. Although model manuals are necessary, the model remains as objective as its user. Therefore, apart from model qualifications, it could do no harm to prescribe user qualifications in the manual.

Acknowledgements

The Horst and Hupsel studies were carried out by Delft Hydraulics under contract of the Ministry of Transportation and Public Works in the Netherlands. We wish to express our gratitude to Dr. J. Uunk of the Ministry for his cooperation.

References

Aalst, R.M. van and Diederen, H.S.M.A. 1983. *The role of nitrogen-oxides and ammonia in the acidification of the Netherlands* (in Dutch). MT/TNO. R 83/42.

Abrahamse, A.H., Baarse, G. and van Beek, E. 1982. *PAWN,* Vol. 12: Model for Regional Hydrology, Agricultural Water Demand and Damages from Drought and Salinity. RAND/ Delft Hydraulics.

Beasley D.B. and Huggins, L.F. 1980. *ANSWERS* (Areal Nonpoint Source Watershed Environment Simulation)- User's Manual. Dept. of Agr., Eng., Purdue Univ., West Lafayette, IN.

Bergström, L. and Johansson, R. 1991. Soil processes and chemical transport. Leaching of nitrate from monolith lysimeters of different types of agricultural soils. *J. Environ. Qual.,* 20, 801-807.

Bhat, K.K., Flowers, T.H. and O'Callaghan, J.R. 1980. A Model for the Simulation of the Fate of Nitrogen in Farm Wastes Land Application. *J. Agric. Sci. Cambr.,* 94, 183-193.

Bingner, R.L. *et al.,* 1992. Predictive Capabilities of Erosion Models for Different Storm Sizes. *Trans. Am. Soc. Agric. Eng.,* 35, 505.

Bruyn, P.J. de, Hopstaken, C.F., Wesseling, J.W. and Wit, P. 1986. Water-Agriculture-Environment. *Model-Design and Application on the North-West Veluwe,* The Netherlands (in Dutch). Delft Hydraulics, T0111.

Central Bureau of Statistics (CBS). 1983. *The use of manure and sewage sludge in agriculture* (in Dutch). CBS 83/2.

Cooper, A.B. *et al.,* 1992. Predicting Runoff of Water, Sediment, and Nutrients from a New Zealand Grazed Pasture Using CREAMS. *Trans. Am. Soc. Agric. Eng.,* 35, 105.

Dijk, T.A. van, 1984. *Manure surplus in the Netherlands* (in Dutch). Institute for Soil Fertility (IB-Haren), Nota 84-6.

Dillon, P.J. 1989. Models of the nitrate transport at different space and time scales for groundwater quality management. In: Jousma *et al.* (ed.), *Groundwater Contamination: Use of models in Decision Making,* pp.272-284. Dordrecht, Kluwer Academic Publishers.

Donigian, A.S., Baker, J.L., Haith, D.A. and Walter, H.F. 1983. Application guide for Hydrological Simulation Program Fortran (HSPF). Environmental Research Lab., U.S. EPA, Athens, GA.

Donigian, A.S., Baker, J.L., Haith, D.A. and Walter, H.F. 1983. *HSPF Parameter Adjustment to Evaluate the Effects of Agriculture; Best Management Practices.* Env. Res. Lab. EPA-600/3-83-066.

Engelenburg, M.A.L. and van den Eertwegh, G.A.P.H. 1988. *Waterquantity and waterquality in the Hupsel research area: a model approach* (in Dutch). Agricultural University Wageningen.

Facchino, F., de Leeuw, A.M., Ruijgh, E.F.W. and Streng, J.M.A.1993. Water quality analysis of the Po river basin. *Application of a modeling framework for nitrogen transport in soil and groundwater.* Delft Hydraulics/IRSA, T690.

Gerrits, H.J., van Amstel, H., Helmerhorst, T., de Leeuw, A. and Ruijgh, E. 1993. *Effects of earlier implementation of equilibrium-fertilization on phosphorus loads to Lake Eem and Lake Gooi* (in Dutch). H2O (26) 600-605.

Goeller *et al.*, 1983. Summary Report, *Policy Analysis of Water Management in the Netherlands.* RAND-Cooperation/Delft Hydraulics.

Grashoff, P.S., Menke, M.A., van Belois, Ch. and Ruijgh, E.F.W. 1989. *Gathering and calculation of hydrological and agricultural inputdata for waterquality calculations* (in Dutch). Delft Hydraulics, T420.

Grashoff, P.S, Adriaanse, B., Maaten, R., Sprengers, C.J. and Tacoma, J. 1990. CERES: coupling DEMGEN with waterquality models (in Dutch). Delft Hydraulics, T332.

Hopstaken, C.F., Wesseling, J.W., Wit, P., de Bruyn, P.J. and Klomp, R. 1986. Excessive Manuring and its Effects on the Nitrogen and Phosphorus Concentrations in Soil, Ground- and Surfacewater of the North-West Veluwe (The Netherlands). *Proc. Joint Dutch-Hungarian Seminar on Flatland and Inland Drainage and Irrigation.* Delft Hydraulics Communication No. 363.

Hopstaken, C.F., Wesseling, J.W., Wit, P. and de Bruyn, P.J. 1987. Modeling the Environmental Impact of Agriculture with Respect to Surface and Groundwater Quality. The Nitrogen Soil Model (NITSOL) applied to the North-West Veluwe (The Netherlands) *Wat. Supply,* 6, 319-326.

Hopstaken, C.F., Ruijgh, E.F.W., Grashoff, P.S. and Menke, M.A. 1988. Water-Agriculture-Environment. Model Application on the Horst Catchment Area (in Dutch). Delft Hydraulics, T297.

Hopstaken, C.F., Ruijgh, E.F.W., Grashoff, P.S. and Menke, M.A. 1989. The environmental impact of agriculture on surface and groundwater quality and the use of Nitsol/Phosol in decision-making. In: Jousma *et al.* (eds), *Groundwater Contamination: Use of models in Decision Making*, pp.378-382. Kluwer Academic Publishers, Dordrecht.

Hopstaken, C.F., Ruijgh, E.F.W., Streng, J.M.A., Sweerts, J-P.R.A, de Ridder, A.C., Gerrits, H.J., de Leeuw, A.M. and Sprengers, C.J. 1991. *Water quality analysis of six Emilia Romagna river basins.* Delft Hydraulics, T659.

Knisel, W.G., 1980. *CREAMS: a Field Scale Model for Chemicals, Runoff and Erosion from Agricultural Management Systems.* Rep. No. 26, U.S. Dep. of Agric.

Kroeze, C., H.G. van Faassen and P.C. de Ruiter, 1989. Potential Denitrification Rates in Acid Soils under Pine Forest. *Neth. J. Agric. Sc.,* 37, 345-354.

Konikow, L.F. and Bredehoeft, J.D. 1992.Groundwater models cannot be validated. *Adv. Water Resour.,* 15, 75.

Moore, L.W. *et al.*, 1992. Modeling of best management practices on North Reelfoot Creek, Tennessee. *Water Environ. Res.,* 64, 241.

Project Eutrophication Borderlakes (PER), 1982. *Eutrophication of borderlakes Wolderwijd and Nuldernauw 1976-1979* (in Dutch). Ministry of Transport and Public Works, Lelystad.

Reddy, K.R., Khaleel, R., Overcash, M.R. and Westerman, P.W. 1979. A Nonpoint Source Model for Land Areas Receiving Animal Wastes: 1. Mineralization of Organic Nitrogen. *Trans. ASAE,* 22,863-874

Riemsdijk, W.H. van, Bomans, L.J.M and de Haan, F.A.M. 1984. Phosphate sorption by soils: a model for phosphate reaction with metal oxides in soil. Soil Sci. Soc. Am. J., 537-541.

Rijtema, P.E. 1980. Nitrogen emission from grassland farms: a model approach. *ICW-technical bulletin,* 119, ICW-Wageningen.

Ruijgh, E.F.W. and Hopstaken, C.F. 1990. Districtwatermodule DIWAMO (in Dutch). Delft Hydraulics, T568.

Ruijgh, E.F.W., Hopstaken, C.F. and Grashoff, P.S. 1990. *Effects of the phosphorus saturated soils control on nutrientloads to surface water* (in Dutch). Delft Hydraulics, T695.

Ruijgh, E.F.W., Hopstaken, C.F., Grashoff, P.S. and Witte, J.P.M. 1990. *Verification of DEMNIP on data from the Hupselse beek research basin.* TNO-CHO Proc. and Info. No. 44. pp 261-270.

Ruijgh, E.F.W. 1991. *Research of the nitrogen and phosphorus content of the soil in the Hupsel research basin* (in Dutch). Delft Hydraulics, T771.

Ruijgh, E.F.W. 1992. *User-manual Nitsol/Phosol* (in Dutch). Delft Hydraulics, T904.

Ruijgh, E.F.W., H.J. Gerrits, M.R., de Leeuw, A.L. and Menke, M.A. 1993. *Analysis of nutrientloads to Lake Veerse from the surrounding polders* (in Dutch). Delft Hydraulics, T680

Shiromohammadi, A. *et al.,* 1992. Model Simulation and Regional Pollution Reduction Strategies. *J. Environ. Sci. Health,* Part A, 27, 2319.

Tuckwell, S.B. and Knight, M.S. 1988. Guidelines for the agricultural use of water supply catchments to minimize leaching of nitrate. *Wat. Supply,* 6, 295-302.

United States Environmental Protection Agency (US-EPA), 1992. *The National Rural Clean Water Program Symposium - 10 Years of Controlling Agricultural Nonpoint Source Pollution.* EPA/625/R-92/006., Office of Water, Washington, D.C.

United States Geological Survey (USGS), 1988. *A modular three dimensional finite difference groundwater flow model.* Book 6. Modeling techniques.

Vasak. L., Krayenbrink, G.W.J. and Appelo, C.A.G. 1981. *The Spatial Distribution of Polluted Groundwater from Rural Centers in a Recharge Area in the Netherlands.* Free University of Amsterdam.

Veen, J.A. van 1977. The Behavior of Nitrogen in Soil. Thesis. Free University of Amsterdam.

Vries, I. de and Hopstaken, C.F. 1984. Nutrient Cycling and Ecosystem Behavior in a Salt Water Lake. *Neth. J. Sea Res.,* 18 (3/4), 221-245.

Vuuren, W. van, 1990. *The Calibration and Verification of the Agrohydrological Model DEMGEN in Two Experimental Areas in the Netherlands.* TNO-CHO Proc. and Info. No. 44. pp 261-270.

Willigen, P. de and Neeteson, J.J. 1985. Comparison of Six Simulation Models for the Nitrogen Cycle in the Soil. *Fertilizer Research,* 8, 157-171.

Wijnands, J.H.M., de Kruif, F.F., de Zodder, K. and Leusink, H.M. 1983. *The use of fertilizer in agriculture in 1979/1980.* LEI-report 3.125 (in Dutch).

Witte, J.P.M. 1987. *Water quality research in the areas Hupsel and Sleen.* DBW-RIZA report 87.078x (in Dutch).

Wright, J.A., *et al.,* 1992. Water Table Management Practice Effects on Water Quality. *Trans. Am. Doc. Agric. Eng.,* 35, 823.

7

Interactive modeling of groundwater contamination: visualization and intelligent user interfaces

Kurt Fedra[1], **Hans-Jörg Diersch**[2] **and Frank Härig**[3]

[1]*Advanced Computer Applications,International Institute for Applied Systems Analysis (IIASA) A-2361 Laxenburg, Austria*
[2]*WASY Gesellschaft für wasserwirtschaftliche Planung und Systemforschung, D-12526 Berlin, Germany*
[3]*Institut für Wasserwirtschaft, Hydrologie und Landwirtschaftlichen Wasserbau Universität Hannover, DW-3000 Hannover 1, Germany*

Abstract
The management and remediation of groundwater contamination problems is among the more complex tasks in environmental management and technology. Not only is the available information usually sparse and the physical processes very complex, i.e., dynamic and spatially distributed, they are also difficult to analyze, understand, and communicate. At the same time, problems are often of considerable urgency and the economic and political stakes are usually high.

Numerical simulation models are certainly powerful tools with enormous potential, but they are difficult to use and difficult to validate. Their results are difficult to communicate and thus not easily understandable by a non-technical audience.

Putting groundwater models into an interactive framework, with a user-friendly interface, employing computer graphics for problem visualization can help to improve this situation. This makes available numerous support functions, such as transparent data base handling that allows the efficient use of models and at the same time the effective communication of model results. The discussion centers on three basic concepts, i.e.,integration of various sources of information and tools that go beyond the immediate groundwater model, but are important components in a problem-oriented application; examples include coupling to geographical information systems or image processing of remote sensing data, various data base management systems, models for related systems such as surface water or land fills, or multi-criteria decision support tools; interaction, i.e., the ability to run the models interactively, produce dynamic animated output, define scenarios interactively, run, modify, and re-run a problem or set of problems, tune output and display options in real time, etc., to get immediate answers in an exploratory and experimental man–machine dialogue; and visualization, i.e., the graphical and symbolic representation of the system and its behavior in time and space, but also including the entire problem that goes beyond the basic physical system and includes economic and socio-political dimensions and attributes.

This chapter introduces an interactive approach to groundwater contamination modeling, and discusses the approach, its philosophy, and concrete implementation examples based on two 2D finite-element and finite difference simulators respectively.

Combining groundwater flow and transport models with an AI-based and symbolic, graphics user interface, selected functions of geographical information systems (GIS), and transparent data and file handling, the systems are designed to allow the easy and efficient use of complex groundwater modeling technology in a problem- rather than model-oriented style.

Using color-graphics super-microcomputer workstations, the approach provides a problem manager with numerous built-in support functions such as the selection of site-specific, as well as generic, groundwater problems from problem libraries and GIS functions for the comparative analysis of model runs, the preparation of topical background maps, the interactive modification of a problem for extensive scenario analysis, or the complete interactive design of a new problem. Using either satellite imagery (such as LANDSAT or SPOT) or Digital Line Graph (DLG) standard vector maps as a background, problems can be edited and modified and then simulated under interactive user control. The results are presented as dynamic graphical output, that can be interactively controlled and configured. Repeated simulations, comparing alternative assumptions on parameter values or management options can be compared in a variety of analysis functions such as GIS-based overlay techniques.

The systems also feature extensive problem-editing capabilities, ranging from the interactive location of wells or sources of pollution to a Computer Aided Design (CAD) component for the interactive design of a complete problem geometry. This allows the user to design very efficiently, and parametrize, a new problem from the very beginning, using a map or an auxiliary grid as a background for the definition of problem geometry and hydrogeological parameters.

Building some of the knowledge of an experienced groundwater modeler into the software system and its interface through the rule bases driving, e.g., numerous transparent, error correction functions and an automatic mesh generator, allows for a very fast and efficient formulation of a new problem, even by an inexperienced user. Freed from the task of very detailed and demanding formalization of the computer representation of the problem, the user can thus concentrate on, e.g., the management, regulatory, economic, or technological aspects of his problem. Designed for complex applications such as hazardous waste management, site evaluation, and groundwater contamination problems, ease of interactive use, responsiveness, and efficiency in problem definition and comparative analysis are the most important characteristics of this interactive approach to groundwater simulation.

Introduction

Groundwater resources are an increasingly important component of socio-economic development. Municipal water supply, in many countries, is for the most part based on groundwater, *e.g.*, between 40-50% in the USA, 60% in the FRG, or 98% in Denmark. Furthermore, groundwater is frequently the most important source for agricultural, as well as industrial, water supply.

The causes and consequences of qualitative changes in groundwater regimes can be separated by decades or even centuries. Once contaminated, groundwater resources may be permanently impaired. The widespread use of chemical products in developed countries, the intentional disposal of large volumes of waste materials, hazardous wastes from the chemical process industry, and accidental leaks and spills of waste materials during transportation are the actual and potential sources of extensive groundwater contamination.

The contamination of surface water resources poses a threat to groundwater resources both in the case of infiltration under natural conditions (especially in the case of floods), and infiltration forced by groundwater extraction for drainage and water supply (bank filtration).

Effective techniques have to be designed then, to mitigate groundwater contamination. An integrated risk assessment, which attempts to estimate the probabilities of groundwater contamination together with their likely consequences, in a technical, as well as a socio-economic framework, could be one such technique. In any case, as one of the major inputs, the prediction of the distribution patterns of groundwater pollutants is required, in terms of concentrations as well as mass flow, under various conditions of groundwater management.

Computer models to predict such patterns do exist (Bachmat, Bredenhoeft and Andrews*et al.*, 1980). In fact, a large amount of formal, mathematical and computational methods have been developed in the area of groundwater resources planning and management, and the field has a considerable history in the use of computers. However, their use and interpretation requires considerable effort and technical expertise.

Many of the available computer-based models and methods certainly are potentially very useful; to turn a potentially useful method into one actually used however, requires a number of special features, as well as an approach that takes psychological and institutional aspects together with scientific and technical ones into account.

The examples used to illustrate some concepts that can make groundwater models easier to use are the prototype finite-element groundwater contamination model system FE FLOW/FEMCAD, which was designed for the assessment of waste management technologies and facilities such as landfills and dump sites (Fedra and Diersch, 1989). One of the main application areas of the system is in the analysis of mitigation options. The second example is a finite-difference model of groundwater flow and contaminant transport, that was integrated into a city-level environmental information systems framework and coupled to a GIS that facilitates the interactive generation of topical background maps and the display and analysis of related data and model results.

Smart software

Software and computer-based tools are designed to make things easier for the human user; and they improve the efficiency and quality of information processing tasks. In practice, only very few programs do that. They make things possible that would not be possible without the computer, but they rarely make it easy on the user.

As with the better mouse trap, one would expect to see increased demand for such techniques; however, simply doing things faster – once all the input has been painstakingly collected and entered, or solving a more complex version of the problem – and then leaving it to the user to extract the meaning from a flood of output and translate into his problem description language may not rate as a better mouse trap in the eye of the practitioner. Every tool, and models in particular, has to be seen as an integrated part in a much more complex information processing and decision-making procedure, that involves not only just running the model, but certainly preparing its inputs over and over again, interpreting and communicating its results, and making them fit the usually rather formalized framework of the existing procedures.

There are several important aspects that need to be addressed: computer-based tools, information and decision support systems as a rule imply a change in personal work habits, institutional procedures, and thus, institutional culture. While they may or may not change what is done, they most certainly change the way how things are done – if they are used.

There is a trade-off between the efficiency and ease of use and the flexibility of a system. The more options are predetermined and available from a menu of choices, the more defaults are provided, the easier it becomes to use a system – for an increasingly smaller set of tasks.

There also is a trade-off between the ease of understanding and the (at least numerical) precision of results. Providing a visual graphical or symbolic representation changes the quality of the information provided from a quantitative, and thus at least apparently precise format, to a qualitative format. The latter, however, certainly is more appropriate for displaying patterns and complex interdependencies.

Mathematical models for groundwater contamination

Although the emphasis of this chapter is on interaction, visualization, and therefore the user interface of the two models described below, a brief ntroduction to the models provides a better basis for the discussion of their specific interactive implementation. To provide the context for this presentation, a short description of basic formulations in numerical modeling is included.

Basic formulations

An unified theoretical basis of the model's governing equations capable of describing all important mechanism in groundwater pollution problems has been systematically developed from conservation principles using averaging theory for

multiphase/multispecies (chemically reactive) porous media systems, *e.g.,* Hassanizadeh and Gray, 1979; Nguyen,Gray, Pinder *et al.,* 1982; Diersch, 1984. As a result, macroscopic balance equations for different levels of complexity can be grouped roughly into two categories:
 (1) transport of solved constituents as a contaminant plume and
 (2) displacement and distribution of a nonaqueous contaminant phase.

The former are typical of a wide range of applications where the groundwater pollutants can be considered as water-soluble fractions. These models are widely used in practice, *e.g.,* Pinder, 1973; Diersch and Kaden, 1984; Frind, 1987; Diersch, 1988; and Kinzelbach and Rausch, 1989.

Nonaqueous phase liquid (NAPL) contamination by organic compounds leads to models based on multiphase transport of immiscible fluids. If the contaminant can be assumed nonvolatile and insoluble those models which have found widespread use in petroleum reservoir modeling would be appropriate *e.g.,* Aziz and Settari, 1979.

More recently, models for both miscible and immiscible contaminant transport have been developed and applied for selected cases *e.g.,* Abriola and Pinder, 1985; Forsyth, 1988; Kaluarachchi, Parker and Lenhard, 1990; and Diersch, 1990. The derivation of the governing model equations starts from microscopic balance laws for mass, species concentration, momentum, and, to close the system with constitutive relationships, for energy and entropy (Hassanizadeh and Gray, 1979).

Jump conditions at phase interfaces of the multiphase porous media system complete the set of model equations. Averaging procedures are used to develop representative macroscopic balance equations for the porous media system consisting of different phases (solid, water, liquid contaminant, gas *etc.*) and chemical species (ions, soluble organics *etc.*) Diersch, 1984.

Further theoretical considerations are not pursued here as of interest is more the handling of these basic model equations for classes of groundwater contamination problems in the interactive modeling context via numerical solution strategies, discussed below.

Numerical approaches
The most common numerical techniques to solve the governing model's equations for contamination processes in subsurface water resources are the finite difference method (FDM) and the finite element method (FEM) (Pinder and Gray, 1977 and Huyakorn and Pinder, 1983). Other related methods, such as the boundary element method (BEM), can be of value for specific tasks (*e.g.,* for free-surface seepage through dams).

In the past, the FDM had found widespread application in groundwater hydraulics. Its advantages are simplicity and effectiveness for rectangular study areas which are often representative of model flow situations. However, if mass (or heat) transport equations have to be solved, there is a desire for very accurate fluid velocities and modeling sharp fronts. Accordingly, classic finite difference discretization techniques have been found to be inappropriate, unless generalized meshing techniques are used. This has led to the development of volume-difference methods,where balance-improved formulations are incorporated and more flexibility in (Neumann-or Cauchy-type) boundary conditions occur. In groundwater hydraulics an alternative solution strategy is sometimes preferred to avoid problems in the numerical approximation of the transport equations. In principle, only the flow equations are then solved in a well-known manner, however, followed by a post-processing technique for evaluating the convective paths of particles introduced in the flow field, so-called particle tracking (Kinzelbach and Rausch, 1989). These methods work fast as long as the number of particles is not too high. Another advantage is that they circumvent numerical difficulties associated with oscillatory results or false damping measures when convective-dominated transport equations have to be solved. However, their price is that they are very restrictive in the application and unable to solve more complex transport phenomena, for instance fluid-density coupling, phase transitions, reacting processes and capillary effects. On the other hand, if coupled with random walk techniques (Kinzelbach and Rausch,1989), effects of hydrodynamic dispersion are modeled on a statistical basis. Examples of packages are MODFLOW (McDonald and Harbaugh, 1984) and ASM (Kinzelbach and Rausch, 1989).

Interactive groundwater models

The two software systems for interactive modeling of groundwater contamination have been designed for different purposes: while FEFLOW/FEMCAD is a more general purpose system, designed to cover a large range of possible applications, GWSIM has been designed for a single-site application for, typically, a city administration. The resulting choice of techniques and ranges of functionality must be understood with this distinction in mind.

FEMCAD offers access to a large number of problems and examples, both site-specific as well as generic, and includes an option for the interactive definition of a new problem from scratch. GWSIM, in contrast, works with a given site and thus a fixed basic grid;it only allows in its current version the positioning of wells and sources of contamination, and, for display, the zooming into any sub-area of the main grid.

FEMCAD makes it easy to select and run, but also to build a new model; GWSIM is designed to assist in the definition of a problem scenario within a well-defined set of problems, and then to run it with minimum user effort to analyze the particular problem. In both cases, integration of GIS technology is provided for generating a background to relate model results to as well as, in the case of GWSIM, store and analyze model results in conjunction with other topical maps.

In summary, the common features can be grouped as follows:
a fully menu-driven, largely symbolic and graphics-oriented user interface;
specific interactive problem definition and editing options;
the coupling to dedicated GIS features;
the use of dynamic color graphics for visualization;
transparent data handling.
The systems have a number of distinct features:

First, the complexity of a tightly integrated model, data bases, pre-, main-, and post-processors, is completely hidden from the user, who is only exposed to a simple menu-driven and graphics-oriented user interface. The complexities of model selection, preparation of the necessary input data, communication between model system components, and different levels of detail and resolution, are all taken care of by the system's intelligent control and interface modules. As part of the various modules, small rule bases control the system's operation and guide the user with features such as automatic problem correction, warning and explanatory messages, and transparent context-dependent reconfiguration and enforcement of constraints. All the user has to do is to select appropriate options from the menus, thereby specifying his problem to a degree sufficient for the system to act upon, and, in the case of the CAD module, select or position spatially distributed objects with a graphical input device.

Second, the systems are hierarchically structured. The level of detail required, and the implied trade-offs between level of treatment, the information the user has to

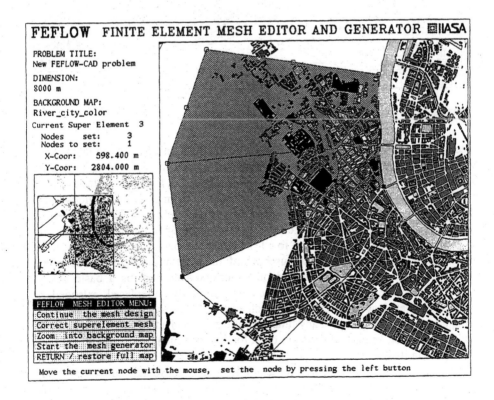

Figure 1 *An integrated framework for interactive modeling.*

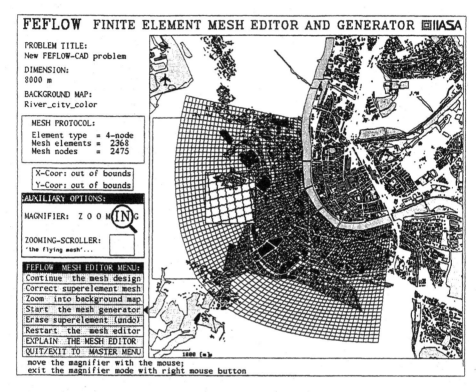

Figure 2 *Defining super–elements in the mesh editor*

specify, and the response time of the system, are interactively determined by the user. Most options provide a hierarchically organized system of defaults and standard procedures, which the user can optionally refine, traversing the levels of the system's problem representation hierarchy. As a particular example of this hierarchical structure, GWSIM offers a two level interface: a top level that is largely driven by symbolic menu buttons and allows only very few numerical problem definitions, and an expert level that provides access to all the numerical parameters of the model for the expert modeler and experienced user. Third, the systems are highly modular and built with an open architecture and object oriented concept as a basic design guideline. Modules are functional entities that can easily be replaced, extended, or added and deleted, without changing much of the rest of the system. Thus, while being an operational system at any time, modifications and adjustments can easily be made at low cost, keeping the system open for adaptive modifications with growing experience in its use.

Integration with other modules, for example the interactive map editor that is part of the GWSIM framework system, or a data base of potentially contaminated industrial sites in the same system, is made easy through this modular design

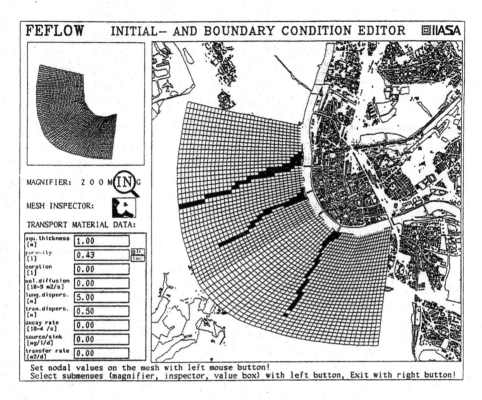

Figure 3 *Mesh generation: local magnification of the mesh .*

philosophy. While these modules can be integrated, the core system can work just as well without them, using only the current default map, for example, as a background.

Similarly, FEMCAD operates on a current problem data set, which could, in principle, have been generated by hand just as well as by the system's own CAD module for interactive problem definition.

As a basic design philosophy, the user should not have to be bothered with the technicalities of specifying the model, which is usually a highly technical problem. The user should be allowed to concentrate on specifying the problem, not the methods to be used to solve it.

Consequently, the user interface in our approach is always menu driven *i.e.*, the user is prompted to select options from menus of possible options the system offers. Wherever possible, the options are specified in a symbolic *i.e.*, linguistic or graphical format, *i.e.*, as icons.

The selection of options start with the problem selection, identifying and loading a complete ready-to-run site-specific or generic problem from a problem library, or the selection of input data from a listing of alternative data sets, initial or boundary conditions, that can be identified from an appropriate listing and loaded as the basis for further problem specification and a model run.

Selection of numerical data to be changed *i.e.*, the pumping rates of wells in the system, or any of the geohydraulic parameters used, is again possible by identifying the respective value on the display screen. The value can be changed by typing a new value via the keyboard or by using a three-button mouse, where the buttons are used to decrease, set, or increase the selected value. Alternatively, sliders that permit a value to be set within a predefined range are used and offer graphical feedback.

Changes are only allowed in a certain range fixed by the input data, and possibly modified by other changes specified by the user. In other words, the system maintains context dependent bounds on input and control variables to guarantee consistent and feasible problem descriptions. For an interactive system, it is extremely important to assist the user to stay with his assumptions not only within plausible ranges (from the problem perspective), but also within the ranges over which the methods to be used are valid and meaningful.

FEFLOW: a finite-element based system

In principle, the movement of contaminants in subsurface water represents an unsteady 3D mass transport problem. Taking into account the extensive numerical calculations required to solve such problems, and the accuracy of data available for model quantification (transmissivities, retardation coefficients, *etc.*), in most practical cases a simplification to a 2D problem becomes necessary and is reasonable. This may be either horizontal-plane problems (*e.g.*, areal distribution of a spill or leakage, pumpage of bank filtered water for water supply), or vertical-plane problems (*e.g.*, deep-well injection). The basis of our prototype model system is provided by a 2D finite-element simulator: the FEFLOW model. However, the supporting software and graphical interface could be adapted to any similar 2D, or extended for a 3D, groundwater simulation model.

Basic equations

The finite element solution is based on the governing equation system describing both horizontal aquifer-averaged and vertical (axisymmetric) fluid-density coupled flow and contaminant transport in groundwater which is commonly used for these type of problems

$$S\frac{\partial h}{\partial t} + \frac{\partial q_i}{\partial x_i} = Qh \tag{1}$$

$$q_i = \epsilon v_i = -K_{ij}M\left(\frac{\partial h}{\partial x_j} + \frac{\rho - \rho_o}{\rho_o}e_j\right) \tag{2}$$

$$R\frac{\partial C}{\partial t} + q_i\frac{\partial C}{\partial x_i} - \frac{\partial}{\partial x_i}(D_{ij}\frac{\partial C}{\partial x_j}) + (R\lambda + Q_h)C = Q_c \tag{3}$$

to be solved for $h = h\,(x_i, t)$, $q_i = qi\,(x_i,\ t)$ and $c = c\,(x_i, t)$ with the constitutive conditions

$$S = \epsilon_o + S_s M \tag{4}$$

$$\rho = \rho_o(1 + \alpha\frac{C}{C_s}) \tag{5}$$

$$\alpha = [\rho(C_s) - \rho_o]/\rho_o \tag{6}$$

$$D_{ij} = (\epsilon D_d M + \beta_T V)\delta_{ij} + (\beta_L - \beta_T)\frac{q_i q_j}{V} \tag{7}$$

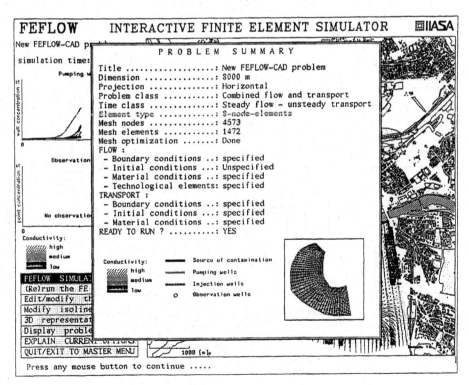

Figure 4 *Interactive definition of parameter values .*

$$R = M[\epsilon + (1 - \epsilon)\chi] \qquad (8)$$

where

$h=$ hydraulic head (L)

$C=$ concentration of chemical species (M L^{-3})

$\epsilon_0=$ drainable or fillable porosity (-)

$\epsilon=$ kinematic porosity (-)

$S=$ aquifer storage coefficient (-)

$\chi=$ sorption coefficient (-)

$S_s=$ specific storage coefficient (L^{-1})

$D_d=$ molecular diffusion coefficient (L^{2}T^{-1})

$M=$ aquifer thickness ($M = 1$ for vertical/axisymmetric problems) (L)

$\beta_L, \beta_T=$ coefficients of longitudinal and transverse dispersivity, respectively (L)

$Q_h=$ source/sink?function (L T^{-1})

$v =(q_i q_i)^{1/2}$, absolute specific flux (L^2 T^{-1})

$K_{ij}=$ hydraulic conductivity (L T^{-1})

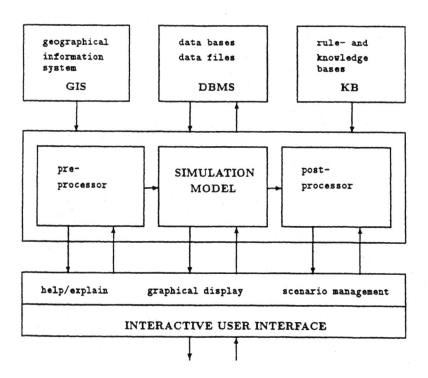

Figure 5 *Status summary page for a site-specific model .*

λ = concentration decay rate (T^{-1})

q_i= Darcy volumetric flux $(L^2 T^{-1})$

Q_c = known source/sink function $(M L^{-3} T^{-1})$

v_i = pore velocity $(L^2 T^{-1})$

D_{ij} = tensor of hydrodynamic dispersion $(L^3 T^{-1})$

ρ, ρ_0= fluid density and reference one $(M L^{-3})$

C_s = maximum concentration $(M L^{-3})$

α = density difference ratio (-)

R = specific retardation factor (L)

e_i = components of the gravitational unit vector (-)

The macroscopic mixing phenomena are described by the Scheidegger-Bear dispersion model (Bear, 1979).

The above equations involve a Henry sorption isotherm and a first order chemical reacting law. Fluid-density effects lead to a nonlinear problem analysis (Pinder and Gray, 1977; Diersch, 1988).

The following initial conditions (I.C.) and boundary conditions (B.C.) are to be specified:

I.C. $\quad h(x_i, O) = h_o(x_i) \qquad C(x_i, O) = C_o(x_i) \qquad$ **first kind**

B.C. $\quad h(x_i, t) = h_1^R \qquad\qquad C(x_i, t) = C_1^R$

$\qquad q_h = q_h^R = -K_{ij} M \frac{\partial h}{\partial x_j} \Upsilon i \quad q_c = q_c^R = -D_{ij} \frac{\partial C}{\partial x_j} \Upsilon i \quad$ **second kind**

$\qquad\qquad\qquad\qquad$ (or) $q_c = C_2^R q_h^R + q_c^R$ for the divergence $\qquad$ (9)

$\qquad\qquad\qquad\qquad$ balance-improved formulation

$\qquad q_h = \Phi_h(h_2^R - h) \qquad\qquad q_c = \Phi_c(C_3^R - C) \qquad$ **third kind**

as well as specific source/sink functions at subregions or well nodes:

$Q_h \;=\; Q_h^a + Q_h^w, \qquad\qquad\qquad Q_c \;=\; Q_c^a + Q_c^w, \qquad\qquad$ **fourth kind**

with $\qquad\qquad\qquad\qquad\qquad\qquad\qquad\qquad\qquad\qquad\qquad\qquad$ (10)

$Q_h^w \;=\; \sum_{m=1}^{r} Q_m^w \delta(x_1 - x_1^m)\delta(x_2 - x_2^m), \;\; Q_c^a \;=\; Q_h^a C^a, Q_c^w = Q_h^w C^w$

where

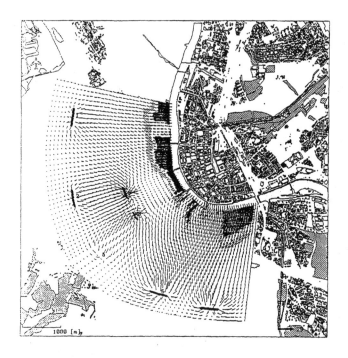

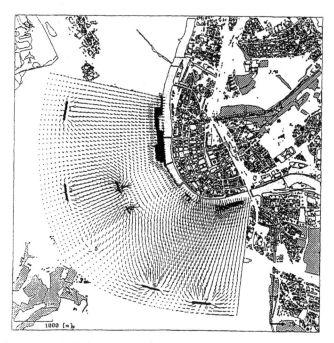

Figure 6 a,b,c, *Animated dynamic model output: subsequent time steps in a bank-filtration example .*

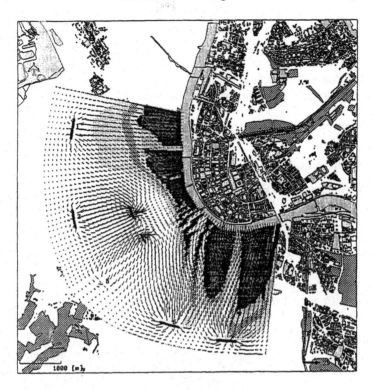

where

h_0 = prescribed initial hydraulic head distribution (L)

h_1^R = prescribed Dirichlet-type boundary hydraulic head (L)

q_h = outward normal Darcy flux $(L^2 T^{-1})$

q_h^R = prescribed Neumann-type normal boundary flux $(L^2 T^{-1})$

Y_i = directional cosines (-)

h_2^R = prescribed Cauchy-type boundary hydraulic head (L)

Φ_h = transfer coefficient (colmation) describing the aquifer-surface water interactions $(L\ T^{-1})$

C_0 = prescribed initial concentration distribution $(M\ L^{-3})$

C_1^R = prescribed Dirichlet-type boundary concentration $(M\ L^{-3})$

q_c = outward normal mass flux (dispersive or total) $(M\ L^{-1}\ T^{-1})$

q_c^r = prescribed Neumann-type boundary mass flux $(M\ L^{-1}\ T^{-1})$

C_2^R = prescribed concentration for convective flux $(M\ L^{-3})$

C_3^R = prescribed Cauchy-type boundary concentration $(M\ L^{-3})$

Q_m^w = the volumetric injection/discharge rate from pumps situated at $(x_i^m)(L^3\ T^{-1})$

r = number of source/sink points (*e.g.*, pumps, positive values imply injection)

Q_h^a, Q_h^w = infiltration rate and well injection/discharge rate, respectively $(L\ T^{-1})$

Q_c^a, Q_c^w= contaminant sources by recharge and well injection, respectively $(M\ L^{-2}\ T^{-1})$

$\Phi_c =$ mass transfer coefficient $(L^2\ T^{-1})$

$\delta()=$ Dirac delta function

$C^w =$ concentration of the pumped fluid at the pump located as above $(M\ L^{-3})$)

$C^a=$ concentration of the recharged/infiltrated fluid $(M\ L^{-3})$

Model implementation

The implementation of FEFLOW described here was designed as a demonstration prototype for the US Bureau of Reclamation in a collaborative project with the University of Colorado's Center for Advanced Decision Support for Water and Environmental Systems (CADSWES). The design objective was to prepare a system that could handle a broad range of groundwater contamination problems, for example, in the context of the US Superfund program. Ease of use by non-modelers, the capability to manage a potentially very large number of cases, and the possibility to work with very limited information on a preliminary screening level were important considerations for the implementation.

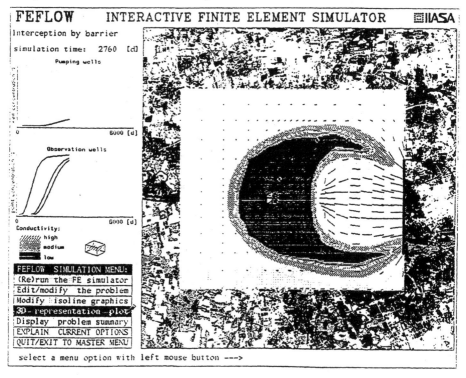

Figure 7 *A color-coded contaminant plume in an interception-barrier example .*

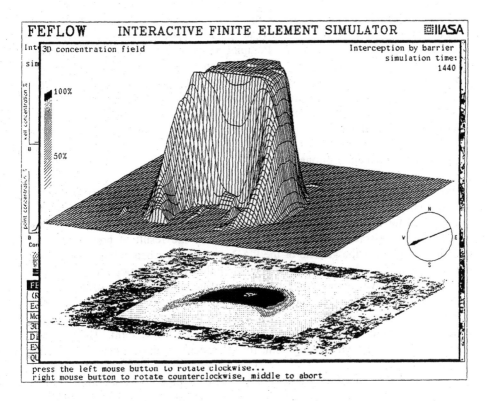

Figure 8 *Pseudo 3D representation of the concentration field.*

From the user's point of view, the model generates predictions of groundwater contamination consequences in terms of concentrations in time and space. The system has to provide estimates to answer questions such as:

How does a contaminant plume migrate from its source (*e.g.*, a waste dump or landfill, or possibly a waste-well injection) through a flowing groundwater field under given hydrogeological conditions.

How will this contamination influence water utilization, *e.g.*, drinking water supply wells ? Will wells or galleries be influenced by wastewater migration, and if so, at what level of contamination?

In a given time, what extent of contaminant distribution can be expected? What is the duration of the contaminant movement?

How can the contamination be stopped or reduced?

Which remediation strategies are recommended and how effective are they for the decontamination of the acquifer?

What is the risk this contamination source or situation poses to the surrounding area, and what can be done to reduce this risk?

To support the experimental nature of the system, each of the control variables determining a problem situation can be modified independently. For example, once a certain problem is defined, the user can run it for several different amounts of substances, or different substances. Pumping rates may be changed, a hydraulic barrier may be introduced, or the dump site can be sealed off. The interactive problem editor with its graphical problem definition tools provides a convenient and efficient means of problem specification with immediate visual feedback.

For the interactive system, three different types of input data sets or problem descriptions are considered:

existing specific sites,

generic problem descriptions,

user-generated problem descriptions.

For existing specific sites, the user can choose from a hierarchically organized description of regions (states) and named sites within this region, currently implemented for the continental US or from the corresponding set of maps. These complete and site-specific problem descriptions are the result of previous model sessions, using either the generic or the CAD options described below to configure and define a new problem. For a given site, more than one description may be stored under different names, representing, for example, alternative clean-up strategies.

However, if only limited information is available, the user can choose generic problem descriptions, either from a list of available generic problem examples or from pictograms representing problems available in a schematic form. They provide a convenient and efficient starting point for the specification of a new site-specific problem with limited data.

The completely specified site-specific or generic problem description is then loaded from a data base with input files for each specific site or generic problem. They are ready-to-run examples, that the user can run with the appropriate menu-option choice; alternatively, he can use them as the starting point for an alternative problem description, using the CAD problem definition and editor module described below. There is also the possibility to store (and retrieve) a problem that is only partially defined, mainly as an interim result of the CAD option. In this case, the system will issue an appropriate warning and invite the user to either complete the problem definition, or select an alternative problem.

For site-specific problems, reference to a background map, either in the form of a raster map (LANDSAT or SPOT), or a vector-based map (in a binary version of the USGS DLG--digital line graph--format), is stored together with the problem description. This graphical background information will then be loaded to provide a geographical reference for the problem in question.

The CAD problem generator and editor

The Computer Aided Design (CAD) problem generator and editor module allows the user to edit a problem description, or define a new problem from scratch. What can be edited depends on the type of problem: for a site-specific problem, only components within the given finite-element representation of the problem can be edited. Examples would be simulation control parameters such as the time step, location of wells and their pumping rates, strength of the source, the decay and adsorption coefficients, initial conditions such as an initial concentration distribution, *etc.* These changes, however, would be temporary, *i.e.,* after the model run, they are lost and no modification of the original data is made unless a new version of the data set is stored under a different name.

For generic examples, the same basic editing options apply; in addition, however, these problems can be rescaled, *i.e.,* the existing finite-element mesh can be rescaled, stretched, and thus adapted to a different geometry.

Also, in the generic cases a large number or super-set of options may be built into the problem descriptions. For example, most cases include a large number of wells at different locations, the majority of which, however, are not active, *i.e.,* pumped but only used as observation wells. By activating/deactivating them through prescribing appropriate pumping rates, alternative well locations can be implemented without necessarily having to modify the geometry of the problem.

If structural changes are made, a new version of the generic problem is created and stored under a new name, possibly as a site-specific problem. Thus, new user-defined problems are either based on modifications and reconfiguration of existing descriptions, starting from any of the generic or specific problem descriptions, or completely generated from scratch.

In the latter case, the user starts by defining the scale of his problem, and by loading a background map, if available. In the next step, the problem area is defined in terms of so-called super-elements, arbitrary quadrilaterals interconnected at their nodal points (Figure 2). They are drawn interactively on the screen, by moving the mouse with a rubber-band display of edges, and setting the nodes by clicking a mouse button. A simple rule-base ensures proper geometry, orientation, and interconnection of super elements with appropriate warning messages and automatic corrections. For the super-element construction, the user can zoom into the map for more detail, and use local magnification and correction for the fine-positioning of individual nodes and arcs.

After a satisfactory definition of the spatial problem domain has been constructed, a finite element mesh can be generated by specifying first the element type (four- or eight-node elements are supported), and then selection of a divisor number for each unconstrained edge (Figure 3).

In the next step, initial and boundary conditions for the flow and the transport component of the problem have to be defined, depending on the type of problem selected. Boundary and initial conditions such as heads, or flows representing wells, material concentrations or flux, or initial concentrations, and geo-hydrological or material parameters such as porosity, storativity, aquifer depth (or width), and dispersion parameters, are then defined, again in an interactive and graphical procedure (Figure 4). The appropriate symbol/description is chosen from an options menu, numerical values (other than the default value suggested by the system) are adjusted, and then set for individual nodes, elements, or globally, depending on the type of parameter. This interactive drawing and editing is supported by auxiliary functions that keep track of what has already been defined, including the option to read back the values set from the mesh by pointing at the respective node on the display. Status summaries and interpolated color overlays over the mesh representing the parameter values defined so far can be called up to guide the editing process.

Upon leaving the editor module, the new problem description is stored by the data base management system and can then be selected like any other site-specific or generic problem for simulation. Alternatively, if the editor was entered from a simulation run, the user will return to the simulation module, and the new and modified problem description will subsequently be used for the next simulation run.

Running the model

Once a problem description has been loaded from the systems problem data bases, a summary information base will inform the user on the status of this particular example (Figure 5). If the example is ready to run, and no further editing is required, the interactive simulation can be started by selecting the appropriate menu option.

In a first step, the model will set up an optimized mesh representation, calculate the flow field, and then start the transport modeling. In the non-stationary case, the flow field will be recalculated as required.

In each time step, the flow field will be displayed with a vector in each node of the mesh together with the concentration field. For the display of the concentration field, a number of options exist: the model can be stopped at any time. It will,

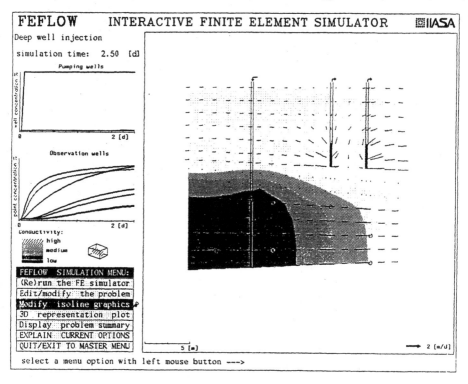

Figure 9 *A vertical representation of deep-well injection in a two-layered acquifer*

however, complete the current time step and its graphical output. If running uninterrupted, the output of subsequent time-steps will result in an animated, dynamic representation of model behavior. Depending on the size and complexity of the example, individual time steps usually only require fractions of a second for computation and display, thus resulting in attractive animation (Figure 6a,b,c).

Once the model is stopped, a number of display options can be used to change the color coding of the concentration field, selecting different resolutions of the output grid, or select an isoline representation (Figure 7). The color coding can also be used to ease direct interpretation, for example using different colors to indicate levels of toxicity or the violation of water quality standards. As another option, the concentration field can be displayed as a pseudo-3D wire-mesh body over the background map and 2D concentration field tilted into the display (Figure 8). The display can be rotated for various viewing angles to provide the most appropriate view of the contaminant plume.

Parallel to the animated 2D representation of the contaminant plume behavior, which can represent both horizontal as well as vertical problems (Figure 9), a set of color coded curves are displayed in separate diagrams, representing concentration over time or break-through curves for the observation wells that may be specified in any example. In addition, concentrations in all the pumping wells of the system are also displayed.

At any point, the user can load the current model into the editor module, and modify the problem definition subject to the constraints listed above. Of course, upon returning to the simulation level, the model will have to be rerun from the start.

The finite difference model

The finite difference model GWSIM (Ground Water Simulation) was implemented as one component in a city-level environmental information system for the city of Hanover, Germany. The basic model, developed at the University of Hanover, was integrated with other components of this system for use by the city's environmental division as well as other branches of the city administration.

Similar to the considerations for the FEFLOW implementation, ease of use and a user friendly interface for users with little or no computer and modeling experience were central to the design. Another objective here was to make the groundwater model output available to several other components of the system through the possibility to export model results in a standardized representation to analysis tools such as a dedicated GIS for overlay analysis.

From the same GIS, topical maps that can incorporate results from other models and data bases in the system can be configured as model background. And the spatially distributed initial and boundary conditions of the model, such as groundwater heads or groundwater recharge, can be treated as any other map set or overlay in the system.

The basic model
Given the use of the model for a single site and by non-modelers, the finite difference

representation with a fixed grid was considered the most appropriate basis for the interactive model. The majority of the input data, including the model geometry but also basic hydrogeological data are given and fixed, so that after the initial testing of the model correct input data can be guaranteed.

The flow model can operate on a rectangular grid in both steady and non-steady modes. Time-variable pumping of wells and variable groundwater recharge can be considered and are no restrictions on the simulation time step. Both confined and unconfined aquifers can be represented.

The basis for the finite difference approach is a diffusion-type partial differential equation (Klenke, 1986). For a large-scale flow problem, where the horizontal dimensions of the aquifer are large compared to the vertical dimension, a two-dimensional simplification is appropriate. Transmissivity and speed of flow are vertically integrated.

$$T_{x\,x} \frac{\partial^2 h}{\partial x^2} + T_{y\,y} \frac{\partial^2 h}{\partial y^2} - Q_g = S \frac{\partial h}{\partial t} \tag{11}$$

where

$h =$ hydraulic head (L)
$T_{x\,x}, T_{y\,y} =$ transmissivity in x and y
$x, y =$ spatial coordinates
$Q_g =$ source/sink function ($L^3\,T^{-1}$)
$S =$ storage coefficient (-)

The aquifer is discretized into rectangular cells, their mid points forming the nodes. Node distances can be variable in both x and y directions. The water budget for a single grid element consists of four horizontal flows Q and a vertical in- or outflow. In a time interval dt in- and outflows will balance with the change in storage in the element:

$$dt(Q_1 + Q_2 + Q_3 + Q_4 + Q_g) = S\,dx\,dy(h_{t=dt} - h_{t=0}) \tag{12}$$

where S is the storage coefficient, relating the amount of water in an element to the hydraulic head. Q_g includes groundwater recharge Q_n, infiltration Q_i, and pumping Q_o.

$$Q_g = Q_n + Q_i - Q_o \tag{13}$$

Horizontal in- and outflows are derived from Darcy's law as

$$Q_1 = dy T_{1\,0} \frac{(h_{i-1,j} - h_{i,j})}{dx}$$

$$Q_2 = dy T_{2\,0} \frac{(h_{i+1,j} - h_{i,j})}{dx}$$

$$Q_3 = dy T_{3\,0} \frac{(h_{i,j-1} - h_{i,j})}{dy}$$

$$Q_4 = dy T_{4\,0} \frac{(h_{i,j+1} - h_{i,j})}{dy} \tag{14}$$

where i,j denote the nodes. $T_{1\,0}$, $T_{2\,0}$, $T_{3\,0}$, and $T_{4\,0}$ are the harmonic mean of transmissivity between the nodes and the neighboring cells, *e.g.,*

$$T_{1\,0} = \frac{2T_{i-1,j}T_{i,j}}{T_{i-1,j} + T_{i,j}} \tag{15}$$

Inserting equation(14) in equation(12), after division by dx, dy yields

$$
\begin{aligned}
Q_o \;+\; &T_{1\,0}\frac{(h_{i-1,j} - h_{i,j})}{dx^2} \\
+\; &T_{2\,0}\frac{(h_{i+1,j} - h_{i,j})}{dx^2} \\
+\; &T_{3\,0}\frac{(h_{i,j-1} - h_{i,j})}{dy^2} \\
+\; &T_{4\,0}\frac{(h_{i,j+1} - h_{i,j})}{dy^2} \\
=\; &\frac{S_{i,j}}{dt}(h_{i,j}(t + dt) - h_{i,j}(t))
\end{aligned}
\tag{16}
$$

A node at an impervious boundary thus has only three inflows; the corresponding transmissivity is set to zero

$$T_1 = 0 \longrightarrow T_{1\,0} = 2\frac{T_1\,T_0}{(T_1 + T_0)} = 0 \tag{17}$$

Boundaries with constant inflow are modeled as impervious with a constant Q_q. Nodes with a constant head are modeled with a very large storage coefficient.

For the simulation of infiltration or exfiltration from surface water bodies the source term Q_g in equation(16) is augmented by another inflow Q_L depending on hydraulic heads:

$$Q_L = L_{i,j}(hr_{i,j} - h_{i,j}) \quad for \quad h_{i,j} > br_{i,j} \tag{18}$$

$$Q_L = L_{i,j}(hr_{i,j} - br_{i,j}) \quad for \quad h_{i,j} < br_{i,j} \tag{19}$$

where

hr = head of the river (L)
br = height of the riverbed (L)
L = the leakage factor:

$$L = \frac{k_r}{d_r} \frac{A}{dx \; dy} \tag{20}$$

where
$\quad k_r$ = transfer coefficient (L T^{-1})
$\quad d_r$ = thickness of the river bed (L)
$\quad A$ = area per element (L^2)
$\quad dx, \; dy$ = dimension of the element (L)

For an unconfined aquifer, the extent of the saturated zone can be calculated in every time step from the hydraulic head and the aquifer base. Flow velocities between nodes are obtained by

$$vx_{i,j} = -kf_x(h_{i+1,j} - h_{i,j}) \tag{21}$$

$$vy_{i,j} = -kf_y(h_{i,j+1} - h_{i,j}) \tag{22}$$

Transport modeling

The simulation of contaminant transport is directly coupled with the flow model. The method used is random-walk particle tracing, which allows the basic fixed grid to be retained independent of local sources of contaminants and consequent concentration gradients that would require local refinement of the mesh in both finite-element and finitedifference approaches.

An instantaneous injection of a mass M of an ideal tracer in $x = 0$ in an infinite aquifer will lead to a concentration distribution of

$$C = \frac{C_0}{\sqrt{4\pi \; \beta_L \; v \; t}} \exp\left(-\frac{(x - v \; t)^2}{4 \; \beta_L \; v \; t}\right) \tag{23}$$

$\quad C_x$ = concentration of chemical species (M L^{-3})
$\quad C_0$ = starting concentration
$\quad \beta_L$ = coefficients of longitudinal dispersivity (L)
$\quad x$ = distance from starting point (L)

which is a normal distribution with mean $x = vt$ and a standard deviation of $\sigma = (2\beta_L vt.)^{1/2}$.

This distribution can be generated stochastically. A number of particles are introduced at $t = 0$ in $x = 0$. The path of a particle consists of the deterministic, convective component dx and the stochastic component xz.

$$x = dx + x_z = v \; t + Z\sqrt{2 \; \beta_L \; v \; t} \tag{24}$$

where Z is a normally distributed random number with mean 0 and a standard deviation of 1. The resulting distances travelled are again normally distributed with a mean of $v\,t$ and a standard deviation of $(2\,b_L\,v\,t)^{1/2}$. This can easily be extended to 2 or 3 dimensions.The resulting concentration in a cell is simply determined by counting the number of particles with mass M_p in the cell.

$$C_{i,j} = \frac{N_{i,j}\,M_p}{dx\,dy\,M_{i,j}\epsilon_o} \tag{25}$$

where

ϵ_o = effective porosity (-)

$M_{i,j}$ = aquifer thickness (L)

Sorption and first-order decay can be included in this description. Also, due to the nature of the source description, arbitrary geometries for sources other than points, lines, circles or rectangles can be simulated. Another advantage of the approach is the possibility offered to superimpose individual solutions on a larger representation for the entire city area.

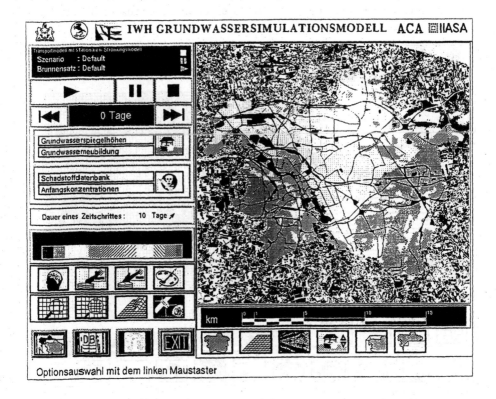

Figure 10 *Icon menu and the background map of the area simulated, showing a combination of a SPOT satellite scene, land use map, and a geological map.*

Model implementation

The FD model has been implemented as a major module of the prototype environmental information system of the city of Hanover. The model is directly accessible from the top-level master menu of the system; if called after a topical (background) map has been constructed in the GIS and map editor component, this map is loaded for use with the groundwater model.

The model itself provides a set of hierarchical iconic menu options, and a large display of the map and model area (Figure 10). The icon menu offers options for problem definition, running the model, and display configuration. Several of the icon options can trigger submenus that again offer a number of logically related options such as a contaminant source editor, a well editor, and an "expert window" that allows the modification of numerical model control parameters. In addition, direct access to the GIS level is provided from the main groundwater menu level.

The latter option provides access to the GIS and map editor, so that an appropriate background map can be constructed or modified even during a model run, responding to the model behavior. The map editor offers a set of topical maps, including satellite images, and themes such as land use, biotopes, or geology, cell grid or raster files that represent pre-processed observation data or possibly the

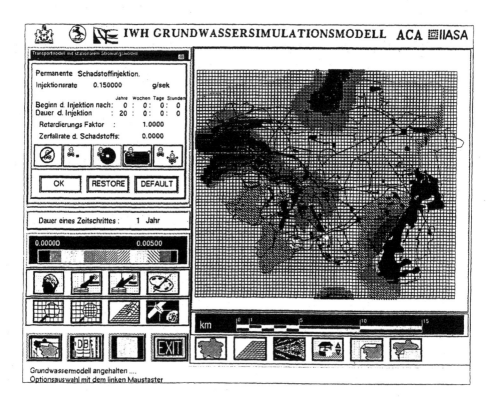

Figure 11 *Interactive editor for the definition and location of contaminant sources.*

output from other models in the overall information system as well as the groundwater model itself. The GIS and map editor not only allows model input data sets (initial and boundary conditions) such as hydraulic head or groundwater recharge to be displayed, but also spatially distributed geohydrological parameters.

From any or all of these maps, selected features can be combined into a new map to provide an appropriate background for a specific simulation run. As an example, it may be of interest to compare a lowering in the groundwater table due to intensive pumping with the location of wetlands and the possibly affected surface water bodies. By loading an appropriate map from a combination of selected features of the land use, biotope, or geological maps, problem-relevant background information can be combined with the model results.

Whenever the model is entered from the master menu level, it is ready to run. In the default case, a combined steady flow and transport model is used. Initial and boundary conditions are set to their default values, and the user can now define a contaminant source in terms of injection rates, substance properties (first-order decay rate), and shape and location of the source (Figure 11). If no source is specified by the user, only the flow model will be run for the default groundwater recharge and

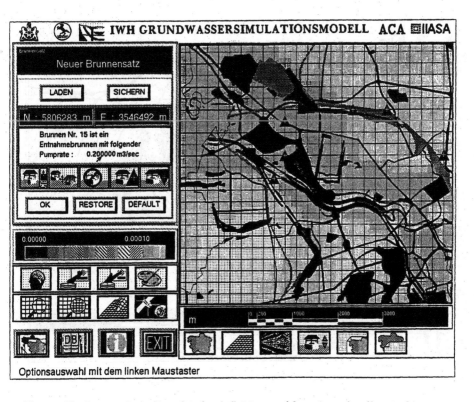

Figure 12 *Interactive editor for the definition and location of wells, working on a zoomed-in sub-area of the model grid.*

well data sets. Selected initial and boundary conditions, such as the hydraulic head or the groundwater recharge, can be displayed as another overlay plane over the map.

A special well editor allows wells to be positioned on the grid, and their pumping rates to be set (Figure 12). Well types are selected from an icon menu, and dragged to their location on the map with the mouse; while moving the well symbol over the map, its current coordinates are displayed and dynamically updated. Alternatively, a well location can be defined by entering its coordinates directly. Existing wells can be picked from the map display for relocation or a modification of their characteristics. Pumping rates can simply be modified with both the mouse buttons or by direct keyboard entry.

To run the model, a stylized tape deck is used, which makes it possible to run the model, pause, halt, rewind, or skip forward (in the case of the non-steady flow model). For example, at any time step, the model can be paused, and the display parameters can be adjusted to provide an optimum representation of the current situation. This can include the selection of an appropriate color gradient, using color ramps from light blue to dark blue, blue to red, back to white, or a contrast enhancing "rainbow", toggling layers of the background map on or off, including displaying or hiding the satellite map, displaying the model grid for reference, or zooming into the

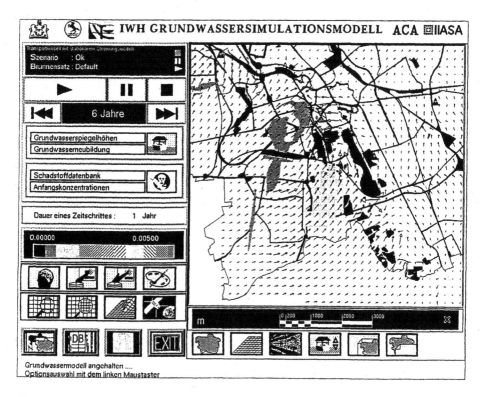

Figure 13 *Model output for a sub-area, showing the flow in a concentration field moving with time.*

display for higher, local resolution. As an additional option, a display of the flow field can be activated (Figure 13).

For the display of a concentration field, the association of the color ramp to the range of concentrations can be modified by interactively setting the upper and lower concentration value for the ramp, thus shifting and scaling it according to the actual current distribution of calculated concentrations. This allows optimization of the color "resolution" in the display.

For the expert modeler, a second layer of editing options is provided with an "expert button" that allows model control parameters to be set and the basic model scenario to be selected. This includes control parameters such as the number of iterations, maximum errors, or the number of particles for the particle tracking method. These parameters can be used to tune the speed versus accuracy performance of the model.

Finally, at any stage, the model output, *i.e.*, hydraulic head and contaminant concentration can be stored on disk as a topical map overlay. These overlays are then available in the GIS system for further analysis in conjunction with other map sets, observation data, or model results from other components in the overall environmental information system.

Discussion

Many computer-based groundwater models and methods of analysis are potentially very useful. A large amount of formal, mathematical and computational methods have been developed in this area of water resources planning and management, and groundwater simulation in particular, and the field has a considerable history in the use of computers. To turn a potentially useful method into one actually used however, requires a number of special features as well as an approach that takes psychological and institutional aspects, as well as scientific and technical ones into account.

Advanced information technology provides the tools to design and implement smart software, where in a broad sense, the emphasis is on the man–machine interface (Fedra,1992;1993a,b). Integration, interaction, intelligence, visualization and customization are key concepts that characterize the systems discussed above.

Tools that are easy to use, equipped with a friendly user interface, use problem-adequate representation formats and a high degree of visualization, customized for an institution and its specific view of problems, and that are developed in close collaboration with the end user, stand a better chance of being used than tools that are based on "only" good science. Good science is a necessary, but certainly not sufficient, condition for a useful and usable information and decision support system; there are definite advantages to user participation, with consideration of questions of maintenance and the update of information requirements from the very beginning.

Integration implies that in any given software system for real-world applications, more than one problem representation form or model, several sources of information or data bases, and finally a multi-faceted and problem-oriented user interface ought to be combined in a common framework to provide a useful and realistic information base. In particular, the integration of transparent data handling, GIS components, and CAD elements with a basic simulation models extends its usefulness and usability considerably.

Interaction is a central feature of any effective man--machine system: a real-time dialogue allows the user to define and explore a problem incrementally in response to immediate answers from the system; fast and powerful systems with modern processor technology can offer the possibility to simulate dynamic processes with animated output, and they can provide a high degree of responsiveness that is essential to maintain a successful dialogue and direct control over the software.

Intelligence requires software to be "knowledgeable" not only about its own possibilities and constraints, but also about the application domain and about the user, *i.e.,* the context of its use. Defaults and predefined options in a menu system, sensitivity to context and history of use, learning, or alternative ways of problem specification, can all be achieved by the integration of expert systems technology in the user interface and in the system itself.

Visualization provides the band-width necessary to understand large amounts of highly structured information, and permits the development of an intuitive understanding of processes and interdependencies, of spatial and temporal patterns, and complex systems in general. Many of the problem components in a real-world planning or management situation are rather abstract: representing them in a symbolic, graphical format that allows visual inspection of systems behavior, and, in general, symbolic interaction with the machine and its software is an important element in friendly and easy-to-use computer-based systems.

Customization is based on the direct involvement of the end-user in systems design and development. It is his view of the problem and his experience in many aspects of the management and decision making process that the system is designed to support. This then must be central to a system's implementation to provide the basis for user acceptance.

The examples presented above share these features and have one common design goal making it easier for the model user to concentrate on the problem, to provide a representation that is directly understandable, and to automate as many as possible of the preparation and interpretation tasks necessary for running a complex mathematical model.

The model itself, its data requirements, the numerical methods used, are largely hidden by the interface. Integration of data handling, interaction and visualization provide a more accessible view of the problem, that lends itself to a more experimental use, but also by a potentially larger group of users.

This, however, introduces a possible problem: how to avoid misuse and misinterpretation of the model result. It is certainly possible to restrict most of the user inputs to reasonable ranges, check specifications for basic consistence and plausibility, and guide the less experienced user in a more tutorial style. We can assume that our user is at least familiar with the nature of the problem and the basic concepts of simulation modeling. Basic background information is also provided in the help and explanatory functions on-line.

For the expert user, ease and efficiency of use are important considerations. The various dedicated editing features and the transparent data handling provide a very efficient tool to set up and refine a problem. At the same time, the graphical representation of the model behavior can provide invaluable diagnostic information: most problems are immediately obvious if one can only see them. The expert user, of course, may want access to model control parameters that the casual user does not want or need to modify. For him two ready-to-run scenario templates such as screening and high-resolution may be sufficient. The expert user, however, may want more degrees of freedom to tune model performance versus precision. Here the two-level interface provides a possible solution. In addition, the default configurations and set-ups are completely data driven, and these input and configuration files can of course be edited by the knowledgeable user, if necessary.

At the other extreme, the integration of more expert systems technology will guide the user in the problem specification process through rule-based deduction options: wherever the system requires a numerical parameter input, it could also use a more refined set of defaults such as porosity taken from a catalog of soil and substrate types, or circumstantial evidence to deduce an appropriate value. First attempts to link such parameter estimation methods to, for example, geological maps, seem quite promising.

As an implementation related consideration, the use of X11 and Xlib for the graphical user interface makes it possible to run the models in a network. While relatively cost-effective machines can display the graphics, the most powerful machine in the system can run the actual numerical model as a compute-server, thus often dramatically increasing the performance of the overall system. This structure clearly lends itself to a more extensive use of distributed and parallel processing for future implementations.

References

Abriola, L.M. and Pinder, G.F. 1985. A Multiphase Approach to the Modeling of Porous Media Contamination by Organic Compounds. 1. Equation Development. 2. Numerical Simulation. *Water Resources Research*, 211., 11–18 and 19–26.

Aziz, K. and Settari, A. 1979. Petroleum Reservoir Simulation. Applied Science Publishers.

Bachmat, Y., Bredenhoeft, J., Andrews, B., Holtz, D. and Sebastian, S. 1980. Groundwater Management: *The Use of Numerical Models Water Resources*, Monograph. 5. American Geophysical Union. Washington, D.C., 127p.

Bear, J. 1979. Hydraulics of Groundwater. McGraw Hill, New York.

Diersch, H-J. 1988. Finite Element Modeling of Recirculating Density-driven Saltwater Intrusion Processes in Groundwater. *Advances in Water Resources*, 111., 25–43.

Diersch, H-J. 1990. *Modeling NAPL Flow and Transport Processes in Porus Media* in preparation..

Diersch, H-J. and Kaden, S. 1984. Contaminant Plume Migration in an Aquifer: Finite Element Modeling for the Analysis of Remediation Strategies: A Case Study. CP-84-11. International Institute for Applied Systems Analysis, A-2361 Laxenburg, Austria. 68p.

Diersch, H.-J. 1984. Modeling and Numerical Simulation of Geohydrodynamic Transport Processes. Post-doctoral Thesis Habilitation.. Academy of Sciences, Berlin. 284p.

Fedra, K. 1992. Intelligent Environment Information Systems. In: Vortragge Wasserbau

Symp., Wintersemster 1991-92. Okologie und Umweltvertraglichkeit. Mitteilungen 85. 22nd Int. Hydrological Engineering Conf., 3-4 January 1992. Technical University, Aachen.

Fedra, K. 1993a. GIS and environmental modeling. RR-94-2. International Institute for Applied Systems Analysis, A-2361, Laxenburg, Austria. Reprinted from Goodchild, M.F., Parks,B.O. and Steyaert, L.T. eds.: *Environmental Modeling with GIS*. Oxford University Press.

Fedra, K. 1993b. Models, GIS and expert systems: integrated water resource models. In: Kovar, K. and Nachtnebel, H.P. eds., *Application of Geographic Information Systems in Hydrology and Water Resources Management*. Proc. Int. Conf. Vienna, Austria, 19-22 April 1993. IAHS Publ. No.211, pp. 297-308.

Fedra, K. and Diersch, H-J. 1989. Interactive Groundwater Modeling: Color Graphics, ICAD and AI. In: Sahuquillo, A., Andreu, J. and O'Donnell, T. eds.., *Proc. Int. Symp. on Groundwater Management: Quantity and Quality*. October 2-5, 1989. Benidorm, Spain. IAHS. Publication No. 188, pp. 305-320.

Forsyth, P.A. 1988. *Simulation of Nonaqueous Phase Groundwater Contamination.*

Frind, E.O. 1987. *TR3--Three Dimensional Advective--Dispersive Transport.* University of Waterloo. Ontario. Canada.

Hassanizadeh, M. and Gray, W.G. 1979. General Conservation Equations for Multiphase Systems. *Adv. Water Res.*, 2, 131–144 and 191–203.

Huyakorn, P.S. and Pinder, G.F. 1983. *Computational Methods in Subsurface Flow.* Academic Press.

Kaluarachchi, J.J., Parker, J.C. and Lenhard, R.J. 1990. A Numerical Model for Areal Migration of Water and Light Hydrocarbon in Unconfined Aquifers. *Advanced Water Resources,* 131., 29–40.

Kinzelbach, W. and Rausch, R. 1989. *ASM–Aquifer Simulation Model. Gesamthochschule.* Kassel Universität, FRG.

Klenke, M. 1986. *Numerische Modelltechnik in der Grundwasserhydrologie.* (Numerical modeling techniques in groundwater hydrology.) Mitteilungen des Institutes für Wasserwirtschaft, Hydrologie und landwirtschaftlicher Wasserbau. Heft 59. Hannover (in German).

McDonald, M.G. and Harbaugh, A.W. 1984. *MODFLOW–A Modular Three-dimensional Finite-difference Groundwater Flow Model.* US Geological Survey. Reston, VA.

Nguyen, V.V., Gray, W.G., Pinder, G.F., Botha, J.F. and Crerar, D.A. 1982. A Theoretical Investigation on the Transport of Chemicals in Reactive Porous Media. *Water Resources Research*, 18(4), 1149–1156.

Pinder, G.F. 1973. A Galerkin Finite Element Simulation of Groundwater Contamination on Long Island. *Water Resources Research*, 96, 1657–1669.

Pinder, G.F. and Gray, W.G. 1977. *Finite Element Simulation in Surface and Subsurface Hydrology.* Academic Press, New York.

8 Modeling contaminant transport and biodegradation in groundwater

Hanadi S. Rifai and Philip B. Bedient

Department of Environmental Science and Engineering, Rice University, PO Box 1892, Houston, Texas 7725, USA

Abstract

The field of groundwater transport modeling has grown tremendously over the past ten years. This is mostly due to the need for quantitative estimates of contaminant mass transport in the subsurface. Transport processes in groundwater include advection, dispersion, adsorption, decay, chemical and biological reactions. This chapter deals with modeling biodegradation, the breakdown of organic contaminants in the subsurface by microorganisms. The main focus of the chapter is on how to incorporate microbially mediated reactions into groundwater contaminant transport models. At the field scale, the transport characteristics of an aquifer limit the delivery of the required nutrients and dissolved oxygen to the microbial population, and therefore it is necessary to incorporate these limitations into a model for biodegradation in the subsurface.

The emphasis in this chapter is on BIOPLUME II, a biodegradation model developed recently. The BIOPLUME II model utilizes a stoichiometric instantaneous reaction between oxygen and the organics to simulate biodegradation. It is assumed that microbial kinetics occur at a much faster time scale than groundwater flow, and that the biodegradation of the organic contaminants is limited by the mass flux of dissolved oxygen into the contaminated plume. The biodegradation input data requirements for the model are simplified and include a quantitative estimate of dissolved oxygen sources in an aquifer. The BIOPLUME II model was developed by modifying an existing two-dimensional transport model from the USGS. The main modifications included simulating the transport of two substrates in the groundwater: oxygen and the organics.

This chapter covers briefly the numerical methods that are utilized in the BIOPLUME II model as well as the input parameters required for the model. Several case studies of modeling biodegradation at field sites are included in addition to a discussion of other models that have been developed to simulate the biodegradation process. The chapter also includes a summary of some of the basic principles that govern the biodegradation of organics in the subsurface and the application of bioremediation as a technology for groundwater clean-up efforts.

Introduction

The field of groundwater transport modeling has grown tremendously over the past ten years. This is mostly due to the need for quantitative estimates of contaminant mass transport in the subsurface. Many articles and books have been written about the science of modeling. The reader is referred to Mercer and Faust (1981), Wang and Anderson (1982) and the National Research Council book on *Groundwater Models: Scientific and Regulatory Applications* (1990) as a starting point.

A groundwater transport model is a tool designed to represent a simplified version of reality (for example, a laboratory column or a field site). There are several types of models. Mathematical models consist of a set of differential equations that describe certain phenomena. The phenomenon which is of a specific interest for this chapter is biodegradation, the breakdown of organic contaminants in the subsurface by microorganisms. Current models of biodegradation in groundwater will be discussed with emphasis on the BIOPLUME II model (Rifai *et al.*, 1988).

The goal of modeling is to predict the value of the unknown variable, for example, the concentration at a certain point in an aquifer. Before a model can be used for predictive purposes, it should be validated and calibrated. Model validation can be accomplished by comparing the model results to an analytical solution of a known problem. Model calibration is quite subjective and involves using the model to simulate observed data from a laboratory experiment or at a field site.

Transport processes
Transport processes in groundwater include advection, dispersion, adsorption, decay, chemical and biological reactions. The first two mechanisms have been analyzed in some detail for both laboratory and field conditions, while the latter four processes are the focus of current research efforts. The incorporation of biological reactions into groundwater model formulations is the main topic for this chapter. Detailed discussions of modeling principles for advection, dispersion, adsorption and chemical reactions can be found in Bredehoeft and Pinder (1973), Fried (1975), Anderson (1979), Bear (1979), Freeze and Cherry (1979), and DeMarsily (1987).

Advection is the movement of a contaminant with the bulk fluid according to the seepage velocity in the pore space (Bedient *et al.*, 1985). Figure 1 shows the breakthrough curve in one-dimension for solute transport due to advection only.

Advection can be quantitatively described using an average seepage velocity, V (L/T)

$$V = -\frac{K}{n}\frac{dh}{dl} \tag{1}$$

where K is the hydraulic conductivity (L/T), n is the porosity and dh/dl is the gradient (L/L).

Dispersion is due mainly to heterogeneities in the medium that cause variations in flow velocities and flow paths. Figure 1 shows the spreading out of the contaminant front due to dispersion in porous media.

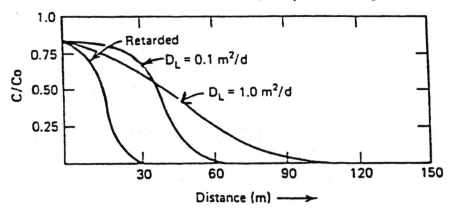

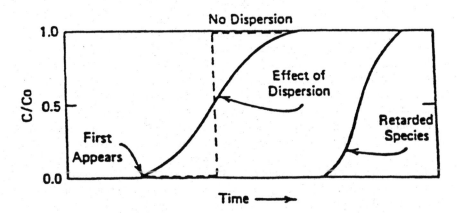

Figure 1 *Breakthrough curves in 1-D showing effects of dispersion and retardation.*

Dispersive flux in a uniform flow field with average velocity V is given by:

$$f_x = -n D_L \frac{\partial C}{\partial x} \tag{2}$$

$$f_y = -n D_T \frac{\partial C}{\partial y} \tag{3}$$

where D_L, D_T are longitudinal and transverse dispersion coefficients (L^2/T), n is the porosity, and C is the concentration of the contaminant. The dispersion coefficients D_L and D_T are given by:

$$D_L = \alpha_L V \qquad (4)$$

$$D_T = \alpha_T V \qquad (5)$$

where α_L and α_T are the longitudinal and transverse dispersivities (L), respectively. The reader is referred to Smith and Schwartz (1980), Gelhar *et al.* (1979) and Anderson (1979) for quantitative estimates of dispersivity.

Sorption is the tendency of organics to adhere to soil particles. The concept of the isotherm is used to relate the amount of contaminant, S, partitioning into the soil to the concentration in solution, C. The freundlich isotherm is one of the most commonly used forms:

$$S = K_d C^b \qquad (6)$$

where S is the mass of solute sorbed per unit bulk dry mass of porous medium, K_d is the distribution coefficient (V/M), and b is an experimentally derived coefficient. If b=1, equation (6) is known as a linear isotherm.

The partition coefficient, K_d, can be measured in static batch experiments or in dynamic column tests; however the "hydrophobic theory" (Karickhoff *et al.*, 1979 and Schwartzenbach and Westfall, 1981) provides a method to estimate this coefficient indirectly, within a factor of 2. For a neutral hydrophobic compound, one that is soluble in water, but more soluble in an organic solvent, a dimensionless partition coefficient, K_{ow}, between water and n-octanol is first measured.

A provisional dimensionless partition coefficient, K_{oc}, is defined for a hypothetical soil made of 100 percent of solid organic material. This coefficient, K_{oc}, is strongly correlated to K_{ow} for a given compound. Schwartzenbach and Westfall (1984) proposed a linear regression of the form:

$$\log K_{oc} = a \log K_{ow} + b \qquad (6a)$$

Values of a and b have been tabulated by Schwartzenbach and Westfall (1984). The distribution coefficient, K_d, is given by:

$$K_d = K_{oc} \cdot f_{oc} \qquad (6b)$$

where f_{oc} is the dimensionless fraction of dry weight of sediment which is made of solid organic carbon compound.

The linear isotherm can be incorporated into the one-dimensional advection-dispersion equation as follows:

$$D_L \frac{\partial^2 C}{\partial x^2} - \overline{V} \frac{\partial C}{\partial x} - \frac{\rho b}{n} \frac{\partial S}{\partial t} = \frac{\partial C}{\partial t} \qquad (7)$$

where ρ_b is the bulk dry mass density (M/V), n is the porosity, and

$$-\frac{\rho b}{n}\frac{\partial S}{\partial t} = -\frac{\rho b}{n}\frac{dS}{dC}\frac{\partial C}{\partial t} \tag{8}$$

For the case of the linear isotherm, $(dS/dC) = K_d$, and

$$D_L\frac{\partial^2 C}{\partial x^2} - \overline{V}\frac{\partial C}{\partial x} = \frac{\partial C}{\partial t}\left(1 + \frac{\rho_b}{n}K_d\right) \tag{9}$$

or finally,

$$\frac{D_L}{R}\frac{\partial^2 C}{\partial x^2} - \frac{\overline{V}}{R}\frac{\partial C}{\partial x} = \frac{\partial C}{\partial t} \tag{10}$$

where R is the retardation coefficient equal to $1 + \rho b/n \cdot K_d$. The effect of retardation on contaminant breakdown is shown in Figure 1.

Governing transport equation

The differential equation for simulating groundwater transport in two dimensions can be written as follows:

$$\frac{\partial}{\partial x_i}\left(D_{ij}\frac{\partial C}{\partial x_j}\right) - \frac{\partial}{\partial x_i}(CV_i) - \frac{C_oW}{nb} + \Sigma R_k = \frac{\partial C}{\partial t} \tag{11}$$

where C is the concentration of the organic (M/L^3), V_i is the seepage velocity (L/T), D_{ij} is the dispersion coefficient (L^2/T), C_o is the solute concentration in source or sink fluid (M/L^3), R_k is the rate of addition or removal of solute due to chemical and biological reactions (M/L^3T) and n is the porosity.

For the purposes of this chapter, the discussion will focus on the term R_k and how it can be formulated to simulate biodegradation. Biodegradation can be incorporated into a transport groundwater model by adding R_k to the transport equations.

Analytical and numerical models

Analytical models are developed by solving the transport equation for certain simplified boundary and initial conditions. Numerous solutions are available in the literature (Bear, 1979; Hunt, 1978; Wilson and Miller, 1978; Cleary and Adrian, 1973; Shen, 1976) for pulse and continuous contaminant sources with boundary conditions ranging from homogeneous confined aquifers to infinite-thickness water table systems. Processes that may be included in these models are advection, dispersion, sorption, and decay. Analytical solutions generally require simple geometries and boundary conditions, but do provide useful insights to many groundwater contaminant problems.

The adverse consequences of contamination can be analyzed if the concentration of a contaminant can be reliably estimated for a given location and time. Two very useful tools for summarizing the variation of a contaminant in space and time are location and arrival time distributions. In a series of papers, Nelson (1977) introduced the concept of arrival distribution and presented a technique for developing those distributions based on determining the travel time along individual flow lines. Variation in the arrival distribution is the result of convergence and divergence of flow lines. Nelson's procedure does not account for dispersion directly, but the arrival distribution concept is useful because it uses the actual flow field to predict transport processes.

The two-dimensional analytical model developed by Wilson and Miller (1978) is one of the simplest to use and can account for lateral and transverse dispersion, sorption, and first order decay in a uniform flow field. Contaminants are assumed to be injected uniformly throughout the vertical axis. The flow velocity must be obtained from a flow model or from detailed field monitoring. The x axis is oriented in the direction of flow.

Numerical or computerized solutions of the transport equation in two dimensions are the most plentiful and commonly used techniques. These solutions are generally more flexible than analytical solutions because the user can approximate complex geometries and combinations of recharge and withdrawal wells by judicious arrangement of grid cells.

The general method of solution is to break up the flow field into small cells, approximate the governing partial differential equations by differences between the values of parameters over the network of time t, then predict new values for time t + Δt. This continues forward in time in small increments Δt. The most common mathematical formulations for approximating the partial differential equation of solute transport are the methods of finite difference, finite element, and characteristics.

The earlier finite-difference methods operate by dividing space into recti-linear cells along the coordinate axes. Homogeneous values within each cell are represented by values at a single node. Partial differentials can then be approximated by differences and the resulting set of equations solved by iteration (Mercer and Faust, 1981; Carnahan *et al.*, 1969; Prickett, 1975). Approximating the differentials by a difference requires neglecting remaining terms, which results in truncation error. Finite-difference models have been developed for a variety of field situations including saturated and unsaturated flow, and for transient and constant pollutant sources. The primary disadvantage of these methods is that the truncation error in approximating the partial differential equations can result in error of the same order of magnitude as does the physical dispersion process (Anderson, 1979). More details about the finite difference method are included below.

The finite element method also operates by breaking the flow field into elements, but in this case the elements may vary in size and shape. In the case of a triangular element, the geometry would be described by the three corner nodes where heads and concentrations are computed. The head or concentration within an element can vary in proportion to the distance to these nodes. Sometimes complex interpolating schemes are used to predict parameter values accurately within an element and thereby reduce the truncation errors common in finite difference procedures. Some numerical dispersion may still occur but is usually much less significant. The use of variable size and shape elements also allows greater flexibility in the analysis of moving boundary problems such as occur when there is a moving water table or when contaminant and flow transport must be analyzed as a coupled problem. A disadvantage of the finite element method is the need for formal mathematical training to understand the procedures properly. Finite element methods generally have higher computing costs (Pinder and Gray, 1977; Pinder, 1973; Wang and Anderson, 1982).

The method of characteristics (MOC) is most useful where solute transport is dominated by advective transport. The most common procedure is to track idealized particles through the flow field. In step one a particle and an associated mass of contaminant is translated a certain distance according to the flow velocity. The second step adds on the effects of longitudinal and transverse dispersion. This procedure is computationally efficient and minimizes numerical dispersion problems (Konikow and Bredehoeft, 1978). The method of characteristics is also discussed in section 5.1.

Biodegradation processes

The biodegradation process is simply a biochemical reaction which is mediated by microorganisms. The biodegradation process and microbial dynamics are described in more detail in the next section . In general, an organic compound is oxidized (loses electrons) by an electron acceptor which in itself is reduced (gains electrons). Several electron acceptors have been identified: oxygen (O_2), nitrate (NO_3^-), sulfate (SO_4^{2-}) or carbon dioxide. The utilization of oxygen as an electron acceptor is termed aerobic biodegradation and that of nitrate is called anaerobic biodegradation. An example of the aerobic biodegradation reaction for benzene is given by:

$$C_6H_6 + 7.5\ O_2 \rightarrow 6\ CO_2 + 3\ H_2O \qquad\qquad (12)$$

The stoichiometric equation (12) is mediated by microorganisms and results in the breakdown of benzene into carbon dioxide and water and the formation of biomass.

The technology is not novel, the biodegradation potential of organic contaminants has been recognized and utilized in the wastewater treatment process for years. Suspended growth processes include activated sludge reactors, lagoons, waste stabilization ponds, and fluidized bed reactors. Fixed film processes include trickling filters, rotating biological discs, and sequencing batch reactors (Lee *et al.*, 1988).

The ultimate goal of the biodegradation process is to convert organic wastes into biomass and CO_2, CH_4, and inorganic salts. Two essential criteria must be in place before biodegradation can occur. First the subsurface geology must have a relatively large hydraulic conductivity to allow the transport of oxygen and nutrients through the aquifer. Second microorganisms must be present in sufficient numbers and types to degrade the contaminants of interest. Early studies such as that presented by Dunlap and McNabb (1973) had verified the presence of active microbial populations in the subsurface. The microbial populations were metabolically active, and often nutritionally diverse.

There are several factors that may limit or inhibit the biodegradation of subsurface organic pollutants, even in the presence of microorganisms. These factors include the concentration of the contaminant, the presence of toxicants, sorption, pH, and temperature (Lee *et al.*, 1988). The lack of oxygen and other inorganic nutrients such as phosphorus and nitrogen has been extensively documented by several researchers to be limiting factors to biodegradation in laboratory and field scale experiments (Alexander, 1980; Lee and Ward, 1985a).

At the field scale, the transport characteristics of the aquifer limit the delivery of the required nutrients to the microbial population. In order to quantitatively estimate the attenuation due to biodegradation, it is necessary to estimate the amount of organics and nutrients that would be available for metabolism by the microorganisms. This can only be achieved through modeling because one needs to simulate the transport of contaminants and nutrients in the subsurface combined with microbial kinetics.

Applications of models

The use of groundwater models has increased tremendously in the past few years due to the complexity of the subsurface environment and the chemical, physical, and biological processes that take place in aquifers. There are two main application areas for which several groundwater models have been developed: (1) resource management and (2) contaminant transport and remediation. Groundwater flow models can be used to manage water supply systems and to predict the effects of pumping on aquifer levels. Groundwater transport models can be used to simulate the migration and fate of contaminants in aquifers. Recently, transport models have been used in a predictive mode to design remediation systems.

Significance of modeling

The significance of models is to:
(1) Test a hypothesis, for example, the BIOPLUME II model was used to test the hypothesis that biodegradation is more efficient in aquifers with relatively high hydraulic conductivities.
(2) Determine the behavior of a system under a variety of conditions, for example, determine the effects of pumping on a given groundwater system.
(3) Understand physical, chemical or biological processes such as the sorption of chemical contaminants to the soil matrix in an aquifer.
(4) Design a remediation system at a field site.
(5) Predict future conditions at a field site using a variety of scenarios for remediation.

The reliability of predictions from a groundwater model depends on how well the model approximates the field situation. In some cases, simplifying assumptions have to be made in order to simulate a complex field situation. For example, many analytical models require that the aquifer be homogeneous and isotropic.

Data requirements for models

The data required for a groundwater flow or transport model are a function of the type of model: analytical or numerical. Analytical models require few input values. The input data in an analytical model do not vary spatially or temporally. The input data for numerical models are more extensive and may vary spatially over the grid system. Numerical models allow more detailed description of a field site and the different processes that may be taking place.

Typical input requirements for numerical flow models include a definition of the geology and boundary conditions at the site. Parameters such as the hydraulic conductivity, thickness of the aquifer, and porosity are required. In addition a vertical definition of the connectivity between different aquifers is required if the numerical model is a three-dimensional model.

For transport models, additional parameters that describe the physical, chemical and biological properties of the contaminant are necessary. A history of the chemical release in terms of when and how much is also an input requirement. Water table conditions and contaminant concentrations at monitoring wells are examples of data that are used in simulating contaminant transport.

As mentioned previously, models have several uses. Good field data are essential when using a model for simulating existing contaminant conditions at a site or when using a model for predictive purposes. However, an attempt to model a system with insufficient data may be useful because it may serve as a method for identifying those areas where detailed field information needs to be collected.

Limitations of models

Mathematical models have several limitations which can be conceptual or application-related. Conceptual limitations are those which relate to representation of the actual process or system with a mathematical model. For example, analytical models are limited by the simplifying assumptions that are required to develop the solution. The analytical models that are available are limited to certain idealized conditions and may not be applicable to a field problem with complex boundary conditions. Another limitation of analytical models is that spatial or temporal variation of system properties such as permeability and dispersivity cannot be handled (Javandel *et al.*, 1984).

Application-related limitations have to do with the solution procedure that is utilized in the model development or with the amount of effort required to implement the solution. For example, the approximation of a differential equation by a numerical solution introduces two types of errors: numerical and residual. The numerical errors are due to the solution method used in solving the differential equations. The residual errors are a result of approximating the differential equation with a series of mathematical expressions.

Numerical models are also limited by their complexity so that the user has to have a certain level of knowledge to be able to apply those models. Achievement of the required familiarity level is time consuming and could be prohibitive when funding is limited or when dealing with a time constraint. Preparation of input data for numerical models often takes a long time. Computer efficiency, *i.e.*, relatively fast computers are necessary when using a numerical model.

Biodegradation of organic contaminants in the subsurface

Prior to developing expressions that can simulate biodegradation in the subsurface, it is important to develop an understanding of the microbial dynamics and their response to the specific environmental factors at a site. The following discussion is a brief overview that outlines some of the principles that should be investigated in more detail using laboratory studies.

Microbial dynamics
The growth of a single micro-organism in an unlimited environment, *i.e.*, an environment which provides all the nutrients required for growth in unlimited quantities, can be described using the following model:

$$\frac{dx}{dt} = \mu x \qquad (13)$$

where μ is the specific growth rate and is a measure of the amount of new biomass (or individual organisms) produced by unit amount of existing biomass in unit time (the units for μ are g new biomass per g of existing biomass per hour), x is the population size and (dx/dt) is the rate of change of population size. Equation (13) can be integrated and rearranged to yield the following expression for estimating xt, the population size x at time t

$$x_t = x_0 \, e^{\mu t} \qquad (14)$$

The value of the specific growth rate depends on the type of micro-organism and the exact physicochemical conditions under which it is growing. In an optimum environment with all the growth requirements in excess, the rate of growth of the population is at a maximum, and so μ becomes the maximum specific growth rate, μ_{max}.

The growth of micro-organisms in a limited environment is depicted in Figure 2. The initial phase of the closed growth cycle, known as the *lag phase*, shows no detectable increase in population size and so the specific growth rate is zero. The *acceleration phase* marks the beginnings of a gradual increase in the specific growth rate. The *exponential phase* may be described by the basic growth equations. The retardation phase is established once the specific growth rate begins to decline. In the *maximum population* phase, the population is metabolically active, but active growth stops. During the *death phase* the numbers of micro-organisms decline.

One of the most used models that describe growth in a closed environment is the logistic model. This model may be used from the beginning of the exponential phase through the maximum population phase:

$$\frac{dx}{dt} = \mu_{max} \cdot x \left(1 - \frac{x}{x_f}\right) \tag{15}$$

where x_f is the population size in the maximum population phase. The term $(1 - x/x_f)$ models the decrease in the rate of increase of the population as the population density increases. When the population density, x, is small at the beginning of exponential

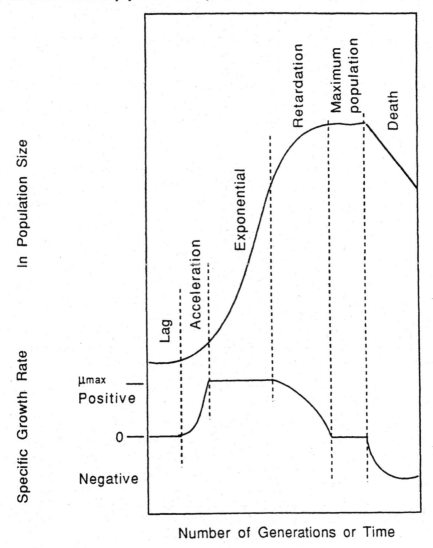

Figure 2 *Growth phases of microorganisms in a limited environment.*

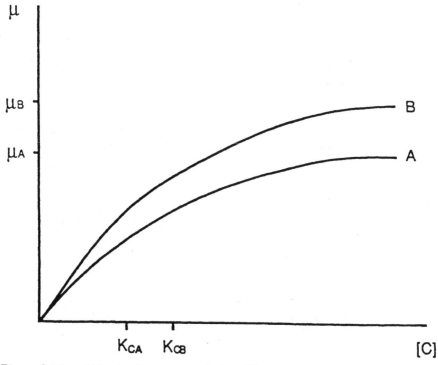

Figure 3 *Monod kinetics for compounds A and B.*

growth, then (x/x_F) is also small. Equation (15) simplifies to the basic growth equation for exponential growth. However, with increasing growth time, x becomes larger and (x/x_F) approaches unity, hence reducing the value of the maximum specific growth rate. At the end of the retardation phase $x = x_f$ and so $f(dx/dt) = 0$ which characterizes the maximum population phase.

The decline in a population's growth rate is related to the depletion of one particular nutrient. Monod (1942) was the first to establish that the form of the relationship was analogous to the Michaelis-Menten enzyme kinetics equation:

$$\mu = \mu_{max} \frac{C}{K_c + C} \qquad (16)$$

where C is the concentration of the growth-limiting substrate. The term K_c is known as the saturation constant and is defined as that growth-limiting substrate concentration which allows the micro-organism to grow at half the maximum specific growth rate (Figure 3). The saturation constant has the same units as substrate concentration. It is a measure of the affinity the micro-organism has for the growth-limiting substrate. The lower the value of K_c, then the greater the capacity of the micro-organism to grow rapidly in an environment with low growth-limiting substrate concentrations. The rate of change in population size is given by:

Table 1 *Biodegradability of selected compounds in laboratory experiments.*

Compound	Conditions	Metabolites	Rate/Extent	Reference
Acenaphthene	aerobic, sandy aquifer		ND, <0.03/wk	Wilson *et al.* 1985
Acetate	methanogenic batch studies & continuous flow fixed film column studies	-	63-93% removal	Bouwer & McCarty 1983b
Acetic acid	aerobic aquifer solids slurries	CO_2 & biomass	$3.79 \times 10\text{-}3$ h-1	Swindoll *et al.* 1988
Alpha-hexachlorocyclohexane	aerobic soil slurry	pentachlorocyclohexane	25 mg/kg soil per day	Bachman *et al.* 1988
	methanogenic soil slurry	monochlorobenzene 3,5-dichlorophenol trichlorophenol isomer	30d acclimation 13 mg/kg soil per day	Bachman *et al.* 1988
Amino Acids	aerobic aquifer solids slurries	CO_2 & biomass	$19.8 \times 10\text{-}3$ h-1	Swindoll *et al.* 1988
m-Aminophenol	uncontaminated soil	-	0.04-0.07% per day	Aelion *et al.* 1987
Aniline	uncontaminated soil	-	.06-.17% per day	Aelion *et al.* 1987
Anilinehydrochloride	aerobic aquifer solids slurries	CO_2 & biomass	$102 \times 10\text{-}3$ h-1	Swindoll *et al.* 1988
Benzene	continuous flow anaerobic aquifer column	-	no degradation	Kuhn *et al.* 1988
	aerobic, sandy aquifer		ND, <0.03/wk	Wilson *et al.* 1986
	aerobic, sandy aquifer		0.1/wk	Barker & Patrick 1986
	water samples		0.006 d-1	Zoeteman 1981
Benzoate	soil and ground water methanogenic sulfate reducing	- -	100% disappearance 100% disappearance	Gibson & Suflita 1985
Bromodichloromethane	anaerobic batch experiments	-	53% removal	Bouwer & McCarty 1983a
	methanogenic batch studies & continuous flow fixed film column studies	-	>99% removal	Bouwer & McCarty 1983b
	aerobic, sandy clay		<0.03-0.02/wk	Wilson *et al.* 1983a
Bromoform	anaerobic batch experiments	-	88% removal	Bouwer & McCarty 1983a
	methanogenic batch studies & continuous flow fixed film column studies	-	>99% removal	Bouwer & McCarty 1983b
Carbon tetrachloride	anaerobic batch experiments	chloroform	92% removal	Bouwer & McCarty 1983a
	methanogenic batch studies & continuous flow fixed film column studies	-	>99% removal	Bouwer & McCarty 1983b
Cellulose	aerobic aquifer solids slurries	CO_2 & biomass	$0.47 \times 10\text{-}3$ h-1	Swindoll *et al.* 1988
Chlorobenzene	uncontaminated soil	-	no metabolism	Aelion *et al.* 1987
	aerobic aquifer solids slurries	CO_2 & biomass	$0.46 \times 10\text{-}3$ h-1	Swindoll *et al.* 1988
	aerobic, sandy clay	-	ND, <0.05/wk not greater than abiotic control	Wilson *et al.* 1983a and b
	aerobic, sand & gravel	-	0.06-0.02/wk	Wilson *et al.* 1986
3-Chlorobenzoate	soil and ground water methanogenic sulfate reducing	benzoate -	100% disappearance 3% disappearance	Gibson & Suflita 1986
4-Chlorobenzoate	soil and ground water methanogenic sulfate reducing	- -	no degradation no degradation	Gibson & Suflita 1986

Table 1 (continued)

Compound	Conditions	Metabolites	Rate/Extent	Reference
4-Chlorobiphenyl	contaminated soil	4'-chloroacetophenone	degradation occurred	Barton & Crawford 1988
Chloroform	anaerobic batch experiments	-	no degradation	Bouwer & McCarty 1983a
	methanogenic batch studies & continuous flow fixed film column studies	-	96% removal	Bouwer & McCarty 1983b
2-Chlorophenol	soil and ground water methanogenic sulfate reducing	- -	100% disappearance loss not above control	Gibson & Suflita 1986
	soil groundwater microcosms		readily biodegraded	Smith & Novak 1987
3-Chlorophenol	soil and ground water methanogenic sulfate reducing	phenol -	100% disappearance loss not above control	Gibson & Suflita 1986
4-Chlorophenol	solid and ground water methanogenic sulfate reducing	phenol -	100% disappearance 26% disappearance	Gibson & Suflita 1986
p-Chlorophenol	uncontaminated soil	-	0.47% per day	Aelion *et al.* 1987
	aerobic aquifer solids slurries	CO2 & biomass	degradation occurred	Swindoll *et al.* 1988
Cinnamic acid	aerobic aquifer solids slurries	CO2 & biomass	0.14 x 10-3 h-1	Swindoll *et al.* 1988
m-Cresol	uncontaminated soil	-	.07% per day	Aelion *et al.* 1987
	aerobic aquifer solids slurries	CO2 & biomass	degradation occurred	Swindoll *et al.* 1988
	water samples	-	0.19 d-1	Zoeteman *et al.* 1981
p-Cresol	contaminated soil batch culture and continuous column reactors	p-hydroxybenzaldehyde & p-hydroxybenzoate	degradation occurred	O'Reilly & Crawford 1989
	water samples	-	0.19 d-1	Zoeteman *et al.* 1981
2,4-D (Aldrich Chemical Co)	soil and ground water methanogenic	2,4-dichlorophenol 4-chlorophenol & phenol	99% disappearance	Gibson & Suflita 1986
	sulfate reducing	-	0% disappearance	
Dibenzofuran	aerobic, sandy aquifer		ND. <0.1/wk	Wilson *et al.* 1985
Dibromochloromethane	anaerobic batch experiments	-	73% removal	Bouwer & McCarty 1983a
	methanogenic batch studies & continuous flow fixed film column studies	-	>99% removal	Bouwer & McCarty 1983b
1,2-Dibromoethane	anaerobic batch experiments	-	no degradation	Bouwer & McCarty 1983a
	aerobic sand & sandy clay	-	ND to <0.012/wk	Wilson *et al.* 1983a and b
1,2-Dibromethane	aerobic soil samples	CO2 & biomass	readily biodegraded	Pignatello 1986
1,3-Dichlorobenzene	aerobic water and soil samples	chloride	degradation occurred	deBont *et al.* 1986
3,4-Dichlorobenzoate	soil and ground water methanogenic sulfate reducing	3- and 4-chlorobenzoate	96% disappearance 15% disappearance	Gibson & Suflita 1986
3,5-Dichlorobenzoate	soil and ground water methanogenic sulfate reducing	3-chlorobenzoate	100% disappearance 5% disappearance	Gibson & Suflita 1986
1,1-Dichloroethane	methanotrophic mixed culture	-	-	Fogel *et al.* 1986

Table 1 (continued)

Compound	Conditions	Metabolites	Rate/Extent	Reference
1,2-Dichloroethane	methanogenic batch studies & continuous flow fixed film column studies	-	no degradation	Bouwer & McCarty 1983b
cis-1,2-Dichloroethylene	methanotrophic mixed culture	-	degradation occurred	Fogel *et al.* 1986
trans-1,2-Dichloroethylene	methanotrophic mixed culture	-	degradation occurred	Fogel *et al.* 1986
2,4-Dichlorophenol	soil and ground water methanogenic sulfate reducing	4-chlorophenol & phenol	100% disappearance loss not above control	Gibson & Suflita 1986
	soil groundwater microcosms	-	readily biodegraded	Smith & Novak 1987
2,5-Dichlorophenol	soil and ground water methanogenic sulfate reducing	3-chlorophenol & phenol	83% disappearance loss not above control	Gibson & Suflita 1986
3,4-Dichlorophenol	soil and ground water methanogenic sulfate reducing	- -	loss not above control loss not above control	Gibson & Suflita 1986
1,3-Dimethylbenzene	anaerobic soil solumn continuous flow	CO_2	>0.45h-1 80% oxidized	Zeyer *et al.* 1986
EDB	uncontaminated soil	-	0.93-1.11% per day	Aelion *et al.* 1987
Ethylbenzene	continuous flow anaerobic aquifer column	-	no degradation	Kuhn *et al.* 1988
	aerobic water samples	-	0.124 mg/l/d	Jamison *et al.* 1976
Ethylene dibromide	aerobic aquifer solids slurries	CO_2 & biomass	103 x 10-3 h-1	Swindoll *et al.* 1988
3-Ethyltoluene	continuous flow anaerobic aquifer column	-	no degradation	Kuhn *et al.* 1988
Fluorene	aerobic, sandy aquifer		ND, <0.1/wk	Wilson *et al.* 1985a
Glucosamine	aerobic aquifer solids slurries	CO_2 & biomass	1.98 x 10-3 h-1	Swindoll *et al.* 1988
Glucose	aerobic aquifer soilds slurries	CO_2 & biomass	0.59 x 10-3 h-1	Swindoll *et al.* 1988
n-Hexane	aerobic water samples	-	0.094 mg/l/d	Jamison *et al.* 1976
1-Methylnaphthalene	continuous flow anaerobic aquifer column	-	no degradation	Kuhn *et al.* 1988
	water samples	-	0-79 d-1	Kappeler & Wuhrman 1978
2-Methylnaphthalene	continuous flow anaerobic aquifer column	-	no degradation	Kuhn *et al.* 1988
	aerobic, sandy aquifer		ND, <0.1/wk	Wilson *et al.* 1985
Naphthalene	aerobic aquifer solids slurries	CO_2 & biomass	1.83 x 10-3 h-1	Swindoll *et al.* 1988
	aerobic, sandy aquifer		ND, <0.01/wk	Wilson *et al.* 1985a
	continuous flow anaerobic aquifer column	-	no degradation	Kuhn *et al.* 1988
	water samples	-	0.006 d-1	Zoeteman 1981
	aerobic water samples	-	1.2 d-1	Kappeler & Wuhrman 1978
	aerobic water samples	-	0.042 mg/l/d	Lee & Ward 1985b
p-Nitrophenol	uncontaminated soil	-	0.72-2.95% per day adaptation period 7-42 days	Aelion *et al.* 1987

Table 1 (continued)

Compound	Conditions	Metabolites	Rate/Extent	Reference
PAHs (polycyclic aromatic hydrocarbons)	nutrient nitrogen limited culture of white rot fungus	-	70-100% disappearance	Bumpus 1989
Pentachlorophenol	soil groundwater microcosms	-	readily biodegraded	Smith & Novak 1987
Phenol	soil and ground water methanogenic sulfate reducing	- -	100% disappearance 99% disappearance	Gibson & Suflita 1986
	uncontaminated soil	-	2.56-2.61% per day	Aelion et al. 1987
	batch culture experiments		Ks=8.5 μM Vmax=466 nmol/min /mg protein	Folsom et al. 1990
	aerobic aquifer solids slurries	CO_2 & biomass	58.10 x 10^{-3} h-1	Swindoll et al. 1988
	subsurface soil slurries	-	5-10,000 μ/kg dry soil	Dobbins et al. 1987
	soil groundwater microcosms	-	readily biodegraded	Smith & Novak 1987
	anaerobic soil samples		0.013-0.062 d-1	Rees & King 1981
Phenoxyacetate	soil and ground water methanogenic sulfate reducing	phenol -	100% disappearance 2% disappearance	Gibson & Suflita 1986
Propylbenzene	continuous flow anaerobic aquifer column	-	no degradation	Kuh et al. 1988
Styrene	aerobic sand/ sandy clay aquifer		0.02-0.04/wk	Wilson et al. 1985
	aerobic, sandy clay aquifer		0.1/wk	Wilson et al. 1983a and b
2,4,5-T (Sigma Chemical Co.)	soil and ground water methanogenic	2,5-dichlorophenoxyacetate dichlorophenol 3-chlorophenol phenol 2-4-D 4-chlorophenol	100% disappearance	Gibson & Suflite 1986
	sulfate reducing	2,4,5-trichlorophenol dichlorophenol 2 unidentified	2% disappearance	
1,1,2,2-Tetrachloroethane	methanogenic batch studies & continuous flow fixed film column studies	1,1,2-trichloroethane	97% removal	Bouwer & McCarty 1983b
Tetrachloroethylene (PCE)	methanotrophic mixed culture	-	no degradation	Fogel et al. 1986
	methanogenic batch studies & continuous flow fixed film column studies	-	86% removal	Bouwer & McCarty 1983b
	aerobic sand & sandy clay	-	ND to <0.03/wk not greater than abiotic control	Wilson et al. 1983a and b
	continuous flow fixed film methanogenic column	TCE, dichloroethylene Vinyl chloride CO_2	24% mineralized	Vogel & McCarty 1985
Toluene	anaerobic soil column continuous flow	CO_2	75% oxidized	Zeyer et al. 1986
	aerobic aquifer solids slurries	CO_2 & biomass	0.50 x 10^{-3} h-1	Swindoll et al. 1988
	continuous flow anaerobic aquifer column	CO_2	97% removal	Kuhn et al. 1988
	aerobic, sandy aquifer		0.009(abiotic) -0.52/wk	Wilson et al. 1983a and b
	aerobic sand/ sandy clay aquifer		>2.5/wk	Wilson et al. 1983a and b

Table 1 (continued)

Compound	Conditions	Metabolites	Rate/Extent	Reference
	aerobic, sandy aquifer		0.1/wk	Barker & Patrick 1986
	aerobic, sandy clay aquifer		0.03/wk, not greater than abiotic control	Wilson *et al.* 1983b
	aerobic, sandy aquifer		0.04/wk, not greater than abiotic control	Wilson *et al.* 1986
	water samples	-	0.019 d-1	· Zoeteman *et al.* 1981
1,2,4-Trichlorobenzene	aerobic water samples uncontaminated soil	-	1.22 d-1 no metabolism	Kappeler & Wuhrman 1978 Aelion *et al.* 1987
Trichlorobenzene	aerobic aquifer solids slurries	CO2 & biomass	4.59 x 10-3 h-1	Swindoll *et al.* 1988
1,1,1-Trichloroethane	anaerobic batch experiments	-	no degradation	Bouwer & McCarty 1983a
	methanogenic batch studies & continuous flow fixed film column studies	-	98% removal	Bouwer & McCarty 1983b
	aerobic sandy clay		ND-<0.05/wk	Wilson *et al.* 1983a and b
Trichloroethylene (TCE)	aerobic unsaturated soil column exposed to a mixture of natural gas in air (0.6%)	CO2	95% removal	Wilson & Wilson 1985
	batch culture experiments	-	0.06-0.02nmol removed/min /mg protein	Harker & Young 1990
	batch culture experiments (cell growth with phenol)		Ks=3μM Vmax=8nmol/min /mg protein	Folsom *et al.* 1990
	mixed methanotrophic cxultures	-	degradation occurred	Little *et al.* 1988
	aerobic subsurface sediment cultures with methanol, methane or propane as energy source	Hydrochloric acid CO2 dichloroethylene vinylidine chloride chloroform	30-99% removal	Fliermans *et al.* 1988
	aerobic sandy clay		ND-<0.01/wk	Wilson *et al.* 1983a and b
	aerobic GW samples simulated with toluene or phenol	-	<0.6-85% TCE remaining	Nelson *et al.* 1988
	methanotrophic mixed culture	CO2 & biomass	degradation occurred	Fogel *et al.* 1986
2,4,5-Trichlorophenol	soil and ground water methanogenic sulfate reducing	- -	loss not above control loss not above control	Gibson & Suflita 1986
	soil groundwater microcosms		readily biodegraded	Smith & Novak 1987
Vinyl Chloride	methantrophic mixed culture	-	degradation occurred	Fogel *et al.* 1986
m-Xylene	continuous flow anaerobic aquifer column	-	95% removal	Kuhn *et al.* 1988
	aerobic, sandy aquifer		ND, <0.05/wk	Wilson *et al.* 1986
	aerobic, sandy aquifer		0.2/wk	Barker & Patrick 1986
o-Xylene	continuous flow anaerobic aquifer column	-	no degradation	Kuhn *et al.* 1988
	aerobic, sandy aquifer		ND, <.03/wk	Wilson *et al.* 1986
	aerobic, sandy aquifer		0.2/wk	Barker & Patrick 1986
p-Xylene	continuous flow anaerobic aquifer column	-	no degradation	Kuhn *et al.* 1988
	aerobic, sandy aquifer		0.2/wk	Barker & Patrick 1986

$$\frac{dx}{dt} = \mu_{max} \frac{C}{K_c + C} \cdot x$$

(17)

In the context of biodegradation, with the organic contaminant as the growth-limiting substrate, an organic compound, B, which is utilized at a maximum rate, μ_{max}, greater than that of compound A, is termed as being more biodegradable than compound A (Figure 3). The half-saturation constant for compound A, K_c, is smaller than that for compound B. The two organics, A and B, might have the same maximum utilization rate, μ_{max}, but compound A could have a smaller half-saturation constant than B (Figure 3). The microbial growth rate from compound B, as shown in Figure 3, is more limited by the concentration of the substrate than that for compound A.

Typical microbial numbers in uncontaminated shallow aquifers range from 10^6 to 10^7 cells g^{-1} dry soil (Wilson et al., 1983b; Ghiorse and Balkwill, 1985; Ghiorse and Balkwill, 1983; Bone and Balkwill, 1986; Webster et al., 1985). Bacteria is the predominant form of microorganism observed ·in the subsurface (Wilson et al., 1983b; Ghiorse and Balkwill, 1985; White et al., 1983). The sampling methods for subsurface microorganisms have been described in detail by Dunlap et al., 1977 and Wilson et al., 1983b. Basically, the method consists of procuring a soil sample with a core barrel, and extruding a sample through a sterile paring device (Lee et al., 1988).

Sinclair et al. (1990) completed a survey of microbial populations in twenty-two aseptically collected sediment core samples. The cores were collected from below the water table (60 to 280 ft deep) at four pristine sites along a major buried-valley aquifer system in northeastern Kansas. Total counts of bacteria varied between 10^6 to 10^8 per gram of dry sediment. A relationship between sediment texture and microbial population density was confirmed statistically by the researchers. Total numbers of bacteria correlated highly with variations in sediment sand and clay content.

Laboratory studies of biodegradation
Laboratory studies of biodegradation are needed to ensure the successful implementation of the bioremediation technology at the field scale. There are several questions that need to be answered in laboratory experiments: is a given compound biodegradable and under what conditions? what numbers and types of micro-organisms are required for biodegradation? would the micro-organisms respond to stimulation if growth-limiting nutrients are added? and at what rates would a given organic chemical degrade.

The biodegradation potential of organic contaminants has long been recognized and utilized in the wastewater treatment process. A common indicator of biodegradation has been the Biochemical Oxygen Demand (BOD) test. BOD is defined as the amount of oxygen required by bacteria while stabilizing decomposable organic matter under aerobic conditions. The BOD test is normally used to determine the pollutional strength of domestic and industrial wastes in terms of the oxygen demand that they will exert if disposed of in natural watercourses. The BOD test is also used to design biological treatment units at wastewater treatment plants.

Similarly, studies of biodegradation of contaminants in fresh waters had been spurned by the Streeter-Phelps report on the effect of pollution and natural purification on the Ohio River (Streeter and Phelps, 1925). They presented a first order function that related water quality (dissolved oxygen) as a function of the amount of sewage discharge, location of discharges, and the condition of the stream (*e.g.* flow rates, depth, re-aeration rate, *etc.*). A common indicator of biodegradation in fresh waters and sediments has been the River-Die-Away test where river waters at utilized to quantitatively determine disappearance rates.

In groundwater, the problem of determining whether a compound is degradable or not and under what conditions is complicated by several factors. The combination of soil and water imposes a mixed-media analysis approach and the complexity of the physical and chemical processes which control the transport and fate of the organic contaminants provide a challenge for designing indicator parameters for biodegradation in the subsurface. The current state-of-the-art does not offer a concise and uniform methodology for laboratory and field quantitative estimation of biodegradation.

To date, the most aerobically biodegradable compounds in the subsurface have been petroleum hydrocarbons such as gasoline, crude oil, heating oil, fuel oil, lube oil waste, and mineral oil (Lee *et al.*, 1987). Other compounds such as alcohols (isopropanol, methanol, ethanol), ketones (acetone, methyl ethyl ketone) and glycols (ethylene glycol) are also aerobically biodegradable. Recent studies have expanded the list of aerobically degraded compounds to include methylated benzenes, chlorinated benzenes (Kuhn *et al.*, 1985), chlorinated phenols (Suflita and Miller, 1985), methylene chloride (Jhaveri and Maocacca, 1983), napthalene, methylnapthalenes, dibenzofuran, fluorene, and phenanthrene (Wilson *et al.*, 1985; Lee and Ward, 1985).

Numerous laboratory studies have also examined the efficacy of anaerobic, methanogenic and sulfate reducing microorganisms in biodegrading a large number of compounds. The reader is referred to Table 1 for a partial listing of the biodegradability of the more common contaminants. Table 1 is by no means a complete list and the reader is encouraged to pursue some of the listed references for more details.

Biodegradation rates

The term "biodegradation rate" of an organic contaminant refers to some measure of substrate mass loss or change in substrate concentration over time. The term has recently been used rather loosely in the literature and has taken on several different interpretations. One encounters different biodegradation "rates" for the same organic contaminant from field and laboratory studies. Differing biodegradation rates for the same organic contaminant in different hydrogeologic environments are not uncommon and are to be expected. Table 1 lists some of the reported rate constants and serves to show the range of variability in some of these constants for several organic contaminants.

Zero and first order kinetics have been utilized frequently in the literature to describe the biodegradation of contaminants in subsurface environments:

$$r_c = - k_1 \qquad \text{zero-order} \qquad\qquad (18)$$

$$r_c = -k_2 C \qquad \text{first-order} \qquad\qquad (19)$$

where k_1, k_2 are rate constants (mg $L^{-1}d^1$ for the zero-order model and day^{-1} for the first order model), and r_c is the reaction rate.

It is noted that the zero-order and first-order rate laws frequently cited in the literature are limits of the Monod (or Michaelis-Menten) kinetics. For $K_c \ll C$, equation (16) reduces to $\mu = \mu_{max}$, a zero-order rate law. Conversely, for $K_c \gg C$, equation (16) becomes a first-order rate law.

There are some distinctions between laboratory and field calculated rates that need to be recognized. In the laboratory, the conditions under which the experiments are conducted may vary significantly from the in-situ conditions in the field. Heterogeneities of the aquifer media that may control transport are difficult to duplicate in laboratory experiments.

Another comment which relates specifically to laboratory determined rate data is that the laboratory rate data are determined under experimental conditions which may be vastly different from one experiment to the next. The parameters include the use of subsurface soil or groundwater, the use of saturated or unsaturated soil, pristine or contaminated soil, pH, temperature, the use of acclimated or non-acclimated microorganisms, the concentration of the substrate, and the amounts of nutrients and oxygen available during the experiment.

The field calculated biodegradation rates for a specific organic compound are subject to the site-specific transport and chemical characteristics of the aquifer, and may not be applicable at another field site. For example, the rates calculated from monitoring well data at a field site might be misleading if oxygen transport was a limiting factor for the specific aquifer. The zero-order or first-order rate constants determined from field data are more likely "fitting" constants and may have no relation to underlying biotransformation processes that mediate the biodegradation of organic contaminants.

The biodegradation rates reported in the literature are at best useful as an indicator of the range of biodegradation rates for different organic compounds, and should not be used as an absolute number in a modeling effort.

Modeling biodegradation in the subsurface

The problem of quantifying biodegradation in the subsurface can be addressed by using models which combine physical, chemical and biological processes. Developing such models is not simple, however, due to the complex nature of microbial kinetics, the limitations of computer resources, the lack of field data on biodegradation, and the need for robust numerical schemes that can simulate the physical, chemical, and biological processes accurately. Several researchers have developed groundwater biodegradation models. The main approaches that have been utilized for modeling biodegradation kinetics include:

(1) First-order decay model
(2) Biofilm models (with kinetic expressions)
(3) Instantaneous reaction model
(4) Dual-substrate monod model

The first-order decay model
The first-order decay model is a simple expression which assumes a constant decay rate for the contaminant:

$$C = C_0 e^{-k_2 t} \tag{20}$$

where C is the biodegraded contaminant concentration, C_0 is the initial contaminant concentration (M/V), k_2 is the first-order decay rate (T) and t is time (T). The first-order decay model is quite useful because it is simple to use and requires very little data. Also, the first-order decay model has a mechanistic basis, enzyme kinetics reduce to first-order at low substrate concentrations (typical of groundwater contaminants), and transport kinetics across the microbial cell membrane are also first-order at low concentrations.

One of the main limitations of the first-order decay model is that the first-order decay rate has to be determined on a case-by-case basis. The expression used in the first-order decay model does not incorporate an oxygen concentration term (or other nutrient limitations), so that the simulated biodegradation using this model may overestimate the amount of biodegradation that would occur in some cases.

The instantaneous reaction model
The instantaneous reaction model used in BIOPLUME II was first proposed by Borden and Bedient (1986). The main assumption in the model is that microbial biodegradation kinetics are fast in comparison with the transport of oxygen, and that the growth of microorganisms and utilization of oxygen and organics in the subsurface can be simulated as an instantaneous reaction between the organic contaminant and oxygen. From a practical standpoint, the instantaneous reaction model assumes that the rate of utilization of the contaminant and oxygen by the microorganisms is very high, and that the time required to mineralize the contaminant is very small, or almost instantaneous.

The advantages of the instantaneous reaction model are that the governing equations are simplified, and that the amount of data that would be required to simulate biodegradation processes in groundwater is minimized. Biodegradation is calculated from the relation:

$$\Delta C_R = -\frac{O}{F} \tag{21}$$

where ΔC_R is the change in contaminant concentration due to biodegradation (M/V), O is the concentration of oxygen (M/V), and F is the ratio of oxygen to contaminant consumed.

This approach has several implications: the instantaneous reaction assumption implies that oxygen transport is the limiting process for biodegradation. Intuitively, the instantaneous reaction assumption seems logical when the organic contaminant is "rapidly degrading", *i.e.*, the rate of the reaction is on the order of seconds or minutes, and the rate of oxygen transport is on the order of days and years. The assumption of an instantaneous reaction may be poor when the rate of oxygen transport is on the order of days, and the rate of the reaction is on the order of years. In this case, microbial kinetics will take on a more significant role in limiting the biodegradation of an organic contaminant in groundwater.

The Monod kinetic model
The dual Monod kinetics model relates the decline in a microbial population's growth rate to the depletion of two substrates (see previous section for more details on Monod kinetics as they relate to the dynamics of microbial growth). For aerobic biodegradation, and assuming that oxygen and the contaminant are the only substrates required for growth, the change in contaminant and oxygen concentrations due to biodegradation is given by:

$$\Delta C = M_t \, \mu_{max} \left(\frac{C}{K_c+C}\right)\left(\frac{O}{K_o+O}\right)\Delta t \tag{22}$$

$$\Delta O = M_t \, \mu_{max} \, F \left(\frac{C}{K_c+C}\right)\left(\frac{O}{K_o+O}\right)\Delta t \tag{23}$$

where C and O = contaminant and oxygen concentration (M/V), M_t = the total microbial concentration (M/V), max = maximum contaminant utilization rate per unit mass microorganisms (T^{-1}), K_C and K_O = contaminant and oxygen half saturation constants (M/V), F = ratio of oxygen to contaminant consumed, and Δt is the time interval being considered (T).

The dual Monod substrate model is more accurate and it accounts for microbial kinetics. The microbial kinetic expression shown above for oxygen and the contaminant can be incorporated into the transport equations as discussed later so that the transport limitations on the amount of contaminant and oxygen in the aquifer can be taken into account. The main limitations of the Monod kinetic model is that it requires more microbial data than the first-order and instantaneous models. The formulation shown in equations (22) and (23) is also more complex to solve numerically.

Current biodegradation modeling studies
McCarty *et al.* (1981) modeled the biodegradation process using biofilm kinetics. It was assumed that substrate concentration within the biofilm changes only in the

direction which is normal to the surface of the biofilm. It was also assumed that all the required nutrients are in excess except the rate-limiting substrate. Three basic processes were used in the model: mass transport from the bulk liquid, biodecomposition within the biofilm, and biofilm growth and decay. McCarty *et al.* (1981) evaluated the applicability of the biofilm model to aerobic subsurface biodegradation using a laboratory column filled with glass beads. The experimental data and the model predictions were relatively consistent.

Kissel *et al.* (1984) developed differential equations describing mass balances on solutes and mass fractions in a mixed-culture biological film within a completely mixed reactor. The model incorporated external mass transport effects, Monod kinetics with internal determination of limiting electron donor or acceptor, competitive and sequential reactions, and multiple active and inert biological fractions which vary spatially. Results of hypothetical simulations involving competition between heterotrophs deriving energy from an organic solute and autotrophs deriving energy from ammonia and nitrite were presented.

Molz *et al.* (1985) and Widdowson *et al.* (1987) presented one-dimensional and two-dimensional models for aerobic biodegradation of organic contaminants in groundwater coupled with advective and dispersive transport. A microcolony approach was utilized in the modeling effort, microcolonies of bacteria were represented as disks of uniform radius and thickness attached to aquifer sediments. A boundary layer of a given thickness was associated with each colony across which substrate and oxygen are transported by diffusion to the colonies. Their results indicate that biodegradation would be expected to have a major effect on contaminant transport when proper conditions for growth exist. Simulations of two-dimensional transport suggested that under aerobic conditions microbial degradation reduces the substrate concentration profile along longitudinal sections of the plume and retards the lateral spread of the plume. Anaerobic conditions developed in the plume center due to microbial consumption and limited oxygen diffusion into the plume interior.

Borden and Bedient (1986) developed a system of equations to simulate the simultaneous growth, decay, and transport of micro-organisms combined with the transport and removal of hydrocarbons and oxygen. Simulation results indicated that any available oxygen in the region near the hydrocarbon source will be rapidly consumed. In the body of the hydrocarbon plume, oxygen transport will be rate limiting and the consumption of oxygen and hydrocarbon can be approximated as an instantaneous reaction. The major sources of oxygen, the researchers concluded, are transverse mixing, advective fluxes and vertical exchange with the unsaturated zone.

Borden *et al.* (1986) applied the first version of the BIOPLUME model to simulate biodegradation at the Conroe Superfund site in Texas. Oxygen exchange with the unsaturated zone was simulated as a first-order decay in hydrocarbon concentration. The loss of hydrocarbon due to horizontal mixing with oxygenated groundwater and resulting biodegradation were simulated by generating oxygen and hydrocarbon distributions independently and then combining by superposition. Simulated oxygen and hydrocarbon concentrations closely matched the observed values.

Widdowson *et al.* (1988) extended their 1986 and 1987 studies to simulate oxygen- and/or nitrate based respiration. Basic assumptions incorporated into the model included a simulated particle-bound microbial population comprised of heterotrophic, facultative bacteria in which metabolism is controlled by lack of either an organic carbon-electron donor source (substrate), electron acceptor (O_2 and/or NO_3), or mineral nutrient (NH_4^{3+}), or all three simultaneously.

Srinivasan and Mercer (1988) presented a one-dimensional, finite difference model for simulating biodegradation and sorption processes in saturated porous media. The model formulation allowed for accommodating a variety of boundary conditions and process theories. Aerobic biodegradation was modeled using a modified Monod function; anaerobic biodegradation was modeled using Michaelis-Menten kinetics. In addition, first-order degradation is allowed for both

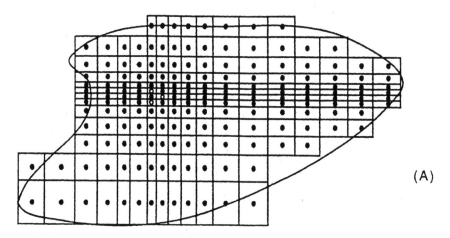

(A)

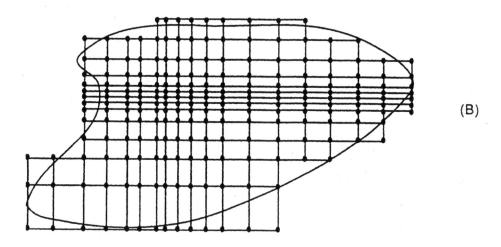

(B)

Figure 4 *Finite difference representation of an aquifer region.*

substances. Sorption can be incorporated using linear, Freundlich, or Langmuir equilibrium isotherms for either substance.

Baek *et al.* (1989) simulated the mitigation of chemical contamination by microbial enrichment in unsaturated soil systems. Their model, BIOSOIL, incorporated the influence of micro-organisms on soil water flow and chemical removal rates. From the modeling study, the authors concluded that the depth of the unsaturated zone seems to be less crucial under bioremediation than it is in waste disposal considerations. The authors also indicated that the microbial distribution in a vertical column of soil is as important as the total population size of soil micro-organisms.

MacQuarrie *et al.* (1990) utilized a similar approach to Borden *et al.* (1986) and Rifai *et al.* (1988) to develop a biodegradation model. The advection-dispersion equation was coupled with a dual-Monod relationship. The system of equations was solved using an iterative principal direction finite element technique. Comparisons of numerical results with the results of a laboratory column experiment showed that the model equations adequately describe the behavior of toluene, dissolved oxygen, and the microbial population, without considering solute diffusion through stagnant fluid layers or biofilms. The authors concluded that in a two-dimensional shallow aquifer setting, an organic plume experiences mass loss, spreading controlled by the availability of dissolved oxygen, and skewing in the direction of groundwater flow.

MacQuarrie and Sudicky (1990) utilized the model developed by MacQuarrie *et al.* (1990) to examine plume behavior in uniform and random flow fields. In uniform groundwater flow, a plume originating from a high-concentration source will experience more spreading and slower normalized mass loss than a plume from a lower initial concentration source because dissolved oxygen is more quickly depleted. Large groundwater velocities produced increases in the rate of organic solute mass loss because of increased mechanical mixing of the organic plume with oxygenated groundwater.

The BIOPLUME II model

The BIOPLUME II model was developed by modifying an existing two-dimensional transport model developed by the USGS and known as the Method of Characteristics (MOC) model (Konikow and Bredehoeft, 1978). Two governing equations are solved in MOC: the groundwater flow equation and the transport equation. The numerical approximation to the flow equation is a finite difference expression that is solved using an alternating-direction implicit procedure. The method of characteristics is utilized to solve the transport equation. A brief description of finite differences and the method of characteristics as applied in MOC precedes a discussion of the development of the BIOPLUME II model.

Finite differences and the method of characteristics
In general, a finite difference model is developed by superimposing a system of nodal points over the problem domain (Figure 4). In the finite difference method, nodes may be located inside cells (block-centered, Figure 4) or at the intersection of grid

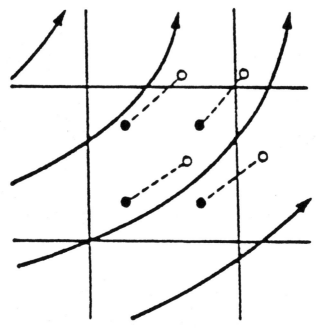

Figure 5 *Part of hypothetical finite-difference grid showing relation of flow field to movement of points.* •, *Initial location of particle.* o, *New location of particle.* →, *Flow line and direction of flow.* ----, *Computed path of particle.*

lines (mesh centered, Figure 4). The MOC/BIOPLUME II model is a block-centered model with a rectangular and uniformly spaced finite-difference grid. Aquifer properties and head values are assumed to be constant within each cell in a block-centered finite difference model. An equation is written in terms of each nodal point in finite difference models because the area surrounding a node is not directly involved in the development of the finite difference equations (Wang and Anderson, 1982).

The finite-difference flow equation is solved numerically in MOC/BIOPLUME II using an iterative alternating-direction implicit (ADI) procedure. The derivation and solution of the finite-difference equation and the use of the iterative ADI have been discussed extensively by Pinder and Bredehoeft (1968), Prickett and Lonquist (1971), and Trescott, Pinder, and Larson (1976).

In general, the basis of the ADI method is to obtain a solution to the flow equation by alternately writing the finite-difference equation, first implicitly along columns and explicitly along rows, and then vice versa. An implicit (or backward) difference formulation is one where the heads (h) are evaluated at time (n). An explicit (or forward) difference formulation is one where the head values are evaluated in terms of known or old values of h. In order to reduce the errors that may result from the ADI method, an iterative procedure is added so that within a single time step, the solution would converge within a specified error tolerance.

The method of characteristics (MOC) is used in MOC/BIOPLUME II to solve the

Table 2 *Aquifer parameters used in simulation runs.*

Porosity	0.3
Aquifer thickness	3.33m
Initial contaminant concentration	100 mg L^{-1}
Retardation coefficient, R_c	1.0
Total microbial concentration, M_t	1.0 mg L^{-1}
Oxygen half-saturation constant, K_o	0.1 mg L^{-1}

solute transport equation. This method is usually used to solve hyperbolic equations. The solute transport equation may closely approximate a hyperbolic partial differential equation if transport is dominated by advection (as is the case in many field problems). The approach used for the method of characteristics solution is to solve an equivalent system of ordinary differential equations.

The method of characteristics used in the MOC/BIOPLUME II model utilizes representative fluid particles that are converted with flowing groundwater. The rate of change in concentration in the groundwater is observed in the aquifer when moving with the fluid particle. The solution of the transport equation is then given by simplified ordinary differential equations called the characteristic curves. The numerical solution of the ordinary differential equations involves introducing a set of moving particles that can be traced within the coordinates of the finite-difference grid. Each particle has a concentration and position associated with it and is moved through the flow field in proportion to the flow velocity at its location.

Intuitively, the method may be visualized as tracing a number of fluid particles through a flow field and observing changes in chemical concentration in the fluid particles as they move. The first step in the method of characteristics involves placing a number of traceable particles or points in each cell of the finite-difference grid. The initial concentration assigned to each point is the initial concentration associated with the node of the cell containing the point.

For each time step every point is moved a distance proportional to the length of the time increment and the velocity at the location of the point (see Figure 5). After all points have been moved, the concentration at each node is temporarily assigned the average of the concentrations of all points then located within the cell. The moving particles simulate advective transport because the concentration at each node of the grid will change with each time step as different particles having different concentrations enter and leave the cell.

The basic concept applied to modify the USGS MOC model and to develop the BIOPLUME II model includes the use of a dual-particle mover procedure to simulate the transport of oxygen and contaminants in the subsurface. The transport equation is solved twice at every time step to calculate the oxygen and contaminant distributions:

$$\frac{\partial(Cb)}{\partial t} = \frac{1}{R_c}\left(\frac{\partial}{\partial x_i}\left(bD_{ij}\frac{\partial C}{\partial x_j}\right)\right) - \frac{\partial}{\partial x_i}(bCV_i) - \frac{C'W}{n} \qquad (24)$$

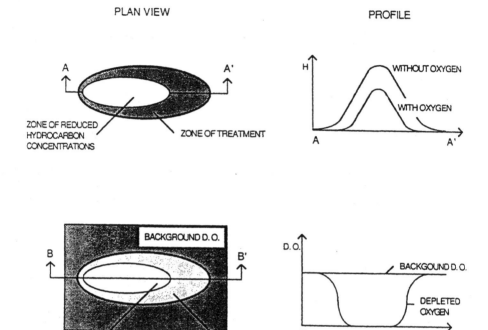

Figure 6 *Principle of superposition for organics and oxygen in BIOPLUME II model.*

$$\frac{\partial(Ob)}{\partial t} = \left(\frac{\partial}{\partial x_i}\left(bD_{ij}\frac{\partial O}{\partial x_j}\right)\right) - \frac{\partial}{\partial x_i}(bOV_i)\right) - \frac{O'W}{n} \tag{25}$$

where C and O = concentration of contaminant and oxygen respectively (M/V), C' and O' = concentration of contaminant and oxygen in a source or sink fluid (M/V), n = effective porosity, b = saturated thickness (L), t = time (T), x_i and x_j = cartesian coordinates, W = volume flux per unit area (V/A), V_i = seepage velocity in the direction of x_i (L/T), R_h = retardation factor for contaminant, and D_{ij} = coefficient of hydrodynamic dispersion (L^2/T).

It is emphasized that the BIOPLUME II model simulates dissolved contaminant concentrations which are vertically averaged over the thickness of the aquifer. Burnett and Frind (1987) have attempted to study the effect of simulating averaged horizontal plane concentrations. Their results indicated that the apparent length of the simulated plume will be much less than that of the equivalent plume defined in terms of peak concentrations. It is not anticipated that the general results from this analysis would change significantly if the analysis were conducted in three dimensions.

The two plumes are combined using the principle of superposition to simulate the instantaneous reaction between oxygen and the contaminants, and the decrease in contaminant and oxygen concentrations is calculated from:

$$\Delta C_{RC} = O/F \; ; \; O = 0 \quad \text{where C} > O/F \tag{26}$$

$$\Delta C_{RO} = C.F \; ; \; C = 0 \quad \text{where O} > C. F \tag{27}$$

where ΔC_{RC}, ΔC_{RO} are the calculated changes in concentrations of contaminant and oxygen, respectively, due to biodegradation.

Figure 6 is a conceptual schematic of the BIOPLUME II model. At the top of the figure, a plan view of the contaminant and oxygen plumes with and without biodegradation are shown. After the two plumes are superimposed, the contaminant plume is reduced in size and concentrations. The dissolved oxygen is depleted in zones of high contaminant concentrations and reduced in zones of relatively moderate contaminant concentrations. The bottom schematics in Figure 6 present transects down the plume centerline and help to illustrate the distributions of contaminant and oxygen concentration with and without biodegradation. It is noted that field data have verified the correlation between oxygen and contaminant concentrations at sites.

Model parameters
There are four types of BIOPLUME II input parameters: physical aquifer parameters, biodegradation parameters, numerical methods related parameters, and input and output control parameters. The emphasis in this section is on the biodegradation parameters so the other parameters will only be discussed briefly. As discussed earlier, it is necessary to superimpose a block-centered finite-difference grid over a site (Figure 4) by specifying the number of cells in the x and y directions (note that the y-axis should be oriented along the main direction of flow at the site). The values of the various parameters in the model can be uniform over the whole domain, or varying over each cell in the domain.

The aquifer specific parameters in MOC/BIOPLUME II include porosity, longitudinal and transverse dispersivity, thickness of the aquifer, transmissivity, and recharge. There are several model parameters which relate to the numerical methods used in MOC/BIOPLUME II. The convergence criteria for the flow equation can be defined using three parameters, the number of iterations, the number of iteration parameters, and the error tolerance allowable in computing the heads at nodes.

Two control parameters are used for the transport equation. The first parameter is used to limit the maximum distance within a cell that a particle can move during a time step. The time step is determined internally in the model and is controlled by four stability criteria that are discussed later in this section. The second parameter is the number of particles in a cell (4, 5, 8 or 9). Using 9 particles per cell is usually adequate. A related parameter is utilized to specify the maximum number of particles in the whole grid. The maximum number of particles is determined from the following relationship:

$$\text{Max. no. of particles} = NX \cdot NY \cdot \text{no. of particles in each cell} \qquad (28)$$

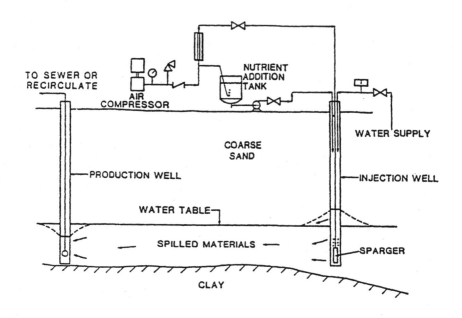

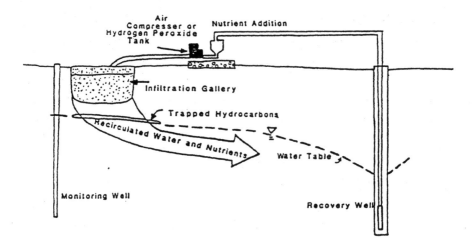

Figure 7 *Schematic for bioremediation using injection wells or infiltration*

Table 3 *Parameters utilized in the modeling study at the Traverse City field site.*

Parameter (1) (2)	Value
Grid size	16 x38*
Cell size	30 m x 30 m
Porosity	0.3
Longitudinal dispersivity	3m
Hydraulic conductivity	50 m day^{-1}
Dissolved oxygen	8.0 mg L^{-1}
Rearation decay coefficient	0.003 day^{-1}
Recharge	1.14E - 3 m day^{-1}

See Figure 12 for grid orientation and flow direction.

where NX is the number of cells in the x-direction and NY is the number of cells in the y-direction.

Several parameters can be used to specify the type of output that is desired. The user is referred to Konikow and Bredehoeft (1978) and Rifai *et al.* (1987) for details on those parameters. The length of time for which modeling is required is specified in MOC/BIOPLUME II using three parameters: the number of pumping periods in the simulation time, the actual time in years for each pumping period, and the number of time steps in a pumping period.

The number of pumping periods is more than one usually in cases where the hydraulic conditions at a site have changed through additional pumping or injection of water into the aquifer. The number of time steps is usually a function of how often is the output desired from the model. The number of time steps has some implications on run-time and numerical errors which will be discussed later. Additional timing parameters are required for transient flow conditions

Source parameters include injection wells, constant concentration cells, and recharge cells. Injection wells and recharge cells basically define a source that leaks into an aquifer, *i.e.*, a source that has a flow rate and a concentration associated with it ($Q = Q_0$ and $C = C_0$). A constant concentration cell ($C = C_0$ boundary condition) simulates a source which adds contaminant mass at natural gradients into the aquifer. The MOC/BIOPLUME II model allows the specification of up to 5 observation wells or monitoring wells at a given site. The history of chemical concentrations in those wells is included in the output from the model.

Boundary conditions in MOC/BIOPLUME II are specified by the user. Types of boundary conditions that can be used include constant head cells or constant flux cells. A constant flux boundary can be used to represent aquifer underflow, well withdrawals or well injection. A constant head boundary can represent parts of the aquifer where the head will not change in time. Constant head boundaries are simulated by using a high leakage term (1.0 s^{-1}). The resulting rate of leakage into or out of the constant head cell would equal the flux required to maintain the head in the aquifer at the specified altitude. If a constant flux or constant head boundary represents a source then the chemical concentration must be specified.

The numerical procedure in MOC/BIOPLUME II requires that a no flow boundary surrounds the modeled site. No flow boundaries simply preclude the flow of water or contaminants across the boundaries of the cell. Initial conditions in the aquifer also have to be specified: the initial water table, and the initial contaminant and oxygen concentrations.

There are two methods that can be used to simulate biodegradation in the currently available version of BIOPLUME II: first-order decay and instantaneous reaction. The Monod kinetics method was utilized in a research mode and was not incorporated as a computational option in BIOPLUME II because it has not been field verified and because there are no kinetic data to support its application.

For the first-order decay model, the reaction rate, k_2, is required as input. Note that the decay rate is only applied to the dissolved components and not to the source materials. The model input parameters required for the instantaneous reaction include the amount of dissolved oxygen in the aquifer prior to contamination, and the oxygen demand of the contaminant determined from a stoichiometric relationship. Modeling the biodegradation of several components such as benzene, toluene, and xylenes (BTX) at a site requires that an average stoichiometric coefficient for the three components be calculated.

Two additional sources of oxygen can be specified in BIOPLUME II. Injection of oxygen in a bioremediation project can be simulated by using injection wells or infiltration galleries (Figure 7. Re-aeration from the unsaturated zone can be simulated in an indirect way by specifying a first-order decay rate for the contaminants at the site.

The numerical solution of the solute-transport equation has a number of stability criteria associated with it. The time step used to solve the flow equation may be subdivided into smaller time steps for calculating transport. Four stability criteria are generally checked internally in the model:

$$\Delta t \quad \leq \text{ Min over grid} \left(\frac{0.5}{\dfrac{D_{xx}}{(\Delta x)^2} + \dfrac{D_{yy}}{(\Delta y)^2}} \right) \tag{29}$$

$$\Delta t \quad \leq \text{ Min over grid} \left(\frac{n\, b_{ijk}}{W_{ijk}} \right) \tag{30}$$

$$\Delta t \quad \leq \frac{\gamma\, \Delta x}{(V_x)_{max}} \tag{31}$$

$$\Delta t \quad \leq \frac{\gamma\, \Delta y}{(V_y)_{max}} \tag{32}$$

where Δt is the transport time step, D_{xx} and D_{yy} are the dispersion coefficients in the x and y directions, Δx, Δy are the cell dimensions, W_{ijk} is the volume flux per unit area of any sources or sinks specified in a given cell, n is the porosity, b_{ijk} is the thickness of the aquifer, γ is the maximum distance within a cell that a particle is allowed to move in a line step, and V_x, V_y are components of the velocity.

The smallest time step calculated using the four criteria is used as the computational time step for the transport equation. This time step determines the number of particle moves that need to be completed and hence directly affects the run-time in the model. The output from the model lists which of the four criteria was

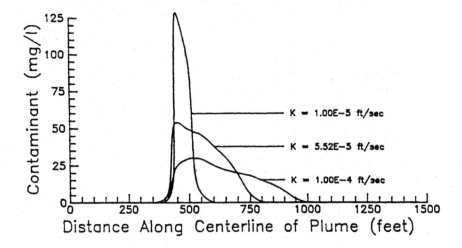

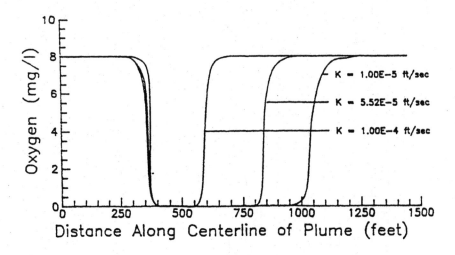

Figure 8 *Variation of contaminant and oxygen concentrations with hydraulic conductivity.*

utilized in determining the transport time step for a given model run. The user can manipulate the input data to decrease the run-time if necessary. It is important to note, however, that the run-time is also a function of the number of pumping periods, the simulation time, and the number of time steps in a pumping period.

The numerical errors in the MOC/BIOPLUME II model are computed for both the flow and transport equations. The flow numerical errors should be less than 1.0 percent and can be decreased by increasing the number of iterations, or decreasing the tolerance, (at the cost of possibly increasing the run-time).

The numerical transport errors are more complex to predict and analyze. The user

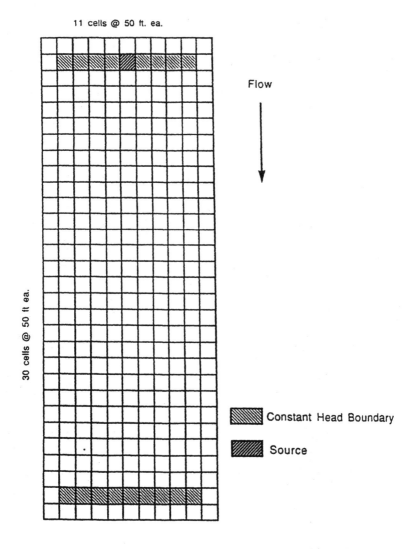

Figure 9 *Problem setup for analysis simulations.*

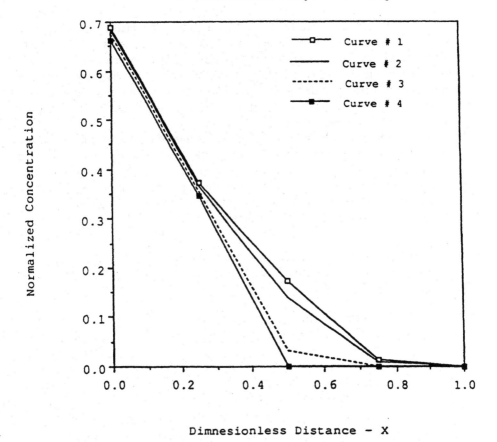

Figure 10 *Normalized concentrations along plume centerline for instantaneous and kinetic model.*

needs to become more familiar with the model and the magnitude of numerical errors for a variety of scenarios. General guidelines are that the transport errors may be higher for flow fields which are influenced by pumping and injection wells. In some cases, the transport numerical errors can be controlled by varying some of the input parameters. The numerical errors are the lowest when 9 particles per cell and a maximum cell distance per particle move of 0.5 are used.

Model sensitivity
The output from the MOC/BIOPLUME II model consists generally of a head map and a chemical concentration map for each node in the grid. Immediately following the head and concentration maps is a listing of the hydraulic and transport errors. If observation wells had been specified, a concentration history for those wells would be included in the output. The user is encouraged to check the echo of the model input printed in the output and the results from the model run.

Model verification for the MOC model had been completed by comparing the model results with two analytical solutions (Konikow and Bredehoeft, 1978). Model verification for the BIOPLUME II model in terms of the biodegradation component was more complicated and consisted of the model comparisons discussed above (the comparison to first-order decay and Monod kinetics) as well as field simulation efforts . There are no analytical solutions which incorporate kinetics that the BIOPLUME II model could have been compared with.

The sensitivity of aerobic biodegradation to some of the model parameters had been analyzed in detail by Rifai *et al.* (1988). Their analyses indicated that biodegradation is mostly sensitive to the hydraulic conductivity (Figure 8). This result verified some field observations about the applicability of bioremediation for systems with relatively large hydraulic conductivities. Biodegradation was not sensitive to the retardation factor or dispersion.

Microbial kinetics and the instantaneous reaction theory
The system of equations for modeling contaminant and oxygen transport coupled with Monod kinetics is given by:

$$\frac{\partial C}{\partial t} = \frac{\nabla(D\nabla C - vC)}{R_c} - M_t \frac{\mu_{max}}{R_c} \left(\frac{C}{K_c+C}\right)\left(\frac{O}{K_o+O}\right) \tag{33}$$

$$\frac{\partial O}{\partial t} = \nabla(D\nabla O - vO) - M_t \mu_{max} F \left(\frac{C}{K_c+C}\right)\left(\frac{O}{K_o+O}\right) \tag{34}$$

where C and O = contaminant and oxygen concentration, v = seepage velocity vector, D = dispersion tensor, M_t = the total microbial concentration, μ_{max} = maximum contaminant utilization rate per unit mass microorganisms, R_C = contaminant retardation factor, K_C and K_O = contaminant and oxygen half saturation constants, F = ratio of oxygen to contaminant consumed.

Rifai and Bedient (1990) have conducted a study to compare biodegradation kinetics to the instantaneous reaction model. In their comparison, it was assumed that the total concentration of microorganisms, M_t, remains constant. The Monod kinetics were incorporated into the BIOPLUME II model for the contaminant and oxygen using an enhanced form of the time-splitting algorithm proposed by Wheeler *et al.* (1987, 1988). The time-splitting algorithm basically accounts for the fact that the reactions have to be calculated on a much smaller time scale than the advection and dispersion components of the transport equation.

The following example is presented to illustrate some of the results from the two models. Consider a simulation domain of 183.33 m x 500 m as shown in Figure 9. The cells in the domain are 16.67 m on each side (11x30 cells in the grid). The aquifer parameters and the kinetic parameters for the simulation are listed in Table 2. A constant source strength (subscript of "0") applied as a Dirichlet boundary

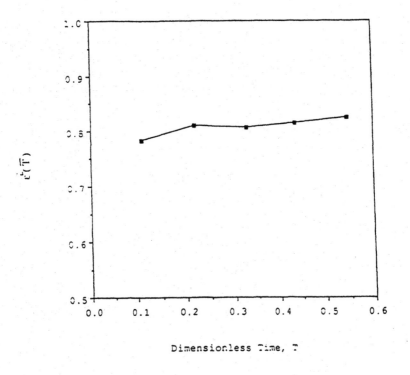

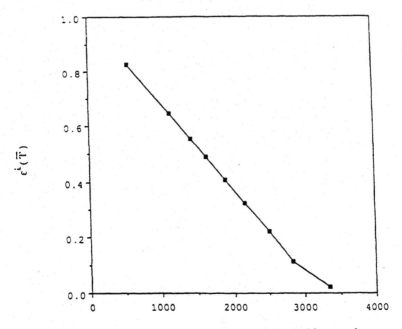

Figure 11 *Error as a function of dimensionless time and Damköhler number.*

condition at the upstream constant head boundary was simulated as shown in Figure IX. The Damköhler number is given by (Boucher and Alves, 1959): $\dot{D}_{a1} = \mu_{max} L/v$

Figure 10 presents the solution for this problem setup for four different scenarios. The normalized concentrations shown in curve # 1 are those for the case of a conservative tracer which does not biodegrade ($D_{a1} = 0$), while the concentrations

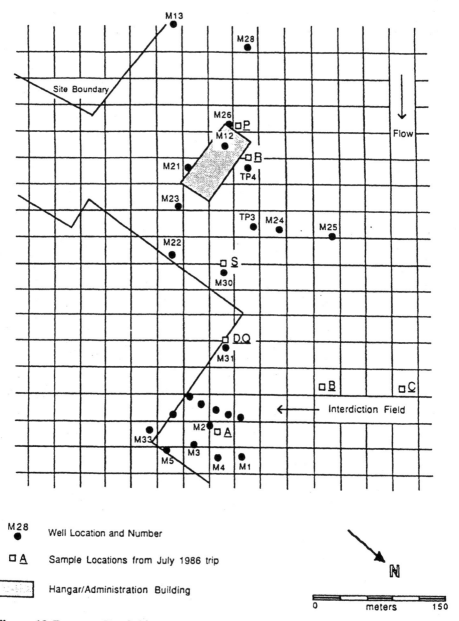

Figure 12 *Traverse City field site map.*

shown in curve numbers 2, 3, and 4 are those for the case of a biodegradable organic (D_{a1} = 1,134, 3,802 and ∝, respectively).

The first observation from Figure 10 is that curve number 1 represents the maximum attainable concentrations while curve number 4 represents the minimum attainable concentrations in the aquifer. The concentrations calculated from the kinetic model for different values of the Damköhler number will fall between the two extremes. The second observation is that the smaller the Damköhler number, then the closer the solution would be to that of a conservative tracer. Similarly, the larger the Damköhler number, the closer the solution would be to that of the instantaneous model.

The comparison between the instantaneous reaction model and the kinetic reaction rate model is complicated by several factors. The major difficulty is that the differences between the two models may be a function of space and time. If one evaluates the two models at arbitrary points in space and time, the conclusions may not be general enough. Another complicating factor is that unequal degrees of deviation between the two models would be expected for chemicals with different rates of biodegradation. Moreover, from a numerical aspect, it seems that one would need to separate the differences between the two models due to numerical errors from differences due to modeling concepts.

In this effort, a measure of the differences between the two models is defined as follows:

$$\epsilon^i(\overline{T}) = \frac{M_k - M_i}{M_d - M_i} \tag{35}$$

where M_k is the dissolved mass in the kinetic model, M_i is the dissolved mass in the instantaneous model, M_d is the dissolved mass in the aquifer for a conservative tracer, and $(\overline{T})$ is an arbitrary time. It is noted that as $\epsilon^i(\overline{T})$ approaches 0, the differences between the two models would get smaller. For slow kinetic reaction rates, $\epsilon^i(\overline{T})$) approaches 1.

For the problem setup described in the previous section (no dispersion for this case), Figure 11 answers a fundamental question: what are the requirements to guarantee that reaction mechanisms are not kinetically limited? The error measure, $\epsilon^i(\overline{T})$) is clearly a function of the kinetic rate of the reaction, but not a function of time. For a specific Damköhler number (D_{a1} = 567)), the error measure does not change with dimensionless time as shown in Figure 11 (the slight variation is due to numerical errors in the computations). This is a logical conclusion since the total number of microorganisms does not change with time. If the total number of microorganisms were to increase with time for instance, then the error would decrease with time.

The error, $\epsilon^i(\overline{T})$), decreases as the Damköhler number increases as shown in Figure 11. This conclusion is particularly interesting because one can select the maximum allowable error, and specify the magnitude of the Damköhler numbers necessary to guarantee that the instantaneous reaction model yields a satisfactory

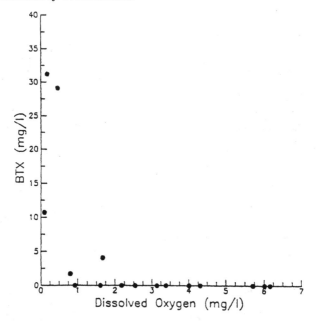

Figure 13 *Total BTX versus dissolved oxygen at Traverse City.*

approximation. If Damköhler numbers are less than the selected threshold Damköhler number, kinetic modeling would be necessary, and one has to determine the kinetic rates of biodegradation for the organic contaminant in question.

Application to a field site - Traverse City, Michigan

Site description

The Traverse City field site is a US Coast Guard Air Station located in Grand Traverse County in the northwestern portion of the lower peninsula of Michigan (Figure 12). The groundwater at the site is contaminated with organic chemicals from a source near the Hangar/Administration building. In the upper reach of the plume, hydrocarbons occur at the surface of the water table and move downward in the aquifer as the plume migrates toward east Grand Traverse Bay (Twenter *et al.* 1985). The soils at the site are generally free of contamination except near the source area. The major contaminants at the site include BTX.

The contaminant plume ranges from 50- to 120-m wide, and is about 1,300-m long. An interception well field (pump and treat, see Figure 12) was installed in April 1985 to prevent the migration of the plume off the site. The captured contaminated water is renovated by carbon adsorption.

The U.S. Geologic Survey performed an extensive study of the geology and hydraulic characteristics of the site (Twenter *et al.* 1985). glacial deposits, as much as 100-m thick, underlie the site. The upper 10 to 40 m is a sand and gravel unit that is underlain by impermeable clay that is about 30-m thick. The water table varies from 3 to 6.5 m below land surface, with groundwater velocities ranging from 1 to 2 m per

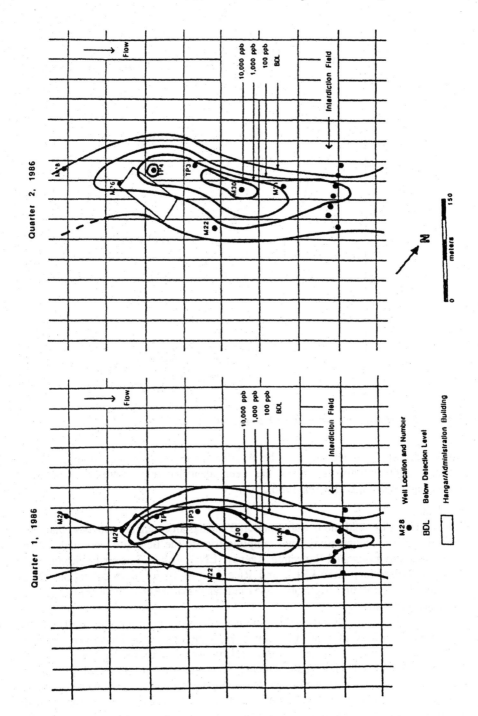

Figure 14 *BTX Plume at Traverse City.*

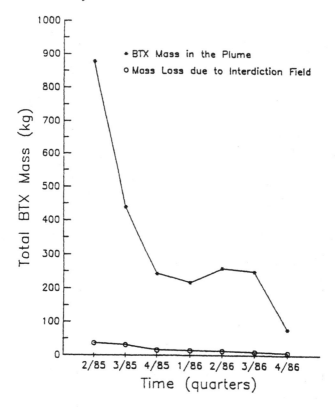

Figure 15 *Variation in total BTX with time at Traverse City.*

day. Hydraulic conductivities in the aquifer range from 28 to 50 m per day. Biodegradation activity at the site has been identified and discussed by several researchers (Wilson, B.H. *et al.* 1986).

Field data

The Traverse City field site is well monitored and has an extensive data base that was utilized in the modeling study. Two sampling trips to the field site were undertaken by the U.S. environmental Protection Agency (EPA), Robert S. Kerr Environmental Research Laboratory (RSKERL), and Rice University for data collection. Data collected during the first trip (July 1985) indicated that aerobic biodegradation was occurring at the site.

The data in Figure 13 indicate the absence of oxygen in monitoring wells where high concentrations of total alkylbenzenes were observed. Dissolved oxygen concentrations were elevated in monitoring wells with negligible levels of hydrocarbons. This phenomenon, which is characteristic of biodegradation sites, seems to indicate that it is appropriate to assume hydrocarbons react with dissolved oxygen for modeling purposes. It was noted that pristine zones in the aquifer had background levels of 8.0 mg L^{-1} of dissolved oxygen.

The second field trip (June 1986) involved more extensive sampling longitudinally and along a transect across the lower portion of the plume (Figure 12, locations A,B,C,P,R,S,D, and Q). The data collected indicated that anaerobic activity was taking place in the center of the plume near locations S,D,Q, and A. Methane concentrations at locations D and Q were higher than at S and A. No methane was detected near the source (locations P and R) or at the outer edge of the plume (locations B and C).

In addition to the above mentioned field trips, the U.S. Coast Guard collected data on a regular basis from a large number of wells. The data collected from April 1985 through 1986 from approximately 25 selected wells (Figure 12) were analyzed to: (1) Define the areal extent of the contaminant plume; and (2) estimate the amount of mass biodegraded over the period of sampling.

The total BTX from monitoring well data were averaged over three-month intervals beginning with the second quarter of 1985. The data were also averaged vertically. This was a necessary approach due to the large number of data points at each well.

A statistical analysis was performed with the averaged data to determine the mean and standard deviation for each well over each quarter. Data from wells with relatively high concentrations of BTX show a small amount of spreading, whereas some of the wells with contaminant levels less than 5.0 mg L^{-1} exhibit more variation. It is noted that some of the variation correlates well with rainfall and snowmelt events and water table fluctuations in the aquifer. Overall, it was concluded that the average values over each three-month period were representative values for the period in question. Concentration contours were developed from the averaged data and are shown in Figure 14.

It can be seen that BTX plume is undergoing significant changes with time: (1) The plume exhibits a significant loss of mass; and (2) the interception field has succeeded in halting the migration of the contaminant plume across the site boundaries.

In general, three basic mechanisms could have contributed to the observed mass loss: biodegradation; volatilization; and dispersion. The flow system at the site is highly advective, so it is unlikely that dispersion could account for the observed loss of mass. A sensitivity analysis on the dispersion coefficients at the site supported this conclusion.

Estimating the rate of mass loss due to volatilization is more difficult. Chiang *et al.* (1987) calculate mass loss due to volatilization using Henry's Law for benzene to be five percent of the total mass loss at a site similar to the U.S. Coast Guard facility. It was reasonable to conclude that biodegradation is the most important mechanism that accounts for the observed mass loss at the site.

The contour plots shown in Figure 12 were utilized to estimate the mass of contaminants in the system for each time period. It is noted that these computations are very rough estimates since the data have been averaged over time in the vertical direction. The contour level that represents concentrations less than 0.01 mg L^{-1} was not included in these computations because there was less certainty and more interpolation in its development. The mass loss due to biodegradation for each

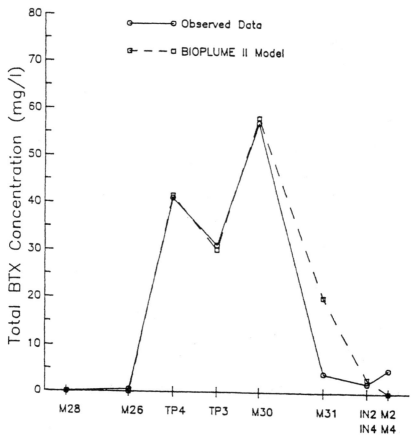

Figure 16 *Calibrated versus observed BTX concentrations at Traverse City.*

quarter was calculated by comparing the BTX mass in the system for successive quarters and subtracting the BTX mass captured by the interception field during the quarter.

The data in Figure 15 show the change in total BTX mass with time in the system. The data also show the mass removed from the interception field. It is obvious that biodegradation had a much more significant effect than the pumping well field on removing contaminant mass. The data indicated a slight increase in mass for the second quarter of 1986. This could be due to the leaching of contaminants from the soil due to seasonal recharge.

The modeling effort at the site was performed in order to calibrate to the April 1985 conditions, the time the interception field was turned on.

Calibration results

The simulation grid used in this effort extends well beyond the site to cover the area downstream of the interception field and into the Grand Traverse Bay. The aquifer parameters utilized are presented in Table 3. The aquifer thickness data at the finite difference grid cells were obtained from well logs, and were used to calculate the transmissivity assuming a constant hydraulic conductivity.

The source was represented with seven injection wells near the hangar building (Figure 12) at a combined flow rate of 6.85 m^3 day^{-1}, which is equivalent to half the annual rate of precipitation. It is noted that this source definition is rather arbitrary, and can account for some of the calibration error that is presented later. The source concentrations varied for the seven wells with an average concentration of about 1,380 mg L^{-1} The interception field was simulated with pumping wells at the specified flow rates.

Three sources of oxygen into the contaminated aquifer were simulated: (1) Mixing with uncontaminated water containing dissolved oxygen of 8.0 mg L^{-1}; (2) natural recharge of dissolved oxygen from the upgradient constant head boundary at the site (also at 8.0 mg L^{-1}); and (3) re-aeration from the unsaturated zone. The re-aeration coefficient was estimated from computations performed at another site (Borden *et al.* 1986), and in this case would be another source of error in calibration results.

The boundary conditions were developed using constant head cells and diffuse recharge. Constant head cells and diffuse recharge. Constant head cells simply mean that the water table elevation will not change with time for that part of the aquifer, and diffuse recharge refers to infiltration due to rainfall. The constant head data were obtained from water table measurements and from water levels in the Grand Traverse Bay. Diffuse recharge was assumed to be 50 percent of the average annual precipitation except at cells along the ditch on the northern side of the site, where the recharge rate was increased to match the hydraulics. No recharge was simulated at the source cells since the infiltration rate was already accounted for with the injection rate.

The calibrated water table elevations matched the observed data reasonably well for most wells except M28, M30 and TP3. The differences between calibrated and observed water table elevations are attributed to variations in seasonal recharge. In the calibration process, recharge was assumed to be constant for the simulation period.

The simulated plume matched the observed plume in shape, extent, and levels of contamination reasonably well. The simulation results are discussed in more detail along the centerline to point out some of the significant differences between the calibrated plume and the observed plume.

The data in Figure 16 show the results of the transport simulation along the centerline of the plume. It can be seen that the model predictions match the observed concentrations at the monitoring wells reasonably well. BIOPLUME II predictions overestimate the observed plume in the vicinity of well M31. This is due to the fact that the simulations to this point have not accounted for anaerobic biodegradation processes.

Data from the June 1986 sampling trip show elevated levels of methane near well M31, but the coefficient of anaerobic decay at the site has not yet been estimated. It should be noted that anaerobic biodegradation activity at the site was verified in laboratory microcosms, and the reader is referred to Wilson, B.H., *et al.* (1986) for details of those studies.

The mass balance computations for the calibrated plume indicate that the total BTX dissolved mass in the system for the April 1985 conditions is 1,470 kg. This is higher than the BTX mass computed from the contour plots (878.0 kg for the second quarter of 1985). The difference is partly due to the anaerobic biodegradation component which was not simulated at this point, and partly due to the approximation in the mass computations presented earlier.

Other field demonstrations

Researchers from Suntech, Inc. are amongst the earliest pioneers who utilized bioremediation at sites contaminated with gasoline. Two field experiments are discussed by Jamison *et al.* (1975), Raymond *et al.* (1977) and Raymond *et al.* (1978). The first field study was at a site in Ambler, Pennsylvania. A leak in a gasoline pipeline had caused the township to abandon its groundwater supply wells. The free product was physically removed prior to the initiation of biodegradation studies at the site. Laboratory studies showed that the natural microbial population at the site could use the spilled high-octane gasoline as the sole carbon source if sufficient quantities of the limiting nutrients, in this case, oxygen, nitrogen and phosphate, were supplied. Pilot studies which were carried out in the field in several wells confirmed the laboratory findings.

The second field experiment completed by the Suntech, Inc. group was at a site in Millville, New Jersey. Again laboratory studies utilizing sands from the aquifer confirmed the ability of the natural microorganisms to degrade the spilled gasoline at the site. The main difference between the Ambler site and the Millville site was the hydrogeologic nature of the media. The aquifer at the Millville site consisted of medium to coarse well sorted quartz sand while that at Ambler was a highly permeable limestone aquifer. The results from the field effort supported the feasibility of bioremediation at sites with consolidated and unconsolidated sands.

In a controlled field experiment at the Canada Forces Base Borden site, two plumes of gasoline contaminated groundwater were introduced into the aquifer. Immediately upgradient of one plume, groundwater spiked with nitrate was added so that a nitrate plume would overtake the organic plume (Berry-Spark *et al.* 1986). The success of the field experiment was limited (Berry-Spark and Barker 1987). The dissolved organic contaminant mass (BTEX) decreased rapidly due to residual oxygen concentrations in the aquifer prior to the nitrate overlap. Insufficient organic mass left in the aquifer was not adequate to evaluate anaerobic biotransformation.

Semprini *et al.* (1988) presented the results from a field evaluation of in-situ biodegradation of trichloroethylene (TCE) and related compounds. The method that was used in the field demonstration relied on the ability of methane-oxidizing bacteria to degrade these contaminants to stable, non-toxic, end products. The field site is located at the Moffett Naval Air Station in Mountain View, California and the test zone is a shallow confined aquifer composed of coarse grained alluvial sediments. Results from the biotransformation experiments at the site indicates that biodegradation of TCE was on the order of 30 percent of the mass injected.

Major *et al.* (1988) investigated the biodegradation of benzene, toluene and the isomers of xylene (BTX) in anaerobic batch microcosms containing shallow aquifer material. BTX loss occurred with the addition of either nitrate or oxygen. Denitrification was confirmed by nitrous oxide accumulation. When a limiting amount of nitrate was added, there was a corresponding limit to the loss of BTX and a limited amount of nitrous oxide production.

Borden and Bedient (1987) conducted a three well injection-production test at the United Creosoting Company (UCC) site in Conroe, Texas, to evaluate the significance of biotransformation in limiting the transport of polycyclic aromatics present in the shallow aquifer. During the test, chloride, a non-reactive tracer and two organic compounds, naphthalene and paradichlorobenzene (pDCB), were injected into a center well for 24 hours followed by clean groundwater for six days. Groundwater was continuously produced from two adjoining wells and monitored to observe the breakthrough of these compounds. A significant loss of naphthalene and pDCB attributed to biotransformation processes was observed during the test .

Chiang *et al.* (1989) characterized soluble hydrocarbon and dissolved oxygen in a shallow aquifer beneath a field site by sampling groundwater at 42 monitoring wells. Results from 10 sampling periods over three years showed a significant reduction in total benzene mass with time in groundwater. The natural attenuation rate was calculated to be 0.95 percent per day. Spatial relationships between DO and total benzene, toluene, and xylene (BTX) were shown to be strongly correlated by statistical analyses and solute transport modeling.

Bioremediation as a clean-up technology

The practical application of biodegradation is termed bioremediation (sometimes referred to as in-situ bioremediation because it is carried out without physically removing the contaminants from the aquifer). The main concept behind bioremediation is to introduce oxygen and nutrients (for aerobic biodegradation) into the subsurface and enhance the microbial growth and activity so that they would metabolize more of the contaminants in a shorter period of time. A brief discussion of the technology and its limitations is presented. It is important to keep in mind that any bioremediation project at a field site needs to be preceded by laboratory experiments of microbial stimulation and modeling studies of nutrient transport to ensure efficient performance of the system.

Discussion of bioremediation

The main purpose of bioremediation is the treatment of subsurface pollutants by stimulating the native microbial population. The advantage of biological treatment is that it offers partial or complete breakdown of the contaminant instead of simply transferring the contaminant from one phase in the environment to another. Most of the other alternatives for remediation do not have the potential to reduce contaminants to the required levels at a reasonable economic cost. For example, in the case of a gasoline spill, a significant portion of the original product can remain even after all the free product had been removed. The gasoline held at residual saturation in the aquifer would act as a continuous source of dissolved compounds into the groundwater for a long time.

Bioremediation is not without its problems, however. The most important being the lack of well documented field demonstrations that show the effectiveness of the technology and what if any are the long term effects of this treatment on groundwater systems. The lack of documented field studies may be attributed in part to the significant commercial potential of the technology which basically means that the information is proprietary. Other problems include the possibility for generating undesirable intermediate compounds during the biodegradation process which are more persistent in the environment than the parent compound.

Enhanced aerobic bioremediation for a petroleum spill is essentially an engineered delivery of nutrients and oxygen to the contaminated zone in an aquifer. The main constraints on the rate of delivery of nutrients and oxygen are: (1) hydrogeologic; and (2) oxygen sources. Bioremediation is difficult in aquifers with an average hydraulic conductivity (K) which is less than 10^{-4} cm s^{-1} (Thomas and Ward, 1990). Oxygen sources include air, pure oxygen (gaseous and liquid forms), and hydrogen peroxide. Sparging the groundwater with air and pure oxygen can supply only 8 to 40 mg L^{-1} of oxygen depending on the temperature of the injection fluid (Lee *et al.*, 1988).

Hydrogen peroxide, which dissociates to form water and 1/2 molecule of oxygen, is infinitely soluble in water (Thomas and Ward, 1990); however, hydrogen peroxide can be toxic to microorganisms at concentrations as low as 100 ppm. A stepping-up procedure is usually utilized to allow the microorganisms to adapt to the higher concentrations of the oxidant. Other problems have to do with the stability of hydrogen peroxide. The key to success in using hydrogen peroxide as an oxygen source is to add a relatively large quantity to water and have oxygen released in a controlled manner as it advances through the aquifer. If hydrogen peroxide is destabilized, oxygen will come out of solution as a gas, and the process becomes less efficient (Hinchee *et al.* 1987). Proprietary techniques have been developed to stabilize hydrogen peroxide.

In severe cases, gas production (both O_2 and CO_2) can lead to a reduction in hydraulic conductivity. One undesirable effect of using hydrogen peroxide is that other redox reactions may also be enhanced. For example, the oxidation of ferrous iron (Fe^{2+}) to Fe_2O_3 (s), this solid precipitates and can plug up the pore system at places where the oxygen is being added. Spain *et al.* (1989) discussed the excessive bacterial decomposition of H_2O_2 during enhanced bioremediation at a jet fuel site in Florida. The uncontrolled and rapid decomposition was catalyzed by a combination of chemical and biological reactions at the injection point. A rather important drawback for the use of hydrogen peroxide may be its cost; at 1987 prices, the cost of 35 percent H_2O_2 was $ 4.2 per gallon. Depending on the size of the spill and the required amount of oxygen, the cost of materials could make bioremediation an expensive alternative.

The basic steps involved in an *in-situ* biorestoration program (Lee *et al.*, 1988) are:

(1) site investigation

(2) free product recovery

(3) microbial degradation enhancement study

(4) system design

(5) operation

(6) maintenance.

It is important to define the hydrogeology and the extent of contamination at the site prior to the initiation of any in-situ effort. The parameters of interest include the direction and rate of groundwater flow, the depths to the water table and to the contaminated zone, the specific yield of the aquifer, and the heterogeneity of the soil. In addition, other parameters such as hydraulic connections between aquifers, potential recharge and discharge zones, and seasonal fluctuations of the water table should be considered. The pumping rate that can be sustained in the aquifer is an important consideration because it limits the amount of water that can be circulated in the system during the bioremediation process.

After defining the hydrogeology, recovery of free product, if any, at the site should be completed. The pure product can be removed using physical recovery techniques such as a single pump system that produces water and hydrocarbon or a two-pump, two-well system that steepens the hydraulic gradient and recovers the accumulating hydrocarbon. Physical recovery often accounts for 30 to 60 percent of the hydrocarbon before yields decline (Lee *et al.* 1988).

Prior to the initiation of a bioremediation activity, it is important to conduct a feasibility study for the biodegradation of the contaminants present at the site. First, contaminant-degrading microorganisms must be present, and second the response of these native microorganisms to the proposed treatment method must be evaluated. In addition, the feasibility study is conducted to determine the nutrient requirements of the micro-organisms. These laboratory studies provide a reliable basis for performance at the field level only if they are performed under conditions that simulate the field.

The chemistry of a field site will affect the types and amounts of nutrients that are required. Limestone and high mineral content soils, for example, will affect nutrient availability by reacting with phophorous. Silts and clays at sites may induce nutrient sorption on the soil matrix, and hence decrease the amount of nutrients available for growth. In general, a chemical analysis of the groundwater provides little information about the nutrient requirements at a field site, it is mostly the soil composition that is of significance.

Nutrient requirements are usually site specific. Nitrogen and phosphorus were required at the Ambler site (Raymond *et al.*, 1976); however, the addition of ammonium sulfate, mono- and disodium phosphate, magnesium sulfate, sodium carbonate, calcium chloride, manganese sulfate, and ferrous sulfate was required at other sites (Raymond *et al.* 1978; Minugh *et al.* 1983). The form of the nutrient may also be important; ammonium nitrate was less efficient than ammonium sulfate in one aquifer system.

Feasibility studies can be completed using several different techniques. Batch culture techniques are used to measure the disappearance of the contaminant, electrolytic respirometer studies are utilized to measure the uptake of oxygen. Tests which are designed to measure an increase in microbial numbers are not sufficient indicators of metabolization of the contaminant in question, instead, studies which measure disappearance of the contaminant or mineralization studies which confirm the breakdown of the contaminant to carbon dioxide and water need to be conducted. Controls to detect abiotic transformation of the pollutants and tests to detect toxic effects of the contaminants on the microflora should be included (Flathman *et al.* 1984).

A system for injection of nutrients into the formation and circulation through the contaminated portion of the aquifer must be designed and constructed (Lee and Ward, 1985a). The system usually includes injection and production wells and equipment for the addition and mixing of the nutrient solution (Raymond, 1978). A typical system is shown in Figure 9. Placement of injection and production wells may be restricted by the presence of physical structures. Wells should be screened to accommodate seasonal fluctuations in the level of the water table. Air can be supplied with carborundum diffusers (Raymond *et al.*, 1975), by smaller diffusers constructed from a short piece of DuPont Viaflo tubing (Raymond *et al.*, 1978), or by diffusers spaced along air lines buried in the injection lines (Minugh *et al.*, 1983). The size of the compressor and the number of diffusers are determined by the extent of contamination and the time allowed for treatment (Raymond, 1978). Nutrients also can be circulated using an infiltration gallery (Figure 9); this method provides an additional advantage of treating the residual gasoline that may be trapped in the pore spaces of the unsaturated zone. Oxygen also can be supplied using hydrogen peroxide, ozone, or soil venting (see section on alternative oxygen sources).

Well installation should be performed under the direction of a hydrogeologist to ensure adequate circulation of the groundwater (Lee and Ward, 1985a). Produced water can be recycled to recirculate unused nutrients, avoid disposal of potentially contaminated groundwater, and avoid the need for makeup water. Inorganic nutrients can be added to the subsurface once the system is constructed. Continuous injection of the nutrient solution is labor intensive but provides a more constant nutrient supply than a discontinuous process. Continuous addition of oxygen is recommended because the oxygen is likely to be a limiting factor in hydrocarbon degradation.

The performance of the system and proper distribution of the nutrients can be monitored by measuring the organic, inorganic, and bacterial levels (Lee and Ward, 1985a). Carbon dioxide levels are also an indicator of microbial activity in the formation (Jhaveri and Mazzacca, 1985). Depending on the characteristics of the nutrients and soil, nutrients can be removed from solution by sorption onto soil (Brubaker and Crockett, 1986). About 90 percent of the ammonium and phosphate and 70 percent of the hydrogen peroxide added to a sandy soil with low calcium, magnesium, and iron was recovered.

After passage of a nutrient solution through a column packed with a clay soil that had high calcium and magnesium but low iron and chloride levels, 100, 66 and 25% of the ammonium, phosphate, and hydrogen peroxide were recovered, respectively. However, after passage of a nutrient solution through a column packed with a clay soil high in calcium, magnesium, and chloride, but low in iron, 75, 100, and 15 % of the ammonium, phosphate, and hydrogen peroxide, respectively, were recovered.

Both soil and groundwater samples should be collected and analyzed to fully evaluate the treatment effectiveness. Raymond *et al.* (1975) reported that the most difficult problem in optimizing microbial growth in the Ambler reservoir was the distribution of nutrients, which was made difficult by the heterogeneity of the dolomite formation.

Summary and conclusions

Biodegradation of organic contaminants is an important attenuation mechanism that may provide a remediation alternative at some contaminated site. The process involves breaking down of the contaminants into harmless components by naturally occurring microorganisms. In order to evaluate the effectiveness of biodegradation, it is necessary to use groundwater models that combine physical, chemical and biological processes of contaminant transport.

Several models of biodegradation have been proposed in the literature. One of the simpler models is the BIOPLUME II model which utilizes an instantaneous reaction assumption between the organics and dissolved oxygen in groundwater. The modeling approach in BIOPLUME II limits the biodegradation data input requirements to those that deal with the mass flux of oxygen into the subsurface. The BIOPLUME II model simulates the transport of the organics and oxygen in groundwater and superimposes the two plumes to calculate the loss of the organics due to biodegradation.

The BIOPLUME II model can be used to quantify biodegradation at a field site by specifying the oxygen concentrations in the groundwater prior to contamination. More important, however, is the ability to simulate oxygen injection and enhanced biodegradation. The time required for cleanup in a bioremediation project and under a variety of injection schemes can be estimated. The modeling activity needs to be supplemented with laboratory studies of microbial growth enhancement.

Overall, bioremediation seems to be a promising technology that can be used to clean up sites contaminated with gasoline products and certain solvents. More research is needed to implement the technology more effectively in the field and to develop more sophisticated models that can simulate the complex microbial growth dynamics in groundwater.

References

Aelion, C.M., Swindoll, C.M. and Pfaender, F.K. 1987. Adaptation to and Biodegradation of Xenobiotic Compounds by Microbial Communities from a Pristine *Aquifer. Appl. Environ. Microbiol.*, 53, (9), 2212-2217.

Alexander, M. 1980. Biodegradation of Chemicals of Environmental Concern. *Science*, 211, 132.

Anderson, M.P. 1979. Using Models to Simulate the Movement of Contaminants Through Groundwater Flow Systems. *Crit. Rev. Environ. control*, 9, 97-156.

Bachmann, A., Walet, P., Wijnen, P., de Bruin, W., Huntjens, J.L.M., Roelofsen, W. and Zehnder, A.J.B. 1988. Biodegradation of Alpha- and Beta-Hexachlorocyclohexane in a

Soil Slurry under Different Redox Conditions. *Appl. Environ. Microbiol.,* 54, (1), 143-149.

Baek, N.H., Clesceri, L.S.and Clesceri, N.L. 1989. Modeling of Enhanced Biodegradation in Unsaturated Soil Zone. *J. Envir. Engrg.* 115, (1), 150-172.

Barker, J.F.and Patrick, G.C. 1986. Natural Attenuation of Aromatic Hydrocarbons in a Shallow Aquifer. *Proc. NWWA/API Conf.Petroleum Hydrocarbons and Organic Chemicals in Groundwater--Prevention, Detection and Restoration,* Houston, Texas, November 13-15, 1985.

Barton, M.R. and Crawford, R.L. 1988. Novel Biotransformations of 4-Chlorobiphenyl by a *Pseudomonas* sp. *Appl. Environ. Microbiol.,* 54, (2), 594-595.

Bear, J. 1979. *Hydraulics of Groundwater.* McGraw-Hill, New York.

Bedient, P.B., Borden, R.C. and Leib, D.I. 1985. Basic Concepts for Groundwater Transport Modeling. Offprints from *Groundwater Quality,* 28, 513.

Berry-Spark, K., Barker, J.F., Major, D. and Mayfield, C.I. 1986. Remediation of Gasoline-Contaminated Groundwaters, A. Controlled Field Experiment. *Proc. of Petroleum Hydrocarbons and Organic Chemicals in Groundwater, Prevention, Detection and Restoration,* Houston, Texas. National Water Well Association, pp. 613-623.

Berry-Spark, K.and Barker, J.F. 1987. Nitrate Remediation of Gasoline Contaminated Groundwaters, Results of a Controlled Field Experiment. *Proc_of the NWWA/API* Conf. *on Petroleum Hydrocarbons and Organic Chemicals in Groundwater--Prevention, Detection and Restoration,* Houston, Texas. National Water Well Association, pp. 127-144.

Bone, T.L and Balkwill, D.L. 1986. Improved Flotation Technique for Microscopy of *in situ* Soil and Sediment Microscopy. *Appl. Environ. Microbiol.,* 51, 462.

Borden, R.C. and Bedient, P.B. 1986. Transport of Dissolved Hydrocarbons Influenced by Oxygen-Limited Biodegradation 1. Theoretical Development. *Water Resour. Res.,* 22, (13), 1983-1990.

Borden, R.C., Bedient, P.B., Lee, M.D., Ward, C.H., Wilson, J.T. 1986. Transport of Dissolved Hydrocarbons Influenced by Oxygen-Limited Biodegradation 2. Field Application. Water Resources Research 22 (13), 1983-1990.

Borden, R.C. and Bedient, P.B. 1987. *In situ* Measurement of Adsorption and Biotransformation at a Hazardous Waste Site. *Water Resources Bulletin,* 23(4), 629-636.

Boucher, D.F. and Alves, G.E. 1959. Dimensionless Numbers for Fluid Mechanics, Heat Transfer, Mass Transfer, and Chemical Reaction. *Chemical Engineering Progress,* 55 (9), 55-64.

Bouwer, E.J and McCarty, P. 1983a. Transformations of Halogenated Organic Compounds Under Denitrification Conditions. *Appl. Environ. Microbiol.,* 45(4), 1295-1299.

Bouwer, E.J. and McCarty, P. 1983b. Transformations of 1- and 2-Carbon Halogenated Aliphatic Organic Compounds Under Methanogenic Conditions. *Appl. Environ. Microbiol.,* 45(4), 1286-1294.

Bredehoeft, J.D. and Pinder, G.F. 1973. Mass Transport in Flowing Groundwater. *Water Resources Research,* 9(1), 194-210.

Brubaker, G.R. and Crockett, E.L. 1986. *In situ* Aquifer Remediation Using Enhanced Bioreclamation. *Proc.* of HAZMAT 86, June 1986, Atlantic City, NJ, in press.

Bumpus, J.A. 1989. Biodegradation of Polycyclic Aromatic Hydrocarbons by *Phanerochaete chrysosporium. Appl. Environ. Microbiol.,* 55 (1), 154-158.

Burnett, R.D. and Frind, E.O. 1987. Simulation of Contaminant Transport in Three Dimensions, 2, Dimensionality Effects. *Water Resour. Res.,* 23, 695-705.

Carnahan, B., Luther, H.A. and Wildes, J.O. 1969. Applied Numerical Methods. Wiley, New York.

Chiang, C.Y., Chai, E.Y., Salanitro, J.P., Klein, C.L. and Colthart, J.D. 1987. Effects of Dissolved Oxygen on the Biodegradation of BTX in a Sandy Aquifer. *Proc. of NWWA/API Conference on Petroleum Hydrocarbons and Organic Chemicals in Groundwater, Prevention, Detection and Restoration*, pp.451-469.National Water Well Association, Houston, TX.

Chiang, C.Y., Salanitro, J.P., Chai, E.Y., Colthart, J.D. and Klein, C.L. 1989. Aerobic Biodegradation of Benzene, Toluene, and Xylene in a Sandy Aquifer--Data Analysis and Computer Modeling. *Groundwater,* 27(6), 823-834.

Cleary, R.W. and Adrian, D.D. 1973. New Analytical Solutions for Dye Diffusion Equations. *J. Environ. Eng. Div.*, *ASCE, 99*, 213-227.

de Bont, J.A.M., Vorage, M.J.A.W., Hartmans, S. and van den Tweel, W.J.J. 1986. Microbial Degradation of 1, 3-Dichlorobenzene. *Appl. Environ. Microbiol.* 52(2), 677-680.

de Marsily, G. 1986. Quantitative Hydrogeology, Groundwater Hydrology for Engineers. Harcourt Brace Jovanovich, Orlando, FL, 440 p.

Dobbins, D.C., Thornton-Manning, J.R., Jones, D.D. and Federle, T.W. 1987. Mineralization Potential for Phenol in Subsurface Soils. *Journal of Environmental Quality,* 16, 54-58.

Dunlap, W.J. and McNabb, J.F. 1973. Subsurface Biological Activity in Relation to Groundwater Pollution. EPA-660/2-73-014. U.S. Environmental Protection Agency, Ada, OK. 60.

Dunlap, W.J., McNabb, J.F., Scalf, M.R. and Cosby, R.L. 1977. *Sampling for Organic Chemicals and Microorganisms in the Subsurface.* EPA-600/2-77-176, U.S. Environmental Protection Agency, Ada, OK

Flathman, P.E. and Githens, E.G. 1984. *In situ* Biological Treatment of Isopropanol, Acetone, and Tetrahydrofuran in the Soil/Groundwater Environment. In: Nyer, E.K. (ed.), *Groundwater Treatment Technology*, p.173. Van Nostrand Reinhold, New York.

Fliermans, B., Phelps, T.J., Ringelberg, D., Mikell, A. and White, D.C. 1988. Mineralization of Trichloroethylene by Heterotrophic Enrichment Cultures. *Appl. Environ. Microbiol.,* 54(7), 1709-1714.

Fogel, M.M., Taddeo, A.R. and Fogel, S. 1986. Biodegradation of Chlorinated Ethenes by a Methane-Utilizing Mixed Culture. *Appl. Environ. Microbiol.,* 51(4), 720-724.

Freeze, R.A. and Cherry, J.A. 1979. *Groundwater* Prentice-Hall, Englewood Cliffs, NJ.

Fried, J.J. 1975. *Groundwater Pollution.* Elsevier, Amsterdam.

Gelhar, L.W., Gutjahr, A.L. and Naff, R.L. 1979. Stochastic Analysis of Macrodispersion in a Stratified Aquifer. *Water Resour. Res., 15*, 1387-1397.

Ghiorse, W.C. and Balkwill, D.L. 1983. Enumeration and Morphological Characterization of Bacteria Indigenous to Subsurface Environments. *Dev. Ind. Microbiol.*, 24, 213.

Ghiorse, W.C. and Balkwill, D.L. 1985. Microbial Characterization of Subsurface Environments. *Groundwater Quality,* p.387 John Wiley & Sons, New York.

Gibson, S.A. and Suflita, J.M. 1986. Extrapolation of Biodegradation Results to Groundwater Aquifers, Reductive Dehalogenation of Aromatic Compounds. *Appl. Environ. Microbiol.,* 52 (4), 681-688.

Harker, A.R. and Young, K. 1990. Trichloroethylene Degradation by Two Independent Aromatic-Degrading Pathways in *Alcaligenes_eutrophus* JM.P.134. *Appl. Environ. Microbiol.*, 56 (4), 1179-1181.

Hinchee, R.E., Downey, D.C. and Coleman, E.J. 1987. Enhanced Bioreclamation Soil Venting and Groundwater Extraction; A. Cost-effectiveness and Feasibility Comparison.

Proc. NWWA/API Conference on Petroleum Hydrocarbons and Organic Chemicals in Groundwater--Prevention, Detection, and Restoration, Dublin, Ohio, pp.147-164.

Hunt, B. 1978. Dispersive Sources in Uniform Ground-Water Flow. *J. Hydrol. Div. ASCE,* 104, 75-85.

Jamison, V.W., Raymond, R.L.and Hudson, Jr., J O. 1975. Biodegradation of High-Octane Gasoline in Groundwater. *Dev. Ind. Microbiol.,* 16, 305.

Javandel, I., Doughty, C.and Tsang, C.F. 1984. *Groundwater Transport, Handbook of Mathematical Models.* American Geophysical Union, Washington, DC. 228 p.

Jhaveri, V. and Mazzacca, A.J. 1983. Bio-reclamation of Ground and Groundwater. A. Case History. *Proc. 4th Natl. Conf. on Management of Uncontrolled Hazardous Waste Sites,* Washington, DC. 242 p.

Jhaveri, V. and Mazzacca, A.J. 1985. Bio-reclamation of Ground and Groundwater by In-Situ Biodegradation, A. Case History. *Proc. 6th Natl. Conf. on Management of Uncontrolled Hazardous Waste Sites,* Washington, DC. 239 p.

Kappeler, T. and Wuhrmann, K. 1978. Microbial Degradation of the Watersoluble Fraction of Gas Oil--II Bioassays with Pure Strains. *Water Res.,* 12, 335-342.

Karickhoff, S.W., Brown, D.S. and Scott T.A. 1979. Sorption of Hydrophobic Pollutants on Natural Sediments. *Water Res.* 13, 241-248.

Kissel, J.C., McCarty, P.L. and Street, R.L. 1984. Numerical Simulation of Mixed-Culture Biofilm. *J. Environ. Eng. Div. ASCE.,* 110, 393.

Konikow, L.F. and Bredehoeft, J.D. 1978. Computer Model of Two-Dimensional Solute Transport and Dispersion in Groundwater, Automated Data Processing and Computations. *Techniques of Water Resources Investigations of the U. S. G. S.,* Washington, DC. 100 p.

Kuhn, E.P., Colberg, P.J., Schnoor, J.L., Wanner, O., Zehnder, A.J.B. and Schwartzenbach, R.P. 1985. Microbial Transformations of Substituted Benzenes during Infiltration of River Water to Groundwater, Laboratory Column Studies. *Environ. Sci. Technol.,* 19, (10), 961-967.

Kuhn, E.P., Zeyer, J., Eicher, P. and Schwartzenbach, R.P. 1988. Anaerobic Degradation of Alkylated Benzenes in Denitrifying Laboratory Aquifer Columns. *Appl. Environ. Microbiol.,* 54 (2), 490-496.

Lee, M.D., Jamison, V.W. and Raymond, R.L. 1987. Applicability of *in-situ* Bioreclamation as a Remedial Action Alternative. *Proc. of Petroleum Hydrocarbons and Organic Chemicals in Groundwater, Prevention, Detection and Restoration,* Houston, Texas, National Water Well Association, pp. 167-185.

Lee, M.D., Thomas, J.M., Borden, R.C., Bedient, P.B, Wilson, J.T, Ward, C.H. 1988. Biorestoration of Aquifers Contaminated with Organic Compounds. *CRC Critical Reviews in Environmental Control, 18, (1),* 29-89.

Lee, M.D.and Ward, C.H.. 1985a. Biological Methods for the Restoration of Contaminated Aquifers. *J. Environ. Toxicol. Chem.,* 4, 743.

Lee, M.D.and Ward, C.H.. 1985b. Microbial Ecology of a Hazardous Waste Disposal Site, Enhancement of Biodegradation. *Proc. Second International Conference on Groundwater Quality,* pp. 25-27 OSU University Printing Services, Stillwater, OK.

Little, C.D. Palumbo, A.V, Herbes, S.E., Lindstrom, M.E., Tyndall, R.L.and Gilmer, PJ. 1988. Trichloroethylene Biodegradation by a Methane-Oxidizing Bacterium. *Appl. Environ. Microbiol., 54*(4), 951-956.

MacQuarrie, K.T.B. and Sudicky, EA. 1990. Simulation of Biodegradable Organic Contaminants in Groundwater, 2, Plume Behavior in Uniform and Random Flow Fields. Paper 89WRO1448. *Water Resources Research,* 26(2), 223-240.

MacQuarrie, K.T.B., Sudicky, E.A.and Frind, EO. 1990. Simulation of Biodegradable Organic Contaminants in Groundwater, 1. Numerical Formulation in Principal Directions. Paper 89WR01449. *Water Resources Research,* 26(2), 207-222.

Major, D.W., Mayfield, C.I. andBarker, J.F. 1988. Biotransformation of Benzene by Denitrification in Aquifer Sand. *Groundwater,* 26(1), 8-14.

McCarty, P.L., Reinhard, M. and Rittmann, B.E. 1981. Trace Organics in Groundwater. *Environ. Sci. Technol.,* 15(1), 40-51.

Mercer, J.W. and Faust, C.R. 1981. *Ground-Water Modeling.* National Water Well Association, Worthington, OH.

Minugh, E.M., Patry, J.J., Keech, D.A. and Leek, W.R. 1983. A. Case History, Cleanup of a Subsurface Leak of Refined Product. *Proc. 1983 Oil Spill Conf. – Prevention, Behavior, Control, and Cleanup,* San Antonio, TX, 397 p.

Molz, F.J., Widdowson, M.A.and Benefield, L.D. 1985. Simulation of Microbial Growth Dynamics Coupled to Nutrient and Oxygen Transport in Porous Media. *Water Resour. Res.,* 22 (8), 1207-1216.

Monod, J. 1942. *Recherches sur la Croissance des Cultures Bacteriennes.* Herman & Cie, Paris.

National Research Council. 1990. *Groundwater Models, Scientific and Regulatory Applications,* National Academy Press, Washington, DC. 303 p.

Nelson, M.J.K., Montgomery, S.O. andPritchard, P.H. 1988. Trichloroethylene Metabolism by Microorganisms That Degrade Aromatic Compounds. *Appl. Environ. Microbiol.,* 54(2), 604-606.

Nelson, R.W. 1977. Evaluating the Environmental Consequences of Groundwater Contamination. #1. An Overview of Contaminant Arrival Distributions as General Evaluation Requirements. *Water Resour. Res.,* 14, 409-415.

O'Reilly, K.T. and Crawford, R.L. 1989. Kinetics of p-Cresol Degradation by an Immobilized *Pseudomonas* sp. *Appl. Environ. Microbiol.,* 55(4), 866-870.

Pignatello, J.P. 1986. Ethylene Dibromide Mineralization in Soils under Aerobic Conditions. *Appl. Environ. Microbiol.,* 51(3), 588-592.

Pinder, G.F. 1973. A. Galerkin Finite Element Simulation of Groundwater Contamination on Long Island, NY. *Water Resour. Res.,* 9, 1657-1669.

Pinder, G.F. and Bredehoeft, J.D. 1968. Application of the Digital Computer for Aquifer Evaluation. *Water Resources Research,* 4(5), 1069-1093.

Pinder, G.F. and Gray, W.G. 1977. *Finite Element Simulation in Surface and Subsurface Hydrology.* Academic Press, New York.

Prickett, T.A. 1975. Modeling Techniques for Groundwater Evaluation. In: *Advances in Hydroscience,* Vol. 10, pp.1-143. Academic Press, New York.

Prickett, T.A. and Lonnquist, C.G. 1971. Selected Digital Computer Techniques for Groundwater Resource Evaluation. Illinois *Water Survey Bull.,* 55, 62 p.

Raymond, R.L., Hudson, J.O.and Jamison, V.W. 1977. American Petroleum Institute Project No. 307-76. *Final Report, Bacterial Growth in and Penetration of Consolidated and Unconsolidated Sands Containing Gasoline.* Washington, DC.

Raymond, R.L., Jamison, V.W.and Hudson, J.O. 1975. Committee on Environmental Affairs. Final *Report on Beneficial Stimulation of Bacterial Activity in Groundwater Containing Petroleum Products.* American Petroleum Institute, Washington, DC.

Raymond, R.L., Jamison, V.W. and Hudson, J.O. 1976. Beneficial Stimulation of Bacterial Activity in Groundwaters Containing Petroleum Products. *AIChE Symp. Ser.,* 73, 390.

Raymond, R.L., Jamison, V.W., Hudson, J.O., Mitchell, R.E and Farmer, V.E. 1978.

American Petroleum Institute Project No. 307-77. *Final Report, Field Application of Subsurface Biodegradation of Gasoline in a Sand Formation.* Washington, DC.

Rees, J.F. and King, J.W. 1981. The Dynamics of Anaerobic Phenol Biodegradation in Lower Greensand. *J. Chem. Tech. Biotechnol.,* 31, 306-310.

Rifai, H.S., Bedient, P.B.. 1990. Comparison of Biodegradation Kinetics With an Instantaneous Reaction Model for Groundwater. *Water Resources Research,* 26(4), 637-645.

Rifai, H.S., Bedient, P.B., Borden, R.C. and Haasbeek, J.F. 1987. BIOPLUME II - Computer Model of Two-Dimensional Transport under the Influence of Oxygen Limited Biodegradation in Groundwater, User's Manual, Version 1.0, Rice Univ., Houston, TX. 100 p.

Rifai, H.S., Bedient, P.B., Wilson, J.T, Miller, K.M.and Armstrong, J.M. 1988. Biodegradation Modeling at an Aviation Fuel Spill Site. *ASCE J. Envir. Engrng, ,* 114(5), 1007-1029.

Schwartzenbach, R.P. and Westall, J. 1981. Transport of Non-polar Organic Compounds from Surface Water to Groundwater. Laboratory Sorption Studies. *Environ. Sci. Technol.* 15, 1360-1375.

Schwartzenbach, R.P.and Westall, J. 1984. Sorption of Hydrophobic Trace Organics in Groundwater Systems. *Proc. Symp. Degradation, Retention Dispersion Pollutants Groundwater,* 1984, Univ. Copenhagen, Denmark. 39-55.

Semprini, L., Roberts, P.V., Hopkins, G.D.and Mackay, D.M. 1988. *A. Field Evaluation of in-situ Biodegradation for Aquifer Restoration.* EPA Agency Project Summary 600/S2-87/096. 7p.

Shen, H.T. 1976. Transient Dispersion in Uniform Porous Media Flow. *J. Hydraul. Div. ASCE,* 102, 707-716.

Sinclair, J.L., Randtke, J.E., Hathaway, L.R. and Ghiorse, W.C. 1990. Survey of Microbial Populations in Buried-Valley Aquifer Sediments from Northeastern Kansas. *Groundwater,* 28(3), 369-377.

Smith, J.A. and Novak, J.T. 1987. Biodegradation of Chlorinated Phenols in Subsurface Soils. *Water, Air, and Soil Pollution,* 33, 29-42.

Smith, L. and. Schwartz, FW. 1980. Mass Transport, 1. A. Stochastic Analysis of Macroscopic Dispersion. *Water Resour. Res.,* 16, 303-313.

Spain, J.C., Milligan, J.D., Downey, D.C. and Slaughter, J.K. 1989. Excessive Bacterial Decomposition of H_2O_2 During Enhanced Biodegradation. *Groundwater,* 27(2), 163-174.

Srinivasan, P. and Mercer, J.W. 1988. Simulation of Biodegradation and Sorption Processes in Groundwater. *Groundwater,* 26(4), 475-487.

Streeter, H.W. and Phelps, E.B. 1925. A. study of the Pollution and Natural Purification of the Ohio River, *Public Health Bulletin* No. 146, Public Health Service, Washington, DC.

Suflita, J.M. and Miller, GD. 1985. Microbial Metabolism of Chlorophenolic Compounds in Groundwater Aquifers. *Environ. Toxicol. Chem.,* 4, 751.

Swindoll, C.M., Aelion, C.M. and Pfaender F.K.. 1988. Influence of Inorganic and Organic Nutrients on Aerobic Biodegradation and on the Adaptation Response of Subsurface Microbial Communities. *Appl. Environ. Microbiol.,* 54(1), 212-217.

Thomas, J.M. and Ward, C.H. 1990. In situ Biorestoration of Organic Contaminants in the Subsurface. *Environ. Sci. Technol.,* 23(7), 760-766.

Trescott, P.C., Pinder, G.F. and Larson, S.P. 1976. Finite-Difference Model for Aquifer Simulation in Two Dimensions with Results of Numerical Experiments, *U. S. Geological Survey Techniques of Water-Resources Investigations,* Book 7, Chapter C1, 116 p.

Twenter, R.F., Cummings, T.R and Grannemann, N.G. 1985. U.S. Geologic Survey. Groundwater Contamination in East Bay Township, Michigan. Water Resources Investigation Report 85-4064.

Vogel, T.M and McCarty, P.L. 1985. Biotransformation of Tetrachloroethylene to Trichloroethylene, Dichloroethylene, Vinyl Chloride, and Carbon Dioxide under Methanogenic Conditions. *Appl. Environ. Microbiol.,* 49(5), 1080-1083.

Wang, H.F.and Anderson, M.P. 1982. *Introduction to Groundwater Modeling, Finite Difference and Finite Element Methods.* W. H. Freeman and Co., San Francisco, CA. 237 p.

Webster, J.J., Hampton, G.H., Wilson, J.T. Ghiorse, W.C.and Leach, F.R. 1985. Determination of Microbial Cell Numbers in Subsurface Samples. *Groundwater,* 23, 17.

Wheeler, M.F., Dawson, C.N.and Bedient, P.B. 1988. Numerical Modeling of Subsurface Contaminant Transport with Biodegradation Kinetics - BIOPLUS, *Proc. NWWA/API Conference on Petroleum Hydrocarbons and Organic Chemicals in Groundwater, Prevention, Detection, and Restoration,* pp. 471-490 Houston, TX, National Water Well Association, Dublin, OH.

Wheeler, M.F., Dawson, C.N., Bedient, P.B., Chiang, C.Y., Borden, R.C. and Rifai, H.S. 1987. Numerical Simulation of Microbial Biodegradation of Hydrocarbons in Groundwater, *Proc. AGWSE/IGWMC Conference on Solving Groundwater Problems with Models,* Feb. 1987, Denver, CO, National Water Well Association, Dublin, OH, 92-109.

White, D.C., Smith, G.A., Gehron, M.J., Parker, J.H., Findlay, R. H., Martz, R.F. and Fredrickson, H.L. 1983. The Groundwater Aquifer Microbiotia, Biomass, Community Structure, and Nutritional Status. *Dev. Ind. Microbiol.,* 24, 204.

Widdowson, M.A., Molz, F.J. and Benefield, L.D. 1987. Development and Application of a Model for Simulating Microbial Growth Dynamics Coupled to Nutrient and Oxygen Transport in Porous Media, in *Proc. AGWSE/IGWMCH Conference on Solving Groundwater Problems with Models,* pp.28-51 Denver, CO, National Water Well Association, Dublin, OH.

Widdowson, M.A., Molz, F.J, Benefield, L.D. 1988. A. Numerical Transport Model for Oxygen- and Nitrate-Based Respiration Linked to Substrate and Nutrient Availability in Porous Media. Paper 7W5057. *Water Resources Research,* 24(9), 1553-1565.

Wilson, BH, *et al.* 1986. Biological Fate of Hydrocarbons at an Aviation Gasoline Spill Site. *Proc. NWWA/API Conference on Petroleum Hydrocarbons and Organic Chemicals in Groundwater – Prevention, detection and restoration, National Water Well Association,* 78-89.

Wilson, J.L. and Miller, P.J. 1978. Two-Dimensional Plume in Uniform Groundwater Flow. *J. Hydraul. Div. ASCE,* 104, 503-514.

Wilson, J.T., McNabb, J.F., Wilson, B.H. and Noonan, MJ. 1983a. Biotransformation of Selected Organic Compounds in Groundwater. *Dev. Ind. Microbiol.,* 24, 134-142.

Wilson, J.T., McNabb, J.F.andBalkwill, D.L.. 1983b. Enumeration and Characterization of Bacteria Indigenous to a Shallow Water-Table Aquifer. *Groundwater,* 21, 134-142.

Wilson, J.T., McNabb, J.F., Cochran, J.W., Wang, T.H. Tomson, M.B. and Bedient, P.B. 1985. Influence of Microbial Adaption on the Fate of Organic Pollutants in Groundwater. *Environ. Toxicol. Chem.,* 4, 721-726.

Wilson, J.T., Miller, G.D., Ghiorse, W.C. and Leach, F.R. 1986. Relationship Between the ATP Content of Subsurface Material and the Rate of Biodegradation of Alkylbenzenes and Chlorobenzene. *J. Contaminant Hydrol.,* 1, 163-170.

Wilson, J.T. and Wilson, B.H. 1985. Biotransformation of Trichloroethylene in Soil. *Appl. Environ. Microbiol.*, 49(1), 242-243.

Zeyer, J., Kuhn, E.P. and Schwartzenback, R.P. 1986. Rapid Microbial Mineralization of Toluene and 1, 3-Dimethylbenzene in the Absence of Molecular Oxygen. *Appl. Environ. Microbiol.* 52(4), 944-947.

Zoeteman, B.C., De Greef, E. and Brinkman, F.J.J. 1981. Persistency of Organic Contaminants in Groundwater. Lessons from Soil Pollution Incidents in The Netherlands. *Sci. Total Environ.*, 21, 187-202.

9   *In situ* detection of contaminant plumes in groundwater by remote spectroscopy

W. Rudolf Seitz

Department of Chemistry, University of New Hampshire, Durham, NH 03824, USA

Abstract

Groundwater contaminants can be detected in situ by making spectroscopic measurements through fiber optics. In addition to direct measurements, it is possible to couple fiber optics with chemical indicators that interact with the contaminants to enhance their detectability. Direct fluorescence measurements have been used to sensitively detect aromatic hydrocarbons in fossil fuels. Direct Raman measurements are also possible but can only detect relatively high concentrations (>0.1%). Parts per billion levels of nitroaromatics and halogenated hydrocarbons can be detected using indicators that react to form colored products. The rate at which the absorbance of the colored product increases is proportional to concentration. Refractive index measurements offer a rugged, reversible approach to detecting classes of organic contaminants in the low parts per million range. Other spectroscopic techniques are considered in this chapter but are not practical for in situ groundwater monitoring at this time. A few of the more promising methods are ready for commercial development while many others require considerable further development before they can be reliably used on a routine basis.

Introduction

Present technology for determining the concentrations of contaminants in groundwater requires that the sample be transported back to a laboratory for analysis. This necessarily involves a significant time delay between acquisition of the sample and reporting of the analytical result. This is particularly important when the data are to be used for deciding on the location of sampling wells. Laboratory analysis also requires careful attention to sampling and sample preservation if the analytical results are to accurately reflect groundwater composition.

Because of the intrinsic limitations of laboratory based analysis, there is an urgent need for methodology that can be applied *in situ*. The most promising approach is to apply spectroscopic methods using fiber optics to conduct light between the spectrometer and the sampling site. However, it is important to understand that spectroscopic methods are subject to significant limitations. Even in the laboratory, spectroscopic methods generally lack the selectivity to detect specific compounds in a complex mixture of contaminants. Furthermore, many methods are incapable of detecting concentrations in the low parts per billion range characteristic of many contaminant plumes. When coupled to remote sample locations through fiber optics, the performance of spectroscopic methods may degrade because long lengths of optical fiber attenuate light and may contribute unwanted background to the observed signal. In many cases sensitivity can be increased by using an indicator that reacts with the compound(s) of interest to enhance detectability. However, indicator reactions are rarely specific to single compounds. The net result is that the dream of compound-specific remote *in situ* detection of contaminants in groundwater will only be realized in special cases. More often, the role of *in situ* spectroscopy methodology will be to screen for the presence of classes of compounds, yielding data that are only semiquantitative. Rather than replacing laboratory methods, *in situ* spectroscopic methods will serve to identify those samples which are most in need of laboratory analysis.

This chapter critically reviews the current status of *in situ* spectroscopic sensing methodologies that can potentially be used to detect the following types of plumes in groundwater:

(1) Petroleum fuels (aliphatic and aromatic hydrocarbons)
(2) Chlorinated hydrocarbons
(3) Nitrated organics including explosives
(4) Heavy metal ions

The chapter will emphasize methods which are sensitive to levels in the part-per-billion range. However, less sensitive methods are also considered since they may be useful for monitoring point source contamination.

The potential for using fiber optics for *in situ* spectroscopic measurements in groundwater was recognized several years ago (Hirschfeld *et al.*, 1983; Hirschfeld *et al.*, 1984). More recent progress in developing systems for *in situ* spectroscopic analysis is summarized in review articles (Angel, 1987; Seitz, 1988; Wolfbeis, 1988). The rationale for using fiber optics for *in situ* environmental analysis has been considered from the perspective of the Environmental Protection Agency (Eccles and Eastwood, 1989). The field of fiber optic sensors is covered in a recent two-volume book (Wolfbeis, 1991). The principle strategies for coupling chemical indicators to optical fibers for *in situ* measurements have been reviewed (Arnold, 1992).

Fiber optics and spectroscopy

Traditionally, most spectroscopic methods of analysis have been laboratory oriented. The sample is brought back to the laboratory and presented to the spectrometer in a controlled manner. Often the sample is treated, *e.g.*, pH adjustment or addition of color-forming reactants, to convert analyte to a form that is more readily measured spectroscopically.

Recently, developments in fiber optic technology have stimulated interest in applying spectroscopy *in situ* using optical fibers to conduct light between the sample and the spectrometer. This allows the spectrometer to remain in a controlled environment while coupled to remote sampling sites without serious losses in light intensity.

Plastic-clad silica is the preferred optical fiber for most intensity-based measurements in the visible and ultraviolet regions of the electromagnetic spectrum (Skutnik *et al.*, 1988). It is available with numerical apertures as high as 0.48 and transmits in the ultraviolet down to 220 nm. Attenuation by the fiber increases drastically as wavelength decreases, greatly complicating remote measurements in the ultraviolet below 300 nm. Plastic-clad silica fibers are commonly available with core diameters of 0.20, 0.60 and 1.0 mm. The larger diameter fibers allow for enhanced light transmission but are significantly more expensive. Fibers with 0.60 mm core diameters provide a good compromise between cost and optical throughput and have been used for remote laser induced fluorescence measurements of aromatic hydrocarbons in water (Chudyk *et al.*, 1985).

Glass fibers developed for communications applications have extraordinarily high transmission in the near-infrared region of the spectrum but have low numerical apertures and do not transmit below 380 nm. Interferometric methods may be implemented with single mode optical fibers designed to transmit light without loss of coherence.

There is considerable current interest in developing fibers that conduct light at

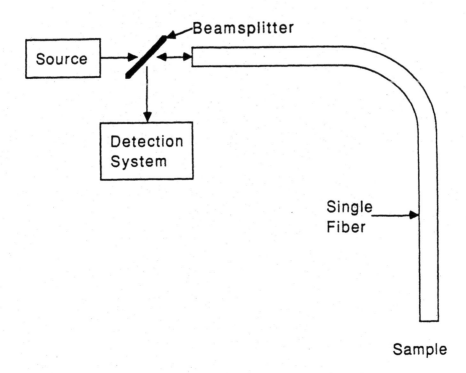

Figure 1 *Schematic of arrangement for single fiber measurements.*

longer wavelengths, extending further into the infrared region of the spectrum (Drexhage and Moynihan, 1988). However, while developments in this area are enhancing the capabilities of infrared spectroscopy, this technique is not sensitive enough to detect contaminant plumes in groundwater, even when a polymer coating is used to preconcentrate analyte (Krska *et al.*, 1992). Furthermore, the attenuation of presently available infrared transmitting fibers is too high to allow measurements over distances greater than three to five meters.

A single fiber can be used to transport light both to and from the sample. Optical arrangements for implementing this are shown in Figure 1. However, while this arrangement conserves fiber and provides a well-defined reproducible geometry, it is

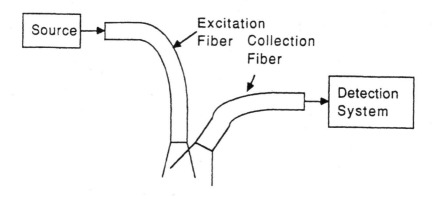

Schematic of system using separate fibers to conduct light to and from the sample. The two fibers are at an angle relative to each other to maximize overlap between the zone illuminated by the excitation fiber and the zone collected by the collection fiber.

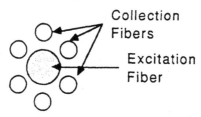

Schematic of arrangement in which sample is illuminated by a single large excitation fiber surrounded by several collection fibers.

Figure 2 *Separate fibers used to conduct light to and from the sample.*

subject to high levels of stray light arising from reflections at interfaces where there is a change in refractive index, *e.g.* the points where light enters and exits the fiber. Fluorescence can be resolved from stray light using a monochromator or filter because it occurs at longer wavelengths than the stray light. However, the quality of the required optics is higher when fluorescence has to be measured in the presence of high levels of stray light. Furthermore, fluorescence and Raman scattering from the fiber itself produce an unavoidable background signal.

For *in situ* groundwater monitoring, it is more practical to use separate fibers to conduct light to and from the spectrometer as shown in Figure 2a. An experimental comparison of single vs. dual fiber arrangements has shown that fluorescence intensities observed for a single fiber measurement are not significantly greater than those observed using two fibers (Louch and Ingle, 1988). As shown in Figure 2b, collection efficiencies are maximized during emission spectroscopy by holding the two fibers at an angle relative to each other such that there is maximum overlap between the cone of radiation excited by the fiber from the source and the cone of radiation accepted by the fiber leading to the detection system (Plaza *et al.*, 1986). Intensity can be further enhanced by surrounding the excitation fiber with several return fibers, as shown in Figure 2b (Schwab and McCreery, 1984; Plaza *et al.*, 1986).

Direct spectroscopy

All spectroscopic methods that can be implemented in the ultraviolet, visible and near infrared regions of the electromagnetic spectrum can be coupled to optical fibers. Direct spectroscopy can be applied if a contaminant or class of contaminants can be detected at the concentrations that occur in groundwater. This form of detection is preferred when applicable since it can provide rapid and continuous results. Various types of spectroscopy are considered below.

Near infrared spectroscopy

Near infrared spectra are due to weak overtone and combination vibrational bands. Although they appear weak and featureless, these spectra can be recorded at very low noise levels. This makes it possible to use multivariate statistical methods to enhance small spectral differences. This type of measurement is readily implemented through optical fibers (Archibald *et al.*, 1988; Foulk and Gargus, 1987; Weyer *et al.*, 1987). However, it is limited to major and minor constituents (components present at levels > 1%) and requires some *a priori* knowledge of the sample composition. This method is not suitable for *in situ* plume monitoring although it may be useful to detect certain types of concentrated point sources.

Raman spectroscopy

Raman spectroscopy is most frequently implemented at wavelengths well out in the visible where optical fibers transmit efficiently. Although Raman spectra have been measured *in situ* through fibers (Archibald *et al.*, 1988; Leugers and McLachlan, 1989; Lewis *et al.*, 1988; Schwab and McCreery, 1984), the detection limits are on the order of 1%, not nearly low enough to detect typical contaminant levels in plumes. Detection limits for phenols in the mg L^{-1} range have been obtained using conventional Raman spectroscopy (Marley *et al.*, 1985). A special cell with a 1-m path length coupled to fiber optics has been used to increase the *in situ* sensitivity for normal Raman scatters to approximately 1 mg L^{-1} (Schwab and McCreery, 1987). Resonance Raman spectroscopy is capable of significantly lower detection limits (Schwab and McCreery ,1987). However, resonance Raman spectroscopic detections of groundwater contaminants would be difficult technically and expensive because it requires laser excitation in the ultraviolet.

Interaction with a metal surface greatly enhances the Raman effect and allows compounds to be detected at much lower concentrations than possible with conventional Raman spectroscopy (Carrabba *et al.*, 1987). Because this technique, known as surface enhanced Raman spectroscopy, requires interaction between the analyte and a surface, it is considered below as a spectroscopic technique coupled to a chemical indicator. The "indicator" is the surface that enhances the Raman effect.

The advantages of Raman spectroscopy for *in situ* detection are that it can be applied to aqueous solutions and can detect any organic contaminant. Dow Chemical Company has developed *in situ* Raman spectroscopy for process control applications (Leugers and McLachlan, 1989).

While sensitivity is likely to remain a problem for conventional Raman spectroscopy, new technology developments are greatly simplifying this technique while improving its performance. The new instrumentation is much more easily implemented in the field (Angel *et al.*, 1993). The most important development is the use of highly sensitive, multichannel charge coupled device detectors in place of the usual double or triple monochromator with a high sensitivity photomultiplier. Other developments include the use of solid state lasers for excitation and high quality band reject filters to greatly reduce Rayleigh scattering without affecting the Raman intensities. Already these advances are being coupled to fiber optic sampling (Allred and McCreery, 1990; Newman *et al.* 1991; Carraba *et al.*, 1993).

It is also possible to enhance the performance of Raman spectroscopy through optical fibers by using filters to block out the Raman band from the excitation fiber, thus reducing a significant source of background (Myrick and Angel, 1990). Another possibility is to focus light emerging from the fiber at a point away from the surface of the fibers (Myrick and Angel, 1990, Carraba *et al.*, 1993). This allows the ends of the optical fibers to be separated from the sampling point by a window.

Laser induced fluorescence

Laser-induced fluorescence (LIF) is the most promising of the direct methods of spectroscopy that can be implemented through fiber optics. Fluorescence methods are inherently sensitive, capable of measuring concentrations in the part-per-billion range. A laser provides intense, highly collimated excitation radiation that illuminates a precise area and can be efficiently coupled into an optical fiber.

Of the four types of contaminant plumes considered in this chapter, LIF is only applicable to petroleum fuels, since aromatic hydrocarbons fluoresce with high efficiency. Heavy metals, chlorinated and unchlorinated aliphatic hydrocarbons, and nitroaromatic hydrocarbons do not fluoresce. Chlorinated aromatics may fluoresce weakly but the presence of chlorine substituents reduces the quantum efficiency of emission. At this point, the possibility of *in situ* fluorimetric detection of chlorinated aromatics remains to be demonstrated.

Benzene, toluene and xylenes are the principal aromatic components of petroleum fuels. Naphthalenes and other higher aromatics may occur in lesser amounts but still at levels readily detected by fluorescence. However, because higher aromatics are not as soluble in water as other components of petroleum fuels, it is to be expected that they will not migrate as rapidly in groundwater. Dyes added to oils would also migrate at their own unique rate. Therefore, LIF techniques based on detection of higher aromatics or dyes may not adequately represent plume distribution. The author's opinion is that LIF techniques for mapping petroleum plumes are properly focussed on detection of single-ring aromatics even though instrumentally this is far more challenging than detecting higher aromatics which fluoresce with greater efficiencies at longer wavelengths.

The feasibility of detecting single-ring aromatics by LIF coupled to fiber optics has been demonstrated (Chudyk *et al.*, 1985). The source was a frequency quadrupled Nd-YAg laser emitting at 266 nm. Radiation was coupled into a 0.60 mm core diameter plastic-clad silica fiber. Separate fibers at an angle of 22° were used to conduct light to and from the sample. Fiber lengths up to 25 m were used. At the excitation wavelength, 266 nm, the attenuation is approximately 500 dB km^{-1} which corresponds to an absorbance of 0.050 m^{-1} Experimentally, the transmittance of 25 m of 0.60 mm core diameter fiber was found to be about 0.04.

Detection limits depend on the particular compound and the measurement conditions. However, they are well down in the part-per-billion range using a filter with maximum transmittance at 320 nm to resolve fluorescence. Typically, measurements required 3 to 5 minutes corresponding to approximately 5,000 laser pulses to build up a large integrated intensity value. However, this could easily be shortened, although it would necessarily involve some degradation in the detection limit.

Intensities measured by LIF have been compared to the total aromatic content of gasoline measured by gas chromatography at a series of sites including 12 gasoline stations, two manufacturing companies and one chemical company (Chudyk *et al.*, 1989). The correlation between the two methods was poor, with differences exceeding an order of magnitude on many samples. In particular, LIF gave higher results on samples on which gas chromatography measured low aromatic concentrations. The origin of the discrepancies is uncertain. Because the aromatic content of gasoline varies from sample to sample, the two methods would not be expected to give identical results. However, the pattern of the data suggests that there may be background signal due to several possible sources including natural fluorescence in the sample, increases in stray light levels resulting from scattering in the sample, and memory effects associated with incomplete removal or aromatics between samples. Recent laboratory measurements indicate that crosstalk between the excitation and emission fibers and the highly nonlinear calibration curves contribute to the difficulty of getting a good correlation with other methods (Chudyk *et al.*, 1990).

Although LIF shows promise for direct *in situ* detection of hydrocarbon plumes, actual performance needs to be improved.

Field LIF measurements made to date involve a single intensity measurement. One way of enhancing the information content is to measure intensities at multiple wavelengths. Research to develop the instrumentation technology to do this is currently underway.

Refractive index detection

A variety of methods may be used to optically sense refractive index (Bobb *et al.*, 1989). One approach is to measure the intensity of light reflected at the end of a fiber (Meyer and Eesley 1987). Refractive index measurements potentially provide a convenient low-cost method for identifying pockets of pure contaminant existing at a highly polluted site. They may also be useful for detecting contaminants at very high concentrations.

Spectroscopic techniques coupled to chemical indicators

Petroleum fuels and other petroleum derived products with aromatic components are the only type of groundwater contaminants amenable to direct spectroscopic detection at the low part-per-billion level. Current research to develop methods for *in situ* detection of other types of contaminants has concentrated on systems involving an "indicator" phase that somehow interacts with the analyte to render it spectroscopically detectable.

The reaction between the indicator and the analyte can be reversible or irreversible. Reversible indicators can potentially be used for continuous sensing. However, the range of concentrations that is detected depends on the equilibrium constant for the analyte/indicator interaction. Reversible indicator reactions with the appropriate equilibrium constant are often not available.

If the indicator reaction is irreversible, the reaction will proceed continuously resulting in the accumulation of product. High sensitivity can be achieved by waiting sufficiently long for a detectable amount of product to form. However, this approach is inherently kinetic and depends on the rate of the indicator/analyte reaction.

Nitro compounds

Amine-containing membranes: Zhang *et al.*, have developed a primary amine containing poly(vinyl chloride) (PVC) membrane that reacts with 2,4,6-trinitrotoluene (TNT) and other polynitroaromatics to form a brown product (Zhang *et al.*, 1989a; Zhang *et al.*, 1989b). The absorption maximum for membranes exposed to TNT is at 510 nm. Because the absorption spectra of the product differs for various polynitroaromatics, the membrane could potentially be used to distinguish different explosives in a mixture.

Formation of the brown product is irreversible. The measured parameter is the rate of color formation. For a typical membrane formulation with a thickness of 0.36 mm, the initial rate of color formation is 0.001 absorbance unit per ppm TNT per minute, when membranes are exposed to aqueous TNT standards (Zhang *et al.,* 1988b). This is independent of membrane area. The detectability depends on the length of time between exposing the membrane to TNT and measuring the membrane absorbance. To get the detection limits below 1 ppm, a typical plume level, it is necessary to wait at least 10 minutes (assuming the minimum detectable change in absorbance is 0.010). Slow diffusion in the PVC membrane is the reason that the TNT-sensitive membrane responds so slowly.

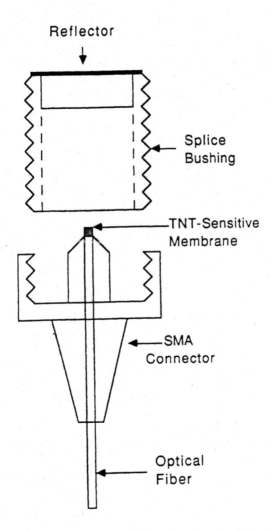

Figure 3 *Arrangement for remote measurements of the absorbance of TNT sensitive membranes.*

The color change occurring in PVC membranes has been measured through a single optical fiber (Zhang and Seitz, 1989; Zhang *et al.*, 1989a). However, this measurement is subject to relatively high levels of stray light due to reflection at interfaces where there is a change in refractive index. For reasons discussed above, the two-fiber arrangement shown in Figure 2a would be preferred.

Figure 3 shows the arrangement used for measuring membrane absorbance. A small piece of membrane is coupled to an optical fiber using Thomas Lubriseal, a grease which serves both as an adhesive and refractive index coupling medium. A reflector behind the membrane helps to redirect a large fraction of the incident light back into the optical fiber.

The system shown in Figure 3 and other remote absorption measurements through fiber optics are all single beam because there is no practical way to construct a reference cell. Instead, intensity at the analytical wavelength is referenced to transmitted intensity at another wavelength which is not absorbed by the sample. This way both the reference and the sample beam traverse the same optical path.

Because the measurement is based on absorption, absolute intensity levels are considerably higher than they are for fluorescence. A field instrument capable of making this measurement could be constructed at modest cost (*ca.* $2,000-5,000) using LEDs or small incandescent bulbs as light sources. The membrane has been shown to respond reliably to TNT in munitions wastewater samples. There is no detectable blank response when membranes are exposed to uncontaminated groundwater. The chemical reaction that is believed to be responsible for formation of the brown product is specific for polynitroaromatics. In short, all available evidence suggests that this membrane is suitable for field use but it would be strongly advisable to test its response on a larger number of samples.

Fluorescence quenching: An alternative approach to the detection of nitrated

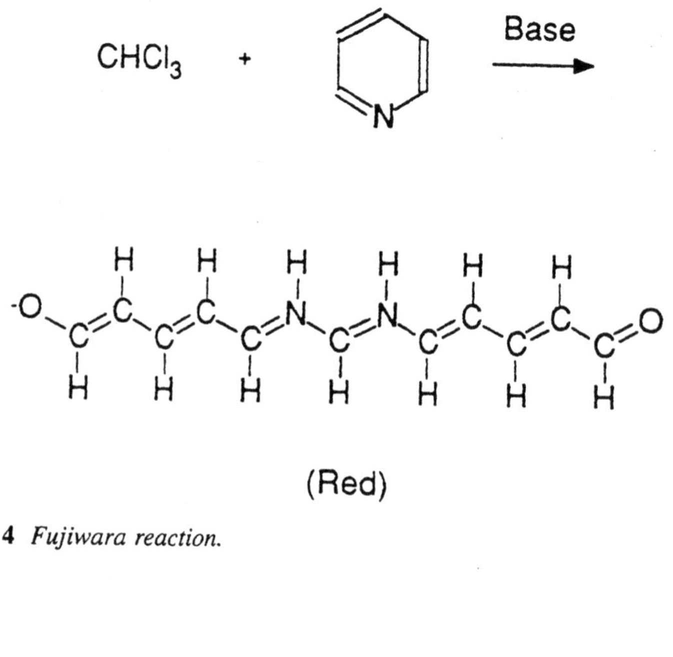

(Red)

Figure 4 *Fujiwara reaction.*

organics based on fluorescence quenching has been investigated (Jian and Seitz, 1990). Pyrenebutyric acid is incorporated into plasticized cellulose acetate membranes. These membranes preconcentrate nitrated organics. The presence of nitro compounds in the membrane quenches pyrenebutyric acid fluorescence. The membrane responds to all nitro compounds with sensitivity depending on the tendency of the nitro compound to partition into the membrane. The fluorescence quenching approach offers the advantage that it responds to hexahydro-1,3,5-trinitro-1,3,5-triazine (RDX), an important nonaromatic explosive that is not detected by the PVC membrane.

Detection limits are in the 1–10 ppm range and response times are on the order of an hour or more. This approach lacks the sensitivity required to map typical contaminant plumes. However, where applicable, it can provide continuous reversible *in situ* response to nitro compounds.

Volatile halogenated organics
Fujiwara reaction: Efforts to detect halogenated organics have centered on the Fujiwara reaction shown in Figure 4. Halogenated organics react with basic pyridine

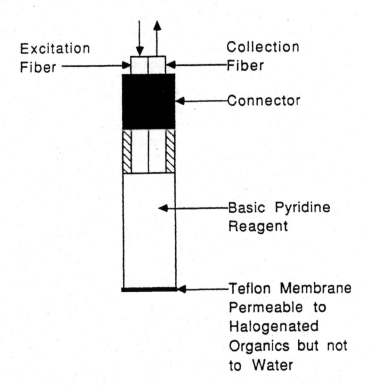

Figure 5 *Schematic of indicator system for detecting halogenated organics. The Teflon membrane serves as a reflector to redirect excitation light back towards the collection fiber.*

to form a fluorescent red product. For sensing applications, the reagents are held at the end of an optical fiber and prevented from direct contact with the sample by a hydrophobic membrane that is permeable to volatile halogenatd organics. The reaction is irreversible. When response reaches a saturation level, the indicator reagents need to be renewed. The reaction is selective in that different reagent formulations react at different rates with various halogenated organics. Formulations that are selective for trichloroethylene (TCE) and chloroform have been developed (Angel *et al.*, 1989; Angel and Ridley, 1990).

The measured parameter is the rate of product formation determined either via the rate of increase in fluorescence or absorbance. Initial studies were based on the rate of increase of fluorescence (Milanovich *et al.*, 1986a; Milanovich *et al.*, 1986b). More recently this type of sensor has been used to determine chloroform concentrations in the headspace of contaminated wells (Herron *et al.*, 1989). However, when the measured parameter is fluorescence, it has been difficult to engineer this system to get reproducible response from sensor to sensor. Fluorescent product initially forms at the interface where halogenated organic first comes in contact with the reagent. This is followed by mixing of the product and the reagent due primarily to convection, which does not occur in a reproducible way. The primary cause of poor reproducibility is variation in the position of the fluorescent product relative to the end of the fiber. This affects both the excitation intensity and the efficiency with which the fluorescence is collected. Some improvement in reproducibility can be achieved by adding polymers to the reagent to increase its viscosity and reduce the extent of convective mixing (Herron *et al.*, 1989).

Fluorescence measurements are also subject to saturation due to the "inner filter effect." As red product accumulates, it absorbs a significant fraction of the excitation radiation, which causes the observed fluorescence to be attenuated.

The performance of Fujiwara reaction based sensors for volatile halogenated organics is significantly improved by measuring the increase in the absorbance of the red product using separate fibers to conduct light to and from the Fujiwara reagent (Angel *et al.*, 1989; Angel and Ridley, 1990). The optical arrangement of the reagent phase is shown in figure 5. In this arrangement the Teflon membrane serves as a reflector to redirect the incident radiation back toward the fiber leading to the detection system. The absorbance approach offers several important advantages. Sensor-to-sensor variability is greatly improved. One reason is that, in the absorbance mode, all absorbing molecules in the optical path affect the observed signal equally, independent of how far they are from the end of the fiber. Another reason is that the measured parameter is the ratio of intensity at 530 nm where the red product absorbs strongly to the intensity at 610 nm which is not absorbed. The use of a ratio measurement compensates for variability in source intensity and the optical characteristics of the indicator phase, most notably variations in the position of the reflecting surface relative to the ends of the two fibers. The absorbance measurement also requires much less intensity. This not only minimizes photodegradation as a source of error but also substantially simplifies instrumentation requirements.

The rate of response to chloroform is 0.15 absorbance units per ppm per minute with a lag time of less than a minute (Angel *et al.*, 1989). This is quite rapid, making

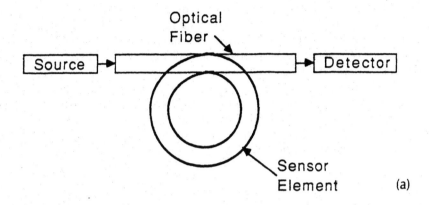

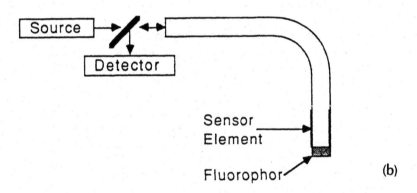

Figure 6 *Schematic of arrangements for refractive-index-based sensors. a, Source and detector at opposite ends of the fiber. b, Source and detector at the same ends of the fiber.*

it possible to detect concentrations in the low parts per billion range in a few minutes. This system has been successfully field tested. The system has been engineered to fit into a cone penetrometer to measure TCE vs. depth as a sampling hole is dug into the ground (Milanovich *et al.*, 1993).

Refractive index: Trichloroethylene (TCE) in water and air has been detected through fiber optics based on a change in refractive index (Oxenford *et al.*, 1989). The cladding is stripped from the core of a fiber and replaced by a coating that has an affinity for TCE. TCE causes a change in the refractive index of the coating, which affects the intensity of light transmitted through the fiber. This approach has been more fully characterized as a method for detecting hydrocarbon contaminants (Klainer *et al.*, 1988). Because it is reversible, it could be left in place in groundwater and used for continuous *in situ* monitoring. However, considerable further work is required to characterize this approach. It is doubtful that detection limits below 1 ppm can be attained.

Another possible approach to *in situ* detection of halogenated hydrocarbons (and other species) is to apply an RF discharge or an electrical spark to a sample at the tip of an optical fiber. Preliminary experiments have demonstrated response to chlorine-containing compounds (Griffin *et al.*, 1989). However, considerable further work is required before the feasibility of this approach can be evaluated.

Hydrocarbons
Refractive index: Hydrocarbons in water and air have been detected based on changes in refractive index (Kawahara *et al.*, 1983). The core of an optical fiber is coated with a layer that interacts with hydrocarbons, undergoing a change in refractive index. This in turn affects the intensity of light propagated through an optical fiber. Maximum sensitivity to small changes in refractive index is achieved by designing the system so that the refractive index of the coating is only slightly lower than the refractive index of the core. Intensities propagated through the fiber can be measured directly by placing a source and a detector at opposite ends of the fiber. As shown in Figure 6a, the coated section of fiber can be coiled to make a relatively small sensing element. Because light penetrates into the coating for only a distance on the order of a couple wavelengths, it is easy to make the coating thick enough so that the light does not directly interact with sample and is not affected by turbidity.

An alternative arrangement for making the measurement is shown in Figure 6b. The coating is placed on a short length of fiber right near the end and a fluorophor is attached directly to the end (Klainer *et al.* 1988; Dandge *et al.*, 1990). The measured parameter is the intensity of fluorescence. The advantage of this arrangement is that the sensing element is at the end of a fiber and can be inserted directly in to the sample.

The sensitivity of this approach to a particular compound depends on the affinity of the coating for the compound and the magnitude of the change in refractive index when the compound partitions into or onto the phase. In general, the less soluble a compound is, the more strongly it will partition into an organic coating and the more sensitively it can be detected. A particular coating will be selective for a class of compounds, *e.g.* aliphatic hydrocarbons, rather than responding to a particular compound. Up to a point, sensitivity can be enhanced simply by coating a longer length of fiber. However, a point will ultimately be reached where sensitivity will be limited by thermal fluctuations in refractive index.

A sensor for oil in water has been developed based on the above principle (Kawahara *et al.*, 1983). The core of a fused silica fiber is reacted with octadecylsilane to produce a lipophilic surface which absorbs hydrocarbons. The sensor is more sensitive to aromatic hydrocarbons because they have higher refractive indices than aliphatic hydrocarbons. The detection limit depends on both the refractive index of a compound and its affinity for the modified surface. The detection limit for crude oil was 3 mg L^{-1}. However, for many other compounds like xylenes it was significantly higher. The measure was made with the transmission arrangement shown in Figure 6a using a helium-neon laser as the source.

Gasoline vapors have been detected using the optical arrangement shown in Figure 6b (Klainer *et al.*, 1988). Response is reversible but quite slow. The response for 24-hour equilibration is considerably greater than for 3-hour equilibration, indicating that the sensor is far from equilibrium after 3 hours. The sensor responds to levels as low as 1 µL gasoline L^{-1} air. It is not clear how this translates to response to hydrocarbons in water.

In general, sensors in which the target contaminants interact with a coating to produce a refractive index change have long response times. However, because response involves partitioning rather than a chemical reaction, these sensors are inherently reversible and can be used for continuous monitoring. However, the sensitivity levels are marginal at best for mapping contaminant plumes in groundwater. Considerable further research and development will have to be done on the preparation and evaluation of such sensors to establish that they could be used for continuous *in situ* groundwater monitoring. Although the fiber optic gasoline sensor is under development by a company with the goal of commercialization, its primary application would be to look for relatively high concentrations of gasoline near a contamination source rather than the lower levels that would be found in a plume.

Environment sensitive fluorescence: A fiber optic sensor for gasoline vapors has been prepared by incorporating an environment sensitive fluorophor in a siloxane polymer at the tip of a single optical fiber (Walt *et al.*, 1989; Barnard and Walt 1991). Gasoline vapors partition into the polymer, modifying the microenvironment of the fluorophor and causing an increase in fluorescence intensity and a shift in the emission maximum to shorter wavelength.

More recently, a similar approach using Nile Red as the environment sensitive fluorophor has been shown to respond to a variety of organic vapors including hydrocarbons (Angel *et al.*, 1993).

The response of these sensors has yet to be characterized in depth. While response is inherently reversible, it has to be established that the fluorophor is sufficiently stable for longterm use. While the concept is interesting, considerable further work is required to determine whether it can be applied in a practical context.

Surface Enhanced Raman Spectroscopy (SERS)

Adsorption on metal surfaces greatly enhances the Raman effect. The possibility of using this effect for *in situ* analysis of groundwater contaminants is under investigation (Carrabba 1988, Carrabba *et al.*, 1988; Carrabba *et al.*, 1987; Vo-Dinh *et al.*, 1988; Bello *et al.*, 1990; Alarie *et al.*, 1992). The most widely applied system consists of an electrochemical cell designed so that analyte adsorbs on a silver electrode positioned so that compounds on the surface are efficiently excited. This approach offers several attractive features. It can be applied to any organic contaminant. In addition to spectroscopic selectivity, electrochemical selectivity can be achieved by taking advantage of variations in the potential dependence of adsorption for different compounds. Sensitivity can be enhanced by allowing time for analyte to accumulate at the electrode. Because electrochemical cycling can be used to renew the surface, it can be used for several sequential measurements *in situ*.

Design aspects of a SERS/electrochemistry system for *in situ* measurements have been addressed, and several proof-of-principle experiments have demonstrated its attractive features (Carrabba *et al.*, 1987). Nevertheless, the challenges of rendering this approach practical in a field context are formidable. While exciting progress has been made in improving detection limits, it still remains to be shown that SERS/electrochemistry can detect contaminants in groundwater at the 10 to 100 ppb level common for many samples. The transition from laboratory measurements of prepared samples to field measurements of contaminants in groundwater will almost certainly compound the sensitivity problem. Samples which fluoresce will be subject to high background signals. Other samples may contain components which adsorb on the electrode and interfere with the electrochemistry.

The time scale of SERS/electrochemistry depends on the desired sensitivity but is on the order of minutes or more. It includes time to allow analyte to accumulate on the electrode surface as well as time to acquire the spectral data with adequate signal to noise.

At this point in time, considerable further development must be completed before SERS/electrochemistry can be considered a serious candidate for practical *in situ* contaminant measurements in groundwater.

Immunosensing

Immunochemical methods are based on the highly selective interaction between an antibody and an antigen. An *in situ* immunochemical method for benzo(a)pyrene has been developed by immobilizing antibody to benzo(a)pyrene on the end of an optical fiber (Vo-Dinh *et al.*, 1987, Tromberg *et al.*, 1988). The measured parameter is the rate of increase in benzo(a)pyrene fluorescence as it binds to antibody. The detection limit was subnanomolar for a 15-minute incubation time.

Nonfluorescent analytes can be detected via the rate at which they displace fluorophor-labeled antigens from antibodies immobilized at the tip of an optical fiber (Anderson and Miller 1988). The measured parameter in this case is the rate at which fluorescence intensity decreases due to displacement. The time scale of these measurements is on the order of minutes. They are specific to particular compounds rather than responding to a class of compounds. The approach is generic for all compounds against which antibodies can be prepared. This includes substituted aromatics like TNT and multiring compounds like benzo(a)pyrene. Current research is directed at developing immunochemical methods for a variety of compounds including tetryl, benezene, dieldrin and para-chlorophenylmethylsulfone in environmental samples.

The instrumentation used for remote measurements has been an argon ion laser as the source with a monochromator to resolve fluorescence from scattered excitation radiation. This instrumentation is relatively complex and expensive. Immunoassays can be implemented in the field using simple color development methods. While these methods can't be implemented *in situ*, they are much less expensive. It's not clear that the *in situ* capability justifies the additional cost for most application contexts.

Detection of metal ions

There have been several reports of optical *in situ* measurements of metal ions (Inman *et al.*, 1989; Lieberman *et al.*, 1987; Saari and Seitz, 1983; Zhujun and Seitz, 1985). The indicator is a ligand with fluorescence characteristics that change upon metal ion binding. Most systems have involved nonfluorescent ligands that form fluorescent complexes. These systems are limited to metal cations that form fluorescent complexes. Of the important metal ion contaminants, Cd(II) can be detected by this approach. Since detection limits are on the order of 10 ppb, the method has sensitivity approaching that required for groundwater analysis. However, most other metal ions including Pb(II), Hg(II) and Cr(III) do not form fluorescent complexes.

Selectivity depends on the relative affinity of the ligand for various metal ions. Response is inherently pH dependent for most systems reported to date, because the metal ion has to displace one or more protons to form a complex with the indicator ligand.

None of the metal ion sensing systems reported to date meets the requirements for detecting groundwater contaminants. However, in principle, it should be possible to find a system for screening for heavy metal ions. It would have to be based on absorbance rather fluorescence in order to get response to the metal ions that do not form fluorescent complexes. Sensitivity can be enhanced by using a ligand with a high affinity for metal ions and allowing time for metal ions to accumulate in the indicator phase. However, as with other indicator systems, there will be a tradeoff between sensitivity and time.

The simplest approach for developing a metal ion sensor is to immobilize the indicator on a solid phase at one end of an optical fiber. A more powerful approach is to design the system to provide a constant supply of indicator to the tip of the fiber. Since the indicator is constantly renewed, photodegradation is much less of a problem. Also, sensors can be designed to respond continuously on a steady state basis. A pressurized membrane indicator system has been used to supply indicator to a fluorigenic fiber optic metal ion sensor system (Inman *et al.*, 1989). Another attractive approach is to use a controlled release polymer to provide a constant supply of indicator to a sensor (Barnard and Walt, 1991). Although this has not been used with metal ion sensors, it has been successfully implemented with pH indicators (Luo and Walt, 1989).

An alternative approach to metal ion sensing based on ion pairing combines high sensitivity with improved selectivity. The reagent phase is a poly(vinyl chloride) membrane containing a neutral lipophilic ionophore selective for the analyte ion and a pH indicator which is neutral in the protonated form. When the ionophore binds analyte it becomes positively charged. This induces the pH indicator to release protons and form an anion which pairs with the cationic ionophore-analyte complex.

This is accompanied by a change in optical properties. The chemistry of this approach is considered in a recent review (Arnold, 1992). This approach has been used to selectively detect lead ion with detection limits below 1 ppb (Lerchi *et al.*, 1992). Because the overall response process involves exchange of a metal ion for a proton, this approach is inherently pH sensitive and would require an independent measurement of pH.

Acknowledgement

The author gratefully acknowledges partial financial support for this project by US Army Cold Regions Research and Engineering Laboratory (USACRREL) through Contract #CRRLRC90472576. Dr. Tom Jenkins of USACRREL provided several helpful suggestions.

References.

Alarie, J.P., Stokes, D.L., Sutherland, W.S., Edwards, A.C. and Vo-Dinh, T. 1992. Intensified charge coupled device-based fiber optic monitor for rapid remote surface-enhanced Raman scattering sensing. *Appl. Spec.*, 46, 1608-1612.

Allred, C.D. and McCreery, R.L. 1990. Near-infrared Raman spectroscopy of liquids and solids with a fiber-optic sampler, diode laser, and CCD detector. *Appl. Spec.*, 44, 1229-1231.

Angel, S.M. April 1987. Optrodes, chemically selective fiber-optic sensors. *Spectroscopy*, 2, 38-48.

Angel, S.M., Katz, L.F., Archibald, D.D., Lin, L.T. and Honigs, D.E. 1988. Near-infrared surface-enhanced Raman spectrscopy. Part I, Copper and gold electrodes. *Appl. Spec.*, 42, 1327-1329.

Angel, S.M., Ridley, M.N., Langry, K., Kulp, T.J. and Myrick, M.L. 1989. New developments and applications of fiber-optic sensors., In: Murray, R.W., Dessey, R.E., Heineman, W.R., Janata, J. and Seitz, W.R. (eds), *Chemical Sensors and Microinstrumentation*. American Chemical Society Series 403, Washington, DC. 345 pp.

Angel, S.M. and Ridley, M.N. 1990. Dual-wavelength absorption optrode for trace-level measurements of trichloroethylene and chloroform. *Proc. SPIE*, 1172, 115-122.

Angel, S.M., Vess, T., Langry, K., Kyle, K. and Kulp, T. 1993. Development of remote sensing methods for environmental and process monitoring applications. *Proc. Symp. Chemical Sensors II*, pp. 625-633. Electrochem. Society, Pennington, NJ.

Anderson, F.P. and Miller, W.G. 1988. Fiber optic immunochemical sensor for continuous, reversible measurement of phenytoin. *Clin. Chem.*, 34, 1417-1421.

Archibald, D.D., Lin, L.T. and Honigs, D.E. 1988. Raman spectroscopy over optical fibers with the use of a near-IR FT spectrometer. *Appl. Spec.*, 42, 1558-1563.

Archibald, D.D., Miller, C.E., Lin, L.T. and Honigs, D.E. 1988. Remote near-IR reflective measurements with the use of a pair of optical fibers and a Fourier transform spectrometer. *Appl. Spec.*, 42, 1549-1558.

Arnold, M.A. 1992. Fiber optic chemical sensors. *Anal. Chem.*, 64, 1015A-1025A.

Barnard, S.N. and Walt, D.R. 1991. Fiber optic organic vapor sensor. *Environ. Sci. Technol.*, 25, 1301-1304.

Barnard, S.N. and Walt, D.R. 1991. Chemical sensors based on controlled-release polymer systems. *Science,* 251, 927-929.

Bello, J.M., Narayanan, V.A., Stokes, D.L. and Vo-Dinh, T. 1990. Fiber-optic remote sensor for *in situ* surface-enhanced Raman scattering analysis. *Anal. Chem.,* 62, 2437-2441.

Bobb, L.C., Krumholtz, H.D. and Davis, J.P. 1989. Optical fiber refractometer. *Proc. SPIE,* 990, 164-169.

Carrabba, M.M. 1988. Surface enhanced Raman spectroscopy SERS., an existing or emerging chemical sensing technology? *Proc. Department of Energy* In situ *Characterization and Monitoring Technologies Workshop* DOE/HWP-62.

Carrabba, M.M., Edmonds, R.B. and Rauh, R.D. 1987. Feasibility studies for the detection of organic surface and subsurface water contaminants by surface-enhanced Raman spectroscopy on silver electrodes. *Anal. Chem.,* 59, 2559-2563.

Carrabba, M.M., Edmonds, R.B., Marren, P.J. and Rauh, R.D. 1988. The suitability of surface enhance Raman spectroscopy SERS. to fiber optic chemical sensing of aromatic hydrocarbon contamination in groundwater. *Presented at Sympoisum on Field Screening Methods for Hazardous Waste Site Investigations,* Las Vegas, NV.

Carraba, M.M,. Edmonds, R.B., Spencer, K.M. and Rauh, R.D. 1993. Fiber optic Raman chemical sensors. *Proc. Symp. Chemical Sensors II* , pp. 634-642. Electrochem. Society, Pennington, NJ.

Chudyk, W.A., Carrabba, M.M. and Kenny, J.E. 1985. Remote detection of groundwater contaminants using far-ultraviolet laser-induced fluorescence. *Anal. Chem.,* 57, 1237-1242.

Chudyk, W., Pohlig, K., Rico, N. and Johnson, G. 1989. Groundwater monitoring using laser fluorescence and fiber optics. *Proc. SPIE,* 990, 4545-4548.

Chudyk, W., Pohlig, K., Wolf, L. and Fordiani, R. 1990. Field determination of groundwater contamination using laser fluorescence and fiber optics. *Proc. SPIE,* 1172, 123-139.

Dandge, D.K., Salinas, T., Klainer, S.M., Goswami, K. and Butler, M. 1990. Fiber optic chemical sensor for jet fuel. *Proc. SPIE,* 1172, 132-139.

Drexhage, M.G. and Moynihan, C.T. 1988. Infrared optical fibers. *Scientific American,* Nov, 110-116.

Eccles, L.A. and Eastwood, D. 1989. Rationale for *in situ* environmental monitoring with fiber optics. *Proc. SPIE,* 990, 30-36.

Foulk, S. and Gargus, A.G. 1989. Fiber optic spectroscopy and multivariate analysis for *in situ* chemical monitoring. *Amer. Lab.,* December, 52-53.

Griffin, J.W. and Olsen, K.B., Matson, B.S., Nelson, D.A. and Eschbach, P.A. 1989. Fiber optic spectrochemical emission sensors. *Proc. SPIE,* 990, 55-68.

Herron, N.R., Simon, S. and Eccles, L.A. 1989. Remote detection of organochlorides with a fiber optic based sensor III, Calibration and field evaluation of an improved chloroform fiber optic chemical sensor with a dedicated portable fluorimeter. *Anal. Instrument.,* 18, 107-126.

Hirschfeld, T., Deaton, T., Milanovich, F. and Klainer, S. 1983. Feasibility of using fiber optics for monitoring groudnwater contaminants. *Opt. Eng.,* 22, 527-531.

Hirschfeld, T., Deaton, T., Milanovich, F. and Klaimer, S. 1984. *Feasibility of using fiber optics for monitoring groundwater contaminants.* EPA Report 600/7-84-067.

Inman, S.M., Stromvall, E.J. and Lieberman, S.H. 1989. Pressurized membrane indicator system for fluorogenic-based fiber optic chemical sensors. *Anal. Chim. Acta,* 217, 249-262.

Jian, C. and Seitz, W.R. 1990. A membrane for *in situ* optical detection of organic nitro compounds based on fluorescence quenching. *Anal. Chim. Acta,* 235, 265-271.

Kawahara, F.K., Fiutem, R.A., Silvus, H.S., Newman, F.M. and Frazar, J.H. 1983. Development of a novel method for monitoring oils in water. *Anal. Chim. Acta,* 151, 315-327.

Klainer, S.M., Dandge, D.K., Goswami, K., Eccles, L.A. and Simon, S.J. 1988. *A fiber optic chemical sensor FOCS. for monitoring gasoline.* EPA Report 600/X-88/259.

Krska, R., Rosenberg, E., Taga, K., Kellner, R., Messica, A. and Katzir, A. 1992. Polymer coated silver halide infrared fibers as sensing devices for chlorinated hydrocarbons in water. *Appl. Phys. Lett.,* 61, 1778-1780.

Lerchi, M., Bakker, E., Rusterholz, B. and Simon, W. 1992. Lead-selective bulk optodes based on neutral ionophores with subnanomolar detection limits. *Anal. Chem.,* 64, 1534-12540.

Leugers, M.A. and McLachlan, R.D. 1989. Remote analysis by fiber optic Raman spectroscopy. *Proc. SPIE,* 990, 88-95.

Lewis, E.N., Kalasinsky, V.J. and Levin, R.W. 1988. Near-infrared Fourier transform Raman spectroscopy using fiber-optic assemblies. *Anal. Chem.,* 60, 2658-2661.

Lieberman, S.H., Inman, S.M. and Stromvall, E.J. 1987. Fiber optic fluorescence sensors for remote detection of cehmical species in seawater. In: Turner, D.R. (ed.), *Chemical Sensors.* The Electrochemical Society 464 p.

Louch, J. and Ingle, J.D. 1988. Experimental comparison of single- and double-fiber configurations for remote fiber-optic fluorescence sensing. *Anal. Chem.,* 60, 2537-2540.

Luo, S. and Walt, D.R. 1989. Fiber-optic sensors based on reagent delivery with controlled-release polymers. *Anal. Chem.,* 61, 174-177.

Marley, N.A., Mann, C.K. and Vickers, T.J. 1985. Raman spectroscopy in trace analysis for phenols in water. *Appl. Spec.,* 39, 628-633.

Meyer MS, Eesley GL 1987. Optical fiber refractometer. *Rev. Sci. Instrum.,* 58, 2047-2048.

Milanovich, F.P., Brown, S.B. and Colston, B.W. 1993. Penetrometer compatible, fiber optic sensor for continuous monitoring of chlorinated hydrocarbons - field test results. *Proc. Symp. Chemical Sensors II,* pp.643-647. Electrochem. Society, Pennington, NJ

Milanovich, F.P., Daley, P.F., Klainer, S.M. and Eccles, L.A. 1986a. Remote detection of organochlorides with a fiber optic based sensor II. A dedicated portable fluorimeter. *Anal. Instrument.,* 15, 347-358.

Milanovich, F.P., Garvis, D.G., Angel, S.M., Klaine,r S.M. and Eccles, L.A. 1986b. Remote detection of organochlorides with a fiber optic based sensor. *Anal. Instrument.,* 15, 137-147.

Myrick, M.L. and Angel, S.M. 1990. Elimination of background in fiber-optic Raman measurements. *Appl. Spec.,* 44, 565-570.

Newman, C.D., Bret, G.G. and McCreery, R.L. 1992. Fiber-optic sampling combined with an imaging specgtrograph for routine Raman spectroscopy. *Appl. Spec.,* 46, 262-265.

Oxenford, J.L., Klainer, S.M., LeGoullon, D.E., Salina, T.M., Dandge, D.K. and Goswami, K. 1989. Development of a fiber optic chemical sensor for the monitoring of trichloroethylene in drinking water. *Proc. SPIE,* 1172, 108-114.

Plaza, P., Dao, N.Q., Jouan, M., Feurier, H. and Saisse, H. 1986. Simulation et optimisation des dapteurs a fibres optiques adjacentes. *Appl. Opt.,* 25, 3448-3454.

Schwab, S.D. and McCreery, R.L. 1984. Versatile, efficient Raman sampling with fiber optics. *Anal. Chem.,* 56, 2199-2204.

Schwab, S.D. and McCreery, R.L. 1987. Remote, long pathlength cell for high sensitivity Raman spectroscopy. *Appl. Spec.,* 41, 126-130.

Seitz, W.R. 1988. Chemical sensors based on immobilized reagents and fiber optics. CRC Crit. Rev. *Anal. Chem.,* 19, 135-173.

Skutnik, B.J., Brucker, C.T. and Clarkin, J.P. 1988. High numerical aperture silica core fibers for biosensor applications. *Proc. SPIE,* 906, 21-25.

Suzuki, K., Tohda, K., Tanda, Y., Ohzora, H., Hishihama, W., Inoue, H. and Shirai, T. 1989. Fiber-optic mangesium and calcium ion sensor based on a natural carboxylic polyether antibiotic. *Anal. Chem.,* 61, 382-384.

Tromberg, B.J., Sepaniak, M.J., Alarie, J.P., Vo-Dinh, T. and Santella, R. 1988. Development of antibody-based fiber-optic sensors for detection of a benzo[a]pyrene metabolite. *Anal. Chem.,* 60, 1901-1908.

Vo-Dinh, T., Alak, A. and Moody, R.L. 1988. Recent advances in surface-enhanced Raman spectrometry for chemical analysis. *Spectrochim. Acta,* 43B, 605-616.

Vo-Dinh, T., Tromberg, B.J., Griffin, G.D., Ambrose, K.R., Sepaniak, M.J. and Gardenhire, E.M. 1987. Antibody-based fiber-optic biosensor for the carinogen benzo[a]pyrene. *Appl. Spec.,* 41, 735-738.

Walt, D.R., Munkholm, C., Yuan, P., Shufang, L. and Barnard, S. 1989. Design, preparation, and aplications of fiber-optic chemical sensors for continuous monitoring. In: Murray, R.W., Dessy, R.E., Heineman, W.R., Janata, J. and Seitz, W.R. (eds), *Chemical Sensors and Microinstrumentation.* American Chemical Society symposium Series 403, Washington, DC. 252 p.

Weyer, L.G., Becker, K.J. and Leach, H.B. 1987. Remote sensing fiber optic probe NIR spectroscopy coupled with chemometric data treatment. *Appl. Spec.,* 41, 786-790.

Wolfbeis, O.S. 1988. Fiber optical fluorosensors in analytical and clinical chemistry. In: Schulman, S.G. (ed.), *Molecular Luminescence Spectroscopy, Methods and Applications- Part II.* Wiley Interscience, New York, NY. 129 pp.

Wolfbeis, O.S. 1991. *Fiber Optic Chemical Sensors and Biosensors* -Vols I and II. CRC Press, Boca Raton, FL.

Zhang, Y. and Seitz, W.R. 1989. Single fiber absorption measurements for remote detection of 2,4,6-trinitrotoluene. *Anal. Chim. Acta,* 221, 1-9.

Zhang, Y., Seitz, W.R. and Sundberg, D.C. 1988a. Development of a membrane for *in-situ* optical detection of TNT. USATHAMA Report CETHA-TE-CR-88021.

Zhang, Y., Seitz, W.R. and Sundberg, D.C. 1989a. Single fiber measurements for remote optical detection of TNT. USATHAMA Report CETHA-TE-CR-89102.

Zhang, Y., Seitz, W.R., Grant, C.L. and Sundberg, D.C. 1989b. A clear, amine-containing polyvinyl chloride. membrane for *in situ* optical detection of 2,4,6-trinitrotoluene. *Anal. Chim. Acta,* 217, 217-227.

Zhang, Y., Seitz, W.R., Sundberg, D.C. and Grant, C.L. 1988b. *Preliminary development of a fiber optic sensor for TNT.* USATHAMA Report AMXTH-TE-CR-87135.

Zhujun, Z. and Seitz, W.R. 1985. An optical sensor for AlIII., MgII., ZnII. and CdII. based on electrostatically immobilized 8-hydroxyquinoline sulfonate. *Anal. Chim. Acta,* 171, 251-258.

SECTION 2
Case Studies

10 Nitrates in groundwater in the southeastern USA

R. K. Hubbard and J. M. Sheridan

US Department of Agriculture, Agricultural Research Service, Southeast Watershed Research Laboratory, PO Box 946. Tifton, GA 31793, USA

Abstract

Nitrogen contamination of groundwater in the nitrate (NO_3-N) or nitrite (NO_2-N) form is of health concern for both humans and animals. Elevated NO_3-N concentrations in drinking water have caused infant death from the disease methemoglobinemia. Formation of potentially carcinogenic nitrosamines in the soil from NO_2-N and secondary amines is also a health concern. Both NO_3-N and NO_2-N have been shown to negatively affect the metabolism of domestic animals.

Movement of NO_3-N and NO_2-N (generally referred to collectively as NO_3-N) to groundwater is of particular concern in the Southeast because of the unique climatic and geohydrologic regimes of the region. The Southeast climatically is characterized by warm temperatures and relatively high rainfall amounts. Because of the length of the growing season multicropping, which requires multiple applications of N, is commonly practiced. Annual rainfall distribution often also requires use of supplemental irrigation. The combination of relatively high N inputs, high rainfall, and use of irrigation for crop production coupled with areas of permeable soils and geologic materials means that the southeastern US has high potential for groundwater NO_3-N contamination.

Current information on NO_3-N leaching to groundwater in the Southeast is rather sparse. This chapter discusses processes and factors affecting NO_3-N leaching, drawing information both from studies conducted in the Southeast and other regions of the USA or world, and also summarizes available information from the Southeast. Topics discussed include principles governing NO_3-N transport to groundwater, physical factors affecting water and NO_3-N transport in the southeastern USA, effects of management on NO_3-N movement, current and potential impact of NO_3-N on groundwater in the southeastern USA, role of models in assessing future impact and for guiding management decisions, and future research needs.

Available research information on NO_3-N movement to groundwater in the Southeast indicates that under presently used N application rates for agriculture, both root zone NO_3-N concentrations and shallow groundwater NO_3-N *concentrations may exceed the 10 mg L^{-1} drinking water standard. Studies in both the Maryland and Georgia Coastal Plain have shown concentrations in excess of 10 mg L^{-1} in shallow groundwater. Much less is currently known about deeper groundwater in the southeastern US, but the research information available from the Midwest and other parts of the US and world clearly indicate the potential for NO_3-N contamination of deeper groundwater. The main conclusion about the Southeast is that the weathered soils of the region require relatively high input of N for adequate crop production, and that this input coupled with high annual rainfall and irrigation results in significant probability for NO_3-N movement to groundwater. Recommendations for minimizing NO_3-N leaching to groundwater primarily are management and land use related. Management decisions include using minimum amounts of inorganic fertilizers, carefully controlling amounts of applied animal wastes, designing septic systems with adequate drain field areas, and minimizing percolation. Winter cover crops may be useful because they minimize nutrient availability for deeper movement during the seasonal wet periods of the winter-spring months, when most subsurface movement and percolation occurs. One major land use tool which may assist in limiting NO_3-N leaching to groundwater is the use of riparian zones to remove NO_3-N from water either by denitrification or by vegetative uptake. Future research needs for the Southeast include new information on the effects of currently used agricultural practices on NO_3-N movement to groundwater, and field, laboratory, and modeling work to better understand factors and processes affecting NO_3-N leaching for this region.*

Introduction

Nitrate nitrogen (NO_3-N) contamination of surface water and groundwater is a concern for both health and environmental reasons. Elevated concentrations of NO_3-N may be reduced to nitrite (NO_2-N) by bacteria in the intestines of newborn infants and thereby cause the disease methemoglobinemia. The bacteria necessary for conversion of NO_3-N to NO_2-N are more likely to exist in infants because the pH of an infant's stomach is more suitable for the growth of NO_3-N reducing bacteria. Once the NO_2-N has combined with the hemoglobin to form methemoglobin, the oxygen carrying capacity of the blood is reduced and methemoglobinemia, characterized by asphyxia and possible death may result (Subbotin, 1961; Smith, 1964). Since the disease was first recognized approximately 2,000 cases of methemoglobinemia have been reported in North America and Europe with a fatality rate of 7 to 8 percent (Winton et al., 1971). Because the condition (methemoglobinemia) is not a reportable disease, it is conceivable that this number is only a small percentage of the actual number of cases. As a safeguard against methemoglobinemia, the United States Public Health Service has set a safety limit of 10 mg L for NO_3-N for municipal water supplies (Lenain, 1967; Abbott, 1949).

The formation of nitrosamines in the soil from NO_2-N and secondary amines is also a health concern (Keeney, 1982). A number of N-nitroso compounds have been shown to induce tumors in test animals on vital tissues; they are also mutagenic and teratogenic (Magee, 1971, 1977; Shank, 1975; Crosby and Sawyer, 1976; NRC,

1978). Although no direct scientifically documented cause-and-effect data have yet been gathered, some circumstantial evidence relating exposure to NO_3-N or NO_2-N to the incidence of cancer is available (NRC, 1978). Domestic animals, including poultry, have also been shown to be sensitive to NO_3-N and NO_2-N (Crabtree, 1972; Case, 1957). These anions have been shown to inhibit iodine and vitamin A metabolism in research animals (Grundman, 1965).

In addition to health concerns, increased NO_3-N levels in ponds or lakes may contribute to eutrophication. If phosphorus is not a limiting factor, nitrogen (N) most often becomes the limiting nutrient for algal growth and hence NO_3-N entering lakes or ponds via surface or shallow subsurface flow may cause eutrophication. In certain physiographic regions, interflow of shallow groundwater emerges in lowland alluvial or riparian areas. Hence, nitrates which have moved in percolating water from upland soils into shallow groundwater may reappear in surface water bodies and become an environmental quality problem.

Contamination of groundwater by NO_3-N or NO_2-N (these generally are lumped together conceptually when discussing nitrate leaching) is of world-wide concern. Regardless of source of N or physiographic area, concentrations of NO_3-N above advisory levels in water which are consumed by humans or animals may result in significant health hazard. The Southeast is climatically unique in the United States. The combination of high rainfall, warm temperatures, and long growing season allows for multiple crop production where landforms and soils are suitable. Along with multiple crop production comes the need for multiple applications of fertilizers. The relatively high rainfall and frequent high intensity storm events in the Southeast provide a potential force for losses of NO_3-N by both surface runoff and leaching. This chapter was prepared with both the recognition that the Southeast has unique potential for NO_3-N movement to groundwater, and to provide currently available information on this subject. However, in reviewing the literature it was discovered that relatively little information on NO_3-N movement to groundwater in the Southeast exists as compared to other sections of the United States. Hence in discussing all of the factors affecting NO_3-N transport to groundwater, it was necessary to draw from information from other parts of the country or other regions of the world. Factors and processes affecting NO_3-N movement are the same world-wide: only rates and interactions of the relevant processes vary based on management and geohydrologic conditions. The primary factors affecting NO_3-N movement which may vary by regions are amount of applied fertilizer, animal waste, or septic tank effluent load; amount and distribution of rainfall or irrigation; permeability of soil and subsoil materials; and management factors such as crop removal or use of buffer strips. Hence where information for the Southeast is currently limited, and results from other regions are cited, it must be recognized that the primary differences are greater rainfall, warmer temperatures, and a longer growing season in the Southeast than in other regions.

Agricultural fertilizers and animal wastes are N sources which impact on NO_3-N leaching to groundwater. In terms of areal extent, agriculture is one of the most widespread human activities that can affect groundwater quality. During the 1960s and 1970s, agricultural fertilizer usage (N, P, and K) increased to a peak of 23 million nutrient tons per year in 1981 (National Research Council, 1989). The usage has since declined because of the large number of hectares withdrawn from production as part of the Conservation Reserve Program and other government programs (National Research Council, 1989). Nationwide, the estimated geographic distribution of N applied to crops closely matches the distribution of corn, wheat, and sorghum growing areas. When applied N exceeds N uptake by crops, the excess N can leach to groundwater.

Numerous studies have documented elevated NO_3-N levels in groundwater

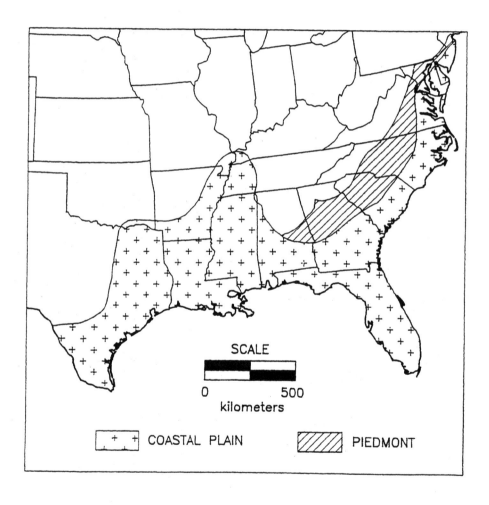

Figure 1 *Location of the Piedmont and Coastal Plain Areas of the southeastern USA.*

associated with agricultural activities. Groundwater in agricultural production areas has been found to contain more than 10 mg L^{-1} NO$_3$-N in Long Island, New York (Meisinger, 1976); Wisconsin (Saffigna and Keeney, 1977); Nebraska (Exner and Spalding, 1979); Arkansas (Wagner *et al.*, 1976); Ontario (Hill, 1982); England (Edmunds *et al.*, 1982; Oakes *et al.*, 1981); Georgia (Hubbard and Sheridan, 1983; Hubbard *et al.*, 1984); and Oklahoma (Naney, *et al.*, 1987; Sharpley *et al.*, 1987). These high concentrations were associated primarily with use of fertilizers and can be attributed to non-point source pollution. In Texas, Kreitler and Jones (1975) determined that agricultural leachates were a major source of NO$_3$-N contamination in groundwater. Wells in or adjacent to corn, wheat, potatoes, or grapes averaged 21.1 ± 3.9 mg L^{-1} NO$_3$-N, whereas public supply and household wells in nearby nonagricultural areas had NO$_3$-N concentrations ranging from 6 to 14 mg L^{-1}. Several investigators have reported large N-fertilizer losses from small grain crops and potatoes grown on coarse-textured soils (Cameron *et al.*, 1978; Endelman *et al.*, 1974; Exner and Spalding, 1979; Gerwing *et al.*, 1979; McNeal and Pratt, 1978; Spalding *et al.*, 1978).

Gormly and Spalding (1979) during 1976-77 found concentrations of NO$_3$-N exceeded the primary drinking standard of 10 mg L^{-1} in 182 of 256 groundwater samples collected from three counties in Nebraska. Comparison of the isotopic ^{15}N and ^{14}N values suggested that the primary source of contamination in most wells was fertilizer. Nitrogen from "native" organic matter is enriched in ^{15}N because

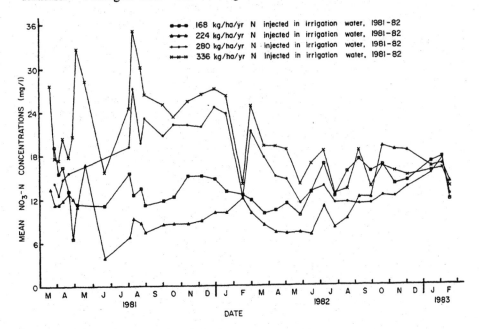

Figure 2 *Mean sampling date NO$_3$-N concentrations at 2.1 m where N was sprinkler applied at rates of 168, 224, 280 and 336 kg ha^{-1}yr^{-1}, March 1981-February 1983. From Hubbard* et al., *1986.*

biological decay processes preferentially use the lighter isotope, [14]N (Gormly and Spalding, 1979). Therefore comparison of ^{15}N versus ^{14}N quantities in water samples indicates whether the source of N is applied fertilizer or native organic matter. In Delaware, Robertson (1977a,b) found that the lowest NO_3-N concentrations were centered around wooded areas while higher values were localized around cultivated fields, poultry houses, farmyards and areas of human population.

Investigations of N levels in central Illinois indicated a correlation between the usage of N fertilizers and levels of NO_3-N found in the rivers of the area (Commoner *et al.*, 1972). Walker's research (1972; Walker *et al.*, 1973) in Illinois also revealed the close correlation that occurs between fertilization and levels of NO_3-N in groundwater. Similarly, the dependency of groundwater NO_3-N levels on the fertilizer management practices for cropping systems in the California area have been indicated (Nightingale, 1972; Feth, 1966; Johnston *et al.*, 1965; Fitzsimmons *et al.*, 1972). Spalding *et al.* (1978) used ^{15}N values to show that the major source of N in groundwater of the Burbank-Wallula area of Washington was agricultural leachates. Fertilizer was suggested as the major groundwater N source by high NO_3-N concentrations (approximately 20 mg L^{-1}) and low ^{15}N concentrations in groundwater adjacent to or underlying fields receiving greater than 170 kg N ha^{-1}.

Nitrate contamination of groundwater from animal waste has been linked to feedlots, grazing animals, and application of animal wastes to land. Notable examples exist of NO_3-N levels in groundwater increasing to critical levels 10 to 20 years after the cessation of animal feeding operations close to the well or monitoring point (Walker and Kroeker, 1982). Studies conducted by Ryden *et al.* (1984) showed that NO_3-N leached below grassland grazed by cattle was 5.6 times greater than that leached below comparable cut but ungrazed grassland. The enhanced NO_3-N movement below the grazed grassland was attributed primarily to the return in urine and feces of as much as 90% of the N in the herbage consumed by cattle.

Liebhardt *et al.* (1979) showed that the concentration of NO_3-N in monitoring wells at 3.0 m was affected by the application of poultry manure, and concluded that application of poultry manure at rates higher than a crop can utilize results in NO_3-N movement to the groundwater. Ritter *et al.* (1986) found that areas with broiler production had higher NO_3-N concentrations in groundwater than areas with only corn and soybean production.

An investigation by Stewart *et al.* (1967) in Colorado revealed that NO_3-N had moved through the soil and into groundwater under both feedlots and irrigated fields, excluding alfalfa. This study showed that larger amounts of NO_3-N were present under feedlots than irrigated lands, but that irrigated lands were contributing more NO_3-N to the groundwater since the ratio of irrigated land to that in feedlot for the study area, 200:1, more than compensated for the difference in concentrations.

Nitrate leaching to groundwater is of greatest concern in sandy soils with direct recharge to drinking water aquifers. Early work in Nebraska (Muir *et al.*, 1976; Muir *et al.*, 1973) concluded that N fertilizers contributed substantially to groundwater NO_3-N contamination only for sites with intensive irrigation development, very sandy soils, and shallow groundwater tables. Nelson (1972) concluded that the occurrence of potentially harmful concentrations of agrichemicals in groundwater, drainage canals, and streams may be greatest in areas where cash crops are grown

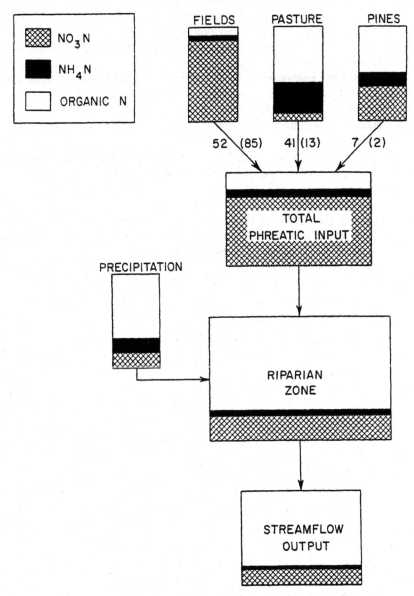

Figure 3 *Change in nitrogen forms from subsurface and precipitation inputs to streamflow outputs. Bars represent relative amounts of nitrate, ammonium, and organic N making up inputs or outputs. Numbers on arrows represent percentages of water discharge and nitrogen in parentheses coming from fields, pastures, and pine forests From Lowrance et al., 1984b.*

mainly on sandy soils which receive frequent applications of irrigation water, insecticides, herbicides, and fertilizers.

A number of investigations have shown substantial losses of fertilizer nutrients in the acid, sandy soils of Florida (Calvert and Phung, 1971; Calvert, 1975; Forbes *et al.*, 1974; and Graetz *et al.*, 1974). Ayers and Branson (1973) observed that NO_3-N concentrations tend to be higher in groundwater located beneath sandy soils which are intensely managed for agricultural use. Research in Missouri indicated that the only water considered contaminated by N fertilizers was located under light sandy alluvial soils along the Mississippi and Missouri rivers (Smith, 1970). More recent research (Hubbard *et al.*, 1987) has shown that NO_3-N will move with time through less permeable materials into shallow groundwater, and hence concerns for NO_3-N contamination of groundwater must also exist for areas with finer textured soils. Also, concern must exist in physiographic regions with karst topography, sinkholes, or breached subsoil materials that may allow more rapid NO_3-N movement towards groundwater.

The major physiographic regions of the southeastern United States are the Coastal Plain and the Piedmont (Figure 1). The Coastal Plain, in general, is characterized by sandy soils through which water, NO_3-N, and other waterborne materials may move rather rapidly, while the Piedmont has soils which generally are higher in clay content. Climatically, the Southeast is characterized by warm

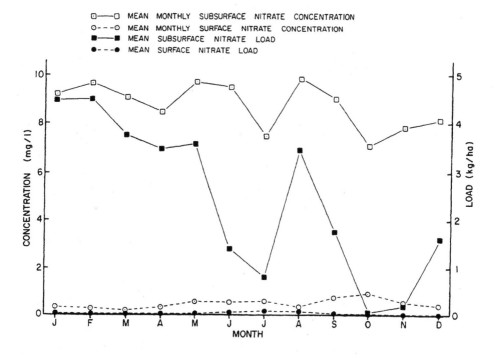

Figure 4 *Mean monthly surface and subsurface nitrate-N concentrations and loads at Station Z, 1969-1978 From Hubbard and Sheridan, 1983.*

temperatures, and relatively high amounts of rainfall. The soils of the southeastern US, in general, are highly weathered and have low organic matter, and consequently require relatively high inputs of N for successful crop production. Because of nonuniform distribution of rainfall and high evapotranspiration associated with high soil and air temperatures, particularly during the summer months, use of irrigation for crop production in the southeastern US has greatly expanded. In the Georgia Coastal Plain for example, irrigation has expanded from 58,575 ha in 1970 to over 495,000 ha by 1989 (Harrison, 1989). The combination of relatively high N inputs, high rainfall, and use of irrigation for crop production consequently means that the southeastern US has high potential for groundwater NO_3-N contamination, particularly in areas with sandy, highly permeable soils. Preservation of the quality of both surface water and groundwater is important for both the Piedmont and Coastal Plain, but groundwater quality is particularly important for the Coastal Plain because it is one of the few remaining areas in the United States that still has an abundant water resource available for crop production. Groundwater is the primary source of drinking water in the Coastal Plain, so preventing contamination from agricultural practices is especially important. Other areas of the country such as the southwest and parts of the Plains states are already depleting their groundwater faster than recharge is occurring (Collins, 1984); therefore, in time the Coastal Plain may be relied upon for a greater share of total US agricultural production. Pressure for increased agricultural production in the Coastal Plain may require increased efforts to protect groundwater quality.

Principles governing NO_3-N transport to groundwater

Role of nitrate in the nitrogen cycle.
Nitrogen exists in soil as nitrite (NO_2-N), nitrate (NO_3-N), ammonium (NH_4-N, and includes NH_3), or organic forms within the soil organic matter fraction. The NO_3-N form of N is particularly important because it is the dominant species and is also highly soluble, which results in much of plant N uptake being in the NO_3-N form, and NO_3-N being the primary N form lost from soils by leaching. Ammonium added to soils is converted to NO_3-N through the process of nitrification.

Nitrification is a process performed by microbes capable of oxidizing reduced forms of nitrogen: NH_4-N is oxidized to NO_2-N, which in turn is oxidized to NO_3-N. Typical nitrifiers are autotrophic aerobic bacteria: Nitrosomonas is a common NH_4-N oxidizer, and Nitrobacter is a common NO_2-N oxidizer. For these organisms the process is related to the energy relations within their cells and is essential to their metabolism. Heterotrophs which can also oxidize NH_4-N include certain bacteria, actinomycetes, and fungi (Alexander, 1965). However, the most significant nitrification appears to be accomplished by the autotrophs.

Nitrifying populations are sensitive to a variety of environmental conditions. The nitrification capacity of soil has been linked to a range of parameters and conditions such as NH_4-N content, pH, oxygen content, moisture, temperature, organic matter, carbon dioxide content, cation exchange capacity, depth of tillage, season of the year, and soil treatment (Alexander, 1965; Mahendrappa, *et al.,* 1966;

Table 1 *Potential pathways of movement, areal extent, and rates of vertical movement of water and associated waterborne materials to regional groundwater aquifers in the southeastern Coastal Plain From Hubbard and Sheridan, 1989.*

Regional Aquifer Condition	Geohydrology	Potential Pathways for Water Movement	Areal Extent	Approximate Rates of Vertical Water Movement	Relative Quantity of Recharge Moving to Regional Groundwater System	Role of Fluviatile Swamps on Leachate Quality	Comments
Unconfined	Primary recharge or aquifer out-crop area	Recharge by infiltration/ percolation of precipitation and runoff	Widespread over out-crop areas	Slow (cm/day)	Large volume of recharge	Typically diminished because drainage is predominantly subterranean	Water available for recharge generally exceeds capacity of regional aquifer to accept additional water
		Direct recharge by sinkholes and other karst features or drainage wells	Localized	Very rapid (cubic meters/ second)	Relatively small but significant volume of recharge	Streamflow thru sinkholes and other karst features may pass thru riparian zone	Relatively small volume, but significant to local groundwater supply and quality
		Recharge thru permeable bottoms of sinkhole lakes and ponds	Localized	Intermediate (Varies with characteristics of fill)	Small but significant volume of recharge	Some benefit possible from riparian vegetation common to low, wet areas	Rate of movement relatively slow, however, water available for vertical movement/ percolation for much or all of year
Semi-confined	Area where upper confining unit is thin or breached (< 100 feet)	Recharge by infiltration/ percolation of precipitation and runoff	Varies with characteristics of upper confining member	Extremely slow (cm/year)	Very small volume of recharge	Possible improvement in quality of per-colation below low, wet areas and floodplain alluvial aquifers	Rate of movement extremely slow due to low vertical conductivities of upper confining member

	Direct recharge by sinkholes and other karst features or drainage wells	Localized	Very rapid (cubic meters/second)	Small, however, locally significant volume of recharge	Streamflow thru sinkholes and other karst features may pass thru riparian zone	Rapid direct contribution significant to local groundwater supply and quality
	Recharge thru permeable bottoms of sinkhole lakes and ponds	Localized	Intermediate (varies with characteristics of fill)	Small volume of recharge may be significant locally	Some benefit possible from riparian vegetation common to low, wet areas	Rate of movement relatively slow, however, volume of water available for vertical movement/percolation for much or all of year
Confined	Recharge by infiltration/percolation of precipitation and runoff	Widespread over confined areas	Extremely slow (cm/year)	Very small volume of recharge	Previous research indicates significant improvement in quality of streamflow and subsurface flows moving from low-lying, wet areas	Area is generally considered to be an aquiclude, however, studies indicate vertical movement is possible, albeit extremely slow
Area of relatively thick upper confining unit (> 100 feet)	Direct recharge by sinkholes and other karst features or by drainage wells	Very localized	Very rapid (cubic meters/second)	Small, possibly negligible volume of recharge	Flow passing thru sinkholes and other karst features may first pass thru riparian zone	Generally not found under confined conditions, however, some limited number may exist. Little evidence of significant movement via this route reported in confined areas

Sabey, 1969; Parker and Larson, 1962; Campbell *et al.*, 1970). Nitrification occurs most rapidly under well aerated conditions, with a temperature of 16-30°C and a pH of 6.5 - 7.5 (Keeney and Walsh, 1972). Little nitrification occurs above 40°C (Alexander, 1965) or below 0°C (Campbell, *et al.*, 1970). Nitrification is reduced abruptly as the soil becomes nearly saturated with moisture (Parker and Larson, 1962).

Nitrate transport by water.

Nitrate ions are negatively charged, are repelled by the clay particles in the soil, and generally are not adsorbed within the soil matrix. Hence as water moves through the soil, NO_3-N generally moves freely with the water. The actual movement of NO_3-N through soil lags behind the wetting front due to mixing processes such as diffusion and hydrodynamic dispersion which occur between the resident soil solution and infiltrating water from irrigation or rainfall. Factors influencing these processes may be diffusion caused by the double-layer surrounding the soil particles (Hagan *et al.*, 1967; Edwards and Monke, 1968); electrical potential unbalance created by displacement of the double-layer (Edwards and Monke, 1968); Van der Waals forces (Sears and Zemansky, 1955); ion exchange (Rible and Davis, 1955); miscible displacement or the diffusion of two solutions with different concentrations (Hagan *et al.*, 1967); diffusion across a concentration gradient into areas of lower shear (Edwards and Monke, 1968; Hunter and Alexander, 1963); or mixing created by Brownian movement (Sears and Zemansky, 1955).

The quantity of NO_3-N transported by leaching through the soil is a function of many variables, including soil, climate, biological characteristics, cultural characteristics, and N characteristics (Rowe and Stinnett, 1975). Edaphic characteristics include texture, porosity, structure, consistency, depth of profile, and percolation rates. Climatic characteristics include amount, frequency, duration and time of precipitation, rates of evapotranspiration, and temperature regimen. Presence or absence of plant cover, depth of root zone, N use characteristics of the vegetation, periods of plant growth, levels of organic matter, and microbial and animal populations are included in the biological characteristics affecting NO_3-N leaching. Land use patterns and soil management practices are included in cultural characteristics, while N characteristics include amount and type of fertilizer application. These variables interact to produce a complex, dynamic pattern of N distribution in soil profiles.

Within storm events, critical factors affecting NO_3-N leaching include the soil moisture content at the start of the precipitation event, the amount of water infiltrating during the event, and the soil's porosity and permeability (Preul and Schroepfer, 1968). Edwards *et al.*, (1972), found that NO_3-N moves essentially with the wetting front when the soil is initially air dry, but does not move at the same rate as the water when the soil is initially saturated. Nitrates leach more quickly through sandier soils with high porosity and permeability. Retardation in the leaching process generally is greater in the more clayey soils, unless these soils have large cracks or macropores. Recent research concern has focused on the important role of macropores in solute transport, both for within storm events and for overall solute

Table 2 *Overall mean NO₃-N concentrations by piezometer depth m and slope position where dairy wastes were applied. From Hubbard et al., 1987.*

		Shallow 0.9-1.2m			Piezometer depth Intermediate 2.4m			Deep 3.6m		
Description	n	NO_3-N mgL^{-1}	Standard error	n	NO_3-N mgL^{-1}	Standard error	n	NO_3-N mgL^{-1}	Standard error
Hilltop	41	[x]46.6[b]	2.4	71	[y]26.6[c]	1.6	98	[z]11.8[dc]	1.2
Upslope	81	[x]42.1[b]	1.7	182	[y]31.5[b]	1.0	240	[z]9.2[c]	0.8
Upper Midslope	87	[x]59.2[a]	1.8	144	[y]45.4[a]	1.4	242	[z]22.1[b]	0·8
Lower Midslope	123	[x]33.7[c]	1.4	186	[x]33.0[b]	1.0	228	[y]24.7[a]	0.9
Downslope	108	[y]21.4[d]	1.5	120	[x]25.1[c]	1.3	69	[z]15.2[cd]	1.4
Hill Bottom	57	[x]20.6[d]	3.1	98	[y]16.8[d]	1.4	89	[y]16.1[c]	21.2

Nitrate concentrations prefixed with the same letters (x,y,z) are not significantly different by depth within slope position (rows) by the least squares mean test (SAS Institute, 1982). Nitrate concentrations suffixed with the same letters (a,b,c,d) are not significantly different by depth within slope position (columns) by the least squares mean test (SAS Institute, 1982).

leaching with time. The characteristics of climate, geology, soils, and management practices in the southeastern US determine the level of impact of N sources on groundwater.

Physical factors affecting water and nitrate transport in the southeastern USA

Geology and soils.
The primary physiographic areas of the southeast are the Piedmont and the Coastal Plain. Piedmont soils tend to be loamy to clayey and well-drained, whereas Coastal Plain soils may be poorly to moderately well-drained sands and clays, or low-lying, poorly-drained peaty soils. Seasonally flooded swamp forests border or formerly bordered most Coastal Plain tributaries (Humenik *et al.*, 1980). Coarse-textured upland soils in the southeast are inherently low in organic matter, with the exception of high elevation forest soils and the organic soils along the coast or in marshy areas. Cropland soils generally contain less than 1% organic matter in the plow-layer and the content drops to 0.05 - 0.1% at the bottom of the root zone (Knisel and Leonard, 1989).

Physical factors affecting the vulnerability of aquifers to contamination include the depth to the water table; the permeability of material overlying the water table; the presence of clay and organic matter, which can absorb contaminants; the amount of precipitation, which affects the amount of recharge and the rapidity with which the contaminants are transported down to the water table; and evapotranspiration rates, which reduce recharge (Moody, 1990). Three factors affect the fate and transport of contaminants: the length of the flow path from the contamination source to the aquifer, the residence time of contaminants in the unsaturated zone, and the potential for biodegradation and decomposition of the contaminants (Johnson, 1988). In general, shallow, permeable, unconfined aquifers overlain by thin soils in humid regions are the most susceptible to contamination from the land surface, because short flow paths to the water table and rapid infiltration reduce the opportunity for physical, chemical and biological reaction to attenuate contaminants. In contrast, deep, confined aquifers are less susceptible to contamination from surface sources because longer flow paths increase opportunities for the attentuation of contaminants, and confining beds retard the movement of contaminated water into the aquifer (Moody, 1990).

The soils and geology of the southeastern US are sufficiently variable to result in a complex hydrogeologic environment which makes regional generalizations about NO_3-N leaching potential difficult. In the sandier areas of the Coastal Plain, NO_3-N may move rapidly out of the root zone, but further movement towards deeper groundwater then depends on the underlying materials. For example, the Floridan aquifer system in the southeastern United States is considered highly susceptible to contamination in its unconfined outcrop areas where recharge is high, but it is relatively less susceptible to contamination toward the coast where increasingly thick sections of clay and silt confine and protect the system (Johnson, 1988). Moody (1990) described three Coastal Plain groundwater regions: (1) complexly interbedded seaward dipping aquifers on the Atlantic and Gulf Coastal Plain; (2) sand and clay layers that overlie semiconsolidated carbonate rocks of the southeast Coastal Plain; and (3) the Floridan aquifer which is one of the most productive aquifers in the world and is the principal source of groundwater in the southeast Coastal Plain.

Generalizations about the materials overlying aquifers in the southeast are more complicated in karst regions. Karst regions exist in both Florida and Georgia. Sinkhole or karst topography has formed in areas where limestone aquifers are at or near the surface. The regional artesian aquifer has cavities and solution channels that are probably comparable in size and extent to those in Mammoth Cave, Kentucky (Stringfield, 1966). Water from ponds, rivers, or lakes may recharge regional aquifer systems through open sinks and man-made or naturally occurring wells that drain low-lying, wet areas directly into underlying aquifers.

The effects of underlying geologic materials on solute transport to groundwater in the Coastal Plain region of the southeastern US are indicated in Table 1, taken from Hubbard and Sheridan (1989). This table indicates rates of vertical movement of water and associated waterborne materials for unconfined, semiconfined, and confined regions. The importance of sinkholes is obvious in that regardless of amount of confinement, regional recharge to the aquifer may be quite rapid when flow is through sinkholes.

Both the presence of important aquifers in the Coastal Plain, and the wide range found in properties of surface soils and underlying geologic materials throughout the southeastern US, mean that the physical properties of these materials are very important relative to rate of NO_3-N leaching to groundwater.

Precipitation.

The total amount, intensity, frequency, and timing of precipitation greatly affect NO_3-N leaching. The southeastern United States receives a relatively large amount of rainfall, although the distribution is not uniform throughout the year. That portion of rainfall which is not lost by evapotranspiration or removed from the landscape via streamflow becomes groundwater recharge.

Climatically the southeastern states include regions classified as the subtropical oceanic margins (Atlantic and Gulf), and subtropical interior (Trewartha, 1961). The first region includes all of the Coastal Plain while the second region includes the Piedmont and parts of the Appalachian Mountain areas lying to the north of the Piedmont.

Along the subtropical Atlantic and Gulf margins from southern New Jersey in the northeast to the Texas-Louisiana boundary in the southwest, an abundant annual rainfall of 114 to 152 cm is concentrated in the warm season. The primary maximum rainfall season for this region occurs during the summer. Most of the abundant precipitation in the summer is from localized convective storms which are frequently accompanied by thunder and lightning. Florida and the northern Gulf coast possess the distinction of having one of the highest frequencies of thunderstorm occurrence for any part of the earth (Trewartha, 1961).

The northern Gulf coast, excluding peninsular Florida, has a biannual rainfall profile characterized by a primary maximum in summer (July-August) and a secondary maximum in the cooler months, which generally reaches a peak in March. Inland from the coast, the summer maximum decreases in prominence so that areas have a March peak as high as that of summer (Trewartha, 1961). The area of Virginia and the Carolinas east of the Appalachians also has the biannual rainfall profile with both summer and winter maximums. However, the winter precipitation peak is less than what occurs in the northern Gulf area (Trewartha, 1961).

The relatively high total amounts of rainfall in the southeastern US (114 to 152 cm) result in a high potential for NO_3-N movement to groundwater. Also, distribution patterns affect this because when supplementary irrigation is used, this water also may move NO_3-N towards groundwater, particularly if excessive amounts of irrigation are applied. The amount of water leaving the land as surface runoff or entering as infiltration relates directly to rainfall intensity, with surface runoff typically increasing in response to increased intensity. The frequency of rainfall or irrigation affects NO_3-N leaching in that NO_3-N leaches with percolating water but moves little in a dry soil. Also, wetter soils tend to have greater surface runoff, which may increase the proportion of available NO_3-N extracted from the soil surface and lost prior to leaching. Timing of rainfall and irrigation on southeastern US soils is critical because significant leaching can occur immediately following fertilizer application.

Effects of management on nitrate movement

Nitrogen application rates

A number of studies have measured the effects of different fertilizer application rates on N losses by surface runoff, leaching, tile drainage, or shallow subsurface flow. Field studies in New York showed that N losses and subsurface concentrations were strongly related to application rate (Zwerman *et al.*, 1971). In the Corn Belt, Burwell *et al.*, (1976) showed that average NO_3-N concentrations in subsurface discharge were 5.8 mg L^{-1} and 21.0 mg L^{-1} for continuous corn watersheds fertilized at 168 kg N $ha^{-1}yr^{-1}$ and 448 kg N $ha^{-1}yr^{-1}$, respectively. Work by Miller (1979) in Ontario showed that NO_3-N in the drainage water from mineral soils fertilized at or below recommended rates seldom exceeded 10 mg L^{-1}, whereas those from sites fertilized at rates greater than recommended were seldom below 10 mg L^{-1}.

Legg and Meisinger (1982), in reviewing results of different studies from a number of locations in the US and Canada, reported that NO_3-N losses ranged from 3 to 95% of input N. Frissell (1977) reported leaching percentages of applied N for corn, wheat, and soybeans as 13, 12, and 8% of input fertilizer, respectively. Watts and Martin (1981), using a field calibrated computer model to estimate NO_3-N losses on irrigated sands in the central Great Plains, concluded that for years of normal rainfall, about 30-35 kg ha^{-1} of NO_3-N leaching loss may be expected during the growing season when 168 kg ha^{-1} of N is applied to irrigated corn. However, they cautioned that this estimate assumes careful water control, and losses may double when excess water is applied.

Hubbard *et al.* (1986) studied NO_3-N leaching for different fertilization rates of corn grown on sandy Coastal Plain soils under irrigation. Nitrogen application rates for the study ranged from 87 to 336 kg $ha^{-1}yr^{-1}$, with the recommended rate for corn being 280 kg $ha^{-1}yr^{-1}$ applied N. Mean NO_3-N concentrations in groundwater at 2.1 m for a 13-month period ranged from 8.8 mg L^{-1} under 87 kg $ha^{-1}yr^{-1}$ applied N to 23.7 mg L^{-1} under 336 kg $ha^{-1}yr^{-1}$ applied N. Significant differences in NO_3-N concentrations in water samples from 2.1 m were found between the two highest sprinkler applied N rates (280 and 336 kg $ha^{-1}yr^{-1}$ N) and the two lowest sprinkler applied N rates (168 and 224 kg $ha^{-1}yr^{-1}$ N) beginning four months after the start of the study and lasting into the following spring (Figure 2). During the second year, differences in NO_3-N concentrations at 2.1 m due to N application rate were blurred because of shallow subsurface flow beneath the entire area resulting from wetter conditions. The results clearly indicated, however, that in the Coastal Plain of the southeastern US, as elsewhere, N application rate affects NO_3-N leaching to groundwater. As part of this N rate study, 20-year simulations using the CREAMS model (Knisel, 1980) were made. With 336 kg $ha^{-1}yr^{-1}$ applied N the predicted mean NO_3-N concentration at 2.1 m over a 20-year period was 20.8 mg L^{-1}. This compared to a 20 mg L^{-1} observed mean NO_3-N concentration at 2.0 m where intensive multiple cropping systems had been used with center pivot irrigation on a Coastal Plain sand (Hubbard *et al.*, 1984). Other studies (Baker *et al.*, 1975; Burwell *et al.*, 1976; Calvert, 1975; Chichester, 1976; Chichester, 1977; Ludwick *et al.*, 1976; Pratt *et al.*, 1972; and Zwerman *et al.*, 1972) also have indicated that high fertilization rates result in excessive concentrations of NO_3-N in subsurface drainage water and/or deep in the soil pedon.

The general conclusion regarding N application rates and groundwater quality is that rates should not exceed the assimilative capacity of the crop. Applying only enough N to meet crop requirements is probably one of the most important best management practices (BMPs) for reducing water pollution from fertilizers. Another important BMP to control N losses is timing of application. Split applications of N can reduce the amount of leachable NO_3-N as compared to a single application. Two studies in New York have shown that N should not be applied when crop uptake is low and deep seepage is prevalent (Klausner *et al.*, 1974; Zwerman *et al.*, 1971). Modeling work by Hubbard *et al.* (1985) in the Coastal Plain of Georgia has shown that NO_3-N leaching losses will be reduced by using fertigation rather than conventional N application, because multiple N applications reduce the probability of leaching.

Cropping sequences, tillage practices, and irrigation practices.

Cropping sequences impact on NO_3-N leaching by affecting NO_3-N uptake from the root zone, and by affecting the permeability of the soil. Winter cover crops may reduce NO_3-N leaching by using up residual N from the previous crop. Minimum tillage has been shown to affect NO_3-N leaching although uncertainty exists as to final impact. Smith *et al.* (1987) showed that minimum tillage and intensive grass production enhanced NO_3-N leaching. They concluded that minimum tillage in general provides a wetter, cooler environment that enhances NO_3-N leaching potentials because more undisturbed large pores and burrows may exist than under conventional tillage. Work by Thomas and associates in Kentucky (Thomas *et al.*, 1973; Tyler and Thomas, 1979; and Thomas *et al.* 1981) showed more N leaching below 90 cm in a killed sod no-till system than in a conventional treatment. These workers concluded that there was a potential for greater leaching in conservation tilled systems as compared to conventional till, particularly during the early growing season. However, Kanwar *et al.* (1985) working on a loam soil in Iowa observed much less leaching of N in no-till plots than in moldboard plow plots. Kitur *et al.* (1984) in Kentucky found no difference in NO_3-N leaching between minimum till and conventional till. Research by Hubbard *et al.* (1994) on a clayey Coastal Plain soil showed greater infiltration into and less surface runoff from a no-till system than from other tillage systems. The greater infiltration with no-till potentially may contribute to greater NO_3-N leaching than what may occur with other tillage systems. Gilliam and Hoyt (1987) in their discussion of the effects of conservation tillage on NO_3-N leaching concluded that conservation tillage may increase or decrease leaching in the short term depending upon soil and other factors.

Depth of tillage may affect NO_3-N leaching by affecting both infiltrability and permeability of the soil. Mansell *et al.* (1977) in Florida found the greatest leaching losses under citrus to be from shallow tillage. Average annual losses of NO_3-N from both surface and subsurface drainage from shallow tillage, deep tillage, and deep tillage with limestone were equivalent to 22.1, 3.1, and 4.5% of total N applied as fertilizer.

Studies have shown that irrigation practices significantly impact NO_3-N leaching to groundwater. Excessive irrigation can result in greater NO_3-N contamination than under natural rainfall while carefully controlled irrigation may actually result in reduced NO_3-N leaching. The increased amount of N in groundwater in some regions of California has been attributed to the liberal use of fertilizer in conjunction with irrigation (Nightingale, 1972; Johnston *et al.*, 1965; Pratt, 1972; Shaw and Wiley, 1969). Hallberg (1986a) reported NO_3-N contamination of shallow aquifers in Iowa and pointed out that irrigation is a management practice which affects NO_3-N loadings to groundwater. Hubbard *et al.* (1984) in the Georgia Coastal Plain showed high NO_3-N concentrations in shallow groundwater under multiple intensive cropping systems using center pivot irrigation.

Ritter *et al.* (1986) concluded that NO_3-N leaching may be less under irrigation because of better N management. Many farmers use fertigation techniques to apply N through the irrigation system. By controlling the timing of N application crops receive fertilizer as needed and NO_3-N leaching may be reduced. Bockstadter *et al.* (1984) showed that N fertilizer application by irrigation on sandy soils in Nebraska could be reduced significantly without reducing corn yields. This conclusion was also reached by modeling studies with the CREAMS model using Georgia Coastal Plain cropping, soil, climate, and management factors (Hubbard *et al.*, 1986). Simulations on a sandy soil with 168 kg $ha^{-1}yr^{-1}$ applied N showed a range of 1502 to 1754 kg ha^{-1} cumulative NO_3-N loss over a 20 year period, with the least loss occurring where N was sprinkler applied and a winter cover of ryegrass was grown, and the greatest loss occurring under side-dressed N application and no winter cover. At an N application rate of 336 kg $ha^{-1}yr^{-1}$, cumulative 20-year NO_3-N losses ranged from a minimum of 3,387 kg ha^{-1} with N sprinkler applied and a winter cover, to 3,773 kg ha^{-1} with side-dressed N and no winter cover. The long-term simulations indicated that the combination of sprinkler applied N and winter cover would be the most effective management system for minimizing NO_3-N leaching on these soils. Other modeling work (Hubbard *et al.*, 1985) comparing NO_3-N losses under fertigation or conventional N application from two different Coastal Plain soils also showed that over a 20-year period more NO_3-N will leave the root zone with conventional N application than with fertigation. From a Bonifay sand, the predicted conventional versus fertigated cumulative 20-year NO_3-N losses were 3,091 and 2,666 kg ha^{-1}, respectively, while the losses under the same treatments from a Tifton loamy sand were 2,778 and 2,576 kg ha^{-1}, respectively.

Land use patterns: agriculture, forestry, urban.
Nitrate movement to groundwater in the southeastern United States depends largely on land use. Nitrate may move to groundwater from agricultural nonpoint sources, such as N fertilizers or animal wastes applied to row crops; from agricultural point sources, such as feedlot operations or leakage around fertilizer storage areas; or from surface runoff in urban areas, or leakage from either septic tanks or large scale sewage disposal systems. In urban areas seepage from over-fertilized over-watered lawns may also contribute NO_3-N to surface runoff and groundwater. Nitrogen generally is not applied in forested areas so this land use has a positive effect on groundwater quality both because there is little source of NO_3-N, and because forests may effectively filter NO_3-N entering via surface runoff or subsurface flow from adjacent upland areas. Nitrate leaching to groundwater in urban areas of the

southeastern US must be of concern because the relatively high rainfall of the region may cause significant leaching from overloaded and/or incorrectly designed septic tank systems.

Row crop agriculture is considered to be the major non-point source of NO_3-N leaching to groundwater. Numerous studies have shown NO_3-N moving to groundwater under cropped areas, particularly on sandy soils. Movement of NO_3-N to groundwater from animal wastes (when a high density of animals are kept in a confined area) can actually be a greater localized problem than under row crops. Nitrate contamination of water supplies is common around dairy operations, barn yards, and feedlots. One result of changes in dairy size in the past 20 years is that the quantity of dairy cattle manure handled per dairy farm has increased at a significant rate (Morgan and Keller, 1987). If the amount of land available for utilization and/or disposal of the dairy cattle manure is too small, then NO_3-N contamination of surface and ground waters may result (Hubbard and Lowrance, 1993). In many parts of the country, the most serious problem is with feedlots (Lorimor et al., 1972; Miner and Willrich, 1970). One investigator in Missouri found over 373 kg ha^{-1} NO_3-N at a depth of 4.3 meters under a feedlot operation (Smith, 1965). Poultry manure also has been shown to cause groundwater contamination in certain areas of Delaware (Ritter et al., 1986). Farm manure contains about 4.5 kg of N per ton of dry matter and poultry waste is even higher (Christy et al., 1965). Some authorities have attributed the amount of N waste per animal in a beef feedlot operation at 43 kg year^{-1} (National Research Council, 1972). Although the localized effects of animal wastes may be more severe to groundwater than the effects of inorganic fertilizers, on a regional basis N applied to row crop agriculture is of much greater concern because animal wastes are applied to a relatively small percentage of the landscape.

Nitrate losses to groundwater from unfertilized forests are small. Gold et al. (1990) found flow-weighted NO_3-N concentrations in groundwater under forest to be less than 1.7 mg L^{-1}, while in contrast concentrations under septic systems and silage corn had annual flow-weighted concentrations in excess of 10 mg L^{-1} for at least 1 of 2 years. Annual losses in groundwater ranged from greater than 70 kg ha^{-1} of NO_3-N from silage corn to less than 1.5 kg ha^{-1} from unfertilized home lawns and forests. Hubbard et al. (1984) found NO_3-N concentrations ranging from 0.2 to 1.2 mg L^{-1} in groundwater at a forested site in the Georgia Coastal Plain. Forest land thus plays a major role in enhancing groundwater quality relative to NO_3-N on a regional scale by serving as a section of the landscape which does not directly receive N fertilizer. Forest land also can act as a filter for NO_3-N, and this will be more thoroughly discussed in the section on riparian zones and buffer strips.

Nitrate losses from urban land include surface runoff, movement by seepage from heavily fertilized lawns, and seepage from septic tanks. Runoff from urban land contains N from lawn fertilizers, dust fall, leaves, waste material from pets, household wastes, etc. (Nelson, 1972). Investigations by Keeney and Walsh (1972) showed that the N content of urban runoff is sometimes higher than the N content of sewage. Vegetative litter such as leaves, twigs, grass clippings, seeds, and weeds may make a significant contribution to this N load. Heaney et al. (1975) reported that evergreen and deciduous leaves contain 0.58 -1.25 percent total N (TN), and that the loading values from these percentages are 48 - 361 kg TN ha^{-1}. Overall, Heaney *et al.* (1975) reported that 3,000 kg ha^{-1}yr^{-1} of vegetative litter is typical of urban areas. Sonzogni *et al.* (1980) reported that home fertilizer use can be a major cause of nutrient impact to receiving waters and that such nutrients are usually readily available to noxious organisms. Oberts (1985) indicated that N is always present at elevated levels in urban runoff, but this contribution is generally less than point discharges. Oberts (1982), however, also reported that even though nutrients from urban nonpoint sources might be less than point sources, many of the receiving waters are urban lakes or quiescent streams where eutrophication impacts can occur. Thus NO$_3$-N in urban runoff can enter water bodies that ultimately impact on groundwater. Regarding urban runoff in general, the EPA's Nationwide Urban Runoff Program (NURP) (US Environmental Protection Agency, 1983) concluded that variability in flow and pollutant concentration is so extreme from site to site and from event to event that relying on specific land uses and geographic locations to make pollution projections is not statistically valid.

The disposal of human wastes has long been recognized as a potential source of contamination of groundwater supplies. The average amount of N released by each individual is considered to be about 5.4 kg yr^{-1} (National Research Council, 1972). If only a small portion of the N from human wastes makes its way to the groundwater, the impact on these supplies could be serious (Rowe and Stinnett, 1975). As of 1980 there were about 22 million domestic disposal systems with about one-half million new systems being installed each year (Moody, 1990). Forty-one states have described individual household disposal systems as a major source of groundwater contamination (Moody *et al.*, 1988). In the southeastern US the highest density of household disposal systems occurs in parts of Florida (The Conservation Foundation, 1987).

It is commonly assumed that from 1/3 to 1/2 of existing septic systems are operating improperly because of poor location, design, construction, or maintenance practices (Cantor and Knox, 1984; The Conservation Foundation, 1987). Also, the life expectancy of a septic system is 10^{-1}5 years, but homeowners tend to try to get more years of service and, consequently, groundwater contamination by septic systems may occur (Cantor and Knox, 1984). Even where operating properly, septic systems can be spaced so densely that their discharge exceeds the capacity of the soil to assimilate the pollutant loads. The US Environmental Protection Agency (1980) has estimated that an average of 4 kg capita^{-1} yr^{-1} of N is contributed to a septic system and that about 10% of the N is removed by the septic tank. Gold *et al.* (1990) estimated that 9.5 kg yr^{-1} dissolved inorganic N would be lost from the septic system of a three-person home. They also indicated that without accounting for N loading from lawns, gardens, and domestic animals, densities of five homes ha^{-1} (0.20 ha^{-1} zoning) would generate 47.5 kg ha^{-1}yr^{-1} of N loading, which is comparable to losses from urea-fertilized silage corn with an effective rye crop. Their conclusion was that where sewering is not a viable alternative, zoning densities of less than five homes ha for on-site sewage disposal may be required to insure potable groundwater quality in residential areas (Gold *et al.* 1990).

Contamination of groundwater by septic system effluent is of particular concern to rural homeowners with drinking water wells. A shallow well can withdraw septic effluent if it is located down-gradient from or too close to the septic system's drainage field. The problem may be acute in rural housing developments, especially with a shallow water table. Over time, effluent may be withdrawn and discharged by different households as the groundwater moves down-gradient through the development (Moody, 1990).

Land use patterns in the southeastern US play a major role in the overall transport of NO_3-N to groundwater. The portion of the landscape in forestry is particularly important on a regional basis. Beck *et al.* (1985) came to this conclusion in a study where deep groundwater was sampled from 34 wells in a primary recharge area of the Claiborne aquifer under the Dougherty Plain of the Georgia Coastal Plain. Elevated NO_3-N concentrations (4-6 mg L^{-1}) in the exposed updip portion of the aquifer were found in areas where land was primarily in crops. In forested areas in the exposed portion of the aquifer, NO_3-N concentrations were less than 1 mg L^{-1}.

Role of riparian zones

Much of the southeastern US can be characterized as having upland soils used for agriculture or forestry, which drain into poorly drained, low-lying riparian areas. The Coastal Plain, in particular, has seasonally wet floodplains as prominent features. These areas, which can be described as riparian zones, have been shown to be sinks for NO_3-N through the processes of denitrification and vegetative uptake (Lowrance *et al.*, 1984a, 1984b).

Denitrification occurs under anaerobic conditions. Denitrifiers are facultative anaerobes which carry out anaerobic respiration by substituting NO_3-N or a related nitrogenous compound for oxygen as the terminal electron acceptor (Rowe and Stinnett, 1975). The pathway generally proceeds by reduction from NO_3-N to NO_2-N to N_2O-N to N_2 gas, although some intermediates have been postulated (Rowe and Stinnett, 1975).

A number of factors influence denitrification rate. Denitrification rates are slower in acid soils than in alkaline soils. The relative amounts of N_2O-N and N_2 produced are affected by temperature, with N_2O-N being predominantly formed at lower temperatures, and N_2 at higher ones (Rowe and Stinnett, 1975). Conditions conducive to denitrification are commonly found in fine-textured, water logged soils with high organic content. Water apparently has a direct effect on denitrification; the closer the soil is to saturation, the more denitrification occurs. Little denitrification occurs in soils less than about 60% saturated (Broadbent and Clark, 1965). High organic content is conducive to denitrification, because heterotrophic denitrifiers need oxidizable organic material as a source of carbon for synthesis of protoplasm and as a source of electrons for the reduction of nitrogenous compounds (Rowe and Stinnett, 1975). In addition to the denitrification occurring in riparian zones, it is probable that denitrification may be quite extensive in the anaerobic capillary fringe zone just above water tables, where water is held in soil spaces by capillary action. This especially occurs in and above saturated zones beneath agricultural areas fertilized with NO_3-N or irrigated with waters high in NO_3-N, since NO_3-N is readily leached from upper layers. Lowrance (1992), at a Coastal Plain site, found that denitrification potential was more than two orders of magnitude higher in the top 10 cm of soil than in the top 10 cm of shallow aquifer. Groffman *et al.* (1992) found that hydric surface ($0^{-1}5$ cm) soils (poorly and very poorly drained) consistently had higher denitrification enzyme activity than upland-wetland transition zone (moderately well and somewhat poorly drained) surface soils. Peterjohn and Correll (1984) estimated N dissimulation by denitrification to be 45 kg N $ha^{-1}yr^{-1}$ for riparian forested areas.

The second mechanism by which riparian zones can reduce NO_3-N concentrations in waters arriving from uplands is through vegetative uptake, particularly by the trees in forested riparian zones. Several investigators (Vitousek and Reiners, 1975; Lowrance *et al.*, 1983, 1984b) have suggested that select harvest of "mature" trees in riparian forests is a method of perpetuating vigorous vegetative uptake of soil nutrients. Odum (1969) hypothesized that constant, pulsed, and annually increasing inputs of nutrients may keep riparian forests in a "bloom" state, and the forest may respond by high and vigorous growth and nutrient uptake rates for a considerable period of time; much longer, perhaps, than the age generally considered as forest maturity. Work of Peterjohn and Correll (1984) using a nutrient mass balance approach indicated a net retention by a forested wetland of 75 kg total N $ha^{-1}yr^{-1}$. Nutrient assimilation and long-term storage in wood biomass ranged from 12 to 20 kg N $ha^{-1}yr^{-1}$.

The combined effects of denitrification and NO_3-N uptake in the riparian zone on N concentrations entering from the uplands have been documented by a number of investigators. Concentrations of N in many watersheds have remained nearly constant as loading of N from agricultural nonpoint sources has increased (Tomlinson, 1970; Thomas and Crutchfield, 1974; Hill and Wylie, 1977). Reddy and Graetz (1981) found that shallow reservoirs and flooded organic soils could be used for NH_4-N and NO_3-N removal from waste water. Van Kessel (1977) measured NO_3-N removal rates in ditches, and found them to be as high as commercial treatment of sewage. Robinson *et al.* (1978), Hoare (1979), and Hill (1983) measured significant stream loss of NO_3-N by denitrification which greatly affected stream and watershed budgets. Brinson *et al.* (1981) showed that 75% of NH_4-N and 94% of NO_3-N was removed as floodwater moved through two riverside swamps. Other studies which have shown the role of forested wetlands as partial nutrient sinks include those of Kitchens *et al.* (1975), Boyt *et al.* (1976), Ewel and Odum (1978), Nessel (1978), Mitsch *et al.* (1979), Tuschall *et al.* (1981), Day *et al.* (1981), and Qualls (1984).

Cooper *et al.* (1986) studied the role of riparian areas in treating N in four watersheds in the Coastal Plain of North Carolina. They found that an upland watershed with predominantly well drained soils in the cultivated fields lost an average of 35 kg $ha^{-1}yr^{-1}$ of NO_3-N past the field edge. However, only 5 and 8 kg $ha^{-1}yr^{-1}$ of NO_3-N and total N, respectively, were lost in drainage water leaving the watershed. They found that riparian strips as narrow as 16 m were effective in removing NO_3-N. Studies by Lowrance *et al.* (1984a) and Peterjohn and Correll (1986) also showed that riparian forests in Coastal Plain agricultural watersheds act as important sinks for NO_3-N.

Both direct and indirect approaches have been taken in studying the role of riparian forests in agricultural settings. One direct method is use of transects running from the edge of the agricultural fields through the riparian forest (Doyle *et al.*, 1975; Lowrance *et al.*, 1984a; Peterjohn and Correll, 1984; Lowrance, 1992; Hubbard and Lowrance, 1993; Simmons *et al.*, 1993). Lowrance (1992), using a transect of wells from a row-crop field to a stream in the Georgia Coastal Plain, determined that NO_3-N in groundwater decreased by a factor of 7 to 9 in the first 10 m of forest. Within the next 40 m of forest, mean NO_3-N concentration decreased from 1.80 to 0.81 mg NO_3-N L^{-1}. Simmons *et al.* (1992) assessed the removal of groundwater NO_3-N on a soil drainage sequence ranging from moderately well to poorly drained. To assess NO_3-N removal, the change in groundwater concentrations of NO_3-N relative to the concentration of the conservative tracer Br was observed in monitoring wells located in each soil drainage class. Removal of groundwater NO_3-N was consistently high in the wetland locations, generally in excess of 80% in both growing and dormant seasons. In the transition zones, attenuation was less than 36% during the growing season, and ranged from 50 to 78% in the dormant season. Attenuation in the transition zones was positively correlated with water table elevations. A second direct method is to utilize chemical budgets (Lowrance *et al.*, 1983; Todd *et al.*, 1983; Lowrance *et al.*, 1984b; Peterjohn and Correll, 1984).

More indirect methods have compared the nutrient concentrations in stream water from agricultural watersheds with varying amounts of riparian forest or have analyzed the predictive capability of models which include or exclude the presence or proximity of riparian forest (McColl, 1978; Schlosser and Karr, 1981a; Schlosser and Karr, 1981b; Omernik *et al.*, 1981, Yates and Sheridan, 1983). With the exception of the Omernik *et al.* (1981) study, all studies have reached the same general conclusion that riparian forests effectively reduce the loss of N from agricultural lands to receiving waters.

Lowrance *et al.* (1983), working in the Georgia Coastal Plain, used a nutrient budget approach to determine the effects of riparian zones on water quality. In preparing the schematic shown as Figure 3, they assumed water and nutrient movements from different land uses to be proportional to the length of interface between each land use and the riparian zone. Nitrogen movement from fields (row crops), pastures, and upland pine forest differed in both kind and amount. Of the total water volume moved from the upland mixed cover ecosystem to the riparian ecosystems, 52% came from fields, 41% from pasture lands, and 7% from pine forests. In contrast, 85% of the N moved from fields, 13% from pastures, and 2% from pines. Discharge from upland fields had 85% of the N as NO_3-N, 4% as NH_4-N and 11% as organic N. From pastures, only 9% of the total N moved as NO_3-N, while 31% was NH_4-N and 60% was organic N. Nitrogen losses in streamflow from the riparian zone were considerably smaller, and the forms had changed. While overall N inputs from the uplands were 74% NO_3-N, 8% NH_4-N, and 18% organic N, streamflow outputs were 18% NO_3-N, 2% NH_4-N, and 80% organic N. Lowrance *et al.* (1983) projected that partial conversion of the riparian forests of the Coastal Plain to cropland could increase NO_3-N and NH_4-N loads to streamflow by up to 800%.

Based on the majority of studies to date, it is clear that riparian forested zones need to be protected so that they can continue to serve as nutrient filters, and where feasible, land that was originally forested wetland, but has been drained and put to use for agriculture, should be returned to riparian forest. Development, clearing, and drainage of riparian areas decreases the NO_3-N treatment capabilities. Channelization decreases residence time and contact time with plants and sediments allowing a more direct delivery of sediment and nutrients downstream (Kuenzler *et al.*, 1977). Clearing and development in riparian areas that have served as sinks for sediment and sediment associated nutrients may initiate erosion and the export of sediment deposited over many years (Cooper *et al.* 1986).

Current and potential impact of NO_3-N on groundwater in the southeastern United States

A limited number of studies on the movement of NO_3-N by surface runoff, through the root zone, and into groundwater have been conducted in the southeastern United States. Movement by surface runoff is pertinent because these waters move downslope where they may enter stream systems which can be connected to groundwater, may flow directly into sinkholes connected to groundwater, or may slowly seep through lowland soils towards groundwater.

Research information on NO_3-N movement in the southeastern US is primarily available for the Coastal Plain physiographic region. Jacobs and Gilliam (1985) measured N concentrations and surface water flow for a 3 year period from two watersheds in the Atlantic Coastal Plain of North Carolina. They found losses of NO_3-N and total N in surface water from a Middle Coastal Plain watershed with 1,299 ha of predominantly well to moderately well-drained soils to be 2.5 and 4.5 kg $ha^{-1}yr^{-1}$, respectively. Nitrate and total N losses from a Lower Coastal Plain watershed with 6,998 ha of somewhat poorly to poorly-drained soils were 0.5 and 2.5 kg $ha^{-1}yr^{-1}$, respectively. The values from both watersheds were lower than expected, and the authors concluded that denitrification between the field and stream was the primary reason for these low values. Concentrations of NO_3-N in streams near the field outlets also decreased significantly as N enriched water moved through the transport system, and the authors concluded that further denitrification was the primary loss mechanism in the stream transport system.

Rates of transport of plant nutrients in drainage waters from two steeply sloping, differently fertilized grassed watersheds in western North Carolina, over a 4-year period, were reported (Kilmer *et al.*, 1974). Nitrate was found to comprise 70% of the total loss in discharge from the two watersheds. The winter and spring months accounted for the highest losses of all nutrients. Average annual NO_3-N concentrations in drainage effluent during the 4-year period were 1.2 and 3.9 mg L^{-1} for the two watersheds, with the NO_3-N concentration exceeding 10 mg L^{-1} only once during the four years. This work was done in the mountain physiographic area, and the authors concluded that steeply sloping pastures, judiciously fertilized, are not an important source of nutrients in surface and groundwaters.

The topic of nutrient movement in Florida was discussed by Campbell (1978).

He indicated that sandy soils, along with high rainfall and warm temperatures, create a condition conducive to rapid nutrient movement under certain conditions, but that the variability of sandy soils in Florida adds confusion to any analysis of the potential for nutrient losses in runoff. Sandy soils in some locations of Florida are deep, with corresponding water table depths of $9^{-1}2$ m or more, while in nearby flatwoods areas, spodic layers and clay layers in the sandy soil may cause a shallow water table fluctuating from 2 m deep up to the soil surface (Campbell 1978). Campbell (1978) concluded that groundwater pollution in Florida is of particular concern because of the state's many lakes, and the ready exchange of surface and groundwater.

Brinsfield and Staver (1990) indicate that elevated NO_3-N concentrations in groundwater have been detected both throughout the Atlantic Coastal Plain (Bachman, 1984), and in other areas of the United States and Europe (Hallberg, 1986b). Concerns were expressed particularly for the Delmarva Peninsula, the Chesapeake Bay drainage basin, and the Maryland Coastal Plain. In the Chesapeake Bay region, shallow aquifers are connected to surface water bodies, and lateral subsurface transport of NO_3-N can make significant N contributions to surface water. Hydrologic studies in the Chesapeake Bay suggest that groundwater flow may be a major mechanism for NO_3-N transport into the Bay (Staver *et al.*, 1989). In the Maryland Coastal Plain, long-term studies on nutrient transport have indicated high rates of N loss associated with corn production, with the major transport pathway being the leaching of NO_3-N from the root zone into groundwater (Staver *et al.*, 1988). The most intense groundwater recharge period in this area is during the winter months; consequently, the most critical factor determining annual NO_3-N leaching rates is the soil NO_3-N concentration following the primary growing season (Brinsfield and Staver, 1990). Brinsfield *et al.*, (1988) and Brinsfield and Staver (1989) recommended use of winter cover crops planted soon after corn grain harvest to reduce the availability of NO_3-N in the soil. Staver and Brinsfield (1990) investigated soil NO_3-N concentrations in non-irrigated corn production systems in the Coastal Plain region of the Chesapeake Bay drainage basin, and observed the lowest NO_3-N concentrations in the upper soil profile (0-30 cm) during the winter months. They also found that under presently recommended fertilizer rates for their area, NO_3-N concentrations in groundwater in the root zone exceeded 10 mg L^{-1} following corn harvest.

Staver and Brinsfield (1990) indicated that for the Coastal Plain agricultural regions of Maryland, corn production uses the highest rates of N fertilizer and hence has the greatest potential for NO_3-N leaching. They concluded that poultry and livestock also are potential groundwater polluters for this region. The saturated conditions necessary for leaching in Maryland generally do not occur from the time of the major N application (30-40 days post-planting) until after the completion of the growing season in nonirrigated corn production systems (Staver *et al.*, 1989). During the growing season major rainfall events tend to be high-intensity short-duration convective storms that may produce surface runoff but rarely deep percolation. Also, in Atlantic Coastal Plain corn production systems, evapotranspiration exceeds the average rate of precipitation during the middle and latter part of the growing season (Hubbard *et al.*, 1984; McLaughlin *et al.*, 1985;

Schepers, 1988; Urban *et al.*, 1989). Consequently, the primary factor that dictates the magnitude of NO_3-N leaching losses to groundwater is the availability of NO_3-N, in the upper soil profile after crop harvest and when rainfall exceeds evapotranspiration demand (Staver and Brinsfield, 1990). The Maryland Coastal Plain showed a sharp decline in NO_3-N in the plow layer during the winter recharge period, which along with groundwater NO_3-N concentrations of 10 to 20 mg L^{-1} in the shallow aquifer underlying the area, suggest that plow-layer NO_3-N is being lost to groundwater (Staver and Brinsfield, 1990). Staver and Brinsfield (1990) also found that a rye cover crop planted immediately following corn harvest reduced NO_3-N levels to less than 50% of those observed in the no-cover areas.

Work by Hickman (1987) on the tidal Potomac River and Estuary involved a drainage area falling within two geomorphic provinces: the Piedmont province to the west of the Fall Line, and the Coastal Plain province to the east. Data from Montgomery County, MD indicated that relatively high concentrations of NO_3-N in groundwater were impacting the river. Groundwater samples taken since 1946 from 48% of the wells in Montgomery County showed concentrations of NO_3-N greater than 1.3 mg L^{-1} (Woll, 1978), the low-flow concentration of NO_3-N observed in streamflow by Hickman (1987).

Considerable research on the hydrologic cycle, NO_3-N movement, and nutrient cycling has been conducted in the Tifton upland area of the Coastal Plain by the Southeast Watershed Research Laboratory of the Agricultural Research Service. The fate of water percolating through the root zone in plinthic soils (Plinthic Paleudults) has been well documented by studies on a 0.34 ha watershed (Station Z) instrumented for both surface runoff and subsurface flow (Jackson *et al.*, 1973; Hubbard and Sheridan, 1983; Hubbard *et al.*, 1991). Surface runoff from the area is directed through a 0.30-m, H-type flume with a sloping floor while shallow subsurface flow moving from the area is intercepted by a gravel-packed terra-cotta tile drain with flow outletting at a V-notch weir. It was determined that on plinthic soils underlain by the Hawthorn Formation (low permeability), lateral subsurface flow accounts for about 80 percent of total runoff and from 17 to 41% of total annual rainfall. Of NO_3-N losses measured in surface runoff and shallow subsurface flow, 99% moved with the subsurface flow. Average NO_3-N concentrations in surface runoff and shallow subsurface flow ($0.5^{-1}.0$ m depth) were 0.47 and 8.75 mg L^{-1} respectively (Hubbard and Sheridan, 1983). Nitrogen lost as NO_3-N in surface runoff and shallow subsurface flow was about 20 percent of the fertilizer N input to the area. Figure 4 shows the mean monthly surface runoff and shallow subsurface flow NO_3-N concentrations and loads from Station Z for a 10-year period.

A 1988-89 study using the same watershed facility (Station Z) also examined transport of NO_3-N by surface runoff, shallow subsurface flow, and leaching, with the focus being on what happens in the field during the first few months after N application (Hubbard *et al.*, 1991). Nitrate concentrations were measured in surface runoff, shallow subsurface flow, soil samples collected from the upper 30 cm of the root zone, and water samples collected from shallow and deep wells. The study showed that most of the NO_3-N leached from the upper 30 cm of the root zone within 1.5 months after N application, given a total rainfall of 200 mm or more. Surface run-off NO_3-N concentrations and loads were small, with monthly loads not exceeding 0.3 kg ha^{-1}. Nitrate concentrations in shallow groundwater (0.9 to 1.8 m) during the sweet corn growing season reflected applied N, with values ranging from 11 to 19 mg L^{-1}. Applied N was also reflected in shallow subsurface flow during the spring months. Mean NO_3-N concentrations in subsurface flow ranged from 3.9 to 14.1 mg L^{-1}, while monthly loads ranged from 0.1 to 9.2 kg ha^{-1}. The study indicated that some NO_3-N is being lost by vertical water movement in plinthic Coastal Plain soils towards deeper groundwater, and that this loss pathway should be of environmental concern.

On a deep sandy soil (Bonifay sand) in the Tifton upland, significant NO_3-N concentrations were found in shallow groundwater where N was applied to a variety of crops (Hubbard *et al.*, 1984). The Bonifay sand is quite sandy to a depth of about 2.7 m, at which depth significant plinthic and/or other water restrictive layers are encountered. A four year study using recommended N rates from 20 to 650 kg ha^{-1}yr^{-1}, showed shallow groundwater NO_3-N concentrations ranging from less than 1 mg L^{-1} to 133 mg L^{-1}, with an overall mean concentration of 20 mg L^{-1}. This contrasted greatly with NO_3-N concentrations observed in wells in an adjacent pristine forest area, where NO_3-N concentrations ranged from less than 0.1 mg L^{-1} to just over 1 mg L^{-1} (Hubbard *et al.*, 1984).

Nitrate leaching research at Tifton using dairy waste applications to forage crops has also indicated contamination problems in unconfined shallow groundwater (Hubbard *et al.*, 1987). Nitrate concentrations were measured in wells placed at 1.2 m, the interface of the overlying sandy material and the plinthite or denser subsoil materials; at 2.4 m, well within the plinthite and dense subsoil materials; and at 3.6 m, where there is a more permanent year-round water table. The observed mean NO_3-N concentrations at these three depths under high rates of dairy wastewater application were 45.4, 36.3, and 16.2 mg L^{-1}, respectively. Relative to landscape position under the center pivot used to apply the wastewater, NO_3-N concentrations were much higher in the shallow and intermediate wells in the upper and middle landscape positions (Table 2). Nitrate concentrations in the deep wells were highest in the middle part of the landscape. At the lowest part of the landscape, which included the edge of a riparian zone, NO_3-N concentrations dropped because of dilution, denitrification, and NO_3-N uptake by vegetation. In a control area used as part of this same study, where inorganic N was applied to Coastal bermudagrass, mean NO_3-N concentrations were 33.7, 28.5, and 31.2 mg L^{-1} at the three depths, respectively. The high NO_3-N concentrations in shallow groundwater from this control area indicated that NO_3-N moved to shallow groundwater through the dense subsoil materials of the Tifton upland regardless of source. More recent work at the same site has shown significant NO_3-N concentrations (33-44 mg L^{-1}) in tile drainage even with reduced N application rates. This work indicates that once a site has been heavily loaded with animal wastes, contamination of shallow groundwater may continue for a long time even with reduced N application rates. Another finding from this study was the relatively high NO_3-N concentrations within and beneath soil and geologic materials considered to be nearly impermeable. In a newer study, water samples collected from a monitoring well network installed in 1991 on a Coastal

Plain site having the same high clay subsoils as the dairy waste site had NO_3-N concentrations of $12^{.}5$ mg L^{-1} at 3 m (Vellidis *et al.*, 1993). This site had never received animal waste and had not been fertilized with inorganic fertilizer in the previous 12 months, which implies that the observed levels were due to inorganic fertilizer applications in previous years. Results from both the dairy waste study (Hubbard *et al.*, 1987) and the background NO_3-N levels observed in the newer study in 1991 (Vellidis *et al.*, 1993) indicate potential for deeper movement of NO_3-N even in less permeable portions of the landscape, and imply that long term concern about NO_3-N contamination of groundwater in these regions is imperative.

Overall, currently available research on movement to groundwater in the southeastern USA indicates that under presently used N application rates both root zone NO_3-N concentrations and shallow groundwater NO_3-N concentrations may exceed the 10 mg L^{-1} drinking water standard. Much less is currently known about deeper groundwater in the southeastern US, but the research information available from the Midwest and other parts of the US and world clearly indicate the potential for NO_3-N contamination of deeper groundwater.

Role of models in assessing future impact and for guiding management decisions

Computer simulation models have become a major tool of both researchers and planners since the mid 1970s. There are a number of types of models, ranging from ones which simulate processes at small scale to those which examine nutrient cycling through mass balance approaches on a regional scale. Some solute transport models have been developed specifically for pesticides, while others include both pesticides and nutrients. Models developed strictly for pesticides include the Pesticide Transport and Runoff (PTR) model of Crawford and Donigian (1973), the Cornell Pesticide Model (CPM), Steenhuis and Walter (1980); and the Pesticide Root Zone Model (PRZM) by Carsel *et al.* (1984). Models having both pesticide and nutrient components include the Agricultural Runoff Management (ARM) model of Donigian and Crawford (1976), the Agricultural Chemical Transport Model (ACTMO) model of Frere *et al.* (1975), the Water-Sediment-Chemical (WASCH) model of Bruce *et al.*, (1975) and the Chemicals, Runoff, and Erosion from Agricultural Management Systems (CREAMS) model (Knisel, 1980). More recent models include the Groundwater Loading Effects of Agricultural Management Systems (GLEAMS) model (Leonard *et al.*, 1987), which was based on the earlier CREAMS model but makes predictions for individual soil layers, the Root Zone Water Quality Model (RZWQM) of DeCoursey *et al.* (1989), and the Riparian Ecosystem Management Model (REMM) of Altier, *et al.* (1993).

Agricultural mass balance models (Frink 1969, Gilliam and Terry, 1973, Fried *et al.*, 1976; Frissel 1978; Scheuler and Kemp 1979; Correll 1981) differ from the process oriented models in that these methods attempt to identify all inputs, storages, and outputs of a particular nutrient within a defined system, assuming the storage pool (soil system, forest system) is not changing (Frissel, 1978). For N, the mass balance approaches consider nutrient inputs to be fertilizer, precipitation, and

N-fixation by legumes, while the principal outputs are denitrification, harvested plant biomass, surface runoff, and leaching to groundwater. As an example, Kuenzler and Craig (1986), using a mass balance approach, made the assumptions that (1) the amount of fertilizer applied annually to a specific crop is the average amount recommended by the state's agricultural extension agency; (2) the amount of N in precipitation falling on a watershed in the Coastal Plain is about 8.7 kg ha^{-1} yr^{-1}; (3) the N-fixation rates for soybeans and peanuts are about 105 and 112kg ha^{-1} yr^{-1}, respectively (Frissel, 1978); (4) denitrification is about 15% of the applied fertilizer N (Frissel, 1978); (5) the amount of N in harvests can be obtained by multiplying the N content of each crop by its average annual crop yield kg ha^{-1} yr^{-1} (Romaine, 1965); and (6) the potential amount of N in surface runoff and leaching is the difference between these outputs and inputs.

Another modeling approach to the determination of the impact of NO$_3$-N leaching on groundwater quality is to develop aquifer vulnerability indexes. Lemme *et al.* (1990) used land use data and soil and geologic information to calculate contamination vulnerability indexes. They used soil and geologic data from a 7.5 minute topographic quadrangle with land use data to formulate surface and aquifer vulnerability index values. For the surface water vulnerability index an abbreviated form of the soil loss equation which included rainfall in j ha^{-1} (R), soil erodibility in mt j^{-1} (K), slope length (L), and slope gradient (S) was used to calculate an index value for each grid point from soil survey data. A Kriging procedure was used to integrate between grid points, and the output was expressed as an iso value map. Four surface water vulnerability classes were used according to the RKLS values: low vulnerability class, RKLS values of 0^{-1}0; medium vulnerability, RKLS of 11-25; high vulnerability, RKLS of 26-32; and very high vulnerability, RKLS greater than 32.

For their aquifer vulnerability index, Lemme *et al.* (1990) used two primary inputs: soil organic matter, because of its ability to retain potential pollutants, and permeability, because of its relationship to the effectiveness of the soil and geologic material to perform as a filter. They used the low end of the permeability ranges (found on the soil interpretation record), and the thickness of material overlying a sand and gravel body (to a depth of 18.3 m) in their equation for aquifer vulnerability. Their index equation resulted in a range of values from 0 to 10, with the aquifer vulnerability categories being: very low vulnerability, index value of 0-2.0; low vulnerability, index of 2.1-6.0; medium vulnerability, index of 6.1-8.0; and high vulnerability, index greater than 8.1.

A methodology developed recently by the US Environmental Protection Agency for evaluating groundwater pollution potential in varied hydrologic settings uses the acronym DRASTIC (Aller *et al.*, 1985). This methodology incorporates the major hydrogeological factors that affect and control groundwater movement, including depth to the water table (D), net recharge (R), aquifer media (A), soil media (S), topography (T), impact of the vadose zone (I), and hydraulic conductivity (C) of the aquifer. The DRASTIC score for a given site is computed by multiplying the rating determined for each of the seven factors by a weight that reflects the importance of each factor to the overall contamination potential. The products are summed to yield the overall DRASTIC score for the given site. Huang et al. (1990) used DRASTIC scores on a national basis and concluded that 34 million hectares were highly vulnerable to pesticide contamination. The DRASTIC index indicates that those Atlantic and Gulf Coastal Plain hydrogeologic settings that are at the surface or that exhibit a high degree of connectedness with the surface, such as the shallow surficial aquifers, floodplain alluvial aquifers, and solution limestone aquifers, show a high groundwater pollution potential (Aller et al., 1985). Where the regional aquifer system is confined and exhibits a lower degree of connectedness with the surficial zone, the model indicates a low groundwater pollution potential.

Conclusions

The southeastern USA is a broad region including physiographic areas which are quite varied in geohydrologic conditions. The commonality of the region includes generally warm temperatures, particularly during the summer months, and high rainfall. The range in soil and geologic materials makes generalizations about NO_3-N leaching to groundwater in the southeastern US difficult. The main generalization is that the weathered soils of the region require relatively high input of N for adequate crop production, and that this input coupled with high annual rainfall results in significant probability for NO_3-N movement to groundwater. Consequently N management decisions, both in agricultural and in urban areas in the southeast, must examine potential effects on groundwater quality.

The primary management tool for reducing the levels of NO_3-N reaching aquifer systems is to keep fertilizers, animal wastes, or septic tank leachates in the root zone as long as possible, so that they can be taken up by plants or degraded by microbial processes. This objective requires using minimum necessary amounts of fertilizers, designing septic systems with adequate drain field areas, and minimizing percolation. Use of the minimum necessary amount of fertilizer increases the percentage of N taken up from that applied, and hence decreases the amount of NO_3-N available for leaching. For animal wastes careful accounting of nutrient contents is needed so that excess amounts are not applied. Recent extension publications from Wisconsin (Bundy et al., 1992; Good et al., 1991; and Wolkowski, 1992) are good examples of information for dairy producers which show how to credit manure applications for nutrient management and protection of water quality. Irrigation should be as efficient as possible, and N fertilizers should be applied only when needed and in amounts that crops can use (Bouwer, 1987). Cropping and management practices in the southeastern US, including use of winter cover crops, need to minimize nutrient availability for deeper movement during the seasonal wet periods of the winter-spring months, when most subsurface movement and percolation occurs. One major land use tool which may assist in limiting NO_3-N leaching to groundwater is the use of riparian zones to remove NO_3-N from water either by denitrification or by vegetative uptake. Such areas are particularly important in places where confined and semiconfined conditions result in shallow lateral subsurface movement of water and nutrients draining from upland agricultural sites into adjacent riparian forests.

The Council for Agricultural Science and Technology (1985) recommends specifications to reduce NO_3-N loss to groundwater from agricultural sources, including N applications based on soil or plant tissue tests; smaller, more frequent N applications; use of slow-release fertilizers; use of chemical inhibitors to delay formation of NO_3-N from NH_4-N and other forms of applied N; avoiding fall application of N fertilizers for succeeding-season crops; use of plant-applied urea solution sprays rather than soil-applied N fertilizers; and increased use of legumes to provide supplemental N for cropping systems. For animal wastes, a chemical analyses is essential in order to estimate application rates appropriate for both crop nutrient and environmental quality goals.

The second key element relative to limiting NO_3-N contamination of groundwater in the southeastern USA is to have technical information on, and understanding of groundwater recharge and NO_3-N movement patterns for the specific area. Necessary information includes rainfall patterns, and infiltration and percolation rates relative to soils and geology. Given that different areas have different potentials for groundwater contamination, management strategies for N fertilizer application rates and timing, for irrigation application and scheduling, and for cropping systems should include consideration of the potential for local groundwater contamination. Agricultural management objectives must include not only concerns for efficient crop production, soil conservation, and surface water quality, but also an awareness of the need to protect and preserve the quality of regional groundwater resources. When evaluating management strategies relative to pollution potential, it must be kept in mind that even in less permeable segments of the landscape, NO_3-N contamination can occur over time. Use of models is one aid for management decisions which may result in minimizing NO_3-N leaching to groundwater. Whether the model is a process-oriented computer simulation model, a mass balance model, or a vulnerability index model, the intent is that it serve as a tool in the management decision process.

The impacts of land management on movement of NO_3-N and other solutes to groundwater in the southeastern USA are happening now, and will happen in the next 10-20 years. Observations of high NO_3-N in drain tiles in the midwestern part of the US in the early 1970s have been followed by reports of high NO_3-N concentrations in deep groundwater in the late 1980s in such places as Iowa and Nebraska. With proper management and understanding by farmers and others in agribusiness and industry, and by the general public concerning the value of water resources and the need to maintain the quality of these resources, the scenario that has occurred in the midwestern US hopefully will be prevented in the southeastern USA.

Future research needs

The literature search in preparation of this book chapter showed a sparsity of information on NO_3-N movement to groundwater in the Southeast as compared to available information from other regions of the US and world. Consequently a major future research need is for new studies which better document current quality of groundwater in the Southeast and, where NO_3-N concentrations are elevated, a determination of the source of NO_3-N contamination. Along with general studies of water quality in the Southeast there is a research need for obtaining basic information on factors affecting NO_3-N movement processes in this region, both so that the factors and processes will be better understood and so that predictive models can be improved and/or verified.

Nitrogen sources applied to the soil surface are subject to movement via surface runoff and infiltration. Detachment and transport of soil particles by erosion processes influence NO_3-N movement. Processes such as surface sealing/crusting affect rainfall partitioning and detachment and transport of sediment and hence affect NO_3-N movement. New research is needed to better define surface runoff, erosion, and infiltration transport processes and mechanisms in southeastern soils.

Transport of NO_3-N vertically through soil and subsoil materials depends on water movement. Soil physical properties such as particle size distribution and porosity largely control water movement rates. Preferential flow through macropores may also greatly influence the rate and extent to which NO_3-N moves to groundwater. More research is needed on water and solute transport pathways through the root zone and vadose zone of southeastern soils. Better understanding of water movement both at the local and regional scale will aid in devising management strategies which protect groundwater quality.

Denitrification is an important loss pathway for NO_3-N. Past research indicates that this process is important in wet soils, particularly in the riparian zones commonly found in the Coastal Plain (Lowrance *et al.*, 1984b). Recent research (Obenhuber and Lowrance, 1991) also has shown that denitrification is an important NO_3-N loss pathway in unconsolidated aquifer material when energy sources for the denitrifiers are available. More information is needed on the amount of denitrification occurring and factors affecting denitrification in southeastern soils.

While denitrification is an important NO_3-N loss pathway in riparian zones, uptake and cycling of NO_3-N by riparian vegetation may be significant. New basic research on southeastern riparian zones is needed to understand the role of all processes involved in NO_3-N transport and cycling in this portion of the landscape. Along with information on basic processes is the need to develop management strategies and predictive models for using riparian zones and adjacent buffer strips to improve quality of water entering from the uplands. Field and laboratory research and modeling efforts concerning riparian zones are needed both now and in the future for the Southeast.

One final area of needed future research pertains to management effects on NO_3-N movement to groundwater. New research is needed on the effect of conservation tillage on NO_3-N leaching in the Southeast, since some studies show increased leaching, others decreased leaching, and still others no difference as compared to conventional tillage. The impact of animal waste application on NO_3-N movement to groundwater also is not clearly understood, and research is needed to address this issue. Finally, the whole research area of N use efficiency, i.e. getting maximum utilization of applied N by crops within acceptable yield goals, while at the same time having minimum negative effects on water quality, needs additional work for the southeastern USA

References

Abbott, G.A. 1949. High Nitrates in Drinking Waters and Their Toxicity to Infants. In: *Annual Proc., North Dakota Academy of Science*, pp. 51-54.

Alexander, M. 1965. Nitrification. In: Bartholomew, W.V. (ed.), *Soil Nitrogen, Agronomy* Monograph 10, pp. 309-346. Madison, American Society of Agronomy:

Aller, L., Truman, B., Lehr, J.H. and Petty, R.J. 1985. *DRASTIC: A standardized system for evaluating groundwater pollution potential using hydrogeologic settings.* EPA/600/2-85/018. U.S. Environmental Protection Agency, Washington, D.C. 163 p.

Altier, L.S., Lowrance, R.R., Williams, R.G., Sheridan, J.M., Bosch, D.D., Hubbard, R.K., Mills, C.M. and Thomas DL. 1993. An ecosystem model for the management of riparian areas. *Proc. Riparian Ecosystems in the U.S. Conf.* (in press).

Ayers, R.S. and Branson, R.L. (eds). 1973. *Nitrates in the Upper Santa Ana River Basin in Relation to Groundwater Pollution.* Bulletin No. 861, California Agricultural Experiment Station, Division of Agricultural Sciences, University of California.

Bachman, L.J. 1984. The Columbia aquifer of the eastern shore of Maryland, Part I. *Hydrogeology. Rpt. Investigations*, No. 40. Md. Geol. Serv., Annapolis. 34 p.

Baker, J.L., Campbell, K.L., Johnson, H.P. and Hanway, J.J. 1975. Nitrate, phosphorus and sulfate in subsurface drainage water. *J. Environ. Qual.*, 4, 406-412.

Beck, B.F., Asmussen, L. and Leonard,R. 1985. Relationship of geology, physiography, agricultural land use, and groundwater quality in southwest Georgia. *Ground Water*, 23,627-634.

Bockstadter, T., Bourgh, C., Buttermore, G., Eisenhauer, D., Frank, K. and Krull D 1984. *Nitrogen and Irrigation Management.* Final Report Hall County Water Quality Project. Nebraska Cooperative Extension Service. Nebraska Department of Environmental Control.

Bouwer, H. 1987. Effect of irrigated agriculture on groundwater. *J. Irr. Drain. Div., Am. Soc. Civil Eng.*,113(1), 4-15.

Boyt, F.L., Bayley, S.E. and Zoltec, J. 1976. Removal of nutrients from treated municipal wastewater by wetland vegetation. *J. Wat. Pollut. Control Fed.*, 49, 789-799.

Brinsfield, R.B., Staver, K.W. and Magette, W.L. 1988. The role of cover crops in reducing nitrate leaching to groundwater. In: *Agricultural Impacts on Groundwater.* Nat. Well Water Assoc., Dublin, Ohio.

Brinsfield. R.B and Staver, K.W. 1989. Cover crops: A paragon for nitrogen management. In: *Groundwater Issues and Solutions in the Potomac River Basin/Chesapeake Bay Region.* Nat. Well Water Assoc., Dublin, Ohio.

Brinsfield, R.B. and Staver, K.W. 1990. Addressing groundwater quality in the 1990 farm bill: Nitrate contamination in the Atlantic Coastal Plain. *J. Soil Water Conserv.*, 45(2), 285-286.

Brinson, M.M., Bradshaw, H.D. and Kane, E.S. 1981. *Nitrogen cycling and assimilative capacity of nitrogen and phosphorus by riverine wetland forests.* Rep. No. 167, Wat. Resour. Res. Inst. Univ. North Carolina. Raleigh, NC. 90 p.

Broadbent, F.E. and Clark, F. 1965. Denitrification. In: Bartholomew, W. V. (ed.), *Soil Nitrogen, Agronomy,* Monograph 10, pp. 347-362. Madison, American Society of Agronomy.

Bruce, R.R., Harper, L.A., Leonard, R.A., Snyder, W.M. and Thomas, A.W. 1975. A model for runoff of pesticides from small upland watersheds. *J. Environ. Qual.*, 4(4), 541-548.

Bundy, L.G., Kelling, K.A., Schulte, E.E., Coombs, S., Wolkowski, R.P., Sturgul, S.J., Binning, K. and Schmidt, R. 1992. *Nutrient Management: Practices for Wisconsin corn production and water quality protection.* A3557. University of Wisconsin-Extension, Cooperative Extension, Madison, WI.

Burwell, R.E., Schuman, G.E., Saxton, K.E. and Heineman, H.G. 1976. Nitrogen in subsurface discharge from agricultural watersheds. *J. Environ. Qual.*, ,5, 325-329.

Calvert, D.V. and Phung, H.T. 1971. Nitrate-nitrogen movement into drainage lines under different soil management systems. *Soil Crop Sci. Soc. Florida Proc.*, 31, 229-232.

Calvert, D.V. 1975. Nitrate, phosphate and potassium movement into drainage lines under three soil management systems. *J. Environ. Qual.*, ,4, 183-186.

Cameron, D.R., DeJong, R. and Chang, C. 1978. Nitrogen inputs and losses in tobacco, bean, and potato fields in a sandy loam watershed. *J. Environ. Qual.*, 7,545-550.

Campbell, C.A., Ferguson, W.A. and Warder, F.G. 1970. Winter changes in soil nitrate and exchangeable ammonium. Canadian *J. Soil Sci.*, 50,151-162.

Campbell, K.L. 1978. *Pollution in runoff from nonpoint sources.* Publication No. 42. Florida water resources research center. Research project technical completion report. OWRT Project Number B-023-FLA. 49 pp.

Cantor, L. and Knox, R.C. 1984. *Evaluation of septic tank effects on ground-water quality.* EPA-600/2-284-107. U.S. Environmental Protection Agency, Washington, D.C. 259 p.

Carsel, R.F., Smith, C.N., Mulkey, L.A., Dean, J.D. and Jowise, P. 1984. *Users manual for the Pesticide Root Zone Model PRZM.* Release 1. U. S. Environmental Protection Agency. Athens, Georgia.

Case, A.A. 1957. Some aspects of nitrate intoxication in livestock. *J. Am. Vet. Med. Ass.*, 130(8), 323-328.

Chichester, F.W. 1976. The impact of fertilizer use and crop management on nitrogen content of subsurface water draining from upland agricultural watersheds. *J. Environ. Qual.* 5, 413-416.

Chichester, F.W. 1977. Effects of increased fertilizer rates on nitrogen content of runoff and percolate from monolith lysimeters. *J. Environ. Qual,.* 6, 211-217.

Christy, M., Brown, J.R. and Murphy, L.S. 1965. *Nitrate in soils and plants guide,* pp. 9804-9806.. Columbia, University of Missouri Extension Division.

Collins, M.A. 1984. *Water for the 21st Century: Will it be there?* Conference papers from April 3-5, 1984. Center for Urban Studies School of Engineering and Applied Science, Southern Methodist University, p. 3

Commoner, B., Shearer, G. and Kohl, D. 1972. *Origins of nitrates in surface water, first year progress report for a study of certain ecological, public health, and economic consequences of the use of inorganic N fertilizer.* Center for the Biology of Natural Systems, Washington University, St. Louis, Missouri. May 5, 1972.

Cooper, J.R., Gilliam, J.W., Jacobs, T.C. 1986. Riparian areas as a control of nonpoint pollutants. In: Correll, D.L. (ed..) *Watershed Research Perspectives*, pp.166-192. Smithsonian Institution Press, Washington, DC 20560.

Correll, D.L. 1981. Nutrient mass balance for the watershed, headwaters intertidal zone, and basin of the Rhode River Estuary. *Limnol. Oceanogr.* 26(6), 1142-1149.

Council for Agricultural Science and Technology, 1985. *Agriculture and groundwater quality.* Rpt. No. 103. Ames, Iowa. 62 p.

Crabtree, K.T. 1972. Nitrate and nitrite variation in ground water. Department of Natural Resources, Madison, Wisconsin. *Technical Bulletin* No. 58. p. 2-22.

Crawford, N.H. and Donigian, A.S., Jr. 1973. *Pesticide transport and runoff model for agricultural lands.* U. S. Environmental Protection Agency, Environmental Protection Technology Series, EPA-660/2-74-013.

Crosby, N.E. and Sawyer, R. 1976. N-Nitrosamines: A review of chemical and biological properties and their estimation in foodstuffs. *Adv. Food Res.*, 22, 1-71.

Day, J.W., Jr., Schlar, F.H., Hopkinson, C.S., Kemp, G.P. and Conner, W.H. 1981. *Modeling approaches to understanding and management of freshwater swamp forests in Louisiana USA*. Coastal Ecology Laboratory. Center for Wetland Resources. Louisiana State University, Baton Rouge.

DeCoursey, D.G., Rojas, K.W. and Ahuja, L.R. 1989. *Potentials for non-point source groundwater contamination analyzed using RZWQM*. ASAE Winter Meeting Paper #SW892562.

Donigian, A.S., Jr. and Crawford, N.H. 1976. *Modeling pesticides and nutrients on agricultural lands*. U. S. Environmental Protection Agency, Environmental Protection Technology Series, EPA-600/2-76-043.

Doyle, R.C., Wolf, D.C. and Bezdicek, D.F. 1975. Effectiveness of forest buffer strips in improving the water quality of manure polluted runoff. In: *Managing Livestock Wastes*, pp. 299-302. Proc. of the 1975 Int. Symp. on Livestock Wastes, Am. Soc. of Agric. Eng., St. Joseph, Mich.

Edmunds, W.M., Bath, A.H. and Miles, D.L. 1982. Hydrochemical evolution of the East Midlands Triassic sandstone aquifer, England. *Geochim. Cosmochim. Act,.* 46, 2069-2081.

Edwards, D.M. and Monke, E.J. 1968. Electrokinetic studies of porous media systems. *Trans. ASAE,* 11(3), 412-415.

Edwards, D.M., Fischback, P.E. and Young, L.L. 1972. Movement of nitrates under irrigated agriculture. *Trans ASAE,* 15(1), 73-75.

Endelman, F.J., Keeney, D.R., Gilmour, J.T. and Saffigna, P.G. 1974. Nitrate and chloride movement in the Plainfield loamy sand under intensive irrigation. *J. Environ. Qual.*, 3, 295-298.

Ewel, K.C. and Odum, H.T. 1978. Cypress swamps for nutrient removal and wastewater recycling. In: Odum, H.T. and Ewel, K.C. (prin. invest.), *Cypress Wetlands for Water Management, Recycling, and Conservation*, pp. 16-34.. Fourth Ann. Rept. to N.S.F. and Rockefeller Found. Center for Wetlands, Univ. of Florida, Gainesville.

Exner, M.E. and Spalding, R.F. 1979. Evolution of contaminated groundwater in Holt County, Nebraska. *Water Resour. Res.,* 15, 139-147.

Feth, J.H. 1966. N Compounds in Natural Waters--A Review. *Water Resources Research,* 21., 41-58.

Fitzsimmons, D.W., Lewis, G.C., Naylor, D.V. and Busch, J. 1972. Nitrogen, phosphorus and other inorganic materials in waters in a gravity irrigated area. *Trans. ASAE,* 15(2)., 292-295.

Forbes, R.B., Hortenstine, C.C. and Bistline, E.W. 1974. Nitrogen, phosphorus, potassium and soluble salts in solution in a Blanton fine sand. *Soil Crop Sci. Soc. Fla. Proc.*, 33, 202-205.

Frere, M.H., Onstad, C.A. and Holtan, H.N. 1975. *ACTMO, an Agricultural Chemical Transport Model*. U. S. Department of Agriculture, Agricultural Research Service, Headquarters, ARS-H-3.

Fried, M., Tanji, K.K. and Van Depol, R.M. 1976. Simplified long-term concept of evaluating leaching of nitrogen from agricultural land. *J. Environ. Qual.* 52, 197-200.

Frink, C.P. 1969. Water pollution potential estimated from farm nutrient budgets. *Agron. J.*, 61(1), 550-553.

Frissel, M.J. 1977. Cycling of mineral nutrients in agricultural ecosystems. *Agro-Ecosystems*, 4, 1-354.

Frissel, M.J. 1978. *Cycling of mineral nutrients in agricultural ecosystems.* Elsevier Press, New York.

Gerwing. J.R., Caldwell, A.C. and Goodroad, L.L. 1979. Fertilizer nitrogen distribution under irrigation between soil, plant, and aquifer. *J. Environ. Qual.,* 8, 281-284.

Gilliam, J.W. and Hoyt, G.D. 1987. Effect of conservation tillage on fate and transport of nitrogen. In: Logan, T.J. *et al.* (eds.), *Effects of Conservation Tillage on Groundwater Quality: Nitrates and Pesticides,* Lewis Publishers, Inc., Chelsea, Michigan 48118.

Gilliam, J.W. and Terry, D.L. 1973. *Potential for water pollution from fertilizer use in North Carolina.* N. C. Agr. Ext. Serv. Circ. 550.

Gold, A.J., DeRagon, W.R., Sullivan, W.M. and Lemunyon, J.L. 1990. Nitrate-nitrogen losses to groundwater from rural and suburban land uses. *J. Soil Wat. Conserv.,* 45(2)., 305-310.

Good, L.W., Madison, F., Cates, K. and Kelling. K. 1991. *Nitrogen credits for manure applications.* Nutrient and Pest Management Guide A3537. University of Wisconsin-Extension, Cooperative Extension, Madison, WI.

Gormly, J.R. and Spalding, R.F. 1979. Sources and concentrations of nitrate-N in groundwater of the central Plattre region, Nebraska. *Ground Water,* 17, 291-294.

Graetz, D.A., Hammond, L.C. and Davidson, J.M. 1974. Nitrate movement in a Eustis fine sand planted to millet. *Soil Crop. Sci. Soc. Fla. Proc.* 33, 157-160.

Groffman, P.M., Gold, A.J. and Simmons, R.C. 1992. Nitrate dynamics in riparian forests, microbial studies. *J. Environ. Qual.,* 21, 666-671.

Grundman, J.E. 1965. Nitrates danger for humans, too. *Missouri Ruralist.* p. 24-25.

Hagan, R.M., Haise, H.R. and Edminister, T.W. 1967. *Irrigation of agricultural lands.* Chapter 14, Number 11 In The Series Agronomy, American Society of Agronomy, Madison, Wisc.

Hallberg, G.R. 1986a. Nitrates in groundwater in Iowa. *Proc. Nitrogen and Groundwater Conf., pp.1-36.* Iowa Fertilizer and Chemical Association.

Hallberg, G.R. 1986b. From hoes to herbicides: Agriculture and groundwater quality. *J. Soil Water Cons.,* 41, 357-364.

Harrison, K.A. 1989. *Georgia 1989 irrigation survey.* Coop. Ext. Serv., Univ. Ga., Tifton.

Heaney, J.P., Huber, W.C., Sheikh, H., Medina, M.A., Doyle, J.R., Peltz, W.A. and Darling, J.E. 1975. *Urban Stormwater Modeling and Decision Making.* U.S. EPA Report, EPA 670/2-75-022.

Hickman, R.E. 1987. Loads of suspended sediment and nutrients from local nonpoint sources to the tidal Potomac River and Estuary, Maryland and Virginia, 1979-81 Water Years: A water-quality study of the tital Potomac River and Estuary. *U. S. Geological Survey Water-Supply Paper* 2234-G. 42 p.

Hill, A.R. and Wylie, N. 1977. The influence of nitrogen fertilizer on stream nitrate concentrations near Alliston, Ontario, Canada. *Prog. Water Technol.,* 8, 91-100.

Hill, A.R. 1982. Nitrate distribution in the ground water of the Alliston region of Ontario, Canada. *Ground Water,* 20, 697-702.

Hil, A.R. 1983. Denitrification: Its importance in a river draining an intensively cropped watershed. *Agric. Ecosyst. Environ.* ,10, 47-62.

Hoare, R.A. 1979. Nitrate removal from streams draining experimental catchments. *Prog. Wat. Technol.* ,11,4-6, 303-313.

Huang, W., Algozin, K., Ervin, D. and Hickenbotham T 1990. Using the conservation reserve program to protect groundwater quality. *J. Soil Water Conserv.,* 45(2), 341-346.

Hubbard, R.K., Asmussen, L., and Allison, H.D. 1984. Shallow groundwater quality beneath an intensive multiple-cropping system using center pivot irrigation. *J. Environ. Qual.,* 13, 156-161.

Hubbard, R.K., Gascho, G.J., Hook, J.E. and Knisel, W.G. 1986. Nitrate movement into shallow groundwater through a Coastal Plain sand. *Trans. ASAE,.* 29, 1,564-1,571.

Hubbard, R.K., Hargrove, W.L., Lowrance, R.R., Williams, R.G. and Mullinix, B.G. 1994. Physical properties of a clayey Coastal Plain soil as affected by tillage. *J. Soil Water Conservation* 49(3), 276-283.

Hubbard, R.K., Knisel, W,G. and Gascho, G.J. 1985. A comparison of nitrate leaching losses under N applied conventionally or by fertigation. In: *Proc. Third Nat. Symp. on Chemigation* Univ. GA., Athens, pp. 51-57.

Hubbard, R.K., Leonard, R.A. and Johnson, A.W. 1991. Nitrate transport on a sandy Coastal Plain soil underlain by plinthite. *Trans. ASAE* 34(3), 802-808.

Hubbard, R.K. and Lowrance, R.R. 1993. Dairy Cattle Manure Management. In: *Animal Waste Management Section of USDA Conservation Research Report on Waste Management and Agriculture.* (in press).

Hubbard, R.K. and Lowrance, R.R. 1993. Spatial and temporal patterns of solute transport through a riparian forest. *Proc. Riparian Ecosystems in the U.S. Conf.* (in press).

Hubbard, R.K. and Sheridan, J.M. 1983. Water and nitrate-nitrogen losses from a small, upland, Coastal Plain watershed. *J. Environ. Qual.,* 12, 291-295.

Hubbard, R.K. and Sheridan, J.M. 1989. Nitrate movement to groundwater in the southeastern Coastal Plain. *J. Soil Water Conserv.* 44(1), 20-27.

Hubbard, R.K., Thomas, D.L., Leonard, R.A. and Butler, J.L. 1987. Surface runoff and shallow groundwater quality as affected by center pivot applied dairy cattle wastes. *Trans. ASAE,* 30, 430-437.

Humenik, F.J., Bliven, L.F., Overcash, M.R. and Kohler, F. 1980. Rural nonpoint source water quality in a southeastern watershed. *J. Wat. Pollut. Control Fed.,* 52, 29-43.

Hunter, R.J. and Alexander, A.E. 1963. Surface properties and flow behavior of kaolinite. Parts I, II and III. *J. Coll. Sci.,* 18, 820-862.

Jackson, W.A., Asmussen, L.E., Hauser, E.W. and White, A.W. 1973. Nitrate in surface and subsurface flow from a small agricultural watershed. *J. Environ. Qual.,* 2, 480-482.

Jacobs, T.C. and Gilliam, J.W. 1985. Headwater stream losses of nitrogen from two Coastal Plain watersheds. *J. Environ. Qual.,* 14(4), 467-472.

Johnson, R.H. 1988. Factors affecting groundwater quality. In: Moody, D.W., Carr, J.: Chase, E.B. and Paulson, R.W. (eds.), *National Water Summary 1986--Hydrologic Events and Ground-Water Quality. Water Supply* Paper 2325, pp. 71-86. U. S. Geol. Surv., Reston, VA.

Johnston, W.R., Ittihadieh, F., Daum, R.M. and Pillsbury, A.F. 1965. Nitrogen and phosphorus in tile drainage effluent. *Soil Science Society of America Proc.* 29(3), 287-289.

Kanwar, R.S., Baker, J.L. and Laflen, J.M. 1985. Nitrate movement through the soil profile in relation to_tillage system and fertilization application method. *Trans. ASAE,* 28, 1802-1807.

Keeney, D.R. and Walsh, L.W. 1972. Available nitrogen in rural ecosystems: Sources and fate. *Hortscience:* 73., 219-223.

Keeney, D.R. 1982. Nitrogen management for maximum efficiency and minimum pollution. In: Stevenson, F.J. (ed.), *Nitrogen in Agricultural Soils. Agronomy* Monograph No. 22. pp. 605-649.

Kilmer, V.J., Gilliam, J.W., Lutz, J.F., Joyce, R.T. and Eklund, C.D. 1974. Nutrient losses from fertilized grassed watersheds in western North Carolina. *J. Environ. Quality,* 3(3), 214-219.

Kitchens, W.H., Dean, J,M,, Stevenson, L.H. and Cooper, J.H. 1975. The Santee Swamp as a nutrient sink. In: Howell, F.G., Gentry, J.B. and Smith, M.H. (eds.), *Mineral Cycling in Southeastern Ecosystems,* pp.349-366. ERDA Symposium Series.

Kitur, B,K,, Smith, M.S., Blevins, R.L. and Frye, W.W. 1984. Fate of N-depleted ammonium nitrate applied to no-tillage and conventional tillage corn. *Agron. J.,* 76, 240-242.

Klausner, S.E., Zwerman, P.J. and Ellis, D.F. 1974. Surface runoff losses of soluble nitrogen and phosphorus under two systems of soil management. *J. Environ. Qual.,* 3, 42-46.

Knisel, W.G. 1980. CREAMS: A field-scale model for chemicals, runoff, and erosion from agricultural management systems. *Cons. Res. Rpt. No.* 26. U.S. Dept. Agr., Washington, D.C. 640 p.

Knisel, W.G. and Leonard, R,A, 1989. Irrigation impact on groundwater: Model study in humid region. *J. Irrig. and Drain. Eng.,* 115(5), 823-838.

Kreitler, C.W. and Jones, D.C. 1975. Natural soil nitrate: the cause of the nitrate contamination of ground water in Runnels County, Texas. *Ground Water,* 13, 53-61.

Kuenzler, E.J., Mulholland, P.J., Ruley, L.A. and Sniffen, R.P. 1977. *Water quality in North Carolina Coastal Plain streams and effects of channelization.* Rep. No. 127. Wat. Resour. Res. Inst. Univ. North Carolina, Raleigh. 160 pp.

Kuenzler, E.J. and Craig, N.J. 1986. Land use and nutrient yields of the Chowan River Watershed. In: Correll, D.L. (ed.), *Watershed Research Perspectives,* pp.77-107. Smithsonian Institution Press, Washington, DC 20560.

Legg, J.O. and Meisinger, J.J. 1982. In: Stevenson, F.J. (ed.), *Soil nitrogen budgets. Nitrogen in Agricultural Soils,* pp.503-566. Madison, Wisconsin, American Society of Agronomy.

Lemme, G., Carlson, C.G., Dean, R. and Khakural, B. 1990. Contamination vulnerability indexes: A water quality planning tool. *J. Soil Water Conserv.,* 45(2), 349-351.

Lenain, A.F. 1967. The impact of nitrates on water use. *J. Am. Water Works Association.,* 59, 1050-1051.

Leonard, R.A., Knisel, W.G. and Still, D,A, 1987. GLEAMS: Groundwater loading effects of agricultural management systems. *Trans. ASAE,* 30(5), 1403-1418.

Liebhardt, W.C., Golt, C. and Tupin, J. 1979. Nitrate and ammonium concentrations of ground water resulting from poultry manure applications. *J. Environ. Qual. ,*8, 211-215.

Lorimor, J.C., Mielke, L.N., Elliott, L.F. and Ellis, J.R. 1972. Nitrate concentrations in groundwater beneath a beef cattle feedlot. *Water Res. Bull.,* 8(5), 999-1005.

Lowrance, R.R., Todd, R..L. and Asmussen, L.E. 1983. Waterborne nutrient budgets for the riparian zone of an agricultural watershed. *Agriculture, Ecosystems Environment .*10, 371-384.

Lowrance, R.R., Todd, R.L. and Asmussen, L.E. 1984a. Nutrient cycling in an agricultural watershed: I. Phreatic movement. *J. Environ. Qual.,* 13, 22-27.

Lowrance, R.R., Todd, R.L., Fail, J. Jr, Hendrickson, O. Jr, Leonard, R. and Asmussen, L. 1984b. Riparian forests as nutrient filters in agricultural watersheds. *BioScience,* 34(6), 374-377.

Lowrance, R. 1992. Groundwater nitrate and denitrification in a Coastal Plain riparian forest. *J. Environ. Qual.,* 21, 401-405.

Ludwick, A.E., Reuss, J.O. and Langin, E.J. 1976. Soil nitrates following four years continuous corn and as surveyed in irrigated farm fields of Central and Eastern Colorado. *J. Environ. Qual.,* 5, 82-86.

Magee, P.N. 1971. Toxicity of nitrosamines: Their possible human health hazards. *Food Cosmet. Toxicol.,* 9, 207-218.

Magee, P.N. 1977. Nitrogen as a health hazard. *Ambio*, 6, 123-125.

Mahendrappa, M.D., Smith, R.L. and Christiansen, A.T. 1966. Nitrifying organisms affected by climatic region in western United States. *Soil Science Society of America Proc.*, 30(1), 60-62.

Mansell, R.S., Calvert, D.V., Stewart, E.H., Wheeler, W.B., Rogers, J.S., Graetz, D.A., Allen, L.H., Overman, A.R. and Knipling, E.B. 1977. *Fertilizer and pesticide movement from citrus groves in Florida flatwood soils.* EPA-600/2-77-177. Office of Research and Development. U.S.-EPA. Nat. Tech. Information Service, Springfield, VA 22161. 134 p.

McColl, R.H.S. 1978. Chemical runoff from pasture: the influence of fertilizer and riparian zones. *N. Z. J. Marine Freshwater Res.*, 12, 371-380.

McLaughlin, R.A, Pope. P.E. and Hansen, E.A. 1985. Nitrogen fertilization and ground cover in hybrid poplar plantation. Effects on nitrate leaching. *J. Environ. Qual.*, 14, 241-245.

McNeal, B.L. and Pratt, P.F. 1978. Leaching of nitrate from soils. In: Pratt, P.F. (ed.), *Proc. Natl. Conf. on Management of Nitrogen in Irrigated Agriculture*, pp.195-230. University of California, Riverside, Calif.

Meisinger, J.J. 1976. *Nitrogen application rates consistent with environmental constraints for potatoes on Long Island*, 6, 1-9. Cornell University Agricultural Experiment Station, Cornell University, Ithaca, NY. SEARCH Agriculture.

Miller, M.H. 1979. Contribution of nitrogen and phosphorus to subsurface drainage water from intensively cropped mineral and organic soils in Ontario. *J. Environ. Qual.*, 8(1), 42-48.

Miner, J.R. and Willrich, T.L. 1970. *Livestock operations and field spread manure as sources of pollutants*, pp. 231-240. Agricultural Practices and Water Quality. Ames, Iowa State University Press.

Mitsch, W.J., Dorge, C.L. and Wiemhoff, J.R. 1979. Ecosystem dynamics and a phosphorus budget of an alluvial cypress swamp in southern Illinois. *Ecology*, 60, 1116-1124.

Moody, D.W., Carr, J.E., Chase, E.B., and Paulson, R.W. (eds.) 1988. *National Water Summary 1986--hydrologic events and ground-water quality.* Water Supply Paper 23-25, U. S. Geol. Surv., Reston, VA 560 p.

Moody, D.W. 1990. Groundwater contamination in the United States. *J. Soil Water Conserv.*, 45(2), 170-179.

Morgan, R.M. and Keller, L.H. 1987. *Economic comparisons of alternative waste management systems on Tennessee Dairy Farms.* Bulletin 656. University of Tennessee Agricultural Experiment Station.

Muir, J., Seim, E.C. and Olson, R.A. 1973. A study of factors influencing the nitrogen and phosphorus contents of Nebraska waters. *J. Environ. Qual.*, 2, 466-470.

Muir, J., Boyce, J.S., Seim, E.C., Mosher, P.N., Diebert, E.J. and Olson, R.A. 1976. Influence of crop management practices on nutrient movement below the root zone in Nebraska soils. *J. Environ. Qual.*, 5, 255-259.

Naney, J.W., Kent ,D.C., Smith, S.J. and Webb, B.B. 1987. *Variability of ground-water quality under sloping agricultural watersheds in Oklahoma*, pp.189-197. Symposium on monitoring, modeling, and mediating water quality.

National Research Council. 1972. *Accumulation of nitrate*, pp.5-23. Washington, DC, National Academy of Sciences.

National Research Council. 1978. *Nitrates: An environmental assessment.* Natl. Acad. Sci., Washington, D.C.

National Research Council. 1989. *Alternative agriculture.* Nat. Acad. Press, Washington, DC. 448 p.

Nelson, L.B. 1972. Agricultural chemicals in relation to environmental quality: chemical fertilizers, present and future. *J. Environ. Qual.*, 1(1), 2-6.

Nessel, J.K. 1978. Distribution and dynamics of organic matter and phosphorus in a sewage enriched cypress swamp. Master's thesis. Dept. Envir. Engrg. Sci., Univ. Fla., Gainesville.

Nightingale, H.I. 1972. Nitrates in soil and ground water beneath irrigated and fertilized crops. *Soil Science*, 114(4), 300-311.

Oakes, D.B., Young, C.P. and Foster, S.S.D. 1981. The effects of farming practices on ground water quality in the United Kingdom. *The Science of the Total Environment*, 21, 17-30.

Obenhuber, D.C. and Lowrance, R. 1991. Reduction of nitrate in aquifer microcosms by carbon additions. *J. Environ. Qual.*, 20(1), 255-258.

Oberts, G.L. 1982. *Water Resources Management; Nonpoint Source Pollution Technical Report.* Metropolitan Council Publication No. 10-82-016, St. Paul, Mn.

Oberts, G.L. 1985. Magnitude and problems of nonpoint pollution from urban and urbanizing areas. *Proc. of Non-point Pollution Abatement Symp.*, Milwaukee, WI. Organized by Div. of Cont. Ed., Marquette University, p. K-III-1 to K-III-19.

Odum, E.P. 1969. The strategy of ecosystem development. *Science,* 164: 262-270.

Omernick, J.M., Abernathy, A.R. and Male, L.M. 1981. Stream nutrient levels and proximity of agricultural and forest land to streams: some relationships. *J. Soil Water Conserv.*, 36: 227-231.

Parker, D.T. and Larson, W.E. 1962. Nitrification as affected by temperature and moisture content of mulched soils. *Soil Science Society of America Proc.*, 26(3), 238-242.

Peterjohn, W.T. and Correll, D.L. 1984. Nutrient dynamics in an agricultural watershed: observations on the role of a riparian forest. *Ecology,* 65, 1466-1475.

Peterjohn, W.T. and Correll, D.L. 1986. The effect of riparian forest on the volume and chemical composition of baseflow in an agricultural watershed. In: Correll, D.L. (ed.), *Watershed Research Perspectives,*pp. 244-262. Smithsonian Institution Press, Washington, DC 20560.

Pratt, P,F,, Jones, W.W. and Hunsaker, V.E. 1972. Nitrate in deep soil profiles in relation to fertilizer rates and leaching volume. *J. Environ. Qual.*, 1, 97-102.

Pratt, P.F. 1972. *Nitrate in the unsaturated zone under agricultural lands.* Environmental Protection Agency. Publication Number EPA 16060 DOE04/72. 45 p.

Preul, H.C. and Schroepfer, G.S. 1968. Travel of nitrogen in soils. *J. Water Pollution Control Federation,* 40, 30-48.

Qualls, R.G. 1984. The role of leaf litter nitrogen immobilization in the nitrogen budget of a swamp stream. *J. Environ. Qual,.* 13, 640-644.

Reddy, K.R. and Graetz, D.A. 1981. Use of shallow reservoir and flooded organic soil systems for waste water treatment: Nitrogen and phosphorus transformations. *J. Environ. Qual.,* 10, 113-119.

Rible, J.M. and Davis, L.E. 1955. Ion exchange in soil columns. *Soil Science,* 79, 41-47.

Ritter, W.F., Chirnside, A.E.M. and Scarborough, R.W. 1986. *Effect of agricultural activities on ground-water quality.* ASAE Summer Meeting Paper #86-2025.

Robertson, F.N. 1977a. *The quality and potential problems of the ground water in Coastal Sussex County, Delaware.* Water Resources Center, University of Delaware, Newark, DE.

Robertson, F.N. 1977b. *The effect of agricultural activity of groundwater quality near Millsboro, Delaware.* Coastal Sussex Water Quality Management Program Report. Contract No. 208-0-77.

Robinson, J.B., Kaushik, N.K. and Chatarpaul, L. 1978. *Nitrogen transport and transformations in Canagigue Creek.* International Joint Commission, Windsor, ON.

Romaine, J.D. 1965. When fertilizing, consider plant food content of your crops. *Better Crops Plant Food*, pp. 1-8.

Rowe, M.L. and Stinnett, S. 1975. *Nitrogen in the subsurface environment.* EPA Report 660/3-75-030 Grant No. R801381. Corvallis, Oregon: National Environmental Research Center, Office of Research and Development. U. S. Government Printing Office, Washington, D.C. 20402

Ryden, J.C., Ball, P.R. and Garwood, E.A. 1984. Nitrate leaching from a grassland. *Nature,* 311, 50-53.

Sabey, B.R. 1969. Influence of soil moisture tension on nitrate accumulation in soils. *Soil Science Society of America Proc.* 33(2), 263-266.

Saffigna, P.G. and Keeney, D.R. 1977. Nitrate and chloride in groundwater under irrigated agriculture in central Wisconsin. *Ground Water,* 15, 170-177.

Schepers, J.S. 1988. Role of cropping systems in environmental quality: Groundwater nitrogen. In: *Cropping strategies for efficient use of water and nitrogen.* Spec. Publ. No. 51. Am. Soc. Agron., Madison, Wisc.

Scheuler, T.R. and Kemp, W.M. 1979. An agricultural mass balance model of nitrogen movement in the Choptank River basin. In: Kemp, W.M., Boynton, W. and Stevenson, J.W. (eds.), *Submerged Aquatic Vegetation in the Chesapeake Bay. Ann. Rep. EPA/CBF.* Vol. I. UMCEES. Ref. No. 80-68-HPEL.

Schlosser, I.J. and Karr, J.R. 1981a. Water quality in agricultural watersheds: Impact of riparian vegetation during base flow. *Water Res. Bulletin,* 17, 233-240.

Schlosser, I.J. and Karr, J,R, 1981b. Riparian vegetation and channel morphology impact on spatial patterns of water quality in agricultural watersheds. *Environ. Manag.,* 5, 233-243.

Sears, F.W. and Zemansky, M.W. 1955. *University physics.* Addison-Wessley Publishing Company, Reading, Mass.

Shank, R.C. 1975. Toxicology of N-nitroso-compounds. *Toxicol. Appl. Pharmacol.,* 31, 361-368.

Sharpley, A.N., Smith, S.J. and Naney, J.W. 1987. Environmental impact of agricultural N and phosphorus use. *J. Agric. Food Chem.* Sept/Oct. 1987, 812-817.

Shaw, E.C. and Wiley, P. 1969. Nitrate ion concentration in well water. *California Agriculture,* 23, 11.

Simmons, R.C., Gold, A.J. and Groffman, P.M. 1992. Nitrate dynamics in riparian forests: Groundwater studies. *J. Environ. Qual.,* 21, 659-665.

Smith, G.E. 1964. Nitrate problems in plants and water supplies in Missouri. Presented at *Second Annual Symposium on the Relation of Geology and Trace Elements to Nutrition,* 92nd Annual Meeting of the American Public Health Association. New York City. October 7, 1964. 21 p.

Smith, G.E. 1965. Why nitrates in water supplies? *Hoard's Dairyman,* 110(18), 1048-1049.

Smith, G.E. 1970. Nitrate pollution of water supplies. In: *Trace Substances Environmental Health-3,* pp273-287. Proc. of University of Missouri's 3rd Annual Conference.

Smith, S.J., Naney, J.W. and Berg, W.A. 1987. Nitrogen and ground water protection. In: *Groundwater quality and Agricultural Practices,* pp. 367-374. Lewis Publishers, Chelsea, Michigan.

Sonzogni, W.C., Chesters, G., Coote, D.R., Jeffs, D.N., Konrad, J.C., Ostry, R.C. and Robinson, J.B. 1980. Pollution from land runoff. *Envir. Sci. Tech.,* 14(2), 148-153.

Spalding, R.F., Gormly, J.R., Curtiss, B.H. and Exner, M.E. 1978. Nonpoint nitrate contamination of ground water in Merrick County, Nebraska. *Ground Water,* 16, 86-95.

Staver, K.W., Brinsfield, R.B. and Stevenson, J. 1989. The effect of best management practices on nitrogen transport into Chesapeake Bay. In: *Toxic Substances in Agricultural Water Supply and Drainage*, pp.163-180. U. S. Committee on Irrigation and Drainage. Denver, Colo.

Staver, K.W., Brinsfield, R.B. and Magette, W.L. 1988. *Nitrogen export from Atlantic Coastal Plain soils*. Paper No. 88-2040. Am. Soc. Agr. Eng., St. Joseph, Mich.

Staver, K.W. and Brinsfield, R.B. 1990. Patterns of soil nitrate availability in corn production systems: Implications for reducing groundwater contamination. *J. Soil Water Conserv.*, 45(2), 318-322.

Steenhuis, T.S. and Walter, M.R. 1980. Closed form solution for pesticide loss in runoff water. *Trans. ASAE*, 23, 615-620, 628.

Stewart, B.A., Viets, F.G. Jr., Hutchinson, G.L., Kemper, W.D. and Clark, F.E. 1967. *Distribution of nitrates and other water pollutants under fields and corrals in the Middle South Platte Valley of Colorado. Agricultural Research Service.* Publication Number ARS 41-134. 206 p.

Stringfield, V.T. 1966. *Artesian water in tertiary limestone in the Southeastern states.* Prof. Paper 517, U. S. Geol. Surv., Reston, VA. 226 pp.

Subbotin, F.N. 1961. Nitrates in drinking water and their effect on the formation of methemoglobin. *S. M. Kirov Order of Lenin Army Medical Academy USSR.* 25(2), 63-67.

The Conservation Foundation 1987. *Groundwater protection--groundwater, saving the unseen resource, and a guide to groundwater pollution problems, causes, and government responses.* Washington, D.C. 240 p.

Thomas, G.W. and Crutchfield, J.D. 1974. Nitrate-nitrogen and phosphorus contents of streams draining small agricultural watersheds in Kentucky. *J. Environ. Qual.*, 3, 46-49.

Thomas, G.W., Blevins, R.L., Phillips, R.E. and McMahan, M.A. 1973. Effect of a killed sod mulch on nitrate movement and corn yield. *Agron. J.*, 63, 736-739.

Thomas, G.W., Wells, K.L. and Murdock, L. 1981. Fertilization and liming. In: Phillips, R.E. *et al.* (eds), *No Tillage Research: Research Reports and Reviews*, pp. 43-54. Univ. KY, Lexington, KY.

Todd, R., Lowrance, R., Hendrickson, O., Asmussen, L., Leonard, R., Fail, J. and Herrick, B. 1983. Riparian vegetation as filters of nutrients exported from a coastal plain agricultural watershed. In: Lowrance,R., Todd, R. L., Asmussen, L. E. and Leonard, R. A. (eds.), *Nutrient Cycling in Agricultural Ecosystems*, p. 485-493. Special Publication Number 23, University of Georgia, College of Agriculture, Experiment Stations, Athens, Georgia.

Tomlinson, T.E. 1970. Trend in nitrate concentrations in English rivers and fertilizer use. *Water Treat. Exam.*, 19, 277-289.

Trewartha, G.T. 1961. *The earth's problem climates.* University of Wisconsin Press. Madison, WI 53701. 334 p.

Tuschall, J.R., Brezonik, P.L. and Ewel, K.C. 1981. Tertiary treatment of wastewater using flow-through wetland systems. In: *National Conference of American Society of Civil Engineers*, July 8-10, Atlanta, Georgia.

Tyler, D.D. and Thomas, G.W. 1979. Lysimeter measurement of nitrate and chloride losses from conventional and no-tillage corn. *J. Environ. Qual.*, 6, 63-66.

Urban, J.B., Gburek, W.J. and Schnabel, R.R. 1989. Upland agricultural watershed contributions of nitrate to streamflow via subsurface flow. In: *Groundwater Issues and Solutions in the Potomac River Basin/Chesapeake Bay Region.* Nat. Well Water Assoc. Dublin, Ohio.

US Environmental Protection Agency US EPA. 1983. *Results of the Nationwide Urban Runoff Program:* Vol. I-Final Report. US EPA Report, Water Planning Division.

US Environmental Protection Agency 1980. *Design Manual: Onsite wastewater treatment and disposal systems.* Publ. No. 625/1-80--12. Office Water Program Operations and Office of Res. and Development, Washington, D.C. 391 pp.

Van Kessel, J.F. 1977. Removal of nitrate from effluent following discharge on surface water. *Water Res.,* 11, 533-537.

Vellidis, G., Lowrance, R., Smith, M.C. and Hubbard, R.K. 1993. Methods to assess the water quality impact of a restored riparian wetland. *J. Soil Water Cons.,* 48(3), 223-230.

Vitousek, P.M. and Reiners, W.A. 1975. Ecosystem succession and nutrient retention: A hypothesis. *BioScience,* 25, 376-381.

Wagner, G.H., Steele. K.F., MacDonald, H.C. and Coughlin, T.L. 1976. Water quality as related to linears, rock chemistry and rain water chemistry in a rural carbonate terrain. *J. Environ. Qual.,* 5, 444-451.

Walker, W.H. 1973. Ground water nitrate pollution in rural areas. *Ground Water* 11(5), 19-22.

Walker, W.H., Peck, T.R. and Lembke, W.D. 1972. *Farm ground water nitrate pollution--A case study.* Presented at American Society of Chemical Engineers Annual and National Environmental Engineering Meeting. Houston, Texas. October 16-22, 1972. 25 p.

Walker, W.E. and Kroeker, B.E. 1982. *Nitrates in groundwater resulting from manure applications to irrigated croplands.* EPA 600/2-82-079 Contract Grant No. R-804827.70.

Watts, D.G. and Martin, D.L. 1981. Effects of water and nitrogen management on nitrate leaching loss from sands. *Trans. ASAE,* 24(4), 911-916.

Winton, E.F., Tardiff, R.G. and McCabe, L.J. 1971. Nitrate in drinking water. *J. Am. Water Works Association,* 63, 93-98.

Wolkowski, R.P. 1992. *A Step-By-Step Guide to Nutrient Management. Nutrient and Pest Management* A3568. University of Wisconsin-Extension, Cooperative Extension, Madison, WI.

Woll, R.S. (ed.). 1978. Maryland ground-water information--Chemical quality data: Maryland Geological Survey. *Water Resources Basic Data Report,* 10, 126 p.

Yates, P. and Sheridan, J.M. 1983. Estimating the effectiveness of vegetated floodplains/wetlands as nitrate-nitrite and orthophosphorus filters. *Agric. Ecosys. Environ.,* 9, 303-314.

Zwerman, P.J., Bouldin, D.R., Greweling, T.E., Klausner, S.D., Lathwell, D.J. and Wilson, D.O. 1971. *Management of nutrients on agricultural land for improved water quality.* Washington, District of Columbia: US Environmental Protection Agency, EPA 13020 DPB-08171.

Zwerman, P.J., Greweling. T., Klausner. S.D. and Lathwell, D.J. 1972. Nitrogen and phosphorus content of water from tile drains at two levels of management and fertilization. *Soil Sci. Soc. Am. Proc.,* 36, 134-137.

11 Residual soil nitrate after application of nitrogen fertilizers to crops

Jacques J. Neeteson

DLO Research Institute for Agrobiology and Soil Fertility (AB-DLO) PO Box 129, 9750 AC Haren, The Netherlands

Abstract

In Western Europe residual soil nitrate, i.e. nitrate present in the soil in the fall after harvest of the crop, is a major source of nitrate that may possibly leach during the subsequent winter period. At the currently recommended N fertilizer application rates high levels of residual soil nitrate are found after potatoes, most field vegetables, grazed grassland, and silage maize. It is to be expected that in these cases the maximum permissible nitrate concentration in the shallow groundwater layers is exceeded. Measures to reduce the levels of residual soil nitrate include refinement of the current recommendations, reduction of N application rates, and also growing winter crops after the cash crops. Current N fertilizer recommendations generally only take into account the amount of soil nitrate present shortly before fertilizer application. Soil N mineralization rates during the growing season and yield levels, which have a dramatic effect on the N fertilizer requirement of the crop, may vary considerably from field to field. Therefore, N mineralization and expected yield level should be explicitly included in N fertilizer recommendations. This can be done by using simulation models. If even after refinement of the recommendations still too high levels of residual soil nitrate occur, recommended rates should be reduced to prevent leaching of nitrate at an unacceptable level. Obviously, this causes a loss of yield. Future research should aim at developing N fertilizer strategies which are able to supply agricultural crops with sufficient N for optimum growth and development, while keeping losses to the environment at a minimum.

Introduction

In Western Europe, nitrate leaching from the root zone to aquifers mainly occurs in the period from late fall to early spring when precipitation exceeds evapotranspiration. The nitrate leached originates from nitrate present in the soil in the fall after harvest of the crop, *i.e.* residual soil nitrate, from soil organic N mineralized in the fall and winter, and from livestock manures applied in the fall and winter. Leaching of nitrate is undesirable because it increases the nitrate concentration in groundwater, which is an important source of drinking water. The European Economic Community (EEC) has set a maximum permissible concentration of 50 mg NO_3 L^{-1} for drinking water (European Economic Community, 1980). If it is assumed that nitrate is not lost from the groundwater through denitrification, that the drinking water would only be produced from shallow groundwater, and that the annual precipitation surplus is 300 mm (the average value for, *e.g.*, the Netherlands) the EEC standard would be exceeded after leaching of 34 kg N ha^{-1}. Currently, the Dutch government aims at a maximum concentration of 50 mg NO_3 L^{-1} in groundwater at a depth of 2 m below the phreatic level in those areas where the groundwater may be used for drinking water (National Environmental Plan of the Netherlands, 1989).

Since residual soil nitrate is a potential loss to the environment through nitrate leaching during the subsequent winter period, it will be shown to what extent nitrate accumulates in the soil when the current N fertilizer recommendations for agricultural crops are followed. It should be noted that the current recommendations have been valid for some years and aim at obtaining maximum financial return without taking into account environmental side-effects of fertilization. Before presenting results on residual soil nitrate, current N fertilizer recommendations are given in the following section . Most results presented are from the Netherlands, but results obtained from other West-European countries are also shown. In this chapter the emphasis is on residual soil nitrate as a source of nitrate leaching. Nitrate concentrations in groundwater under West-European conditions are extensively reviewed elsewhere (Jurgens-Geschwind, 1989).

Current Dutch N fertilizer recommendations

N fertilizer recommendations for arable crops, field vegetables, and grassland and silage maize are given in this section. Although silage maize can be regarded as an arable crop, the N fertilization of the crop is treated in the section dealing with the recommendation for grassland, because under Dutch conditions the crop generally is grown on dairy farms. The silage is used as roughage for dairy cows.

The amounts of recommended N fertilizer do not only include inorganic fertilizers, but also N applied with animal manures. For the latter an efficiency coefficient for the amount of N available for crop uptake, *e.g.* 60% of total amount of N applied with the manures, should be used to take account of losses during application and organic N not yet mineralized during the growing period of the crop.

Arable crops

Almost 80 years ago, Russell (1914) was probably the first to recognize that soil nitrate affects the N fertilizer requirement of arable crops; after dry winters he found larger amounts of soil nitrate and higher yields of winter wheat (*Triticum aestivum* L.) than after wet winters. More than 40 years later this finding was confirmed by Boyd *et al.* (1957) for sugar beet (*Beta vulgaris* L.). By covering parts of experimental fields during periods of rainfall, Van der Paauw (1962) found that the

Table 1 *Current Dutch N fertilizer recommendations for arable crops (Anonymous, 1989a).*

Crop	Recommendation $N_{rec}{}^a = A\text{-}B.N_{min}b$		Sampling depth for $N_{min}{}^b$ (cm)
	A	B	
Winter wheat (*Triticum aestivum* L.)[c]	140	1.0	0-100
Spring wheat (*Triticum aestivum* L.)[c]	120	1.0	0-60
Winter barley (*Hordeum vulgare* L.)[c]	120	1.0	0-100
Spring barley (*Hordeum vulgare* L.)[c]			
clay and loam soils	100	1.0	0-60
sandy soils	80	1.0	0-60
Rye (*Secale cereale* L.)	100	1.0	0-60
Oats (*Avena sativa* L.)	100	1.0	0-60
Sugar beet (*Beta vulgaris* L.)	220	1.7	0-60
Ware potatoes (*Solanum tuberosum* L.)			
clay and loam soils	285	1.1	0-60
sandy soils	300	1.8	0-30
Starch potatoes (*Solanum tuberosum* L	275	1.8	0-30
Seed potatoes (*Solanum tuberosum* L.)	140	0.6	0-60
Green peas (*Pisum sativum* L.)	0-30	-	-
Italian rye grass (*Lolium multiflorum* Lam.)			
for seed production			
clay soils	165	0.6	0-90
sandy soils	165	1.0	0-90
Colza (*Brassica napus* L.)	200	1.0	0-100

[a]N_{rec} = recommended application rate of N fertilizer in kg N ha^{-1}.
[b]N_{min} = amount of soil mineral N at the end of winter in kg N ha^{-1}.
[c]The recommendation refers to the first dressing only.

weak response of potatoes (*Solanum tuberosum* L.) and rye (*Secale cereale* L.) to added N after 'dry' winter periods could be attributed to the relatively large amount of nitrate present in the soil at the end of winter. After 'wet' winter periods, small amounts of soil nitrate were found, probably because N was lost due to leaching. In the Netherlands these findings resulted in N fertilizer recommendations in the 1960s which were corrected annually for the amount of rainfall between November 1 and March 1. For cereals it was advised to increase the recommended amount of N fertilizer by 10-30 kg N ha^{-1} after wet winters and to decrease it by the same amount after dry winters (Van der Paauw, 1966).

To improve this indirect method of taking soil nitrate into account, research efforts were subsequently aimed at determining the relationship between soil nitrate

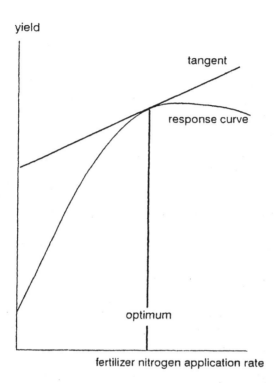

Figure 1 *Schematic representation of the determination of the economically optimum application rate of N fertilizer from a response curve. The slope of the tangent reflects the ratio of the cost of N fertilizer to the price of crop produce.*

at the end of winter and the economically optimum application rate of N fertilizer. For this purpose, many series of field experiments with winter wheat, potatoes and sugar beet were conducted for several years on various soil types (Ris *et al.*, 1981). In these experiments the amount of mineral N, *i.e.* nitrate N + ammonium N, present in the soil at the end of winter was measured and various amounts of N fertilizer were applied to determine the economically optimum application rate of N fertilizer. For each individual trial the optimum was determined from the yield response curve to which a tangent was drawn, the slope of which depended on the ratio between the cost of N fertilizer and the price of crop produce (Figure 1).

As an example, the relationship between the amount of soil mineral N at the end of the winter and the economically optimum application rate for potatoes on clay and loam soils is shown (Figure 2). The relationship is based on 77 trials, which were conducted in various regions in the Netherlands from 1973 to 1982. Although the relationship was found to be weak ($r^2 = 0.22$), the linear regression line was then decided to represent the recommendation. For potatoes on sandy soils and other arable crops the same procedure was followed. All current recommendations for arable crops are summarized in Table 1. Except for green peas (*Pisum sativum* L.), which need only a starter dressing of about 30 kg N ha^{-1}, all N fertilizer recommendations are based on the amount of soil mineral N already present in the soil. After this so-called N_{min}-method was initiated in the Netherlands, research was also done in other West European countries, *e.g.* Belgium, Germany, Sweden and Denmark, to evaluate and further develop the method (Greenwood, 1986). It has now been adopted in these countries.

Field vegetables
As for arable crops the current Dutch N fertilizer recommendations for most field vegetables are given according to the N_{min}-method (Table 2). It should be noted,

Table 2 *Current Dutch N fertilizer recommendations for field vegetables (Anonymous, 1989a).*

Crop	Recommendation N_{rec}[a] = A-B.N_{min}[b]		Sampling depth for N_{min}[b] (cm)
	A	B	
Seed onions (*Allium cepa* L.)	180	1.0	0-60
Brussels sprouts (*Brassica oleracea* L.)	240	1.0	0-60
Chicory (*Chichorium intybus* L.)			
clay soils	70	1.0	0-60
sandy soils	60	1.0	0-60
Leeks (*Allium porrum* L.)	270	1.0	0-60
Asparagus (*Asparagus officinalis* L.)	80	1.0	0-60
Cauliflower (*Brassica oleracea* L)	300	1.0	0-60
Carrot (*Daucus carota* L.)	80	1.0	0-60
Winter carrot (*Daucus carota* L.)	90	1.4	0-30
Strawberry (*Fragaria* ssp.)	120	-	-
White cabbage (*Brassica oleracea* L.)	350	1.0	0-60
Spinach (*Spinacia oleracea* L.)			
1st crop			
clay and loam soils	290	1.4	0-30
sandy soils	240	1.4	0-30
peaty soils	190	1.4	0-30
2nd crop			
clay and loam soils	225	1.0	0-30
sandy soils	200	1.0	0-30
peaty soils	150	1.0	0-30
Broad beans (*Vicia faba* L.)	0-50	-	-
Celeriac (*Apium graveolens* L.)	210	1.4	0-60
Salsify (*Tragopogon porrifolius* L.)	90-140	-	-

[a]N_{rec} = recommended application rate of N fertilizer in kg N ha^{-1}.
[b]N_{min} = amount of soil mineral N at the end of winter in kg N ha^{-1}.

Table 3 *Current Dutch N fertilizer recommendations for grassland and silage maize (Anonymous, 1989b).*

Crop	Recommendation N_{rec}^a = A - B.N_{min}^b		Sampling depth for N_{min}^b (cm)
	A	B	
Grassland			
well-drained peaty soils	250	-	-
other soils	40	-	-
Silage maize (*Zea mays* L.)			
permanent cropping	150	-	-
in crop rotation	200	-	-

[a]N_{rec} = recommended application rate of N fertilizer in kg N ha^{-1}.
[b]N_{min} = amount of soil mineral N at the end of winter in kg N ha^{-1}.

Table 4 *Residual soil nitrate after arable crops.*

Crop	N_{rec}^a (kg ha^{-1})	Sampling depth for N_{res}^b (cm)	N_{res}^b (kg ha^{-1}) ON[c]	N_{rec}^a	Ref[d]
Winter wheat	200-210	0- 90	13-14	15-24	i
(*Triticum aestivum* L.)	140	0-100	18	30	ii
Winter barley	170	0- 40	-	7-43	iii
(*Hordeum vulgare* L.)					
Spring barley	80	0- 90	15	25	iv
(*Hordeum vulgare* L.)					
Sugar beet	120	0- 90	21	23	v
(*Beta vulgaris* L.)	133	0- 60	19	23	vi
Ware potatoes	250	0- 90	45	75	iv
(*Solanum tuberosum* L.) -		0- 60	25-50	-	vii
	250-300	0- 60	30[e]	60-75[e]	viii
Starch potatoes	207	0-100	50	95	iv
(*Solanum tuberosum* L)					

[a]N_{rec} = recommended application rate of N fertilizer.
[b]N_{res} = residual soil nitrate.
[c]ON = without application of N fertilizer.
[d]Ref = reference; i = Dilz, 1988; ii = Machet and Mary, 1989; iii = Smith *et al.*, 1988; iv = Neeteson and Wadman, 1991; v = Lindén, 1987; vi = Neeteson and Ehlert, 1989; vii = Zandt *et al.*,1986; viii = Neeteson *et al.*, 1989.
[e]Model calculations

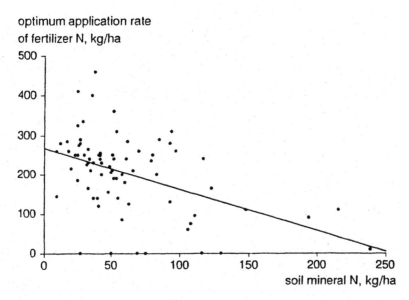

optimum application rate
of fertilizer N, kg/ha

Figure 2 *Relationship between the amount of soil mineral N in the 0-60 cm layer at the end of winter and the economically optimum application rate of N fertilizer for potatoes (Solanum tuberosum L.) on clay and loam soils in the Netherlands.*

however, that the number of field trials on the basis of which the relationship between soil mineral N and the economically optimum application rate of N fertilizer for the various crops was established, was limited. This was due to the small acreages of these crops and the large diversity of crops.

Grassland and silage maize

The current N fertilizer recommendations for grassland and silage maize (*Zea mays* L.) are summarized in Table 3. Contrary to the recommendations for arable crops and field vegetables, the recommendations do not take into account soil nitrate. The recommendations for grassland refer to intensively used grassland on farms without a surplus of roughage. The total amount recommended is to be split over the various cuts during the growing period. If the total amount is 400 kg N ha^{-1}, a common pattern is 80 kg for the first (grazing) cut, 100 kg for the second (mowing) cut, and 60, 60, 60, and 40 kg N ha^{-1} for the third, fourth, fifth and sixth (grazing) cut, respectively. Obviously, the actual number of cuts depends on weather conditions during the growing period.

In practice, the recommendations for silage maize given in Table 3 are rarely followed. Since silage maize is grown on farms with intensive animal husbandry, and the crop does not respond negatively to very high fertilizer application rates, the crop generally is heavily fertilized with slurries.

Table 5 *Residual soil nitrate after field vegetables.*

Crop	N_{rec}^a (kg ha^{-1})	Sampling depth for N_{res}^b (cm)	N_{res}^b (kg ha^{-1}) ON^c	N_{rec}^a	Ref^d
Brussels sprouts (*Brassica oleracea* L.)	240	0-90	-	20- 40	i
White cabbage (*Brassica oleracea* L.)	300	0-90	25	50	ii
Seed onions (*Allium cepa* L.)	120	0-90	50	75	ii
Cauliflower (*Brassica oleracea* L.)	180-300	0-90	-	58-210	i, iii
Celeriac (*Apium graveolens* L.)	50-150	0-90	-	51-227	i
Leeks (*Allium porrum* L.)	100-150	0-90	-	124-200	i
Spinach	215-290	0-90	105-170	212-222	iv
(*Spinacia oleracea* L.)	250	0-60	-	159	v

^{a}N$_{rec}$= recommended application rate of N fertilizer.
^{b}N$_{res}$= residual soil nitrate.
^{c}ON = without application of N fertilizer.
^{d}Ref = reference(s); i = Anonymous, 1982; ii = Neeteson and Wadman, 1991; iii = Anonymous,1983; iv = Van der Boon and Pieters, 1981; v = Wehrmann and Scharpf, 1989.

Table 6 *Amount of N present in residues of field vegetables after application of recommended rates of N fertilizer.*

Crop	N in crop residues (kg ha^{-1})	Ref^a
Spinach (*Spinacia oleracea* L.)	25	i
Celeriac (*Apium graveolens* L.)	60	i
Cauliflower (*Brassica oleracea* L)	80	i
White cabbage (*Brassica oleracea* L.)	100	i
	53-172	ii
Brussels sprouts (*Brassica oleracea* L.)	190	i

^{a}Ref = reference; i = Wehrmann and Scharpf, 1989; ii = Unpublished results DLO Research Institute for Agrobiology and Soil Fertility (AB-DLO).

Residual soil nitrate

Since under West-European conditions residual soil mineral N, *i.e.* soil nitrate N + soil ammonium N, is an important source of nitrate leaching during the subsequent winter period, in this section it will be shown to what extent mineral N accumulates in the soil when the current N fertilizer recommendations are followed; this will be done on the basis of typical observations in field trials which were performed in Western Europe. It should be noted that under West-European conditions residual soil mineral N is almost entirely made up of nitrate N. In the following residual soil nitrate is therefore regarded being equal to residual soil mineral N. Substantial amounts of ammonium N are only expected to be found after recent applications of slurries and in soils recently treated with soil fumigants.

Arable crops

Data on residual soil nitrate after cereals, sugar beet, and potatoes, the most important arable crops, are summarized in Table 4. Since N fertilizer applied to cereals and sugar beet is taken up efficiently by the crop (Dilz, 1988), virtually no nitrate accumulates in the soil after application of the currently recommended rates of N fertilizer to these crops. Without N fertilizer application the amount of residual soil nitrate is 10-20 kg N ha^{-1} in the 0-90 cm layer; with the recommended rate of N

Table 7 *Residual soil nitrate after grassland and silage maize.*

Crop	N_{rec}[a] (kg ha^{-1})	Sampling depth for N_{res}[b] (cm)	N_{res}[b] (kg ha^{-1}) ON[c]	N_{res}[b] (kg ha^{-1}) N_{rec}[a]	Ref[d]
Non-grazed	400	0-100	5-40	40-80	i
grassland	400	0-100	30	57	ii
	420	0- 60	39	45	iii
	420	0- 60	45	50	iv
	420	0- 90	-	46	v
	400	0- 90	13[e]	56[e]	vi
Grazed grassland	420	0- 90	-	170	v
	400	0- 90	14[e]	183[e]	vi
Silage maize	150	0- 90	64	142	vii
(*Zea mays* L.)	150	0- 90	44	105	viii
	-	0- 60	60-80	-	ix

[a]N_{rec} = recommended application rate of N fertilizer.
[b]N_{res}= residual soil nitrate.
[c]ON = without application of N fertilizer.
[d]Ref = reference; i = Prins, 1980; ii = Groot *et al.*, 1989; iii = Prins *et al.*, 1988; iv = Oenema *et al.*,1989; v = Ball and Ryden, 1984; vi = Goossensen and Meeuwissen, 1990; vii = Schröder,1985b; viii = Schröder, 1985a; ix = Schröder, [e]model calculations1990.

Table 8 *Spatial distribution of residual soil mineral N in grassland soils (Ball and Ryden, 1984). N fertilizer application rate was 420 kg ha^{-1}.*

| Soil sampled | No. of cores | Residual soil mineral N (kg ha^{-1}) | | | |
| | | Nitrate-N | | Ammonium-N | |
		Mean	Range	Mean	Range
Non-grazed plot					
At random	10	38	22- 74	8	6- 11
Grazed plot					
At random	39	160	47- 510	10	4- 57
Urine patches	9	920	200-1,710	258	26-1,090
Camped area	6	400	230- 705	36	4- 64

Table 9 *Residual soil nitrate after application of N fertilizer rates exceeding recommended rates.*

Crop	Excess rate of N applied (kg ha^{-1})	Sampling depth for N$_{res}$[a] (cm)	N$_{res}$[a] (kg ha^{-1})	Ref[b]
Winter barley (*Hordeum vulgare* L)	100	0-40	77-149	i
Celeriac (*Apium graveolens* L.)	34	0-90	156	ii
Leeks (*Allium cepa* L.)	282	0-90	248	ii
Cauliflower (*Brassica oleracea* L.)	197-429	0-90	198-564	ii

[a]N$_{res}$ = residual soil nitrate
[b]Ref = reference; i = Smith *et al.*, 1988; ii = Wehrmann and Scharpf, 1989.

fertilizer amounts found are little higher (10-40 kg ha^{-1}; Table 4). It is, therefore, not to be expected that the current N fertilizer recommendations for cereals and sugar beet will result in heavy nitrate losses due to leaching. Potatoes, however, are rather inefficient users of N (Neeteson, 1989). Without N fertilizer application the amount of residual soil nitrate is relatively high (about 50 kg ha^{-1} in the 0-90 cm layer) and after application of the recommended N fertilizer rates the amount increases to 75-100 kg ha^{-1} (Table 4). Thus, the risk of nitrate leaching after application of the recommended fertilizer rates to arable crops is low, except in the case of potatoes. If the foliage and the heads of the sugar beet, which contain 50-150 kg N ha^{-1} (Neeteson and Ehlert, 1989; Wehrmann and Scharpf, 1989), remain on the field after harvest, they can also be regarded as a potential source of nitrate leaching after decomposition. It is not yet clear to what extent the foliage and heads contribute to nitrate leaching.

Table 10 *Residual soil nitrate after application of various amounts of cattle slurry to silage maize (Zea mays L.). Average values of six-years' research on a sandy soil (after Schröder, 1985a).*

Slurry application rate (t ha^{-1})	Total N application with slurry (kg ha^{-1})	Residual soil nitrate in the 0-100 cm layer (kg ha^{-1})
0	0	69
100	495	208
200	975	337
300	1,422	375

Field vegetables
Data on residual soil nitrate after field vegetables are presented in Table 5. After application of the recommended N fertilizer rates to Brussels sprouts (*Brassica oleracea* L), cabbage (*Brassica oleracea* L) and onions (*Allium cepa* L.) low to moderate levels of residual soil nitrate (20-75 kg ha^{-1}) are found. After application of the recommended rates to other field vegetables, however, high amounts of residual soil nitrate may occur. This is especially the case after crops which are harvested before maturing, *e.g.* spinach (*Spinacia oleracea* L.), where residual soil nitrate may even exceed values of 200 kg ha^{-1} (Table 5). Obviously, large amounts of nitrate will then leach during the subsequent winter period.

Another potential source of nitrate leaching are crop residues. Crop residues of cabbage and Brussels sprouts may contain 100-200 kg N ha^{-1} (Table 6). If these residues remain on the field after harvest and are (partly) decomposed before winter, substantial amounts of N from the decomposed plant material may leach during the subsequent winter period.

Grassland and silage maize
Application of the recommended rates of N fertilizer to non-grazed grassland generally does not result in accumulation of soil nitrate (Table 7). When the recommended rates are applied to grazed grassland, however, high amounts of residual soil nitrate (about 150 kg ha^{-1}) are found (Table 7). This is caused by the extra input of N through animal excreta during grazing. Grazing also causes a larger variation in soil mineral N within a field due to the uneven spreading of urine and dung patches. Whereas the average amount of residual soil mineral N in a grazed field was 160 kg ha^{-1}, it exceeded 1,000 kg ha^{-1} under urine patches (Table 8). After application of the recommended rate of fertilizer N to silage maize (*Zea mays* L.) high values of residual soil nitrate (about 100 kg ha^{-1}) have also been observed (Table 7). Thus, it can be concluded from the results presented in Table 7 that there is a serious risk of nitrate leaching after application of the recommended N fertilizer rates to grazed grassland and silage maize.

Additional comments

It should be emphasized that the values for residual soil nitrate given in Tables 4, 5, and 7 only refer to crops without fertilization or to crops fertilized according to the current N recommendations and that there are no constraints to crop growth.

However, when more N fertilizer is applied than recommended, the level of residual soil nitrate can rise substantially (Table 9). This also applies to crops, *e.g.* winter barley (*Hordeum vulgare* L.), with little accumulation at the recommended rate of N fertilizer. In practice, the N recommendation for silage maize is rarely followed and large quantities of slurries are applied to the crop. Silage maize is generally grown on farms with intensive animal husbandry where a surplus of slurry exists. Since yield and quality of silage maize (Zea mays L.) do not respond negatively to an excessive supply of N, large amounts of slurries are applied to the crop. Obviously, considerable quantities of residual soil nitrate are then likely to be found (*e.g.* Table 10) and substantial amounts of nitrate may subsequently leach.

Higher values of residual soil nitrate can also be found on fields where the mineralization rate of the soil is higher than average (Neeteson *et al.*, 1989) or when uptake of N by the crop ceases early due to drought or disease.

Measures to reduce nitrate leaching

Since in practice high levels of residual soil nitrate are often caused by the application of N rates exceeding the recommended rates, the amounts of nitrate currently leached can be reduced by persuading farmers to apply the N fertilizer recommendations. Agricultural Extension Services should increase the awareness among farmers that the application of excess N is disadvantageous from both the economic and environmental point of view. However, it has been shown that the application of the currently recommended N rates will not always result in low levels of residual soil nitrate and thus in little nitrate leaching. Therefore, the current N recommendations should be reevaluated in such a way that they do not result in substantial nitrate leaching. Recommendations can be refined by taking into account more site specific characteristics than currently is the case. In this way it may be possible to reduce levels of residual soil nitrate to an acceptable level without yield losses. If, however, this is not the case, recommended rates should be simply reduced to prevent too high nitrate leaching levels. The associated yield losses will then have to be accepted.

In this section examples are given of methods to refine the current N recommendations . Also, the effect of lowering the recommended rates on crop yield is discussed and the possibility of decreasing the effect of residual soil nitrate on nitrate leaching by growing winter crops. Other methods to refine current N recommendations include soil and tissue N testing during crop growth, and fertilizer placement.

Refinement of current N fertilizer recommendations

It has been shown that most fertilizer recommendations currently used in Western

Europe are given according to the N_{min}-method which is based on measurement of the amount of nitrate present in the soil shortly before fertilization. Soil N mineralization rates during the growing season and yield levels, which have a dramatic effect on the N fertilizer requirement of a crop, may vary considerably from field to field. Therefore, methods are being developed which explicitly take account of N mineralization and expected yield level: the balance-sheet method and the use of a simulation model.

Balance-sheet method
With this method a balance sheet is drawn up in which the N requirement of the crop is given on one side and the contributions of N fertilizer, soil nitrate at the end of the winter, N mineralized during the growing season, and atmospheric deposition of N during the growing season on the other (equation (1)),

$$Y \cdot b = (N_{rec} + N_{min}) \cdot E_1 + (M_o + M_y) \cdot E_2 + N_a \, E_3 \qquad (1)$$

where Y is the expected yield, b is the N uptake by the entire crop per unit of Y, N_{rec} is the recommended N fertilizer rate, N_{min} is the amount of mineral N already present in the soil at the time of application of N_{rec}, M_o and M_y are the amounts of N mineralized from 'old' and 'young' soil organic matter, N_a, is the amount of N supplied to the soil through dry and wet atmospheric deposition during the growing period, and E_1, E_2, and E_3 are efficiency coefficients.

Originally, the balance sheet method was developed in France for cereals and sugar beet (Rémy, 1981; Rémy and Viaux, 1982). Recently, promising results have also been obtained in Belgium and the Netherlands (Neeteson, 1990). The method has been introduced for commercial farming in France some years ago. It has not yet been adopted in Belgium and the Netherlands due to uncertainties about the values for the N uptake per unit of yield and the rate of mineralization.

Use of a simulation model

Simulation models should allow the calculation, on a daily basis, of the growth and the N uptake of a crop, and the N supply to a crop, using weather data and crop and soil parameters as inputs. A simulation model could be used as a tool to indicate the N fertilizer requirement of a crop at any time during the growing season. A simulation model could thus add the time element to N fertilizer recommendations. Moreover, unlike the N_{min}-method and the balance-sheet method, with simulation models it is possible to identify the environmental side-effects of N fertilizer application when all relevant N processes in soil (*e.g.* nitrate leaching) and crop are included in the models.

Recently, many models have been developed for developing the effect of N fertilizer application on the growth and N uptake of arable crops (Neeteson *et al.*, 1987; Van Keulen and Seligman, 1987; Addiscott *et al.*, 1991; Eckersten and Jansson, 1991; Groot and De Willigen, 1991; Hansen *et al.*, 1991; Kersebaum and

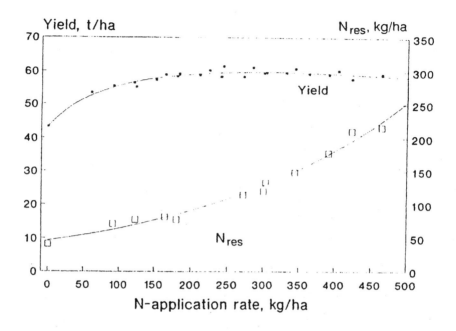

Figure 3 *Relationship between N fertilizer application rate to starch potatoes (Solanum tuberosum L.), and yield and residual soil nitrate (N_{res}) in the layer 0-100 cm (Neeteson and Wadman, 1990).*

Richter, 1991; Mirschel *et al.*, 1991; Vereecken *et al.*, 1991; Whitmore *et al.*, 1991) and grassland (*e.g.* Scholefield *et al.*, 1991). Only results obtained with a recent model for potatoes (Neeteson *et al.*, 1987) will be discussed here as an example. This model was developed for practical purposes rather than to improve scientific understanding.

The potato (*Solanum tuberosum* L.) model was developed and tested in close collaboration between the Horticultural Research International (Wellesbourne, England) and the DLO-Research Institute for Agrobiology and Soil Fertility (Haren, the Netherlands). The model is assuming the potato crop to grow on soil divided into 5-cm layers. The number of layers depends on the maximum rooting depth, which must be given as input parameter. As the crop grows, the roots penetrate more and more of these layers and are able to extract nitrate from them. The model calculates on a daily basis (i) potential increase in total dry weight, (ii) actual increase in total dry weight, (iii) soil organic matter mineralization, (iv) soil moisture content, (v) nitrate leaching, (vi) rooting depth, (vii) potential N uptake, (viii) actual N uptake, (ix) partitioning of dry weight among tubers and foliage, and (x) partitioning of N among tubers and foliage. The field dependent inputs required by the model are (i) time and amount of soil nitrate at the end of the winter, (ii) time and rate of N

Table 11 *N uptake by winter crops (kg N per ha^{-1} in above-ground parts in mid-November) in dependence of sowing time (Van Enckevoort et al., 1990; Elers et al., 1987).*

Crop	Mid-August	Sowing time Early September	Mid-September
Italian ryegrass (*Lolium multiflorum* Lam.)	60-125	35- 80	5-20
Winter rye (*Secale cereale* L.)	100-150	50- 75	20-50
Turnip (*Brassica rapa* L.)	100-160	70-110	20-45
Fodder radish (*Raphanus sativus* L.)	115-175	60-100	25-45

fertilizer application, (iii) time of planting, (iv) expected maximum tuber yield, (v) expected maximum rooting depth, (vi) soil type, (vii) daily values of soil temperature at 10 cm depth, and (viii) daily values of precipitation surplus. It should be noted that all these input parameters can easily be obtained by farmers. Details of the model and its underlying relationships are described by Greenwood *et al.* (1985 a and b) and by Neeteson *et al.* (1987). The model was tested against the results of numerous potato trials. For each trial first the final tuber yield at various N fertilizer application rates was predicted and next the optimum application rate of N fertilizer on the basis of the predicted yields. Although the model properly predicted yield and optimum N application rate (Neeteson *et al.*, 1987), it is not yet used on commercial farms because the description of the mineralization of soil organic matter needs improvement (Neeteson, 1990). Modified versions of the model are used to predict the N response of wheat (Greenwood *et al.*, 1987) and various field vegetables (Greenwood and Draycott, 1989).

Reduction of N fertilizer application rates
We have shown that high levels of residual soil nitrate may be present after application of the currently recommended N fertilizer rates to potatoes, field vegetables, grazed grassland, and silage maize. It was also shown that the levels of residual soil nitrate are generally much lower when these crops do not receive N fertilizer. Therefore, a reduction of the N fertilizer application rates to these crops may lower levels of residual soil nitrate and consequently contribute to a reduction of nitrate leaching during the subsequent winter period. It is not yet clear at which level of residual soil nitrate the maximum permissible nitrate concentration of 50 mg L^{-1} in groundwater will be exceeded. Under West-European conditions 50 mg NO_3 L^{-1} is reached in the shallow groundwater when annually 34 kg N ha^{-1} reaches the groundwater (see Introduction). Due to denitrification losses not all residual soil nitrate will enter the groundwater; levels of residual soil nitrate can therefore be higher than 34 kg ha^{-1}. Recently, it was suggested to the Dutch government to set the maximum value for residual soil nitrate in the 0-100 cm layer at 70 kg ha^{-1}, on the assumption of an average denitrification loss of 50% (Goossensen and Meeuwissen, 1990). Figure 3 gives an example of the relation between the N fertilizer application

rate to starch potatoes (*Solanum tuberosum* L.) and the resulting yield reduction, assuming an upper residual soil nitrate limit of 70 kg ha^{-1}. At the recommended N fertilizer rate of 207 kg ha^{-1} the amount of residual soil nitrate was 95 kg ha^{-1} (see also Table 4) and the yield was 60.0 t ha^{-1}. To avoid the amount of residual soil nitrate exceeding 70 kg ha^{-1}, the N fertilizer application rate should not exceed 120 kg ha^{-1}; at this rate the yield was 56.7 t ha^{-1}; so in this situation avoiding too high residual soil nitrate levels resulted in a yield reduction of about 6%. The same procedure was followed for ware potatoes (*Solanum tuberosum* L.), seed onions (*Allium cepa* L.), and silage maize (*Zea mays* L.); yield reductions were about 2% for ware potatoes and seed onions (Neeteson and Wadman, 1991) and 10% for silage maize (Goossensen and Meeuwissen, 1990). Obviously, a further reduction in the upper limit of the amount of residual soil nitrate will result in higher yield reductions.

Thus, a reduction in N fertilizer application rates to crops with high levels of residual soil nitrate contributes to lowering these levels. It should be taken into account, however, that crop yields will then also be reduced.

Growing of winter crops

Winter crops sown after cash crops are able to take up N and water during the fall. This will result in reduced nitrate leaching during the winter period due to lower levels of residual soil nitrate and a reduced water flow towards the groundwater. Under West-European conditions, however, it is a prerequisite that the winter crops are sown early, preferably before mid-September, so that they can absorb substantial quantities of N before winter. Winter crops sown in mid-August may take up more than 100 kg N ha^{-1}; crops sown in early September 35-110 kg ha^{-1} and crops sown in mid-September 5-50 kg ha^{-1} (Table 11).

Although winter crops contribute to the reduction in nitrate leaching, it can be expected that they will not be sufficiently effective in absorbing N, because many cash crops with high levels of residual soil nitrate are harvested too late to grow a well established winter crop. In Western-Europe potatoes (*Solanum tuberosum* L.) are generally harvested in late September/early October and silage maize (*Zea mays* L.) in late October/early November.

In practice, often about 50 kg N fertilizer ha^{-1} is applied to winter crops shortly before sowing. If this is the case, a winter crop cannot be expected to contribute to a reduction in residual soil nitrate and thus to a reduction in nitrate leaching during winter.

Future research needs

N fertilization strategies should aim at supplying agricultural crops with sufficient N for optimum growth and development, while keeping losses to the environment at a minimum. It has been argued that there is scope to refine the current N fertilizer recommendations and that simulation models may play an important role. Although the currently available models show promising results in describing the N cycle in the crop/soil system (Groot *et al.*, 1991), there are still uncertainties in describing microbiological processes in the soil, *viz.* N mineralization/immobilization and denitrification (De Willigen and Neeteson, 1985; De Willigen, 1991). Net N mineralization rates were found to range from 0.5 to 2.0 kg N ha^{-1} per day (Neeteson *et al.*, 1987), but it is not yet possible to predict the rate in a specific field. Research aimed at finding simple ways to determine mineralization levels by means of incubation or chemical extraction (Stanford, 1981), or on the basis of physical properties of the soil (Van Veen and Kuikman, 1990) or field records (Mary, 1987) should therefore be continued. The phenomenon of temporal immobilization of fertilizer N shortly after its application (Neeteson *et al.*, 1986) also needs clarification. After more knowledge has been gained on N mineralization/ immobilization on a field scale, it should be included in simulation models to describe organic matter dynamics in agricultural soils (*e.g.* Verberne *et al.*, 1990). These models can later be simplified for use as submodels in models for practical purposes, *e.g.* N fertilizer recommendations. The same is true for denitrification. Although there is a proper qualitative understanding of the denitrification process (Leffelaar, 1987; Groffman *et al.*, 1988; Christensen and Tiedje, 1988; Van Cleemput *et al.*, 1988), it is still not yet possible to obtain reliable quantitative information about the process on a field scale. The lack of quantitative data hampers the description of not only the amount of N available to crops during the growing season, but also the amount of residual soil nitrate which denitrifies in fall and winter.

All current N fertilizer recommendations should be reevaluated in such a way that they allow the recommendation of economically and environmentally sound N application rates. When both objectives cannot be met environmental considerations should prevail over economic considerations. Special attention should be given to the inclusion of the effect of grazing cattle in N fertilizer recommendations for grazed grassland and to the possibility of developing environmentally safe N fertilizer recommendations for field vegetables which are harvested before maturity.

References

Addiscott, T.M., Bailey, N.J., Bland, G.J. and Whitmore, A.P. 1991. *Fert. Res.*, 27, 305-312.

Anonymous. 1982. *Gemüsebau Versuchsergebnisse 1981*. Lehr- und Versuchsanstalt für Gartenbau Hannover-Ahlem, Hannover, 138 pp.

Anonymous.1983. *Gemüsebau Versuchsergebnisse 1982*. Lehr- und Versuchsanstalt für Gartenbau Hannover-Ahlem, Hannover, 123 pp.

Anonymous.1989a. *Handbook for the production of arable crops and field vegetables*. Publication 47, Experimental Station for Arable Crops and Field Vegetables, Lelystad. 252 pp. (in Dutch)

Anonymous. 1989b. *Fertilizer recommendations for grassland and fodder crops*. CAD-BWB Veehouderij, Wageningen. 72 pp. (in Dutch)

Ball, P.R. and Ryden, J.C. 1984. *Plant Soil*, 76, 23-33.

Boyd, D.A., Garner, H.V. and Haines, W.B.. 1957. *J. Agric. Sci. (Camb.)*, 48, 464-476.

Christensen, S. and Tiedje, J.M. 1988. In: Jenkinson, D.S. and Smith, K.A. (eds.), *Nitrogen efficiency in agricultural soils*, pp. 295-301. Elsevier Applied Science, London.

De Willigen, P. 1991. *Fert. Res.*, 27, 141-149.

De Willigen, P. and Neeteson, J.J. 1985. *Fert. Res.*, 8, 157-171.

Dilz, K. 1988.In: Jenkinson, D.S. and Smith, K.A. (eds.), *Nitrogen efficiency in agricultural soils,* pp. 1-26. Elsevier Applied Science, London.

Eckersten, H. and Jansson, P.-E. 1991. *Fert. Res.,* 27, 313-329.

Elers, B., Van Brandis, A., Scharpf, H.C. and Hartmann, H.D. 1987. *Gemüse,* 6, 290-292.

European Economic Community. 1980. *Off. J. EEC,* 23, 80/778 EECL, 229, 11-29.

Goossensen, F.R. and Meeuwissen, P.C. 1990. *Recommendations of the Committee on Nitrogen.* DLO, Wageningen, 93 pp. (in Dutch)

Greenwood, D.J. 1986. *Adv. Plant Nutr.,* 2, 1-61.

Greenwood, D.J. and Draycott, A.. 1989. *Fert. Res.,* 18, 153-174.

Greenwood, D.J., Neeteson, J.J. and Draycott, A. 1985a. *Plant Soil,* 85, 163-183.

Greenwood, D.J., Neeteson, J.J. and Draycott, A. 1985b. *Plant Soil,* 85, 185-203.

Greenwood, D.J., Verstraeten, L.M.J., Draycott, A. and Sutherland, R.A. 1987. *Fert. Res.,* 12, 139-156.

Groffman, P.M., Tiedje, J.M. Robertson, G.P. and Christensen, S. 1988. In: Wilson, I.R. (ed.), *Advances in nitrogen cycling in agricultural ecosystems,* pp. 174-192. C.A.B. International, Wallingford.

Groot, J.J.R., and Willigen, P. De. 1991. *Fert. Res.,* 27, 261-271.

Groot. J.J.R.. Willigen, P. De and Wadman, W.P. 1989. *Report 216.* DLO-Institute for Soil Fertility Research, Haren, 24 pp. (in Dutch)

Groot, J.J.R., Willigen, P. De and Verberne, E.L.J. (eds.), 1991. *Nitrogen turnover in the soil-crop system.* Kluwer Academic Publishers, Dordrecht. 386 pp.

Hansen, S., Jensen, H.E.V., Nielsen, N.E. and Svendsen, H. 1991. *Fert. Res.,* 27, 245-259.

Jurgens-Geschwind, S. 1989. In: Follet, R.F. (ed.), *Nitrogen management and ground water protection,* pp. 75-138. Elsevier Science Publishers, Amsterdam.

Kersebaum, K.C. and Richter, J. 1991. *Fert. Res.,* 27, 273-281.

Leffelaar, P.A. 1987. Dynamics of partial anaerobiosis, denitrification and water in soil: experiments and simulation. Thesis, Agricultural University, Wageningen.

Lindén, B. 1987. In: Nielsen, N.E. (ed.), *Proc 3rd meeting NW-European study group for the assessment of nitrogen fertilizer requirement,* pp. 72-86. Royal Vet. and Agric. University, Copenhagen.

Machet, J.M. and Mary, B. 1989. In: Germon, J.C. (ed.), *Management systems to reduce impact of nitrates,* pp, 126-145. Elsevier Applied Science, London.

Mary, B. 1987. *C.R. Acad. Agric. France,* 73, 57-69.

Mirschel, W., Kretschmer, H., Matthäus, E. and Koitzsch, R. 1991. *Fert. Res.,* 27, 293-304.

National Environmental Plan of the Netherlands. 1989. *Tweede Kamer der Staten Generaal, Vergaderjaar 1988-1989,* 21137, no. 1-2. SDU, the Hague, 258 pp. (in Dutch)

Neeteson, J.J. 1989. *Neth. J. Agric. Sci.,* 37, 227-236.

Neeteson, J.J. 1990. *Fert. Res.,* 26, 291-298.

Neeteson, J.J., and Ehlert, P.A.I. 1989. In: *Proc. 2nd winter congress,* pp. 79-91. International Sugar Beet Research Institute, Brussels.

Neeteson, J.J., and Wadman, W.P. 1991. *Report 237.* DLO-Institute for Soil Fertility Research, Haren, 18 pp. (in Dutch)

Neeteson, J.J., Greenwood, D.J. and Habets, E.J.M.H. 1986. *Plant Soil,* 91, 417-420.

Neeteson. J.J., Greenwood, D.J. and Draycott, A.. 1987. *Proc. 262. The Fertiliser Society,* London. 31 pp.

Neeteson, J.J., Greenwood, D.J. and Draycott, A. 1989. *Neth. J. Agric. Res.,* 37, 237-256.

Oenema, O., Postmus, J., Prins, W.H. and Neeteson, J.J. 1989. In: *Proc. XVlth international grassland congress,* pp. 159-160. Nice.

Prins, W.H. 1980. *Fert. Res.,* 1, 51-63.

Prins, W.H., Dilz, K. and Neeteson, J.J. 1988. *Proc. 276. The Fertiliser Society*, London. 27 pp.

Rémy, J.-C. 1981. *C.R. Acad. Agric. France,* 67, 859-874.

Rémy, J.-C., and Viaux, Ph. 1982. In: *Proc. 211, 67-92. The Fertiliser Society,* London.

Ris, J., Smilde, K.W. and Wijnen, G. 1981. *Fert. Res.*, 2, 21-32.

Russell, E.J. 1914. *J. Agric. Sci. (Camb.),* 6, 18-57.

Scholefield, D., Lockyer, D.R., Whitehead, D.C. and Tyson, K.C. 1991. *Plant Soil,* 132, 165-177.

Schröder, J. 1985a. *The effect of heavy application rates of cattle slurry on growth, yield and quality of silage maize and on soil fertility, Heino (sandy soil) 1972-1982. Report 30. Experimental Station for Arable Crops and Field Vegetables, Lelystad,* 151 pp. (in Dutch)

Schröder, J. 1985b. *The effect of heavy application rates of cattle slurry on growth, yield and quality of silage maize and on soil fertility and water pollution, Maarheeze (sandy soil) 1974-1982.* Report 31. Experimental Station for Arable Crops and Field Vegetables, Lelystad, 101 pp. (in Dutch)

Schröder, J. 1990. *Meststoffen* 1/2, 25-32. (in Dutch)

Smith, K.A., Howard, R.S. and Crichton, I.J. 1988. In: Smith, K.A. and Jenkinson, D.S. (eds.), *Nitrogen efficiency in agricultural soils,* pp. 73-84. Elsevier Applied Science, London.

Stanford, G. 1981. In: Stevenson, F.J. (ed.), *Nitrogen in agricultural soils,* pp. 651-688. American Society of Agronomy, Madison.

Van der Boon, J. and Pieters, J. 1981. *Report 97. DLO-Institute for Soil Fertility Research,* Haren, 36 pp. (in Dutch)

Van Cleemput, O., Abboud, S. and Baert, L. 1988. In: Smith, K.A. and Jenkinson, D.S. (eds.), *Nitrogen efficiency in agricultural soils,* pp. 302-311. Elsevier Applied Science, London.

Van Enckevoort, P.L.A., Landman, A., Titulaer, H.H.H. and Jansen, E.L.J. 1990. In: *Themaboeke,* 10, 53-68. Experimental Station for Arable Crops and Field Vegetables, Lelystad. (in Dutch)

Van Keulen, H. and Seligman, N.G. 1987. *Simulation Monographs.* PUDOC Wageningen, 310 pp.

Van der Paauw, F. 1962. *Plant Soil,* 16, 361-380.

Van der Paauw, F. 1966. *Landwirtsch. Forsch.,* 20, 97-105.

Van Veen, J.A. and Kuikman, P. 1990. *Biogeochemistry,* 11, 213-233.

Verberne, E.L.J., Hassink, J., De Willigen, P. , Groot, J.J.R. and Van Veen, J.A. 1990. *Neth. J. Agric Sci.,* 38, 221-238.

Vereecken, H., Vanclooster, M., Swerts, M. and Diels, J. 1991. *Fert. Res.,* 27, 233-243.

Wehrmann, J . and Scharpf, H.C. 1989. In: Germon, J.C. (ed.), *Management systems to reduce impact of nitrates,* pp. 147-156. Elsevier Applied Science, London.

Whitmore, A P., Coleman, K.W., Bradbury, N.J. and Addiscott, T.M. 1991. *Fert. Res.,* 27, 283-291.

Zandt, P.A., De Willigen, P. and Neeteson, J.J. 1986. *Report 9-86. DLO-Institute for Soil Fertility Research, Haren, 57 pp. (in Dutch)*

12 Effect of crops and fertilization of crops and fertilization regimes on the leaching of solutes in an irrigated soil

A. Feigin and J. Halevy[*]

Institute of Soils and Water, ARO, The Volcani Center, PO Box 6, Bet Dagan 50250, Israel

Abstract

Information on the relationship between agricultural practices and solute movement in soil has been collected from the long-term Bet Dagan Permanent Plots field experiment, in which the effect of fertilization on crop yield under Mediterranean climatic conditions has been intensively studied since 1960. Soil samples representing 60 (5 x 4 x 3) NPK fertilizer combinations were analyzed for different chemical properties, mainly the concentration of solutes (NO_3, Cl, SO_4, Na, K, Ca, Mg) in soil saturation extracts. The samples were taken from a depth of 2.1 m in 1986 and 1987, and also from a depth of 12.0 m in some cases in 1987. The mean annual rainfall for the above period was 524 mm, and it was greatly exceeded by the potential evapotranspiration.

The pattern of solute distribution in soil was closely related to crop yield and dry matter production. The higher the yield, the greater the transpiration and the smaller the quantity of rain or irrigation water available for leaching. The Cl level in the soil saturation extract along the 2–12-m profile of the zero fertilizer treatment ranged between 3 and 5 mM, while that found in the optimum fertilization treatment was 6–27 mM. No Cl was added through fertilizer in either case. The distribution of Na and Mg, not originating from fertilizer, was also greatly affected by leaching. However, mainly due to exchange and precipitation- dissolution processes, their distribution pattern differed from that of Cl.

The level of NO_3 in soil was greatly influenced by the plant, both directly through uptake and indirectly through its effect on leaching. The present data show that proper fertilization of irrigated crops can greatly improve yield and slow down solute movement through the soil profile.

---

[*]Dr A. Feigin and Dr J. Halevy passed away a short time after submitting this work for publication. Further correspondence regarding this chapter should be directed to Dr B. Bar-Yosef at the same address.

Introduction

Agricultural activities may affect groundwater quality (*e.g.* Bouwer, 1986; Keeney, 1982; Letey and Pratt, 1984; Sabol *et al.*, 1986). Large quantities of agrochemicals, especially fertilizers and manures, are employed to ensure food and fiber supply to the rapidly increasing world population. The load of fertilizer per area unit of intensively cultivated land is very high. Irrigation and rainfall water transport various solutes derived from soil or originating in fertilizers, waters, manures, sewage effluents and different agrochemicals, through the soil profile to deeper layers and eventually into aquifers. The amounts of irrigation water used under arid and semi-arid conditions often exceed rainfall.

There is ample evidence that optimization of irrigation-fertilization-cropping management increases fertilization efficiency and water use efficiency. For instance, Letey *et al.* (1977) concluded that the control of leaching is the most positive approach to controlling NO_3 movement beyond the root zone of irrigated agriculture. It may also be assumed that another benefit will be environmental quality preservation. A much higher level of NO_3-N was measured below the root zone of cotton plants where irrigation (with sewage effluents) and fertilizer application exceeded crop needs, in comparison with a better controlled management (Feigin *et al.*, 1978).

Well controlled irrigation implies adequate water supply to crops with a minimum flux of solutes beyond the root zone of crops (Bresler *et al.*, 1982). Optimum yield with the lowest acceptable salt flux below the root zone is a desirable goal in irrigated agriculture. Feigin *et al.*, (1982a,b) have shown that NO_3 leaching below the root zone of celery was greatly reduced by the use of well-controlled trickle irrigation.

Efficient use of fertilizers is another important factor. This involves the addition of suitable chemicals in optimum quantities at the correct time and mode of application. Pratt and Jury (1984) described the ideal system of management to maximize N use by the crop and minimize NO_3 leaching, as follows:

(1) Efficient root system for NO_3 uptake.
(2) Efficient system in which there is no increase in NO_3 leaching even though adequate quantities of N fertilizers are added to obtain maximum yields.
(3) Information on the N needs of crops, to ensure maximum yield or maximum economic return, thereby avoiding N excess.
(4) The quantity of water movement through the soil is small during periods when large amounts of NO_3 are available for leaching.

Increasing nitrogen efficiency by means of proper fertilization, irrigation and cropping management has won worldwide attention recently, as indicated by the number of international symposia on this subject (Jenkinson and Smith, 1988; Hargrove, 1988; Follett, 1989).

There are differences in the behavior of various materials involved in ground-water pollution. For instance, both Cl and NO_3 are easily leached beneath the root zone of crops by excessive irrigation or heavy rainfall. On the other hand, the involvement of these ions in biological systems differs greatly (*i.e.* uptake by plants, denitrification, other transformations in soil). Similarly, Na, as well as other cations common in the soil solution (Ca, Mg, K), interacts with soil clay. Consequently, the study of solute behavior involves special attention to the different materials affecting ground water quality.

Reliable field data can contribute greatly to the development of a well-controlled irrigation-fertilization-cropping management aimed at optimizing nutrient uptake and minimizing water and solute flux beneath the root zone of crops. Such information is of value for the preservation of ground water quality under highly intensive agricultural conditions, and can be used by planners to adjust local agricultural activities to different hydrological situations. Suitable information obtained from such long-term studies can be used to verify available models describing solute movement in soil.

Long-term field experiments consisting of different fertilizer and manure treatments given to agricultural crops under irrigation accompanied by adequate recording of chemicals and water inputs, crop yields and nutrient uptake, can be a suitable basis for a field study of the relationship between crop yield and solute movement beyond the root zone of plants.

Hypothesis and purpose of the study

Appropriate fertilization management can lower the potential ground water contamination in the following ways: suitable application of fertilizer satisfies the plant's needs with a minimum amount of residue available for leaching; maximal crop yield can result in enhanced water consumption, thus reducing the probability of excessive leaching of water (Ayers and Westcot, 1985) that is often accompanied by downward movement of nutrients and other solutes; the latter, such as Na and Cl, can be derived from irrigation water and other sources.

The aforementioned assumption is based on well established information indicating a strong linear relationship between transpiration and dry matter production by the plant (Ritchie, 1983). This finding has been presented, for different climatic conditions, by the following equations:

$$Y = mT/E_o \tag{1}$$

where Y is yield of dry matter, m is a crop factor depending on species or variety, E_o is free water (potential) evaporation (Briggs and Shantz, 1914) and T is transpiration. Another equation:

$$Y = nT \tag{2}$$

where n is a crop factor, has been suggested by de Wit (1958) for humid conditions where solar radiation is a limiting factor.

Hanks (1974) suggested the following relationship between yield and transpiration:

$$Y/Y_{max} = T/T_{max} \tag{3}$$

where Y and Y_{max} are actual and potential maximum yield and T and T_{max} are the actual and maximum transpiration under given nutritional conditions, respectively.

The Bet Dagan Permanent Plots Field Experiment consists of 75 NxPxK fertilizer combinations, 60 of which are included in the present study. Besides the possibility of defining the best fertilizer treatments for obtaining maximum yield, this experimental design covering an especially wide range of fertilization treatments and soil fertility levels enables testing of the relation between crop yield and solute distribution through and beneath its root zone. This is particularly true for long-term field studies such as the present one, which has been in existence for close to 30 years.

Proper fertilization management can contribute greatly to the reduced leaching of nutrients (especially NO_3) to ground water. This is also true for other solutes, such as Na and Cl, derived from other sources (*e.g.* irrigation water). Irrigation planned for a certain potential crop yield may exceed plant needs if nutrition becomes the limiting factor. Similarly, uniformly applied irrigation water can result in different leaching fractions in different parts of the same field if, due to nonuniform fertilization, plant development differs; rainfall is usually uniformly distributed over agricultural fields. Thus, nutrient use efficiency can influence the rate of leaching of solutes the origin of which is not fertilizer. Attention should be paid also to interactive effects among nutrients. For instance, P deficiency may greatly reduce crop development and yield, thus lowering the use efficiency of N. The level of the latter, always added to satisfy a higher crop yield, becomes excessive and, where NO_3 build-up occurs, ground water contamination becomes a problem.

The Bet Dagan Permanent Plots Field Experiment

Experimental site and treatments

The Bet Dagan Permanent Plots Field Experiment (Permanent Plots Team 1980) is located in a fine-textured soil - vertisol (Typic chromoxerert) with approximately 50% clay, 1.2% organic matter and 10% $CaCO_3$. The field capacity of the soil is 33 g H_2O/100 g soil.

The experiment was established by the Permanent Plots Team (1980) in 1960 with the main purpose of studying the influence of N, P and K fertilizers, in combination with different manure treatments, on field crops growing in irrigated grumusol (vertisol) clay soil. The experiment incorporates:
(a) Five levels of nitrogen fertilizer (N_0 (no fertilizer), N_1, N_2, N_3 and N_4). (b) Five levels of phosphate fertilizer (P_0 (no fertilizer), P_1, P_2, P_3 and P_4). (c) Three levels of potassium fertilizer (K_0 (no fertilizer), K_1 and K_2). (d) Three organic matter treatments: no organic amendment, farmyard manure ,and green manure.

Each of the 225 combinations (5x5x3x3) was applied on two replicate plots. The layout was designed to favor efficient mechanized cultivation (fertilization and harvest). The dimensions of a single plot are 4 x 27 m, with a total of 450 plots, occupying an area of 7 ha. The experiment was planned as an incomplete block design. The main crops since 1960 were: wheat, cotton, corn, sorghum, sugar beet, chickpea and vetch; more details are given in Table 1.

Table 1 presents the yearly rainfall measured at Bet Dagan during the years 1960 to 1987; the average was 524 mm. The rainfall pattern is typical for a

Table 1 *The sequence of crops grown in the Bet Dagan Permanent Plots field experiment, and the quantity of water and macronutrients applied during the years 1960-1987.*

Crop no.	Year[*]	Crop	Water input (mm) Rainfall	Irrigation	Fertilizer[**] kg ha^{-1} Level 1 of each element, N	P	K
1	1960/61	Wheat	464	0	No fertilizer		
2	1961/62	Chickpea	473	60	20	12.5	62.5
3	1962/63	Sugar beet	94	390	60$^+$	37.5$^+$	106
4	1963	Foxtail millet	0	140	100$^+$	30$^+$	125
5	1963/64	Vetch-Oat	683	0	20$^+$	15$^+$	150
6	1964	Cotton	0	380	30	10	150
7	1965	Sorghum	912	420	30	9	128
8	1965/66	Wheat	357	0	30	9	128
9	1966/67	Chickpea	753	60	19	11	37.5
10	1967/68	Sugar beet	428	250	60$^+$	37.5$^+$	106
11	1968	Corn. silage	0	405	100$^+$	50$^+$	50
12	1968/69	Pea	744	0	No fertilizer		
13	1969	Cotton	0	320	60$^+$	10	150
14	1969/70	Wheat	350	110	30	9	128
15	1970/71	Chickpea	564	0	24$^+$	14$^+$	70
16	1971/72	Sugar beet	666	300	60$^+$	375$^+$	25
17	1972	Corn . silage	0	360	200$^+$	100$^+$	50
18	1973	Cotton	453^{++}	360	No fertilizer		
19	1973/74	Vetch	815	0	No fertilizer		
20	1974/75	Wheat	687	0	36$^\#$	10	50
21	1976	Sorghum	442^{++}	80	No fertilizer		
22	1976/77	Oat	570	0	60$^+$	10	50
23	1977	Corn	0	250	100$^+$	40$^+$	no K
24	1978	Cotton	506^{++}	300	30	10	150
25	1979	Cotton	378^{++}	300	30	10	150
26	1980	Cotton	689^{++}	400	30	10	150
27	1981	Cotton	617^{++}	370	30	10	150
28	1982	Cotton	417^{++}	390	30	10	150
29	1983	Cotton	645^{++}	430	30	10	150
30	1984	Cotton	332^{++}	430	30	10	150
31	1985	Cotton	353^{++}	445	30	10	150
32	1985/86	Wheat	367	190	30	10	150
33	1986/87	Wheat	621	0	30	No P	No K

* Crops grown during winter are marked by a two-calendar-year sign *e.g.* crop no. 1, wheat, 1960/61; the other crops were grown mainly in summer.
** The figures show the quantities of N, P and K kg ha^{-1}. added in the level 1 fertilizer treatments. Level 2 is always equal to 2 x level 1, while levels 3 and 4 unless otherwise stated are equal to 4 x level 1 and 8 x level 1, respectively. Four levels of N and P and two levels of K other than the zero fertilizer level were tested.
+ Fertilizer levels 3 and 4 . N3, N4, P3 and P4. are equal to 3 x level 1 and 4 x level 1, respectively.
# The four levels of N were 36, 72, 144 and 151 kg ha^{-1}, respectively.
++ The rain fell during the winter preceding the main growing season of the relevant crop.

Mediterranean climate, with rain limited to the winter, and falling mainly in December and January. The summer is dry and, as shown in Table 1, irrigation is given to sustain the main summer crop. Irrigation (by sprinklers) was scheduled usually every 14 days.

Crop sequence and total water and fertilizer addition
Thirty-three crops were grown in the Bet Dagan experiment during a period of 28 years (approx. 1.2 crops per year) (Table 1). The most important crops were cotton (11 seasons, grown continuously during the period 1978-1985) and wheat (6 seasons). The main irrigated crops were cotton (between 300 and 445 mm irrigation water/season), sugar beet (250-390 mm/season), and corn (250-400 mm/season).

The quantities of N, P and K fertilizers added to each of the treatments are also listed in Table 1. In addition to the N, P and K levels mentioned above, zero N, P, K treatments were also included in the Permanent Plots study. All fertilizers were broadcasted prior to seeding.

The N fertilizer added was mostly $(NH_4)_2SO_4$ but calcinated ammonium nitrate

Table 2. *Effect of fertilization on the average yield[*] Mg ha^{-1} obtained from the main crops grown in the Bet Dagan Permanent Plots Field Experiment during the years 1960-1987.*

Main fertilizer treatment	Cotton[**]	Wheat[+]	Crop Sugarbeet[++]	Sorghum[#]	Chickpea[@]
N_0	3.26	2.52	8.9	4.15	2.50
N_1	4.17	3.07	11.8	4.85	2.40
N_2	4.89	3.51	13.4	5.40	2.05
N_3	5.05	3.87	14.2	6.45	2.15
N_4	5.01	3.70	12.9	6.95	1.95
P_0	3.92	2.65	7.0	4.95	2.05
P_1	4.63	3.33	13.2	5.80	2.25
P_2	4.74	3.53	13.6	5.65	2.25
P_3	4.74	3.57	13.3	5.75	2.20
P_4	4.72	3.62	14.1	5.80	2.20
K_1	4.48	3.33	12.2	5.50	2.15
K_2	4.48	3.30	12.1	5.55	2.20
K_3	4.59	3.30	12.4	5.66	2.20

[*] The figures show the average yield per growing season obtained from the main crops tested in the presented trial; the values represent the effect of the main N, P, K fertilizer treatments.
[**] Grown for eleven seasons, yield parameter - seed cotton.
[+] Grown for six seasons, yield parameter - grain.
[++] Grown for three seasons, yield parameter - sugar.
[#] Grown for two seasons, yield parameter - grain.
[@] Seasons, yield parameter - grain.

and urea were sometimes used; phosphorus was added as $Ca(H_2PO_4)_2 \cdot CaSO_4$ (superphosphate); and potassium as KCl and K_2SO_4.

The water used for irrigation contained Cl, SO_4, Na, Mg and Ca. Taking into account only small changes in the chemical composition of water during the experiment period, the total quantities of ions added through the irrigation water can be estimated using recent chemical analyses carried out in our laboratory.

The total amount of rainfall during the period of 1960-1987 was 14,680 mm (a mean annual level of 524 mm), but this was exceeded by the potential evapotranspiration and additional irrigation water was necessary to sustain the summer crops (sugar beet, cotton, sorghum, corn). The total quantity of irrigation water applied during the above period was 7,140 mm (Table 1). Considerable quantities of salt and nutrients were added to the soil through the water. The estimated cumulative figures (in kg ha^{-1}) for the 1960-1987 period follow: NO_3-N, 170; Cl, 17,130; SO_4-S, 1,290; HCO_3, 13,780; Na, 8,000; K, 442; Ca, 3,430; and Mg, 2,140. The range of nutrients and salt added through the fertilizer was as follows (in kg ha^{-1}): nitrogen: N_1, 1,310; N_2, 2,680; N_3, 4,640 and N_4, 7,510; phosphorus: P_1, 540; P_2, 1,080; P_3, 1,840; and P_4, 2,990. Sulfate S was added by K_1 fertilizer (K_2SO_4) at a rate of 257 kg ha^{-1}, and by K_2 at a rate of 514 kg ha^{-1}. The quantity of Cl added by the K_1 and K_2 treatments (in KCl fertilizer) was 2,230 and 4,460 kg Cl ha^{-1}, respectively. The quantity of SO_4-S added through N fertilizer $(NH_4)_2SO_4$ was 1,670, 3,340, 5,010 and 6,680 for the N_1, N_2, N_3 and N_4 treatments, respectively, and that applied through the P_1, P_2, P_3 and P_4 superphosphate fertilizer treatments was 825, 1,650, 2,475 and 3,300 kg S ha^{-1}, respectively.

Yield and dry matter determinations

Crop yield was determined for all studied crops. Some of the dry matter data have been obtained by means of direct measurements (vetch, corn), but for the other crops presented in the table, estimates were used to obtain the necessary information. For example, seed cotton yield was always determined, but the yield of stover was calculated using available data on the seed cotton/stover ratio obtained in other field studies carried out in Israel and elsewhere. Information on the grain/straw ratio in wheat, obtained for the 1986/87 yield, enabled us to estimate the total yield of dry matter obtained in previous years. The dry matter yield data were further tested by comparing the results with those obtained in other studies in Israel and elsewhere.

Soil sampling

Soil samples were taken by means of a truck-mounted sampler from 60 fertilizer treatments which received chemical fertilizers only. The samples represented five N treatments x four P treatments (P_0, P_2, P_3, P_4) x three K treatments = 60 treatments, each tested in two replications, bringing the total number of plots tested to 120.

In 1986 and 1987, soil samples were taken after the wheat harvest (during July), from a depth of 0 to 2.1 m. Three cores (5 cm in diameter) were taken from each of the 120 plots. The soil cores were divided into eight different layers: 0-20, 20-40, 40-60, 60-90, 90-120, 120-150, 150-180 and 180-210 cm. The three samples

Table 3 *Effect of N and P fertilizer combinations on the total production of dry matter Mg ha^{-1} season^{-1} [*] obtained during the period of 1960-1987, and during the recent 1980-1987 growing seasons.*

Fertilizer treatments		Cotton	Wheat	All crops
N	P	1980-1985.	1986-1987	1960-1987
0	0	6.16	4.0	4.88
0	2	6.67	5.0	5.56
0	3	6.33	4.5	5.63
0	4	6.33	5.0	5.53
1	0	8.00	5.0	6.34
1	2	9.17	6.5	8.31
1	3	10.00	6.5	7.81
1	4	9.83	7.0	7.88
2	0	10.67	5.0	6.38
2	2	13.33	8.5	9.13
2	3	13.83	9.0	9.38
2	4	13.50	9.5	9.78
3	0	11.00	6.0	10.44
3	2	13.33	9.0	10.50
3	3	13.17	10.0	10.59
3	4	13.33	10.5	10.66
4	0	9.83	5.5	6.91
4	2	12.67	6.5	10.25
4	3	12.17	10.0	10.69
4	4	12.83	10.5	10.69

[*] Average for the three K treatments.

taken from the same depth in each plot were later combined into one representative sample.

Late in the summer of 1987, soil samples were also taken by means of a suitable sampler from the 12.0-m depth. The sampling procedure was as follows: sixty 20-cm soil layers, starting with the top 0-20 cm and ending with the 1,180-1,200 cm layer, were taken separately from ten soil cores representing five different fertilizer treatments (low fertility level - $N_0P_0K_0$ and $N_4P_0K_2$; medium fertility level - $N_2P_2K_0$; and high fertility level - $N_3P_4K_0$ and $N_4P_4K_2$), each in two replicates. Each second 0.2-m-depth soil layer sample (total of 30 samples) was analyzed from each core.

Chemical analysis
The air-dried soil samples were ground to pass a 2-mm sieve and stored in suitable containers. The level of water soluble Na, K, Ca, Mg, Cl, NO_3 and SO_4 was determined in soil saturation extracts. The concentration of NO_3, Cl and SO_4 in the

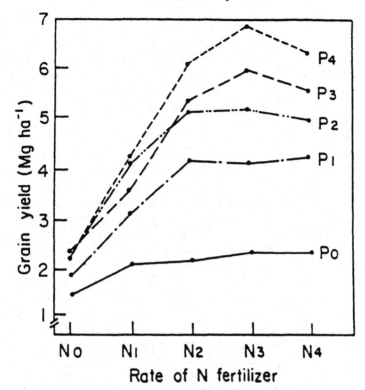

Figure 1 *Effect of N and P fertilizer combinations on the yield of wheat grain harvested in 1987. P_0-P_4 are the P fertilizer levels.*

water extract was measured by means of high performance ion chromatography (using a Dionex HPIC-2000i apparatus), and that of Ca and Mg by means of a 460 Perkin Elmer atomic absorption spectrophotometer; the level of Na and of K was determined by flame photometry. The level of exchangeable NH_4, and NO_3 was determined by means of lM KCl extraction followed by steam distillation.

Crop yield and dry matter production

Data on the effect of the different experimental treatments on crop yield (between 1960 and 1986) have been published elsewhere (Bar-Yosef and Kafkafi, 1972; Blum and Feigenbaum, 1969; Feigin *et al.*, 1974; Halevy, 1979; Kafkafi and Halevy, 1974, and relevant progress reports of the Permanent Plots Team). The 1987 yield data are shown in Figure 1.

The response of crops grown during the period 1960-1987 to the fertilization treatments is summarized in Table 2 and Figure 1. The yield of the main crops was positively affected by proper fertilization and was usually much greater than that obtained under nutrient-deficiency conditions. Similar results were obtained from the other crops. Representative data on the response of wheat yield (the 33rd crop, grown in the 1986/87 season) are shown in Figure 1. Due to the long-term effect of fertilization, and to the depletion of nutrients in the non-fertilized plots, the difference between the treatments is much greater in the 1987 wheat yield than in most of the previous crops. Furthermore, the data in Figure 1 are more striking than those shown by Table 2, since in the latter the average yields of the main treatments only are presented. The effect of N and P fertilizers on the 1987 wheat grain yield is significant, as is the N x P interaction.

Grain yield is responsible for two-thirds of the total dry matter production of wheat. As the relation between transpiration and dry matter production is usually linear, the estimated quantity of water transpired by the high yield plants (Figure 1) was fivefold that of the stunted plants grown in the nutrient-deficient soil.

Data on the total production of dry matter by the crops grown in different treatments of the Bet Dagan long-term experiment provide complementary information to the marketable yield parameters (*e.g.* grain, seed cotton, or sugar yields) that can be used for indirect estimation of other valuable crop parameters, such as transpiration.

The available data on the yield of dry matter obtained from the different crops grown in the Bet Dagan Permanent Plots experiment are given in Table 3. There are large differences between the dry matter production of the different crops as well as great variations in the yield of the same crops in different seasons due to the long-term continuous change in soil fertility. However, certain N and P fertilizer combinations produced the highest quantities of total dry matter.

The highest yields of dry matter were obtained from the N_3P_2, N_3P_3, N_3P_4, N_4P_2, N_4P_3 and N_4P_4 treatments. The average production of dry matter (the 1968-1987 period) was between 10.25 and 10.7 Mg dry matter ha^{-1} $season^{-1}$. Since the difference between the fertilizer treatments has been greatly increased in recent years, the dry matter yield obtained during the years 1980-1987 (cotton and wheat) was also calculated separately (Table 3). The combinations of N_2, N_3 or N_4 with P_2, P_3 or P_4 produced approximately 12-13 Mg of cotton dry matter ha^{-1} $season^{-1}$ during the last seven seasons (1980-1986), while the wheat dry matter was greatest (10-10.5 Mg ha^{-1} $season^{-1}$) in the N_3P_3, N_3P_4, N_4P_3 and N_4P_4 treatments. The P_0 combinations produced the lowest dry matter yields.

Long-term effects of fertilization and dry matter production on the distribution of solutes in soil

Electrical conductivity (EC_e)

The level of the electrical conductivity in soil saturation extracts was greatly affected by the fertilizer treatments. Figures 2 and 3 present data obtained from ten NxP combinations (five N levels x two P levels) in the summer of 1986, 26 years after the commencement of the experiment. Since the specific K effect on the electrical conductivity of the soil saturation extract (EC_e) was negligible, the above data were arranged in ten NxP combinations, each averages of six replications (three K x two original replications). The P_2 and the P_3-N combinations not presented here match well the EC_e data given in the figures. The EC_e pattern of the presented NxP

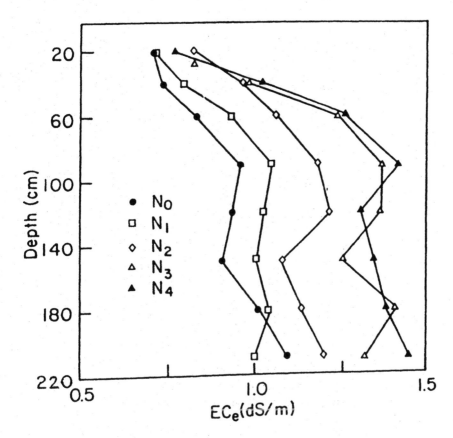

Figure 2 *Effect of five N fertilizer levels. N_0-N_4. combined with the P_0 level on the distribution of electrical conductivity in soil saturation extract EC_c in a 2,1 m-deep soil profile. 1986.*

combinations was similar. The lower values were found in the top soil layers and the highest ones below a depth of 1-1.5 m.

Additional soil samples were taken from the same plots one year later, in the summer of 1987, again after a wheat harvest. The 1987 EC_e data (not presented here) had a similar distribution pattern to those of 1987, but the actual ECe values were lower. This difference may be explained by the greater rainfall in 1987 (621 mm vs 367 in 1986). Thus, the EC_e values of each of the above N treatments (averaged over all soil layers included in the 0-2.1-m-deep profiles for the N_0, N_1, N_2, N_3 and N_4 fertilizer levels) were approximately 0.83, 1.0, 1.15, 1.27 and 1.35 dS m^{-1}, respectively, in 1986. The corresponding figures for 1987 were 0.67, 0.77, 0.84, 0.96 and 1.1 dS m^{-1}.

Further information on the effect of fertilization on the EC_e distribuiton in soil was obtained from the 12-m-deep soil cores taken in 1987. The fertilizer effect on

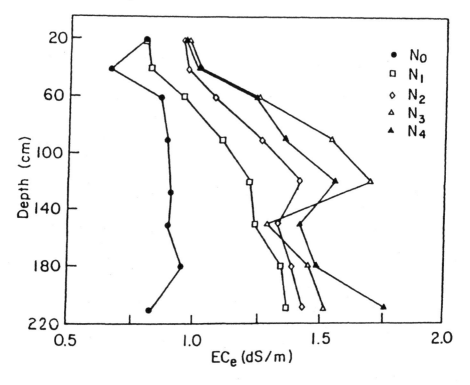

Figure 3 *Effect of five N levels. N_0-N_4. combined with the P_4 level on the distribution of electrical conductivity in soil saturation extract EC_e in a 2.1 m-deep soil profile. 1986.*

the distribution of the EC_e level in these cores is presented in Figure 4. The EC_e values are strongly influenced by the treatments: the lowest levels (approx. 0.4-0.7 dS m^{-1}) were detected in the low fertility ($N_0P_0K_0$) treatment. Much higher values were detected in the fertile soil, up to 2.7 dS m^{-1} in the $N_3P_4K_0$ treatment at the 10-12 m depth. This treatment 'produced' the highest wheat yield in 1987 and was found to be an excellent fertilizer combination for many of the previous crops. Somewhat lower EC_e values were measured in the $N_4P_2K_2$ treatment, from which high crop yields were also obtained. The $N_2P_2K_0$ treatment resulted in intermediate EC_e values.

The EC_e distribution for the fertilized plots has two peaks, one of which is found at the 1-4-m depth and the other at the 10-12 m depth; the second peak is the result of a gradual increase in the EC_e, starting at a depth of about 7 m. This two-peak pattern indicates a different behavior of certain solutes in soil, the combination of which results in the aforementioned EC_e peak pattern.

The effect of N fertilizer on the EC_e level (Figures 2, 3, 4) is clear. The higher the fertilizer level, the higher the EC_e. The N_0P_0 fertilizer treatments resulted in the lowest electrical conductivity values and the highest were detected in the N_3 and N_4 combinations. The high EC_e levels detected at the high N and P fertilizer levels

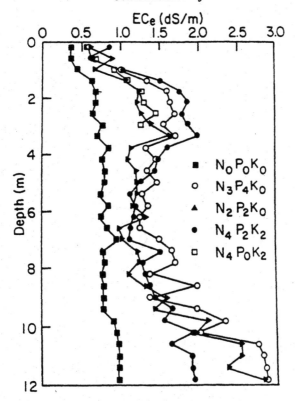

Figure 4 *Long-term effect of N, P and K fertilization on the distribution of electrical conductivity in soil saturation extract EC_e in a 12 m-deep soil profile. 1987.*

could result from each of two reasons: (i) direct contribution of solutes by fertilizer, and (ii) the influence of fertilization on other soil processes involved in the determination of solute distribution in soil. The H_2PO_4 ions added to the calcareous Bet Dagan soil are quickly transformed into the much less soluble Ca compounds containing HPO_4^{2-} or PO_4^{3-} constituents, and eventually result in the formation of sparingly soluble apatite mineral. This was confirmed by the negligible water-soluble P in the relevant soil samples. On the other hand, the ammonium-based fertilizer N was transformed quickly into NO_3, which can be easily leached into deeper soil layers. The data indicate that although the contribution of fertilizer to the increased EC_e values in soil is apparent, the main fertilizer effect on EC_e distribution in soil stemmed from the reduced quantity of water available for leaching. Suitable fertilization results in bigger plants and greater transpiration, and a smaller quantity of water available for leaching.

Chloride
Chloride is a common ion in irrigation water and contributes to the total electrical conductivity of the soil solution. The Cl concentration in the Bet Dagan irrigation

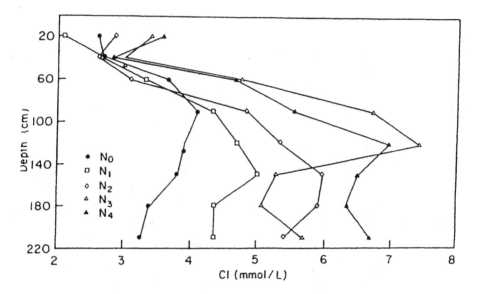

Figure 5 *Effect of five N fertilizer levels, N$_0$-N$_4$ combined with the P$_4$ level on the distribution of Cl⁻ in soil saturation extract in a 2.1 m-deep soil profile. 1986.*

water was 6.2 mM (220 mg L^{-1}). The total quantity of this ion added to the Bet Dagan experiment soil in the irrigation water is estimated to be approximately 17,000 kg ha^{-1}. In treatments K$_1$ and K$_2$ - KCl adding 2,230 and 4,460 kg Cl ha^{-1}, respectively, was included. This is a much smaller quantity of Cl in comparison with that added through the irrigation water.

The Cl concentration determined in the top soil layer in the summer of 1986 was relatively low in all the NxP combinations (Figure 5). This was mainly the result of Cl being leached into deep soil layers by rain water. The peak in Cl concentration was found to be in the 1.0-1.2 m soil layer. The difference in Cl concentration between the N treatments was relatively narrow in the P$_0$ combinations (data not presented). In the P$_4$ combination (Figure 5) a clear effect of N fertilizer on Cl distribution in soil was obtained, indicating a strong effect of plant growth on level of Cl in soil. The only explanation for this phenomenon is the reduced quantities of water available for leaching. The great difference in yield caused by N fertilization in the P$_4$ level was accompanied by a wide range of water consumption. Since the quantity of irrigation water and rainfall is the same in all plots, those carrying smaller plants have greater leaching fractions. Consequently, the distribution of Cl in the 210 cm soil profile, which originated from water, was indirectly affected by plant nutrition.

As in the case of the EC$_e$ samples, the Cl distribution measured in the 1987 soil samples in the 0-2.1 m-deep soil profiles has a similar pattern but the actual values were lower than those measured in 1986. The range of Cl levels in the soil saturation extracts obtained from five N-fertilizer treatments (average values for the four P levels combined with each of the above N treatments, and average for the 0-2.1-m-deep soil profile) was 3.9-5.3 mM in 1986, and 2.3-3.2 mM in 1987.

The Cl distribution in the saturation extract of the 12-m-deep soil profile is shown in Figure 6. The fertilizer treatments had a strong influence on the Cl levels along the soil profiles. The lowest Cl concentration was found in the $N_0P_0K_0$ treatment and the highest in the $N_3P_4K_0$ treatment. The Cl concentration in the other

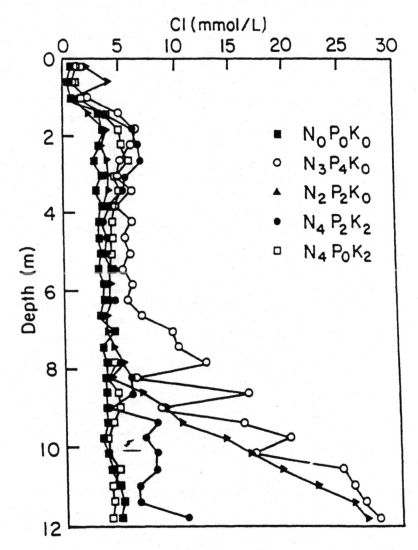

Figure 6 *Long-term effect of N, P and K fertilization on the distribution of Cl⁻ in soil saturation extract in a 12 m-deep soil profile. 1987.*

treatments was also related to yield. The Cl concentration in the low-yielding $N_4P_0K_2$ treatment was similar to that in the $N_0P_0K_0$ treatment, while that in treatment $N_2P_2K_0$ was similar to that of $N_3P_4K_0$ treatment. The Cl concentration in the $N_4P_2K_2$ treatment was intermediary between the two groups, as the grain yield was (Figure 1). The total quantity of Cl added to the $N_0P_0K_0$, $N_2P_2K_0$ and $N_3P_4K_0$ treatments via the irrigation water throughout the 1960-1987 period was 17,100 kg ha^{-1}. The amount of Cl applied through the irrigation water and the K_2 fertilizer (partially added as KCl) was greater (17,100+4,600 = 21,700 kg Cl ha^{-1}). The total amount of Cl detected in the soil in the aforementioned treatments was calculated using the following values: the Cl concentration determined in the soil saturation extract in each soil layer; the ratio of saturation extract to air-dry soil (0.7:1) at a soil depth of 12 m; and average apparent soil density of 1.2 g cm^3. The quantities of Cl in the 12-m profile were 11,290, 35,740, 26,880, 17,510 and 13,620 kg ha^{-1} for the $N_0P_0K_0$, $N_3P_4K_0$, $N_2P_2K_0$, $N_4P_2K_2$ and $N_4P_0K_2$ treatments, respectively. The Cl balance (the Cl found in soil minus the added Cl) for these treatments was as follows: $N_0P_0K_0$, -5810; $N_3P_4K_0$, +18,640; $N_2P_2K_0$, +9,780; N4P2K2, -4,190; and $N_4P_0K_2$, -8,080 kg Cl ha^{-1}. Since the uptake of Cl by crops is relatively small (except for sugar beet, which was grown for three seasons only), and even if it is assumed to be as high as 10% of the total quantity applied, the Cl balance would be virtually unchanged. A negative balance means that part of the added Cl was leached beyond the 12 m soil profile. A positive balance indicates incomplete leaching and/or remains of Cl found in soil prior to 1960. The data show that positive balances were obtained in the high yield treatments, and negative balances were obtained in the low yield treatments.

Data in Table 3 show that the estimated average dry matter production in the high-yielding fertilizer treatments for the 33 growing seasons was approximately 10 Mg $ha^{-1}yr^{-1}$. Assuming a water use efficiency of 2 kg dry matter per 1 m^3 of water (Stanhill 1986), and an average rainfall of 524 mm yr^{-1}, all the rainwater would have been used for dry matter production. As part of the rainwater was lost through evaporation from bare soil and leaching, it was necessary to compensate for this by the irrigation water (whole experiment mean 250 mm yr^{-1}) with the excess leached below the root zone of the crops. As the average yield obtained in the lower-yielding treatments was less than one half, and in recent years even much less than that (Table 3), considerably more water was available for leaching in these treatments. Therefore, whereas a Cl build-up was found below the 6m depth in the $N_3P_4K_0$ and $N_2P_2K_0$ treatments that produced high yields, the Cl was leached to deeper soil layers in the less productive treatments ($N_0P_0K_0$ and $N_4P_0K_2$). No information is available as yet on the Cl distribution below 12 m, but one may assume that a similar Cl build-up occurred in deeper soil layers of the low-yield treatments. Eventually the Cl could reach the ground water, which is more than 30 m below the surface level in the study area.

The Cl concentration in the top layer (0-1 m depth) is much lower than in the deeper soil layers; this Cl pattern was detected also in the 1986 soil samples. This is the result of leaching of Cl to a deeper soil layer. The distribution of Cl in the 1-6m deep soil layer appeared to be uniform with small fluctuations in concentrations. This indicates that the main miscible displacement process took place in the top 1 m, and the Cl carried downwards by the water solution was well mixed. The difference in the Cl level throughout the 1-6-m soil depth is clear and is greater in the $N_3P_4K_0$ than in the $N_0P_0K_0$ treatment, while the Cl values of the other treatments fall between the two extremes.

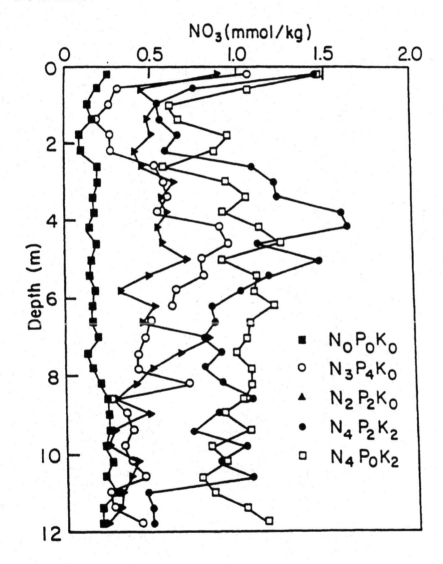

Figure 7 *Long-term effect of N, P and K fertilization on the distribution of KCl-extractable NO_3^- in a 12m-deep soil profile. 1987. To convert to soil saturation extract and to field capacity concentrations mM divided by 0.7 and 0.33, respectively.*

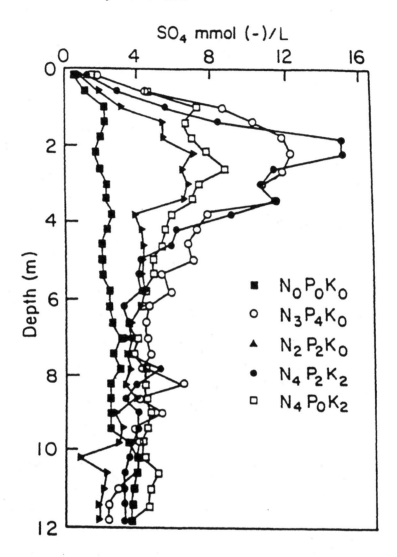

Figure 8 *Long-term effect of N, P and K fertilzation on the distribution of SO$_4^{2-}$ in soil saturation extract in a 12 m-deep soil profile. 1987.*

Nitrate

The NO$_3$-N concentration in the soil was much lower than that of Cl. It was usually around 1 mM in saturation paste (Figure 7), while the Cl concentration ranged between 2 and 30 mM. The main source of NO$_3$-N was N fertilizer. The potential contribution of NO$_3$ by soil organic matter is shown in the N$_0$P$_0$K$_0$ treatment, where the NO$_3$-N profile consists of a low and uniform concentration of about 0.25 mmol kg^{-1}. The NO$_3$-N level in the irrigation water was also low and its contribution to soil NO$_3$-N was negligible. The highest NO$_3$-N levels in soil were recorded in the

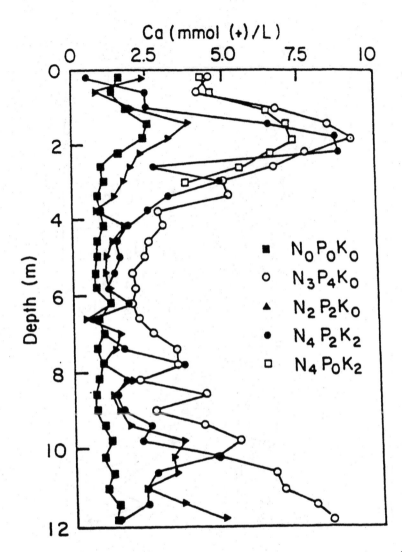

Figure 9 *Long-term effect of N, P and K fertilization on the distribution of Na⁺ in soil saturation extract in a 12 m-deep soil profile. 1987.*

$N_4P_0K_2$ and $N_4P_2K_2$ treatments (0.5 to 1.6 mmol kg^{-1}). This was due to the combined effects of high N application and relatively low yield and N uptake by the crop. The dry matter yield obtained from these treatments was much less than that attained from the $N_3P_4K_0$ treatment, in which the concentration of NO_3-N ranged between 0.3 and 1.0 mmol kg^{-1}. Only a small fraction of the total N added was found as NO_3-N in the 12-m soil profile: the total NO_3-N in the profile was 334, 960,

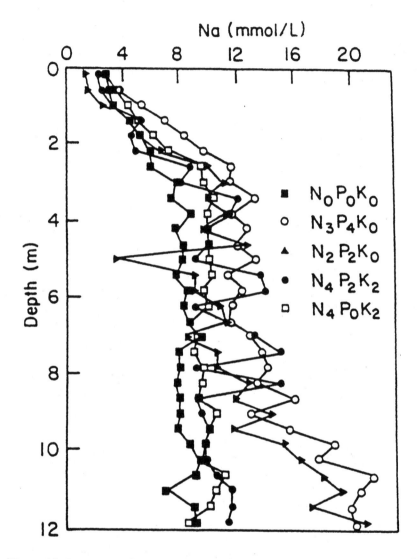

Figure 10 *Long-term effect of N, P and K fertilization on the distribution of K^+ in soil saturation extract in a 12 m-deep soil profile. 1987.*

930, 1,850 and 1,941 kg h^{-1} in the $N_0P_0K_0$, $N_3P_4K_0$, $N_2P_2K_0$, $N_4P_2K_2$ and $N_4P_0K_2$ treatments, respectively.

These data indicate that although it was impossible to prevent NO_3 leaching below the 1 m deep root zone of the crop, improved fertilization management results not only in a better yield - and probably of better quality - but also in greatly reduced NO_3 movement towards the aquifer. Figure 7 shows that the NO_3-N concentration in the deeper soil layers is approximately 1 mmol kg^{-1} in the $N_4P_0K_2$ treatment and less

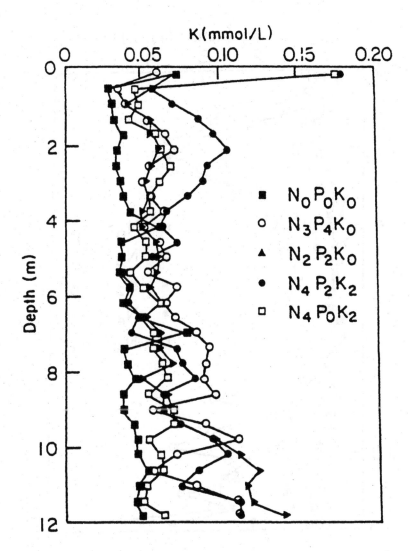

Figure 11 *Long-term effect of N, P and K fertilization on the distribution of* Ca^{2+} *in soil saturation extract in a 12 m-deep soil profile. 1987.*

than 0.5 in the $N_3P_4K_0$ treatment. The NO_3-N levels in the same soil layers were between 0.5 and 1 mmol kg^{-1} in the $N_4P_2K_0$ treatment, and 0.25 mmol kg^{-1} soil in the $N_0P_0K_0$ treatment. As the experimental treatments were not designed for high NO_3 uptake efficiency, a further improvement in N uptake efficiency and reduced residual NO_3 is still expected.

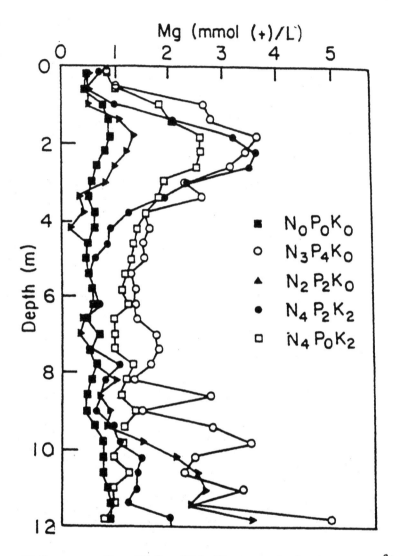

Figure 12 *Long-term effect of N, P and K fertilization on the distribution of* Mg^{2+} *in soil saturation extract in a 12 m-deep soil profile. 1987 .*

Sulfate

The distribution pattern of water soluble SO_4 in the soil differed greatly from that of the monovalent anions (Figure 8). The fertilizer treatments greatly influenced the SO_4 level in soil, since both the N and P fertilizers that were used included SO_4. This was shown also by the 2.1-m soil samples taken in 1986 and 1987 (data not presented).

The concentration of SO_4 ions in the unfertilized soil was relatively low (~1 mM), and uniform above 1 m depth (Figure 8). This concentration is lower than the SO_4 concentration calculated from the K_{sp} of $CaSO_4$ ($K_{sp}=10^{-4.64}$, Lindsay, 1979) assuming a $CaCO_3$ controlled Ca concentration (0.01 M at pH 8 and PCO_2 $3x10^{-4}$ atm), which yields a concentration of 4.6 mM SO_4^-.

The highest concentrations of water-soluble SO_4 were found in the fertilized plots between 1 and 5 m depth, with peaks around a 2–3 m depth. The peak concentrations (12-15 mM) are supersaturated with respect to $CaSO_4$ solubility if assuming a Ca concentration of 0.01 M. They are reasonably close, however, to the SO_4 concentration expected in a water-$CaSO_4$ system ~10 mM SO_4^-). This agreement indicates that SO_4 was accumulated in the soil's peak zone as $CaSO_4$.

Calcium

The distribution of water soluble Ca in the soil had two peaks, which were particularly marked in the treatments which received superphosphate (Figure 9). The upper peak coincided with the SO_4 peak (Figure 8), thus supporting our conclusion that $CaSO_4$ was accumulated in the 1.5-2.5 m soil layer. The deeper peak in Ca concentration coincided with the Cl⁻ peak (Figure 6). As in the case of Cl, Ca concentration below 8 m depth was greater in treatment $N_3P_4K_0$ than in the other fertilizer treatments.

The water extractable Ca concentration in the $N_0P_0K_0$ treatment was about 1 mM(+) throughout the profile (Figure 9). This concentration is appreciably lower than the expected value in a $CaCO_3$-CO_2-pH system at pH 8 and PCO_2 $3x10^{-4}$ atm (~20 mM(+)).

Sodium

The distribution of water-extractable Na in the 12-m-deep soil profile differs from that of Ca (Figure 10). These data match those obtained from the 1986 and 1987 2.1-m-deep soil samples (not presented). The Na content in the top soil layer (0–0.4 m) is the smallest, indicating a strong leaching into the deeper horizons of the soil by rain water. The shape of the Na concentration curve below the 3 – 4 m depth changes greatly among the treatments. It is generally linear and quite constant down to the 12 m depth in the low-yielding treatments ($N_0P_0K_0$ and $N_4P_0K_2$), while it increases gradually, reaching the highest level in the deeper soil horizon (12 m), in the high-yielding treatment ($N_3P_4K_0$ and $N_2P_2K_0$).

The concentration of Na in the deepest part of the soil cores approaches that of Cl (approximately 21 and 28 mM, respectively). This was due to the fact that both ions are not prone to precipitation reactions. The extra Cl concentration balanced the charge of the other cations that were present in the soil bulk solution.

Potassium

There were consistent differences in the water soluble potassium concentration in the soil between the fertilizer treatments (Figure 11) even though the total concentration of K in the soil saturation extract was relatively low (0.05–0.15 mM). The effect of K application was clear at the top soil level and later, to some extent, at the

2-m-depth peak. The increase in the K concentration in the deeper soil layers was greater in the $N_3P_0K_0$ and $N_2P_2K_0$ treatments, in which high yields were obtained, and the greater transpiration and lower leaching was probably responsible for the K movement through the deep soil profile. This result agrees with the general leaching pattern obtained in the present study.

Magnesium

Unlike Ca, Mg in the soil was not derived from the fertilizers. Even so, Mg shows a two-peak distribution pattern as in the case of Ca (Figure 12). The location of the peaks in the soil profile was similar to that of Ca and SO_4, and was probably related to $MgCO_3$ or $MgCa(CO_3)_2$ (dolomite) precipitation. The second peak is probably the result of dissolution and further transport of Mg by water. The height of the upper peak in the $N_3P_4K_0$ treatment was approximately 4 mM(+), and that of the lower peak was between 4 and 5 mM(+). As in the case of Cl, the lower concentration of Mg was found in the low-yield treatments and was probably the result of further leaching of the cation.

General discussion and conclusions

Data obtained during the Bet Dagan Permanent Plot Experiment show that large differences in the solute concentration in soil saturation extracts can develop as the result of direct and indirect effects of fertilization.

The data presented show that the solute distribution in soil is strongly affected by plant growth and yield, which are strongly associated with water consumption by the crop. The mean annual rainfall of the Bet Dagan area for the 1960–1987 period was 524 mm. Rough estimations reveal that this water quantity can support a dry matter production of approximately 10 Mg ha^{-1} yr^{-1} (2 kg dry matter m^{-3} water). This is the average yield of dry matter produced by the above-mentioned high-yielding treatments.

Due to losses through evaporation and leaching outside the soil root zone, supplementary irrigation had to be applied to the crops in order to obtain high yields. Since both rain and irrigation water were added in similar quantities to all the experimental plots, the lower-yielding treatments (in the present case, the N- and P-deficient ones) involved much more water available for leaching. This is clearly indicated by soil data obtained from both the 2.1-m-depth profile considered in 1986 and 1987, and the 12-m-depth soil profile.

The soil data show that fertilization exerted an influence on two groups of solutes: (i) those originating in the fertilizer (NO_3, SO_4 and Ca) and (ii) those contributed by other sources, mainly irrigation water (Cl, Na and Mg). Some solutes (Ca, SO_4) were contributed by both sources. Negligible N was contributed by the irrigation water. The Cl and Na showed a high peak in the soil layer below the 7-m depth. Calcium and Mg were characterized by two peaks - one at the same depth as the monovalent ions (below 7 m depth) and the other close to the soil surface (at a depth of 2–3 m). Sulfate accumulated only at the 2–3 m depth.

A typical solute distribution is depicted by the Cl levels in the soil profile: a low concentration in the top 1 m, which is a result of the fact that during the 1986/7 season rainfall was the only source of water, followed by a higher level of a quite uniformly distributed solute. It seems that miscible displacement homogenizing the Cl in the soil solution took place in the leached part of the profile, and the leachate was later uniformly distributed along a very thick soil layer.

The relatively uniformly distributed Cl level present below the upper leached soil horizon and above the 6 m soil depth indicates that a steady state condition was established within the 1–6-m soil depth. The effect of fertilization on the steady state zone is reflected by the higher Cl level in the high-yielding treatments in comparison with the low-yielding ones. A great accumulation of Cl is found below the 7-m depth in the high-yielding soil, while a low level of Cl is detected in the low-yielding treatments. The present results lead to the conclusion that the high Cl and Na peaks detected below the 7-m soil depth are the result of limited leaching, while greater leaching, resulting from lower transpiration, washed much of these solutes into soil layers deeper than 12 m. It should be noted that even in the high-yielding plots, not all the Cl and Na found in the soil is accounted for, since the peak of these solutes is obtained at the 12-m depth and no information is available about their concentration in deeper layers. Relevant measurements, estimations and calculations indicate that a negative balance in the 12-m-depth soil profile is found in the case of the low-yielding treatments, while a positive balance is found in the high-yielding treatments. This indicates that a considerable quantity of Cl must have been found in the soil before the establishment of the experiment.

The peak of the slowly leached SO_4, and the peaks of Ca and Mg match at 2–3 m below the soil surface. This is probably the average depth which the irrigation or rainwater often reaches under the Bet Dagan conditions. Oversaturation taking place under such circumstances results in the precipitation of Ca, SO_4 and Mg salts in these soil layers.

Downward movement of these salts can result from dissoution of the precipitates in wet years. The solutes are transported by water moving below a depth of 3–4 m. This process may explain the second peak of Ca and Mg found below the 7-m depth. The much greater leaching occurring with the low-yielding crops result in a much deeper movement of solutes not derived from fertilizer.

The NO_3–N distribution in soil was affected mainly by the fertilizer level and uptake by the crop. The present data show that efficient fertilization – irrigation management can greatly reduce the quantity of NO_3–N available for leaching.

The effect of crop growth on leaching depends on the amount of rainfall, and it is therefore appropriate for Israel's coastal plain, where rainfall ranges between 400 and 600 mm yr^{-1}, or for other sites with similar conditions. The effect of fertilization – irrigation – cropping management on the leaching of solutes through its impact on crop transpiration will be less effective under humid conditions.

The Bet Dagan data show that appropriate fertilization and irrigation management can strongly affect crop yield, while considerably reducing the quantity of solutes available for leaching. This is true for solutes originating in both fertilizer and other sources.

The long-term records from the Bet Dagan field study, allow to identify certain fertilization treatments as the most suitable for the local conditions. These treatments (*e.g.* $N_3P_4K_0$) resulted in both high yields and less leaching of solutes.

Acknowledgements

This research was supported by the National Council for Research and Development, Ministry of Science and Development, Israel; and the Commission of the European Communities DG X11.Contribution from the Agricultural Research Organization, The Volcani Center, Bet Dagan, Israel. No. 3122-E, 1990 series.

References

Ayers, R.S. and Westcot, D.W. 1985. *Water quality for agriculture*. Rev Edn FAO Irrig Drain Pap. 29, Rome. 174 p.

Bar-Yosef, B. and Kafkafi, U. 1972. Rates of growth and nutrient uptake of irrigated corn as affected by N and P fertilizers. *Soil Sci. Soc. Am. Proc.*, 36, 931-936.

Bouwer ,H 1986. Effect of irrigated agriculture on ground water. *J. Irrig. Drain Div. ASCE*, 113, 4-15.

Blum, A. and Feigenbaum, S. 1969. The effect of soil fertility on hybrid grain sorghum grown under conditions favoring maximum yield. *Qual Plant Mat. Veg.*, 17, 273-285.

Bresler, E., McNeal, B.L. and Carter, D.L. 1982. *Saline and sodic soils*. Springer-Verlag, Berlin

Briggs, L.J. and Shantz, H.L. 1914. Relative water requirements by plants. *J.Agric. Res.*, 3, 1-63

De Wit, C.T. 1958. Transpiration and crop yields. *Landouwkd Onderzok* No 64, 1-88

Feigin, A., Bielorai, H., Dag, Y., Kipnis, T. and Giskin, M. 1978. The nitrogen factor in the management of effluent-irrigated soils. *Soil Sci.*, 125, 248-254.

Feigin, A., Halevy, J. and Kafkafi, U. 1974. Cumulative effects of manures applied with and without fertilizers, on nutrient availability in the soil and on yields of crops under irrigation. *Hassadeh*, 65, 1351-1384 . Hebrew, with English summary and Table and Figure legends.

Feigin, A., Letey, J. and Jarrell, W.M. 1982a. Celery response to type, amount, and method of N-fertilizer application under drip irrigation. *Agron. J.*, 74, 971-977.

Feigin, A., Letey, J. and Jarrell, W.M. 1982b. Nitrogen utilization efficiency by drip irrigated celery receiving preplant and water applied N fertilizer. *Agron. J.*, 74, 978-983.

Follett, R.F. (ed.), 1989. *Nitrogen management and groundwater pollution*. Developments in agricultural and managed forest ecology 21. Elsevier, Amsterdam, 395 p.

Halevy, J. 1979. Fertilizer requirements for high cotton yields. *Proc. 14th Colloq Int Potash Inst.* Seville, Spain, 2, 1-7.

Hanks, R.J. 1974. Model for predicting plant growth as influenced by evapotranspiration and soil water. *Agron. J.*, 66, 660-665.

Hargrove, W.L. (ed.), 1988. *Cropping strategies for efficient use of water and nitrogen*. ASA Special Publ No 51, 218 p.

Jenkinson, D.S. and Smith, K.A. (eds.), 1988. *Nitrogen efficiency in agricultural soils*. Elsevier Applied Science, London and NY, 450 p.

Kafkafi, U. and Halevy, J. 1974. Rates of growth and nutrient consumption of semidwarf wheat *Proc 10th Int Congr Soil Sci.* Moscow. 4, 137-143.

Keeney, D.R. 1982. Nitrogen management for maximum efficiency and minimum pollution.

In: Stevenson,F.J. (ed.), *Nitrogen in agricultural soils*, 605-647. Amer Soc Agron, Madison, WI.

Letey, J., Blaire, J.W., DeWit, D., Lund, L.J. and Nash, P. 1977. Nitrate-nitrogen effluent from agricultural tile drain in California. *Hilgardia*, 45, 289-318.

Letey, J. and Pratt, P.F. 1984. Agricultural pollutants and ground water quality. .In: Yaron, B., Dagan, G. and Goldshmid, J. (eds.), *Behavior of pollutants in the unsaturated zones*, 211-111. Springer-Verlag, Berlin.

Lindsay, W.L . 1979. *Chemical Equilibria in Soils*. John Wiley, New-York.

Permanent Plots Team . 1980. *The Permanent Plots Experiment Report* No 3, 1970-1977. collated by J Halevy, Agricultural Research Organization, The Volcani Center, Bet Dagan, Israel.

Pratt, P.F. and Jury, W.A. 1984. Pollution of the saturated zone with nitrate. In:Yaron, B., Dagan,G and Goldshmid, J. (eds.), *Behavior of pollutants in the unsaturated zones*, 53-67. Springer-Verlag, Berlin.

Ritchie, J.T. 1983. Efficient water use in crop production, Discussion on the generality of relations between biomass production and evapotranspiration. In: Taylor, H.M., Jordan, W.M. and Sinclair, T.R. (eds.), *Limitations to efficient water use in crop production*. Amer Soc Agron, Madison, WI.

Sabol, G.V., Bouwer, H. and Wierenga, P.J. 1986. Irrigation effects in Arizona and New Mexico. *J. Irrig Drain Eng ASCE*, 113, 30-48

Stanhill, G. 1986. Water use efficiency. *Adv. Agron.*, 39, 53-85.

13 Groundwater pollution in Australian regional aquifers

G. Jacobson[1] **and J.E. Lau** [2]

[1] *Australian Geological Survey Organisation, Box 378, Canberra, A.C.T., Australia, 2601*
[2] *G.C. and J.E. Lau and Associates, Red Hill, A.C.T., Australia, 2603*

Abstract

A total of 144 known groundwater contamination incidents are documented for Australia. A wide range of contaminant sources is involved, including industrial effluent, sewage and landfill leachate. Many cases are of local significance but several important regional aquifers are affected.These regional aquifers include Quaternary sand aquifers of the Perth Basin, in Western Australia; a widespread Tertiary limestone in South Australia and Victoria; Quaternary volcanic rocks and Tertiary sand aquifers near Melbourne, Victoria; and fractured Silurian sedimentary rocks at Canberra. These are all shallow unconfined aquifers that underlie regions of intensive urban, industrial or agricultural development. Remedial measures, including groundwater recovery and treatment, have been successful in a small number of cases. A range of State (as opposed to federal) government legislation is applicable to the control and management of point source contamination, but controls are unevenly implemented. Legislation applicable to diffuse contamination is obscure or non-existent. Surveillance and documentation of the reported incidents is incomplete and more than half of the known incidents are not currently monitored. Remediation of many of the known incidents has not yet been undertaken.

The need for groundwater protection has been recognised in Australia recently, and national guidelines are being developed to assist the formulation of regional policies and controls.

Introduction

Australia has a land area of more than 7 million km^2 of which two-thirds is arid and groundwater dependent. Several hundred small towns and communities use groundwater for drinking water in this region, and groundwater is also used as a supplementary water source in several cities. The total amount of groundwater extracted is about 2,500 million m^3 annually, and this is about 15% of the total water used in the country (Jacobson *et al.*, 1983).

Although groundwater has been developed in Australia for a century or more, the issue of groundwater pollution received little attention prior to 1979 when the Australian Water Resources Council sponsored a national conference on the subject. This resulted in the delineation of specific problems, and recommendations for the development of groundwater quality criteria and for research on particular aspects of groundwater pollution (Lawrence and Hughes, 1981). Subsequently, case studies of groundwater pollution in certain urban and agricultural areas were documented by Bestow (1981), Harvey (1983), Knight (1983), Shugg (1987).

In 1987 we prepared a national inventory of proven groundwater contamination incidents for the Australian Water Resources Council (Jacobson and Lau, 1988). This was intended to form a basis for investigating the extent and seriousness of groundwater pollution in the country, and for assessing national needs for aquifer protection and for monitoring networks. Information was obtained from questionnaires distributed within water agencies, supplemented by a literature search. Details of the incidents were entered into a microcomputer data base and this has recently been updated so that the information reported in this chapter could be generated.

Inventory of groundwater contamination incidents

A total of 144 groundwater contamination incidents have been documented in the inventory; of these, 115 are based on openfile information and 29 are based on questionnaire responses regarded as confidential. The inventory is not exhaustive, and there are undoubtedly other incidents, as yet undocumented or undiscovered. Saline intrusion due to stresses on aquifers or changing land use is a serious problem in Australia, but is not considered in this study. Naturally occurring deleterious substances in groundwater such as nitrate and fluoride, which are common in Australian arid-zone shallow aquifers, are also not considered in the study.

Of the documented incidents, some 38 are described as diffuse and 106 are described as pointsource incidents. Most of the diffuse sources are sewage and agricultural fertilisers and are defined by plumes of high nitrate concentration.

An analysis of contaminant sources for all incidents shows the following distribution:

Contaminant source	Number of incidents
Industrial effluent	46
Sewage	26
Landfill leachate	16
Petroleum products	16
Mining	15
Agriculture	14
Food processing waste	11

This distribution is illustrated graphically in Figure 1. Several of the industrial effluent incidents are directly related to the chemical industry. Contamination from sewage is generally from unsewered suburban areas with a large number of septic tanks. Contamination from petroleum products is not commonly reported because of possible litigation, and there must be many more incidents than the 16 that are documented herein. Incidents relating to mining reflect the legacy of the past century of uncontrolled mining which caused pollution of both surface water and groundwater prior to the introduction of environment protection legislation in the

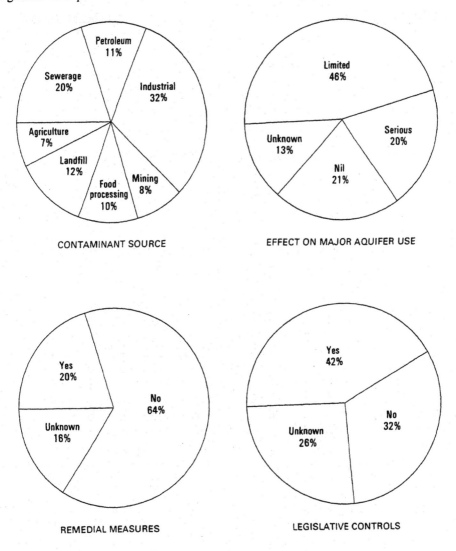

Figure 1 *Australian groundwater contamination incidents: contaminant source, effect on major aquifer use, remedial measures, and legislative controls.*

1970s. In some cases this has required massive publicly funded rehabilitation works, for instance at the former base metals mine at Captains Flat, New South Wales (Jacobson and Sparksman,1988) and at the former uranium and base metals mine at Rum Jungle, Northern Territory (Northern Territory Department of Mines and Energy, 1986).

The major regional aquifers affected, which are all unconfined or semiconfined, are:

Aquifer	Number of incidents
Superficial formations, Perth Basin	27
Gambier Limestone and equivalents, Otway Basin	21
Fyansford Formation and Brighton Group, Port Phillip Basin	11
Cainozoic volcanics, Melbourne	10
Fractured Silurian rocks, Canberra	5

The locations of these important regional aquifers are shown in Figure 2. The remaining 70 incidents are distributed among several other aquifers. The main problem areas, and the increasingly vulnerable areas, are shallow unconfined aquifers which underlie regions of intensive urban, industrial or agricultural development.

The majority of the contamination incidents (101) are described as being continuous over a period of years. Six incidents are described as having occurred once only and five incidents comprise several discrete occurrences. The time frame for groundwater contamination ranges back to the early years of the century, although most incidents are post 1945. Some 28 incidents appear to have started or been discovered in the 1970s and 19 started in the 1980s. The 1970s saw the introduction of environmental legislation in Australia, including licensing of waste disposal, and the growth of public and governmental awareness of water pollution. Clearly also the increasing number of reported cases relates to increasing urban and industrial development.

A total of 57 incidents are described as still occurring, that is the pollution source is still active or leachate is still being generated. Some 49 incidents no longer have active sources, and the situation is unknown in the other 38 cases owing to the lack of surveillance or documentation.

Information on the areal extent of the pollution plume is available for 40 incidents. The most extensive areas of contamination are the diffuse nitrate plumes in the Perth metropolitan area (Figure 3), southeast South Australia (Figure 4), and the Nepean Peninsula, Victoria. Information on the volume of contaminated groundwater is only available for 12 incidents.

A total of 26 cases are described as having serious effects on water use. In general these cases affect drinking water supplies, or have intractable pollutants, or have deleterious effects on surface waters. Another 60 cases are described as having limited effects on water use; 25 cases apparently have no effects on water use; and the effect on water use is unknown in 33 cases (Figure 1).

Remedial measures are known to have been undertaken in 35 of the listed incidents. Where the type of remediation is known it can be categorized as follows:

Remedial measures	Number of incidents
Groundwater recovery and treatment, disposal or reuse	12
Closure of site or changed effluent disposal practice	7
Removal of contaminated ground	2
Elimination of source	2

This is illustrated in Figure 1. The success of remedial measures is described as great in 7 cases, moderate in 10 cases and limited in 3 cases. In the other 15 cases the effectiveness of the remedial measures is unknown. In the majority of documented incidents (109), no remedial measures have been undertaken and this includes several cases described as having serious effects on water use.

Legislative controls apparently pertain in 50 of the documented incidents but have not always been activated. In 38 cases no legislative controls are applicable; this includes some historic contamination incidents that occurred prior to the introduction of legislation. In some 56 cases the situation with regard to legislative controls is unknown or obscure (Figure1). The legislative and administrative framework for control of groundwater contamination is different for each State and Territory in Australia; in some States institutional responsibility is dispersed among several agencies (Australian Water Resources Council, 1990). Administrative controls, including waste disposal licensing and septic tank regulations, appear to be relevant in 42 of the groundwater contamination cases in the inventory.

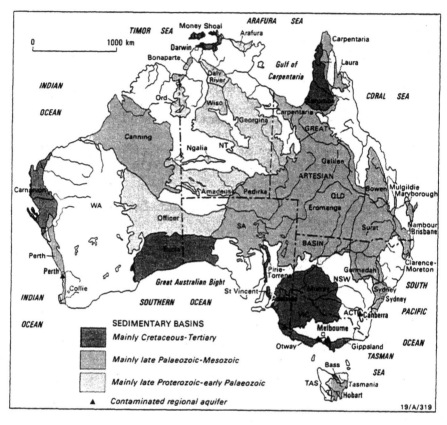

Figure 2 *Australia: major groundwater systems and location of affected aquifers.*

Current monitoring systems are reported for 68 out of the 144 incidents. The monitoring agencies include State and local Government agencies and industry. The frequency of monitoring, and the number of parameters monitored, varies considerably. Significantly, 76 cases are apparently not monitored and these include 11 cases that are described as having serious effects on water use.

Pollution of regional aquifers

Quaternary of the Perth Basin

The Perth Basin (Figure3) contains shallow aquifers which supply about 60% of the city's water supply needs, and also have an important role in the maintenance of wetlands. The population of the city is about one million. The shallow aquifers are mainly Quaternary sands with low absorptive capacity and rapid infiltration rates. Groundwater recharge averages about 12% of the annual rainfall of 870 mm although this proportion varies locally with vegetation type (Smith and Allen, 1987).

Nearly half of the Perth metropolitan area is unsewered, and this part relies on septic tanks. High nitrate concentrations in groundwater are causing algal blooms in wetlands (Appleyard and Bawden, 1987). The nitrogen input to groundwater in the unsewered area is estimated as 260 kg N ha^{-1}yr^{-1}, about three times that of sewered

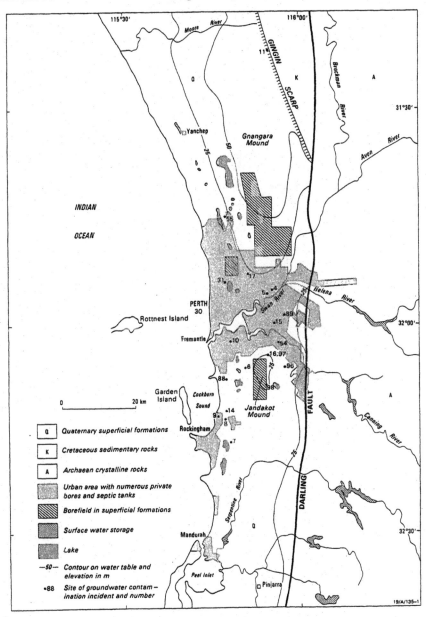

Figure 3 *Groundwater contamination incidents in the Perth Basin, Western Australia.*

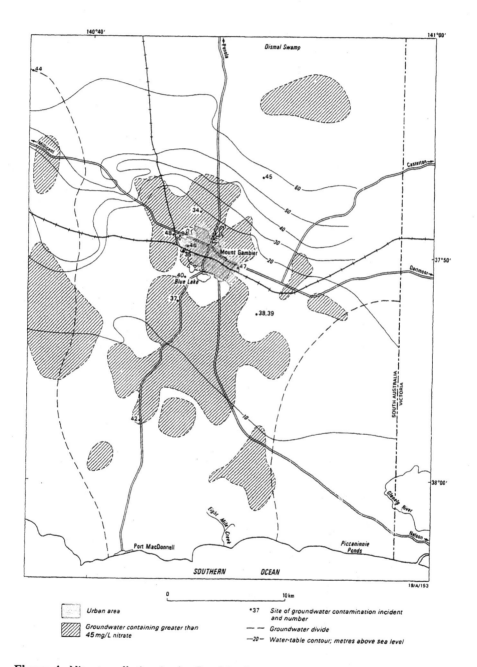

Figure 4 *Nitrate pollution in the Gambier Limestone, South Australia (modified after Waterhouse, 1977).*

ares (Gerritse *et al.*, 1989). The nitrate concentrations in water bores beneath the unsewered areas are 25-80 mg L^{-1} NO$_3$, compared with less than 14 mg L^{-1} NO$_3$ beneath the sewered areas; the limit for drinking water is 45 mg L^{-1}. There is also significant and widespread bacterial contamination of groundwater beneath the urban area (Parker *et al.*, 1981).

About 30 incidents of point source contamination of groundwater have been documented in the Perth basin, but there are believed to be 700 potential sources (Hirschberg, 1989). The industrial area at Kwinana, just south of Perth, has several known pollution incidents, and coalescent pollution plumes cover an area totalling several tens of square kilometres.

A case study of the recovery of highly polluted groundwater in the southern Perth Basin was reported by Whincup *et al.* (1987). At this location effluent from a mineral sands processing plant polluted the unconfined aquifer leading to high sulphate levels in domestic bores and to iron staining in a nearby estuary. Contamination originated from failure of a sulphuric acid storage tank, and leakage from effluent storage and drainage facilities. The total area of contamination was 55 ha, underlain by 0.56 million m^3 of contaminated groundwater and 2,280 t of 100% acid equivalent. Contaminated domestic water supply bores were replaced by deeper artesian bores. Production bores recover 1,800-2,000 m^3 day^{-1} of contaminated groundwater over a 10-year program. Disposal of this water is by several methods: part is included with the factory process water stream; part is discharged to calcareous sand dunes; and the balance is chemically neutralised before reinfiltration to the surficial aquifer.

Gambier Limestone, South Australia

The hydrogeology of the Mt Gambier area, in southeast South Australia, is shown in Figure 4. The Gambier Limestone is a karstic mid-Tertiary unit in the Otway Basin sequence, and forms an unconfined aquifer throughout this area. It is up to 300 m thick, and regional groundwater flow is coastwards, from north to south (Waterhouse, 1977; Forth, 1981). Recharge is about 125 mm yr^{-1} from rainfall of 750 mm yr^{-1} (Allison and Hughes, 1978). The Gambier Limestone supplies water to many small towns and farms; and it overlies confined early Tertiary sand aquifers.

There are several known pollution plumes from point sources in the Gambier Limestone. Smith and Schrale (1982) described a nitrate plume emanating from a cheese factory at Millel, South Australia. For 40 years, until comprehensive new water legislation in 1976, factory waste was discharged underground through a nearby drainage well. Severe pollution of the Gambier Limestone aquifer resulted in relocation of the factory supply wells, and then construction of a much deeper new well to the confined aquifer. Ultimately the plume threatened the town of Mt Gambier, 7 km down-gradient, which relies on groundwater, largely from the unconfined aquifer, for its reticulated supply. The plume contained 0.38 million m^3 of contaminated groundwater, delineated by resistivity survey and monitoring, but the total amount of soluble contaminants was only a small fraction of the total waste discharged underground. Smith and Schrale (1982) suggested that the remainder may have accumulated near the disposal well and be available for remobilization in the future. They proposed recovery of contaminated groundwater by strategically placed irrigation wells.

Recent assessment of diffuse groundwater pollution in southeast South Australia by Dillon (1988) indicates that 27% of this region has nitrate concentrations greater than 45 mg L^{-1}, and that this is the effect of a century of grazing. The main source of nitrate is atmospheric nitrogen fixed by clovers. Infiltration times through the approximately 10 m thick unsaturated zone are 20-30 years.

Newer Volcanics, Melbourne
The hydrogeology of the Melbourne metropolitan area is shown in Figure 5. The population of the city is about 2.5 million. The Newer Volcanics are Quaternary lavas which form plains underlying the western industrial suburbs of the city. They are up to 60 m thick and overlie confined Tertiary sand aquifers. Although Melbourne's reticulated water supply is from surface storages, the volcanics provided a supplementary source for industry. Pollution of the aquifer, largely by dumping of industrial waste in disused quarries, has forced industry to obtain its supplies from the deeper confined sand aquifer.

The basalt and scoria which comprise the Newer Volcanics are closely jointed, with a high though variable permeability (Kenley and Hancock, 1967; Stepan *et al.*, 1981). Recharge is by direct infiltration, and discharge is mainly into surface streams with the remainder directly into Port Phillip Bay (Riha and Kenley, 1978). Several pollution incidents have been documented, particularly in the highly industrialized area between Kororoit Creek and the Maribyrnong River (Leonard, 1982). An example is the disposal of phenolic wastes into an abandoned quarry documented by Riha (1977). This produced a plume of contaminated groundwater 1 km wide, probably extending down-gradient as far as Port Phillip Bay 4.5 km away. After installation of an above-ground treatment plant and a lined sludge pond in 1974 the recharge mound in the quarry declined and there was a corresponding decrease in phenols in groundwater from 608 mg L^{-1} in 1971 to 463 mg L^{-1} in 1974.

Tertiary sands, Melbourne
Upper Tertiary sands of the Fyansford Formation and the Brighton Group underlie the southeast suburbs of Melbourne and also occur near Geelong to the southwest (Figure 5). These deposits are up to 200 m thick and contain aquifers used as a source of water for irrigation and industry. Groundwater pollution from the disposal of wastes in sand pits is a problem on a local level. Regionally, the porous sandy aquifer is capable of attenuating much of the groundwater pollution. However most of the pollution sources are in an area of good quality groundwater which is heavily used (Leonard, 1982).

Fractured Silurian rocks, Canberra
The hydrogeology of urban Canberra, which has a population of about 300,000, is shown in Figure 6. Quaternary alluvium a few metres thick overlies deeply weathered and fractured Silurian mudstone and some limestone in this area. An

ephemeral perched alluvial aquifer overlies a semi-confined fractured bedrock aquifer. Groundwater flows to the south towards an artificial lake. Groundwater recharge occurs through stormwater drains in this paved urban setting, with rapid infiltration after rainfall events. The annual rainfall averages 620 mm.

The affected area of about 20 ha contains more than 20 service stations with a total of about 60 underground gasoline tanks. The tanks are set 3-4 m below ground and are bedded in sand. Pollution from spilled or leaked gasoline has been observed at two localities (Figure 6). At the Center Cinema, a fatal explosion and fire occurred in a basement in 1977 (Jacobson, 1983). This was caused by seepage of gasoline from an up-gradient service station. Remedial action was undertaken by pumping from a deep bore to depress the water table and induce flow of the contaminant to a skimming well. This was only partially successful.

Canberra is a planned city with areas zoned for particular uses. Groundwater pollution in the central city area occurs mainly because the zone of concentration of service stations and motor vehicle workshops has a shallow water table and buildings which extend below it. The fatal accident has led to a tightening of local regulations governing service stations and to the routine checking of underground tanks.

Discussion

Several important Australian regional aquifers are affected by groundwater pollution. These are shallow, unconfined to semi-confined aquifers which underlie sites of urban, industrial or agricultural development. These regional aquifers require vulnerability mapping, pollution transport modelling, and effective monitoring; and the development of groundwater protection strategies to reduce contaminant sources and ameliorate existing pollution. Elsewhere in Australia groundwater pollution is a local scale problem. The effect is to reduce the actual or potential use of groundwater resources and to put at risk surface water bodies located downgradient from polluted groundwater.

The number of point sources of pollution, particularly liquid wastes, has been substantially reduced by administrative controls. Nevertheless, there is a need for improvement in the selection, design and operation of industrial and domestic waste disposal depots, and the prevention of spills. The existing pollution problem, a legacy of past mistakes, is being addressed to varying degrees by State agencies. Vulnerability mapping has been undertaken on a broad scale for one state (New South Wales Department of Water Resources, 1987), and is now being undertaken for the Perth Basin (Appleyard, 1990).

The extent of groundwater contamination in Australia that is documented in this chapter must be regarded as only a partial statement of the problem. Many of the known incidents were discovered by accident and there has been little systematic investigation and monitoring of likely pollution sites. An unknown number of landfills, and districts serviced by septic tanks, are discharging effluent to shallow aquifers. Groundwater contamination is difficult to detect as contaminants are often colourless and the parameters of pollution are not always well defined nor revealed by the onetime only standard inorganic analysis. In many Australian groundwater

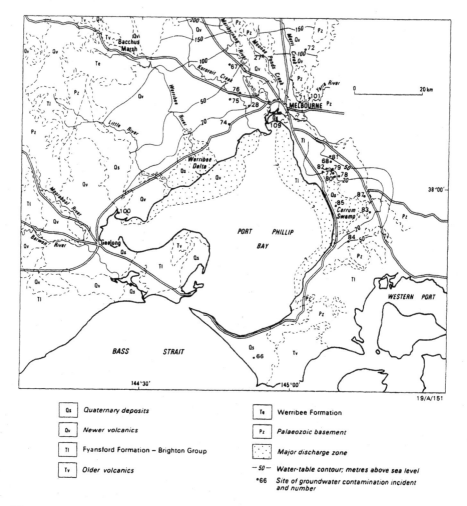

Figure 5 *Groundwater contamination incidents in regional aquifers near Melbourne (modified after Leonard, 1979).*

systems, velocities are low and there may be a long time lag before contamination is detected.

Remedial works for groundwater pollution are costly, timeconsuming and not always successful. Clearly, prevention is better than cure. Groundwater protection programmes are needed in Australia and an important aspect is to provide economic incentives to reduce contaminant sources. Hydrogeological criteria for waste treatment and disposal sites need to be developed.

Little information is available on non-point source pollution of groundwater in Australia and an evaluation is needed which includes pollution from sewage, land

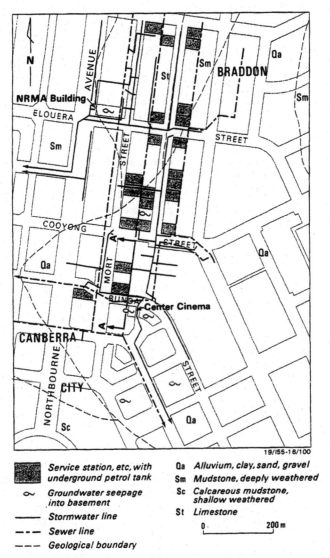

Figure 6 *Hydrogeology of central Canberra.*

use, herbicides and pesticides. Apparently only two studies of groundwater contamination by pesticides have been documented in Australia: in the agricultural Burdekin Delta, Queensland (Brodie *et al.*, 1984); and in urban Perth (Gerritse *et al.*,1990).

Australia has three tiers of government - federal, state and local - and control of groundwater is constitutionally a State government responsibility. All Australian States have groundwater protection laws but they vary and are unevenly implemented. The organizational and administrative structure is different in every State and the control of groundwater contamination is institutionally fragmented in some states. State water legislation was reviewed by Clark (1980) and the regulatory framework for groundwater protection has been reviewed recently by the Australian Water Resources Council (1990). Several states are presently developing groundwater protection strategies (Lane, 1990).

Groundwater constitutes about 15% of Australia's water use. By comparison, groundwater constitutes at least 25% of water use in the United States which has 15 times the population of Australia. In the United States it has been estimated that 12% of usable groundwater is polluted (Pye and Patrick, 1983). Australia's present groundwater pollution problem is thus considered minor compared with that of the United States. Nevertheless, Australia's aridity and growing dependence on groundwater makes conservation and protection of its stored groundwater resources imperative. These resources, including the brackish aquifers, will ultimately be developed to their maximum extent.

Conclusions

(1) A total of 144 groundwater contamination incidents are documented for Australia, with the reservation that there may be many other undiscovered incidents. A range of contaminant sources is involved, especially industrial effluent, sewage and landfill leachate.

(2) The main regional aquifers affected are the Quaternary Superficial Formations of the Perth Basin; the Gambier Limestone and equivalents, in the Otway Basin; the Fyansford Formation and Brighton Group in the Port Phillip Basin; the Newer Volcanics in Melbourne; and fractured Silurian rocks in Canberra. These shallow, unconfined to semiconfined aquifers underlie regions of intensive urban, industrial or agricultural development.

(3) Remedial measures have been undertaken in 35 of the documented cases; the most common measures involve groundwater recovery and treatment, disposal or reuse. The remedial action has been effective in 17 of these cases.

(4) A range of State legislation is applicable to the control and management of pointsource groundwater contamination but is unevenly implemented. Legislation applicable to nonpoint source pollution of groundwater is obscure or nonexistent.

(5) Surveillance of the reported incidents is uneven and the majority of known cases are not currently monitored, including 11 cases that have serious effects on water use.

Acknowledgements

We thank John Bauld and Ray Evans of the Australian Geological Survey organisation for comments on the manuscript. The paper is published by permission of the Executive Director of the Australian Geological Survey Organisation.

References

Allison, G.B. and Hughes, M.W. 1978. The use of environmental chloride and tritium to estimate total recharge to an unconfined aquifer. *Aust. J. Soil Res.*, 16, 181-195.

Appleyard, S.J. 1990. The Perth Basin groundwater contamination vulnerability map. In: Proceedings of the International Conference on Groundwater in Large Sedimentary Basins, pp.267-277. Perth, 1990. *Aust. Wat. Resour. Council, Conf. Ser.* 20.

Appleyard, S.J. and Bawden, J. 1987. The effects of urbanization on nutrient levels in the unconfined aquifer underlying Perth, Western Australia. In: Proceedings of the International Conference on Groundwater Systems under Stress, Brisbane, 1986. *Aust. Wat.Resour.Council Conf. Ser.* 13, 587-594.

Australian Water Resources Council, 1990. The status of groundwater contamination and regulation in Australia. *Aust. Wat. Resour. Council Man. Comm. Occ. Pap.* 1.

Bestow, T.T. 1981. The influence of sanitary landfill on groundwater quality at Hertha Road, Stirling. In: Whelan, B.R. (ed.). Proceedings of the Symposium 'Groundwater resources of the Swan coastal plain', Perth, 1981. Comm.Sci.Ind.Res.Org.Aust., pp.295-312.

Brodie, J.E., Hicks, W.S., Richards, G.N. and Thomas, F.G. 1984. Residues related to agricultural chemicals in the groundwaters of the Burdekin Delta, North Queensland. *Env. Poll. (Ser.B)*, 8, 187-215.

Clark, S.D. 1980. *Groundwater law and administration in Australia.* Aust. Wat. Resour. Council Tech. Pap. 44.

Dillon, P.J. 1988. An evaluation of the sources of nitrate in groundwater near Mt Gambier, South Australia. *Comm. Sci. Ind. Res. Org. Aust. Wat. Res. Ser.* 1.

Forth, J.R. 1981. Modelling of nitrate transport in a regional groundwater system in South Australia. In: Lawrence, C.R. and Hughes, R.J. (eds.), Proceedings of the Groundwater Pollution Conference, Perth, 1979. *Aust. Wat. Resour. Council Conf. Ser.* 1, 101-118.

Gerritse, R.G., Barber, C. and Adeney, J.A. 1989. The impact of residential urban areas on groundwater quality: Swan coastal plain, Western Australia. *Comm. Sci Ind. Res. Org. Aust. Wat. Res. Ser.* 3.

Harvey, P.D. 1983. Groundwater pollution related to land use in a karst area. In Papers of the International Conference on Groundwater and Man, Sydney, 1983. *Aust. Wat. Resour. Council Conf. Ser.* 8, 2, 131-142.

Hirschberg, K.J. B. 1989. Groundwater contamination in the Perth Metropolitan region.In: Lowe,G. (ed.). *Proceedings, Swan coastal plain groundwater management conference,* Perth, 1988. West. Aust. Wat. Resour. Counc. 1/89

Jacobson, G. 1983. Pollution of a fractured rock aquifer by petrol a case study. *Bur. Miner. Resour. J. Aust. Geol. Geoph.*, 8, 313-322.

Jacobson, G. and Lau, J.E. 1988. Groundwater contamination incidents in Australia:an initial survey. *Bur. Miner. Resour. Aust. Rep.* 287.

Jacobson, G. and Sparksman, G.F. 1988. Acid mine drainage at Captains Flat, New South Wales. *Bur. Miner. Resour. J. Aust. Geol. Geoph.*, 10, 391-393.

Jacobson, G., Habermehl, M.A. and Lau, J.E. 1983. Australia's groundwater resources. *Aust.Dept.Resour.Energy. Water 2000 Consult. Rep.*2.

Kenley, P.R. and Hancock, J.S. 1967. The underground water resources.In Geology of the Melbourne district. *Bull. Geol. Surv. Vict.*, 59, 79-82.

Knight, M.J. 1983. Modelling of leachate discharged from a domestic solid waste landfill at Lucas Heights, Sydney, Australia. In Papers of the International Conference on Groundwater and Man, Sydney, 1983. *Aust. Wat. Resour. Council Conf. Ser.* 8, 2, 219-230.

Lane, A.P. 1990. Groundwater pollution-the challenge. *Wat. Miner. Dev.*, 9, 3, 7-12.

Lawrence, C.R.and Hughes, R.J. 1981. Proceedings of the groundwater pollution conference, Perth. 1979. *Aust. Wat. Resour. Council Conf. Ser.*1.

Leonard, J.G. 1979. Preliminary assessment of the groundwater resources in the Port Phillip region. *Geol. Surv. Vict. Rep.* 66.

Leonard, J.G. 1982. Effects to date of uses of groundwaters and/or aquifers in the Port Phillip catchment. *Geol. Surv. Vict. Rep.* 1982/21.

New South Wales Department of Water Resources, 1987. *Groundwater in New South Wales. Assessment of pollution risk.* Map, scale 1:2 000 000.

Northern *Territory Department of Mines and Energy, 1986.* The Rum Jungle rehabilitation project. Final project report.

Parker, W.F., Carbon, B.A. and Grubb, W.B. 1981. Coliform bacteria in sandy soils beneath septic tank sites in Perth, Western Australia. In Proceedings of the Groundwater Pollution Conference, Perth. 1979. *Aust. Wat. Resour. Council Conf. Ser.* 1, 402-414.

Pye, V.I. and Patrick,R. 1983. Groundwater contamination in the United States. *Science,* 221, 713-718.

Riha, M. 1977. Hydrochemical effects of waste percolation on groundwater in basalt near Footscray, Victoria, Australia. *Prog. Wat. Technol.* ,9, 249-266.

Riha, M. and Kenley, P.R. 1978. Investigation of the hydrogeology and groundwater pollution in the basalt aquifers west of Melbourne. *Geol. Surv. Vict. Rep.* 1978/40.

Shugg, A. 1987. Leachate burst phenomena. In Proceedings of the International Conference on Groundwater Systems under Stress, Brisbane, 1986. Aust. Wat. Resour. Council Conf. Ser. 13, 61-71.

Smith, P.C. and Schrale, G. 1982. Proposed rehabilitation of an aquifer contaminated with cheese factory wastes. Water, 9,21-24.

Smith, R.A. and Allan, A.D. 1987.The unconfined aquifer and effects on urbanization, Perth, W.A. In: Proceedings of the International Conference on Groundwater Systems under Stress, Brisbane, 1986. *Aust. Wat. Resour. Council Conf. Ser.,* 13, 575-585.

Stepan, S., Smith, J.R. and Riha, M.1981. Movement and chemical change of organic pollutants in an aquifer. In: Lawrence, C.R. and Hughes, R.J. (eds.), Proceedings of the Groundwater Pollution Conference, Perth, 1979. *Aust. Wat. Resour. Council Conf. Ser.,* 1, 415-424.

Waterhouse, J.D. 1977. *The hydrogeology of the Mount Gambier area.* Geological Survey of South Australia, Report of Investigations 48.

Whincup, P., Bibby, P.A. and Chandler, M.S. 1987. Investigation and recovery of groundwater contamination of SCM Chemicals, Australind, Western Australia. In Proceedings of the International Conference on Groundwater Systems under Stress, Brisbane, 1986. *Aust. Wat. Resour. Council Conf. Ser.* 13, 265-275.

14 Groundwater contamination from municipal landfills in the USA

W. R. Roy

Illinois State Geological Survey, 615 East Peabody Drive, Champaign, Illinois 61820, USA

Abstract

New municipal landfills have become very difficult to site because of public opposition. The lack of new landfills together with the diminishing capacity of landfills now in operation could lead to a major crisis in the United States in 10 years. A major reason for public opposition to new landfills is the fear of groundwater contamination from landfill leachate. However, a review of case studies revealed few serious incidents, particularly among more recent landfills. Also, new federal regulations on landfill operation and design are expected to result in more strictly controlled landfills. The US Environmental Protection Agency concluded that if properly located, designed, and operated, future municipal solid waste landfills will be protective of human health and the environment.

Groundwater contamination by municipal landfills has been an issue for decades. The inorganic and organic chemical composition of landfill leachate and gas is quite variable and is dependent on the age of the landfill, how it was operated, and on the type of wastes present. Pathogens in leachate are not thought to be a major problem. About half of the volatile organic compounds reported in leachate have either a medium or low mobility in groundwater and the other half are classified as highly or very highly mobile. Some of these compounds could readily volatilize from leachate in the unsaturated zone. Chemical oxidation and hydrolysis of volatile organic compounds in landfill leachate may not be significant. Many organic compounds may biodegrade, but a few appear to be resistant.

Although case histories were found where leachate from municipal landfills resulted in serious groundwater contamination, the majority of active landfills did not have any type of groundwater monitoring program. Most problem landfills in the past were located in geologically unsuitable areas, were constructed without an effective liner, or were poorly operated.

Proposed federal regulations mandate the use of soil liners, leachate collection systems, final covers, and other requirements. However, very little documentation is available from which to evaluate the efficacy of these requirements. Soil and plastic liners can potentially control leachate movement, but little research has been done on field-scale soil liners, and there is little actual experience with plastic liners.

In the future, leachate collection and recirculation, travel-time calculations in liner designs, and site impact assessments may be implemented. Short-term remedies to the landfill crisis include on-site expansions, landfill mining, and possibly large regional megafills.

Introduction

In the past, municipal solid wastes have been burned, buried, and dumped indiscriminately in piles and into sinkholes, quarries, coal mines, rivers, lakes, and the ocean. Groundwater contamination resulting from such practices has been recognized for years. For example, Rossler (1950) reported that in 1923, elevated

Capacity problems, public opposition, concerns about the availability of suitable land, and permitting delays

No major capacity problems at present, but concerns about the availability of suitable land, public opposition, or economic problems

No major capacity problems at present or anticipated

?   No relevant information provided or unknown

Figure 1 *Statewide assessment of aggregate municipal landfill capacity based on information in USEPA (1988b) and updated information provided by various state agencies.*

concentrations of hardness (Ca + Mg), Cl, and SO_4 were detected in wells that were 1 km from a refuse site established in 1913. In 1927, contamination was detected 7.6 km away from the site. In 1932, the groundwater beneath an impounding pit had been contaminated with chloride, hardness, and dissolved solids by a "garbage liquor" (Calvert, 1932).

In 1986, about 143,000,000 metric tons of municipal solid wastes were generated in the United States, and about 83% of the wastes were placed into landfills (USEPA, 1988b). A typical community of 500,000 that now produces 1,815 metric tons of trash per day may require a 97-hectare landfill by the year 2010 (Birks, 1989).

Land disposal of municipal solid wastes has been a major environmental and socioeconomic issue for decades. During the first half of this century, "sanitary landfills" were welcomed improvements over the then familiar trash dumps (see McBride, 1958). In the 1960s, the public's reaction to municipal landfills turned negative, at times hostile. The potential for landfills to contaminate groundwater supplies became a major concern to the public. Today because of public opposition to landfills, new facilities have become very difficult to site (Figure 1). Compounding the siting problem is the expensive, time-consuming nature of the permitting process, as well as the lack of suitable land in some areas.

In 1984, between 6,000 and 9,000 municipal landfills existed in the United States (USEPA, 1988b). The exact number is unknown because "municipal landfill" is defined differently throughout the states. Nationally, very few new landfills have opened during the last 5 years, and more than half of the existing facilities are older than 15 years (USEPA, 1988b). The number of landfill-expansion approvals remained fairly constant during the 1980s, but the number of approvals for new sites decreased substantially.

The diminishing aggregate capacity of municipal landfills could lead to a major crisis in the United States by the turn of the century. The USEPA (1988b) estimated that 71% of the nation's landfills have 15 years or less of remaining capacity. Most landfills are designed to be in use for about 20 years, but in many states, filled landfills are not being replaced with new ones at a rate that will meet society's waste disposal needs. Estimates on a statewide basis vary. For example, LaMoreaux et al. (1988) estimated that about 80% of the currently existing landfills in Alabama will be full by 2003. About 72% of the current landfills in Iowa may be full by 2008 (Iowa Department of Natural Resources, 1988). In 1985, the New Jersey Department of Environmental Protection (1985) estimated that the aggregate landfill capacity of that state would not be exceeded until 1991, but it was exceeded sooner, and refuse is currently being exported to nearby states. The remaining capacity in Illinois may last until the late 1990s (Illinois Environmental Protection Agency, 1990); of the state's 102 counties, 24 have no landfills. The aggregate landfill capacity of New York may be exhausted in less than 10 years (Overcast and Heintz, 1990). This trend of diminishing capacity has been called the "vanishing-landfill crisis."

This chapter reviews six broad topics concerning groundwater contamination by municipal landfills: composition of municipal solid wastes, formation and chemical composition of landfill leachate and gas, environmental fate of landfill leachate, case studies on groundwater contamination, minimizing groundwater contamination, and

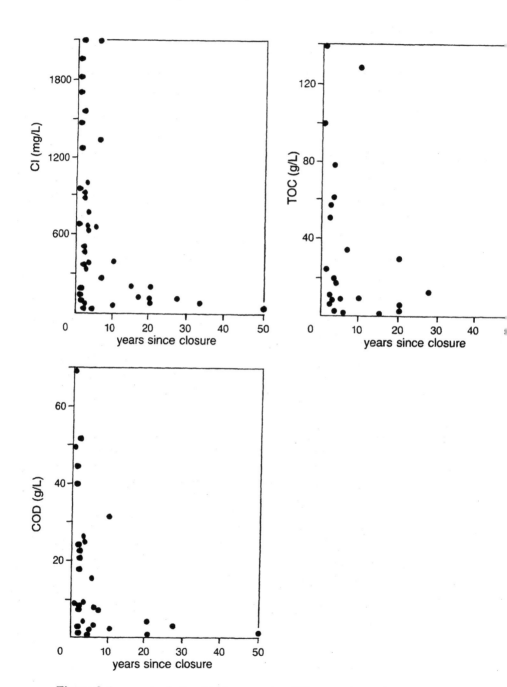

Figure 2 *Apparent relationship between landfill age and the Cl, chemical oxygen demand, and total organic carbon content of leachate (adapted from Lu et al., 1985).*

Table 1 *Inorganic chemical composition of municipal landfill leachate (concentrations are in mg L^{-1} unless noted).*

	Range in literature[a]		Median detected concentrations				
	Low	High	Clark and Piskin (1977)	USEPA (1988b)[b]		Sridharan and Didier (1988)	
Parameter[c]							
pH	1.5	12.5	6.8	6.70	$(1,263)^d$	6.69	$(59)^d$
Eh (mV)	-180	+804	ND[e]	ND	ND	+481	(6)
EC (mmhos cm^{-1})	0.01	990	6.1	5.11	(1,132)	5.60	(55)
Alkalinity[f]	0	57,850	1,225	2,300	(786)	2,650	(29)
Total hardness[f]	0	225,000	1,600	2,000	(999)	1,665	(26)
TDS	0	594,000	5,346	1,396	(172)	4,890	(28)
BOD	1	72,000	1,500	1,710	(777)	2,310	(33)
COD	1.3	97,900	4,490	1,823	(949)	2,800	(52)
TOC	0.3	45,100	ND	1,439	(226)	1,000	(33)
Temperature (°C)	2	122	ND	12	(55)	11	(6)
Constituent							
Al		85	ND	0.60	(5)	2.4	(7)
As		70.2	0	19µg L^{-1}	(83)	13.5µg L^{-1}	(36)
B		133	4.70	4.20	(97)	4.00	(8)
Ba		33.3	2.25	0.32	(59)	0.58	(36)
Be		0.24	ND	0.24	(1)	5µg L^{-1}	(6)
Ca		4,080	430	354	(78)	320	(19)
Cd		17	30µg L^{-1}	20µg L^{-1}	(67)	15.5µg L^{-1}	(31)
Cl		27,100	562	486	(891)	594	(52)
Co		3.40	ND	3.40	(1)	0.08	(2)
Cr		22.5	0.05	0.100	(100)	0.06	(43)
Cu		1,100	0.05	0.05	(76)	0.05	(33)
F		302	0.40	0.36	(37)	0.39	(18)
Fe		57,300	138	73.5	(518)	95.0	(55)
Hg		160µgL^{-1}	3µg L^{-1}	1g L^{-1}	(25)	0.6µg L^{-1}	(16)
K		3,770	150	100	(77)	382	(19)
Mg		15,600	200	91.0	(56)	136	(18)
Mn		1,400	9.20	1.31	(93)	3.7	(43)
Na		8,000	357	360	(251)	693	(37)
Ni		60.0	0.20	0.26	(92)	0.17	(37)
NO$_3$		250	0.10	2.05	(52)	0.22	(31)
P		154	0.59	1.81	(208)	1.4	(14)
Pb		6.60	0.10	0.10	(99)	0.06	(45)
S^{2-}		125.0	ND	1.45	(15)	ND	
Se		1.02	0	14.9g L^{-1}	(28)	20µg L^{-1}	(17)
Si		34.0	15.4	34.0	(1)	ND	
SO$_4$		84,000	153	90.0	(573)	111	(39)
Zn		250	1.7	0.89	(108)	0.68	(50)

[a]Clark and Piskin (1977), James (1977), Fuller (1978), Griffin and Shimp (1978), Kmet and McGinley (1982), Glynn (1985), USEPA (1988b), Sridharan and Didier (1988).
[b]Compiled from 14 sources.
[c]Abbreviations: Eh, oxidation-reduction potential; EC, exchange capacity; TDS, total dissolved solids; BOD, biochemical oxygen demand, COD, chemical oxygen demand, TOC, total organic carbon.
[d]Number of samples where detected. [e]ND, not determined. [f]mg L^{-1} as CaCO$_3$.

future expectations and research. This chapter is intended to be a concise and comprehensive treatise of the topic, a summary of the current thinking of various researchers and regulators in the United States, and a discussion of the technical basis for these views. The impetus for writing this chapter was the vanishing-landfill crisis, which has been fueled in part by concerns about groundwater contamination.

Composition of municipal refuse

The many sources of municipal solid wastes results in a heterogeneous composition. Regional differences in climate, season, and socioeconomic factors contribute to this variability (USEPA, 1988b). Components of garbage have been classified in studies, but these studies often are not comparable because definitions of wastes differ from study to study, as do the measurement techniques used.

Municipal landfills typically receive household refuse and nonhazardous commercial wastes. They may also receive a limited amount of sewage sludge and industrial waste. Municipal wastes can be approximated (by weight) as 36% paper, 20% yard waste, 8% glass, 9% metals, 9% food wastes, 4% wood, 7% plastic, and 7% rubber, leather, textile, and miscellany (Franklin Associates, 1986, 1988). Paper

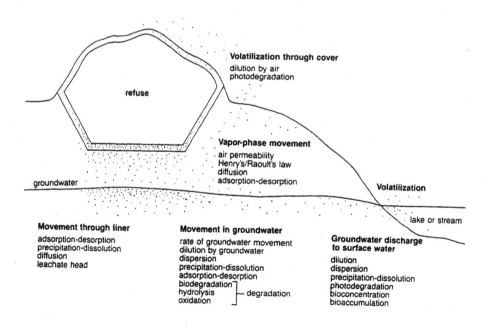

Figure 3 *Possible processes and landfill-subsurface properties that control the environmental fate of inorganic and organic contaminants released from a municipal landfill.*

products are the major component of municipal solid waste, varying from 25 to 61% of the total mass by weight (Mac Namara, 1971; Solid Wastes Management, 1972; Rovers and Farquhar, 1973; Pohland and Harper, 1986). Paper has been the major component of refuse since at least 1938 (Solid Wastes Management, 1972). The composition of the waste placed in landfills has changed over time, reflecting changes in technology and product packaging and marketing. Solid Wastes Management (1972) noted that since the 1930s, the quantity of putrid materials declined, as had ash derived from the domestic combustion of coal and wood; plastics and paper increased in quantity.

Hazardous wastes generated by households can also be a component of municipal wastes. The hazardous waste content of municipal solid waste has been estimated to range from 0.1 to 0.4% by weight (USEPA, 1988a). The quantity of household hazardous waste generated in 1986 in the United States could have been anywhere from 1,800 to 508,000 metric tons. Motor oil, paint, household maintenance items, batteries, and miscellaneous electrical items predominate the household hazardous waste stream (USEPA, 1988b).

Landfill leachate and gas

After refuse is placed into a landfill, it decays at a rate that depends on the composition of the waste, the operation of the landfill, and climatic and hydrogeological factors. The maturation of a landfill can be characterized as occurring in five phases, which may or may not occur simultaneously throughout the site (Pohland and Harper, 1986). In phase 1 (initial adjustment), the wastes are placed into the excavation, the site is closed, and the initial subsidence occurs. In phase 2 (transition), the infiltration of water, either through the landfill cap or from groundwater recharge, exceeds the field capacity of the refuse, and significant quantities of leachate begin to form. At the same time, the soil-chemical environment becomes more reduced, and the oxygen that was trapped in the interstitial voids is replaced by carbon dioxide. Intermediates of microbial degradation, such as volatile organic fatty acids, first appear in the leachate. In phase 3 (acid formation), the pH of the leachate decreases, and hydrogen may be first detected in the landfill gases. In phase 4 (methane fermentation), the intermediary products that appeared during phase 3 are converted into methane and excess carbon dioxide. The pH of the landfill environment increases in response to bicarbonate buffering, and the oxidation-reduction potentials (Eh) are at their lowest levels. In the final phase (final maturation), measurable methane production ceases, oxygen and oxidized species slowly return with a corresponding increase in Eh, and the organic materials that were not readily biodegradable may be slowly degraded.

The rate at which refuse decays is variable. Charnley *et al.* (1988) stated that no landfill has a single "age": each cell and cell section of a landfill exists at a unique stage of progression towards stabilization. Eliassen (1942) concluded that the optimum moisture content for the aerobic degradation of fresh refuse was between 50 and 70% by weight. On the basis of rates of gas generation, the overall half-life of

Table 2 *Organic chemical composition of municipal landfill leachate (concentrations are in g L1 unless noted).*

	Range in literature[a]		Median detected concentration		
	Low	High	Sridharan and Didier (1988)		USEPA (1988)[b]
Acetone	8.0	13.0mg L^{-1}	ND[c]		430
Benzene	1.4	1,630mg L^{-1}	ND		37
Bis(2-ethylhexyl) phthalate	34	7.9mg L^{-1}	1,050	(5/26)[d]	ND
Carbon tetrachloride	3	995	28	(3/50)	202
Chlorobenzene	1	685	25.2	(7/50)	7
Chloroethane	2	860	17	(11/50)	28
Chloroform	4	1,300	7.14	(7/50)	29
Cyanide	4	555mg L^{-1}	80	(63/91)	ND
Di-*n*-butyl phthalate	12	540	28.7	(4/26)	49
1,2-Dichlorobenzene	3	32	ND		12
1,4-Dichlorobenzene	1	250	14	(12/37)	7
1,1-Dichloroethane	BDL[e]	44mg L^{-1}	ND		165
1,2-Dichloroethane	BDL	11mg L^{-1}	ND		10
1,1-Dichloroethylene	BDL	110	ND		ND
1,2-Dichloroethylene	BDL	2,200	ND		330
Diethyl phthalate	3	330	44	(11/26)	83
Endrin	0.04	50	ND		0.25
Ethyl acetate	42	290	ND		86
Ethylbenzene	1	4,900	43.5	(30/52)	58.5
Isophorone	3.18	16mg L^{-1}	76	(13/26)	76
Lindane	0.017	0.023	ND		0.020
Methylene chloride	BDL	58.2mg L^{-1}	483	(25/41)	440
Methyl ethyl ketone	110	37mg L^{-1}	19.6mg L^{-1}	(2/11)	1.55mg L^{-1}
Methyl isobutyl ketone	10	740	ND		270
Naphthalene	4.6	202	33.8	(10/21)	12
Phenanthrene	8.1	1,220	50.7	(5/27)	ND
Phenol	1.1	28.8mg L^{-1}	174	(16/27)	378
Phenolic compounds	0.05	19mg L^{-1}	619	(120/128)	ND
1-Propanol	76	37mg L^{-1}	ND		11mg L^{-1}
2-Propanol	94	41mgL^{-1}	ND		8.45mg L^{-1}
Tannin + lignin	120	264mg L^{-1}	1.94 g L^{-1}	(48/52)	ND
Tetrachloroethylene	1	232	16.3	(10/52)	55.0
Tetrahydrofuran	18	1,400	730	(6/12)	260
Toluene	1	18mg L^{-1}	360	(42/53)	413
1,1,1-Trichloroethane	BDL	13mg L^{-1}	ND		86
Trichloroethylene	1	15mg L^{-1}	19	(12/53)	43
Vinyl chloride	8	3,000	230	(12/42)	40
Xylenes	2.5	320	72.5	(10/13)	71

[a] Kmet and McGinley (1982), Sabel and Clark (1983), Glynn (1985), Friedman (1988), Sridharan and Didier (1988), USEPA (1988b).
[b] Compiled from 13 sources; frequency of detection not reported.
[c] ND, not determined.
[d] Number of detections/total number of samples.
[e] BDL, below detection limit (no value given).

Table 3 *Typical composition of landfill gas (Lang* et al., *1987; USEPA, 1988b).*

Component	Percentage (dry volume basis)
Methane	44 – 60
Carbon dioxide	34 – 60
Nitrogen	2 – 21
Oxygen	0.05 – 1.7
Sulfides, mercaptans disulfides	0 – 1.0
Hydrogen	0 – 0.3
Carbon monoxide	0 – 0.2
Trace organic compounds	0.01 – 0.6

biodegradable-landfill refuse has been estimated at about 24 years (Schumacher, 1983; Rathje, 1990).

Exhumation studies have indicated that some degradable materials may last longer than previously assumed. Much of the organic waste, including grass clippings, bread, and corncobs, was still intact after 2 to 8 years of burial in a landfill in Colorado (Michaels, 1990). In two landfills, between 14 and 68% of the biodegradable material initially present had been transformed in about 8 years (Morelli, 1989): more material degraded at a Florida landfill, where infiltration may have enhanced the process, and less at a landfill in Arizona, where rainfall was much lower than in Florida. At another Arizona landfill, newspapers were still legible after 40 years of burial (Grossman and Shulman, 1990). The current practice of placing household garbage into disposable plastic bags may greatly reduce the rate of decomposition.

Landfill leachate forms primarily from rain water that infiltrates through the refuse. Leachate can mobilize inorganic and organic contaminants from a disposal site into groundwater. The natural dewatering of wet refuse during compaction and settling can also contribute to the volume of leachate. The microbial degradation of biodegradable organics can also generate water that will contribute to leachate generation (Lu *et al.*, 1985). Leachate may also be derived from recharge into the landfill cell if the refuse has been buried beneath the zone of saturation.

Landfill leachate derived from fresh municipal refuse is generally a mildly acidic (pH 5.0) aqueous solution that contains a high content of dissolved solids and biologically degradable organics (Fuller, 1978; USEPA, 1979). The range of hydrogen ion activity in leachate samples reported in the literature spans almost 11 orders of magnitude, and the inorganic chemical composition of landfill leachate has been quite variable (Table 1). The range of values in Table 1 reflects that the samples were collected from many different landfills that were in different climates and at different stages of stabilization and operated by different methods. The composition

Table 4 *Selected volatile organic compounds in landfill gases (from Young and Heasman, 1985; USEPA, 1988b; California Waste Management Board, 1988; air-sample data from LaRegina et al., 1986).*

Component	Range in conc. (mg m⁻³) in gas samples	Median conc.[a] (mg m⁻³)		Geometric mean conc. (μg m⁻³) in surface air samples
Benzene	0 – 127	1.0	(21)[b]	8.57
1,2-Dichloroethylene	0.07 – 28	—		—
Ethylbenzene	0 – 396	6.5	(14)	3.87
Methylene chloride	0 – 812	2.9	(17)	0.92
Trichloroethylene	1.2 – 175	0.66	(19)	0.98
1,1,1-Trichloroethane	0 – 13.3	0.17	(18)	—
Toluene	0 – 1,367	26.0	(20)	70.3
Vinyl chloride	0 – 83.2	5.72	(16)	—
Xylene	0 – 490	0.44	(6)	—

[a] Reported in USEPA (1988b).
[b] Number of samples.

of the waste and the analytical methods used to generate these values also contributed to their variability. Often, no information was available about the type of wastes buried at a given site. Some landfills reported in the literature accepted large but undocumented quantities of industrial hazardous wastes along with municipal refuse.

The first column (Clark and Piskin, 1977) of median values in Table 1 lists values for most of the constituents in leachate samples collected from 54 landfills in Illinois during the 1970s. The second column of median values was compiled by the USEPA (1988b) on the basis of leachate samples collected throughout the United States. Values in the last column, compiled by Sridharan and Didier (1988), were derived from data on 56 landfills in Wisconsin. The three studies were in good agreement, although the methods used to sample the leachate were often unknown.

A plethora of organic chemicals has also been detected in landfill leachate (Table 2). The major fraction of the organics present in leachate is thought to be free volatile fatty acids. Chian and DeWalle (1977) found that between about 20 and 50% of the total organic carbon content of leachate samples was composed of free volatile fatty acids, such as acetic, propionic, butyric, valeric, and formic acid. Other investigators (Burrows and Row, 1975; County of Sonoma, 1973; Mao and Pohland, 1973; Hughes *et al.*, 1971) reported that 25 to 80% of the chemical oxygen demand of leachate samples was attributable to fatty acids.

Municipal landfill leachate probably contains hundreds of different organic chemicals at some finite concentration. The specific organic composition of the leachate of course depends on the type of waste present. For example, toluene concentrations in leachate samples collected at a landfill in New Jersey were as high as 24.0 mg L⁻¹ (Glynn, 1985). This value was regarded as atypical and was not included in Table 2 because about 8 million liters of spent-toluene solvent had been placed into the site. The highest concentrations of 2-bis-dichloroethyl ether and toluene in leachate at the LiPari Landfill in New Jersey were 210 and 22.4 mg L^{-1}, respectively (Gominger and Lyon, 1981). However, about 46,000 55-gallon drums of industrial wastes, including spent solvents, were disposed along with municipal garbage.

The studies of Sabel and Clark (1983), Friedman (1988), and Sridharan and Didier (1988) suggested that the most common organic compounds include phenolics, tannins, lignins, and among the volatile organic compounds, benzene, 1,1-dichloroethylene, ethylbenzene, toluene, trichloroethylene, and xylene. Other commonly occurring compounds include chloroethane, methylene chloride, 1,1,1-trichloroethane, and tetrachloroethylene. Where possible, "typical" or median concentrations are given for these compounds in Table 2.

Little information is available on how long a landfill generates a leachate that is distinctly different from the groundwater. In some studies (Baedecker and Apgar, 1984), leachate did not migrate from the landfill at a continuous rate, but rather as pulses. The duration of leachate generation depends on the composition of the refuse, design and operation of the landfill, and climate, but the limited information available suggests landfill stabilization after closure may be on the order of 15 to 40 years (Miller *et al.*, 1974; Chian and DeWalle, 1977; Lu *et al.*, 1985). The concentrations of some indicator parameters, such as chemical oxygen demand, Cl, and total organic carbon (Figure 2), and metals, such as Cd, Pb, Fe, and Zn, have been found to be greatest at closure, then asymptotically decrease with time to background levels. The fatty acid fraction gradually decreases, whereas organics such as humic carbohydrate-like compounds and fulvic-like materials become more predominant with the increasing age of the fill (Chian and DeWalle, 1977). Analogous data for volatile chlorinated organic compounds have yet to be compiled.

Landfill gas during phase 4 (methane fermentation) consists of mostly methane and carbon dioxide (Table 3). Theoretically, the composition of landfill gas should be 53% methane and 46% carbon dioxide, assuming complete degradation under anaerobic conditions (Ham, 1979). Measurements have shown landfill gases to be variable and that their composition depends on the stage of landfill stabilization and the intensity of microbial activity. Landfill gas generally has a CO_2:CH_4 ratio that is greater than 0.5, whereas most natural gases and coal gases have ratios less than 0.1 (Coleman *et al.*, 1990). Furthermore, Coleman *et al.* (1990) demonstrated that ^{14}C analysis of methane can distinguish landfill gas from natural, coal, and drift gas (microbial gas in organic-rich glacial drift). The combined yield of CO_2 + CH_4 from landfills in the field may vary from 2.2 to 390 L kg^{-1} waste, and yields of 0.47 to 400 L kg^{-1} have been measured in lysimeter experiments (Ham, 1979).

Landfill gases, like leachate, have contained a diverse composition of volatile organic compounds, including alkanes, cycloalkanes, alkenes, aromatic hydrocarbons, organosulfur compounds, esters, halogenated compounds, alcohols, and ethers (Young and Heasman, 1985). The California Waste Management Board (1988) concluded that the volatile organic compounds benzene, ethylbenzene, toluene, vinyl chloride, dichloromethane, trichloroethylene, 1,2-*cis*-dichloroethylene, and tetrachloroethylene are common in landfill gases (Table 4). Vapor-phase organic compounds may also migrate to the surface. LaRegina *et al.* (1986) collected air samples at the surface of a landfill in New Jersey and detected 26 volatile organic compounds. Some of their data is given in Table 4. Toluene and xylene were detected at the highest concentrations; 287 and 46 g m^{-3}, respectively.

Municipal solid waste generally contains a large microbial population and may be heavily contaminated with pathogenic organisms (Gaby, 1975). Ware (1980), however, found no evidence that linked any disease outbreaks to the pathogen content of municipal solid wastes, nor any increased disease rate among workers in the solid waste industry. Although the chemical properties of landfill leachate have been well characterized, the microbiological composition and variability of microbes in leachate has been less studied (Lu *et al.*, 1985). Scarpino *et al.* (1979) found that lysimeter leachates from older landfill refuse contained *Bacillus* sp., *Streptococcus* sp., *Corynebacterium* sp., and yeast. Pathogens included *Listeria monocytogenes*, *Acinetobacter* sp., *Moraxella* sp., and *Allescheria boyii*. These pathogens are Class 2 microbes: agents that can result in urinary infections, abscesses, and skin and eye inflections. In general, intestinal viruses are rarely found in landfill leachate (Lu *et al.*, 1985); apparently landfills pose a harsh environment for the survival of viruses (see Engelbrecht, 1973), and few micro-organisms can survive in the leachate environment (Pahren, 1987). Studies have shown that the populations are greatest in leachate of fresh refuse (Lu *et al.*, 1985; Pahren, 1987). Pathogen survival tends to decrease with an increase in landfill temperature or because of the toxic, antagonistic effects of the leachate (Engelbrecht, 1973). Very little information is available on the occurrence of fungi and parasites in landfill leachate (Lu *et al.*, 1985).

The USEPA (1979) summarized that on the basis of experience with septic tanks and cesspools, extensive migration of pathogens and fecal-indicator organisms as leachate moves in the subsurface is not expected. Pahren (1987) concluded that leachates from properly operated landfills did not constitute an environmental or public health hazard from intestinal viruses. The USEPA (1988a) added that few microbes are transported away from municipal landfills after the waste has been in place for a few months, and that no groundwater impacts associated with the land disposal of infectious wastes have ever been identified (USEPA, 1988b).

Fate of landfill leachate

The potential release pathways of pollutants in landfill leachate and the concomitant mechanisms of attenuation are diverse and complex (Figure 3). Once contaminants reach groundwater, their concentrations can be reduced by dispersion and dilution, depending on the degree of leachate-groundwater mixing. Geochemical interactions in the mixing zone may influence the fate of contaminants. In a study by Nicholson *et al.* (1983), most of the sulfate and Ca in a plume from a landfill appeared to be from gypsum-containing construction debris. The groundwater beneath the landfill was in equilibrium with gypsum and slightly supersaturated with respect to calcite. Ion-exchange reactions between the plume and the subsurface materials resulted in the displacement of Ca and the subsequent precipitation of calcite. Some of the dissolved iron in the plume resulted from the release of iron from ferric iron coating on sand grains under reducing conditions beneath the landfill. The solubility of the iron in the groundwater-leachate mixing zone appeared to be controlled by siderite ($FeCO_3$).

The rate by which the contaminants move away from the landfill of course depends on the hydrogeologic properties of the subsurface, but they may also be retarded by adsorption-desorption interactions between the solutes in solution and the solid phases. Adsorption-desorption reactions are one of the most important geochemical processes affecting the rate of contaminant migration (Olsen and Davis, 1990). The concentration of organic contaminants can be reduced further by volatilization, biodegradation, hydrolysis, and oxidation. For example, dissolved As has been identified in landfill leachate (Table 1) and can be adsorbed from solution. The extent of As adsorption has been correlated with the content of hydrous oxides of aluminum and iron in the adsorbent (Jacobs *et al.*, 1970; Anderson *et al.*, 1976). Griffin and Shimp (1978) found that As was adsorbed from landfill leachate by clays; montmorillonite adsorbed approximately twice as much as kaolinite.

Chloride is a mobile constituent in groundwater because it is not adsorbed significantly by most subsurface materials and because no environmentally significant solid phases control the aqueous concentration of Cl. Consequently, Cl has long been used as an inorganic indicator of groundwater contamination. Griffin *et al.* (1976) generalized that Cl and Na in landfill leachate were not attenuated when passed through laboratory columns containing calcium-saturated clays. Ammonia, K, Mg, Si, and Fe were "moderately attenuated," and heavy metals such as Pb, Cd, Hg, and Zn were "strongly attenuated." Similarly, Fuller (1978) found that Fe, Mn, and Zn in landfill leachate were attenuated when passed through 11 different soils in laboratory columns. The rate of migration of As, Be, Cd, Cr, Hg, Ni, Se, V, and Zn depended on the physicochemical properties of the water-soil system. The extent of Cr (II) adsorption from landfill leachate correlated with the cation-exchange capacity of two clays (Griffin *et al.*, 1977). At pH values greater than 4.5, Cr (II) precipitated as hydrous chromic hydroxide. About 30 to 300 times more Cr (II) was adsorbed by the two clays than Cr (VI). The environmental fate of Cd and Pb can be controlled both by precipitation-dissolution reactions as well as by adsorption-desorption reactions (Griffin and Shimp, 1976). The adsorption of Cd often correlates with the cation-exchange capacity of the adsorbent (Rai *et al.*, 1986).

Farquhar (1977) summarized that adsorption was the most important mechanism in reducing the concentrations of NH_4, Fe, Zn, K, and chemical oxygen demand in landfill leachate based on laboratory column studies. The extent of removal was correlated with the cation-exchange capacity of the soil adsorbents. Griffin *et al.* (1976) and Farquhar (1977) observed the concomitant desorption of Ca and Mg, suggesting that the specific mechanism of attenuation was ion exchange. This phenomena has been observed near municipal landfills. An increase in hardness (i.e., a measure of Ca + Mg) has been detected near landfills in Illinois and was thought to be the result of desorbed Ca and Mg (Cartwright *et al.*, 1977; Griffin and Shimp, 1978). Other studies have also documented the occurrence of "hardness halos" near landfills (Bagchi *et al.*, 1980; Nicholson *et al.*, 1983).

Table 5 Summary of Henry's law constants (K_H), organic carbon-water partition coefficients (K_{OC}), and degradation potential of selected organic compounds in a municipal landfill.

Compound	K_H (atm m^3 mol^{-1}) and reference[a]	K_{OC} (mL g^{-1}) and reference[a]	Degradation potential[b]
Acetone	2.1×10^{-5} (11)	~1 (9)	—[c]
Benzene	5.6×10^{-3} (10)	97 (9)	Biodegrades slowly.
Bis(2-ethylhexyl) phthalate	3.0×10^{-7} (6)	5,000 (2)	Hydrolyses slowly, biodegrades rapidly.
Carbon tetrachloride	3.0×10^{-2} (4)	232 (9)	Biodegrades rapidly. Half-life in soil <7 days
Chlorobenzene	3.9×10^{-3} (1)	318 (9)	Biodegrades slowly. Half-life in soil >9 months
Chloroethane	1.1×10^{-2} (4)	42 (2)	Hydrolyses slowly.
Chloroform	3.7×10^{-3} (4)	34 (4)	Biodegradable. Half-life in soil 5 to 9 months
Cyanide	Very rapidly volatilizes (11)	Can be adsorbed by organic matter, but only weakly (11)	Can biodegrade.
Di-n-butyl phthalate	4.5×10^{-6} (6)	1,820 (2)	Biodegrades rapidly. Hydrolyses very slowly.
1,2-Dichlorobenzene	2.4×10^{-3} (7)	343 (5)	Resistant. Half-life in soil >9 months
1,4-Dichlorobenzene	1.9×10^{-3} (3)	522 (2)	Resistant.
1,1-Dichloroethane	5.6×10^{-3} (4)	43 (2)	Biodegradable. Hydrolyses, oxidizes slowly.
1,2-Dichloroethane	1.3×10^{-3} (7)	36 (2)	Biodegradable. Hydrolyses, oxidizes slowly.
1,1-Dichloroethylene	2.6×10^{-2} (4)	58 (2)	Biodegrades rapidly. Hydrolyses very slowly.
1,2-Dichloroethylene	9.4×10^{-3} (4)	169 (2)	Biodegradable. Hydrolyses very slowly.
Diethyl phthalate	1.2×10^{-6} (6)	132 (2)	Biodegrades rapidly. Hydrolyses very slowly.
Endrin	4.0×10^{-7} (6)	90,000 (2)	Resistant.
Ethyl acetate	1.3×10^{-4} (7)	8 (2)	—[b]
Ethylbenzene	8.7×10^{-3} (7)	622 (9)	Biodegradable.

Table 5 continued

Compound					Comment
Isophorone	5.8×10^{-6}	(3)	26	(2)	Biodegrades rapidly.
Lindane	NA		28,900	(2)	Biodegrades rapidly.
Methylene chloride	2.3×10^{-3}	(7)	25	(9)	Biodegradable. Hydrolyses slowly. Half-life in soil 5 to 9 months
Methyl ethyl ketone	2.7×10^{-5}	(10)	4	(9)	Biodegrades rapidly.
Methyl isobutyl ketone	3.5×10^{-5}	(10)	24	(9)	Biodegrades rapidly.
Naphthalene	4.1×10^{-4}	(7)	1,300	(5)	Biodegrades rapidly.
Phenanthrene	2.5×10^{-5}	(7)	23,000	(5)	Biodegrades rapidly.
Phenol	3.9×10^{-7}	(7)	27	(5)	Biodegrades rapidly.
Propanol	8.5×10^{-6}	(7)	<1	(2)	—b
Tetrachloroethylene	1.5×10^{-2}	(8)	303	(9)	Biodegradability uncertain. Half-life in soil 5 to 9 months
Tetrahydrofuran	?		<1	(2)	?
Toluene	5.9×10^{-3}	(8)	242	(4)	Can biodegrade. Soil half-life >9 months
1,1,1-Trichloroethane	1.7×10^{-2}	(4)	155	(4)	Can biodegrade. Soil half-life <7 days.
Trichloroethylene	9.9×10^{-3}	(8)	152	(4)	Biodegrades slowly. Soil half-life >9 months
Vinyl chloride	2.8×10^{-2}	(4)	66	(2)	Resistant.
o-Xylene	6.1×10^{-3}	(11)	363	(9)	Biodegrades rapidly.

a References: (1) Ehrenfeld *et al.* (1986); (2) estimated from water solubility data using the empirical regression of Hassett *et al.* (1983); (3) estimated from vapor pressure and water solubility data (see Mabey *et al.*, 1982); (4) Gossett (1987); (5) Hassett *et al.* (1983); (6) Mabey *et al.* (1982); (7) Nirmalakhandan and Speece (1988); (8) Roberts and Dandliker (1983); (9) Roy and Griffin (1985); (10) Roy and Griffin (1990); (11) USEPA (1983).

b Comments were derived from references and discussion in Mabey *et al.* (1982), USEPA (1983), and Bennett (1985).

c These compounds can be biodegraded but are themselves potential products of biodegradation.

There is no *a priori* basis for making quantitative predictions of the extent of adsorption of inorganic ionic solutes from landfill leachate or groundwater-leachate mixtures; such measurements must be made experimentally. The adsorption of heavy metals can be influenced by pH, Eh, ionic strength, competitive effects from other ions, and of course the nature of the adsorbent. The adsorption of some nonionic organic contaminants can be dominated by somewhat different mechanisms. The extent of adsorption of sparingly water-soluble organic compounds is often correlated with the organic carbon content of the adsorbent (Hassett *et al.*, 1983). When adsorption is expressed as a function of organic carbon content, an organic carbon-water partition coefficient (K_{oc}) can be estimated. An increase in organic carbon should result in greater adsorption, depending of the K_{oc} of the individual compound. There is also a critical organic carbon concentration below which the inorganic fraction of the adsorbent dominates the process, and K_{oc} concepts may not apply (see Olsen and Davis, 1990). Values of K_{oc} were compiled (Table 5) for the common nonionic compounds reported in landfill leachate (Table 2). These partition coefficients can be used to rank the relative mobilities of each compound in saturated porous media. With the classification system given in Roy and Griffin (1985), about half of the compounds reported in landfill leachate have either a medium or low mobility; their rates of migration through a landfill liner and in groundwater would be expected to be influenced by the amount of organic carbon present. A little less than half of the compounds in Table 2 were classified as highly or very highly mobile; adsorption-desorption processes alone may have little effect in retarding their movement.

A Henry's law constant (H) estimates the tendency of a chemical to partition between leachate and its gas or vapor phase. Values for H for most of the compounds were also compiled (Table 5). About 19% of the compounds in Table 2 are highly volatile, if highly volatile is defined as $H \geq 10^{-2}$ atm m^{3}mol^{-1}. These compounds would be expected to readily volatilize from leachate in the unsaturated zone beneath the landfill. Bonazountas and Wagner (1984) modeled the fate of organic solvents in leachate in a hypothetical scenario. In a simulation where the water table was 10 meters below the surface in a silty loam soil, they estimated that 99.3% of the carbon tetrachloride present would volatilize and that 0.1% would remain as an adsorbed phase. About 0.6% was predicted to migrate down to the water table. If medium volatility is defined as 10^{-4} to 10^{-2} atm m^{3}mol^{-1}, then 15 of the compounds fall into this range; volatilization can be an important mechanism whereby a large fraction of these chemicals are slowly removed from solution, depending on the mole fraction of the components in the aqueous phase, temperature gradients, and the extent of mixing. All of the volatile organic compounds regarded as common in landfill gases (Table 4) are either highly volatile or have a medium volatility. Eleven compounds in Table 5 have a low potential ($<10^{-4}$) for significant vapor-phase movement; they may not partition from leachate in the vadose zone and reach the surface in significant amounts. Although some compounds are mobile in groundwater and may not volatilize quickly from solution, they may degrade. Degradation can be a combination of oxidation, hydrolysis, and biodegradation. Of the organic compounds identified in landfill leachate, chemical oxidation by dissolved oxygen, hydroxyl, or

Table 6 *Concentration ranges of volatile organic compounds detected in groundwater at 6 or more of 13 landfills sampled in Minnesota (Sabel and Clark, 1983).*

Compound	Range (μg L^{-1})	Frequency of detection
Benzene	1.0 – 470	9 of 13 landfills
1,1,2-Trichloroethylene	0.2 – 144	9 of 13
Chloromethane	NQ[a]	9 of 13
Ethylbenzene	1.2 – 590	8 of 13
1,2-Dichloropropane	0.5 – 43	8 of 13
Methyl ethyl ketone	6.8 – 6,200	7 of 13
Methylene chloride	1.0 – 250	7 of 13
1,1-Dichloroethane	0.5 – 1900	7 of 13
1,2-Dichloroethylene	0.5 – 20,000	7 of 13
Acetone	24 – 3,000	6 of 13
Tetrahydrofuran	24 – 3,000	6 of 13
1-Butanol	59 – 17,000	6 of 13
Toluene	1.5 – 8,300	6 of 13
o- and *p*-Xylene	1.3 – 950	6 of 13
Trichlorofluoromethane	0.2 – 150	6 of 13
Chloroethane	NQ	6 of 13
Dichlorodifluoromethane	NQ	6 of 13

[a]NQ, Not quantified

peroxy radicals is not thought to be significant, especially in relatively reduced environments like groundwater. Some of the compounds can hydrolyze in solution, but at slow rates. For example, the half-life of the hydrolysis of tetrachloroethylene may be on the order of 9 months (Mabey *et al.*, 1982). The hydrolysis half-life of methylene chloride has been estimated as 18 months (ATSDR, 1987a). Although these reactions are slow, they may be significant removal mechanisms in slow-moving groundwater near a landfill.

Many organic compounds will be biodegraded by microorganisms to some extent in an aerobic environment, but less information is available for anaerobic environments such as landfills. Of the chemicals in Table 5, 1,2- and 1,4-dichlorobenzene, endrin, and vinyl chloride appear to be resistant to biodegradation. Biodegradation is less likely when the compound has structural features that are not encountered in nature. The potential to biodegrade in Table 5 was based on a static-culture batch-screening procedure (Tabak *et al.*, 1981) and laboratory studies summarized by Bennett (1985). Benzene was identified as a common organic component in landfill leachate, and it apparently will biodegrade slowly under anaerobic conditions (ATSDR, 1987b); its biodegradation half-life in groundwater may be on the order of 110 days (Zoeteman *et al.*, 1981). The microbial metabolism of benzene proceeds through the formation of *cis*-dihydrodiols and with further oxidation to catechols. Tetrachloroethylene is thought to biodegrade only very slowly under anaerobic conditions and may be relatively persistent (ATSDR, 1987c); Roberts *et al.* (1982) estimated that the biodegradation half-life of tetrachloroethylene in groundwater was about 230 days. The literature on *in situ* biodegradation of organic chemicals in landfills is sparse to nonexistent (Bennett, 1985).

Degree of groundwater contamination

In the mid-1960s, the USEPA began to investigate problems related to the management of municipal solid wastes. Early projects on landfills included physical and chemical characterization of leachate, waste decomposition, leachate treatment, gas recovery, composting, source reduction, and related topics. The following review of selected studies helps to illustrate the historical sequence of events that led to modern landfills.

A 1958 survey completed by about 200 landfill operators indicated that the most frequently encountered problem at landfills was blowing paper, followed by flies, fires, rodents, odors, and dust (Weaver, 1961). Water pollution problems were reported in only five cases. In two reports, refuse had been placed in an abandoned gravel pit, and the water table had risen into the fill after a major storm.

Andersen and Dornbush (1967) reported that a landfill located in an inactive gravel pit resulted in a "slight impairment of groundwater quality" immediately downgradient from the fill. Groundwater 370 meters downstream was unaffected. The degree of contamination was based on the distribution of nitrate and chloride concentrations, electrical conductivity, hardness, and pH.

During a 4-year study of five landfills in Illinois, Hughes *et al.* (1971) found that about 50% of the annual precipitation infiltrated the landfills and formed groundwater mounds beneath the sites. They concluded that the fine-textured tills of northeastern Illinois attenuated the leachate and that landfills in that area did not need protective measures to prevent groundwater pollution. In conjunction with these studies, Cartwright and McComas (1968) conducted electrical resistivity and soil temperature surveys. They found that leachate-contaminated groundwater could be detected more than 305 meters from the source. The soil temperature surveys detected a halo of higher temperatures around the landfills and areas of surface recharge.

In his review on the effects of municipal landfills on groundwater quality, Zanoni (1972) found few case histories of serious groundwater contamination that were directly attributable to municipal landfills. Zanoni attributed the relative success of most landfills to subsurface attenuation of leachate. He further remarked that chloride

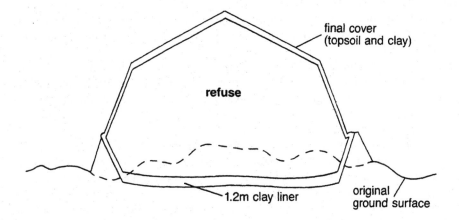

Figure 4 *Schematic illustration of a modern municipal landfill in Wisconsin, the Marathon County Landfill (adapted from Gordon et al., 1984)*

concentrations directly below the landfill were always high, but decreased drastically over a short distance from the landfill. He concluded that landfill failures were the result of poor siting. In the same year, Seitz *et al.* (1972) reported that a landfill placed in thin loess over granite resulted in gross contamination of the groundwater, on the basis of total dissolved solids, hardness, and chloride. However, the zone of contamination was confined to the proximity of the landfill.

In 1974, Miller *et al.* (1974) summarized the extent of groundwater contamination in the northeastern United States by municipal landfills. The authors reported that of the 42 cases in which groundwater contamination was detected, 16 landfills contaminated water-supply wells. They also reported that a pollutant (not identified) had migrated more than 300 meters. No details were given about the extent of contamination, but in 26 of the 42 cases, no action was deemed necessary with respect to the landfill or the contaminated groundwater. Water wells were abandoned in four cases. In six cases, the landfill was closed. The landfills of that era were often placed in marshlands, sand and gravel pits, old strip mines, or sinkholes (Miller *et al.*, 1974).

Robertson *et al.* (1974) and Dunlap *et al.* (1976) documented the occurrence of organic compounds in groundwater from the Norman, Oklahoma, landfill. Qualitative analysis of groundwater samples collected within the boundaries of the landfill area contained more than 40 compounds. A few of the compounds were probably end products of microbial metabolism, but most of them were associated with industrial wastes. The landfill had a long history of unrestricted dumping, regardless of waste type.

Zenone *et al.* (1975) reported on two landfills in Alaska. One landfill had begun operating in 1958, and water-quality data showed that leachate was present in shallow groundwater beneath the landfill. However, the data indicated either that the leachate had not moved beyond the immediate landfill area or that it had been attenuated. The other landfill was located above the water table, and in the 4-year interval after it had begun receiving wastes, no leachate had been detected in monitoring wells.

Each of seven landfills in Marion County, Indiana, was either a former sand and gravel pit or a borrow pit that was used as a landfill in the 1960s and 1970s (Pettijohn, 1977). Although the groundwater at each site was contaminated by leachate, the areal extent of the contamination appeared to be confined to the proximity of the site and to shallow depths.

Lindorff and Cartwright (1977) provided a summary of eight case studies in Illinois where groundwater contamination had been detected. All eight landfills had been operated prior to about 1972. Common to all eight was the unsuitability of the site geology. Two were in floodplains and five were in relatively permeable materials, including three former sand and gravel quarries. Four sites were closed because of environmental problems, and no remedial actions were taken at two of these sites.

Ostenso and Gordon (1980) summarized a case study where leachate run-off was a problem at a landfill in Wisconsin. The site was located on a relatively impermeable soil, but municipal wastes were disposed along with paper mill sludge that had a liquid content of 40 to 60%. Also, an excessive quantity of leachate was ponded on site. The result was the formation of leachate seeps. These problems were attributed to the lack of leachate collection systems and poor soil covers during the operation of the landfill. The excess volume of liquid resulted in the "bathtub effect," in which the landfill fills up with leachate then overflows, forming seeps.

A landfill in Delaware (Army Creek Landfill) was located in a sand and gravel pit, and municipal wastes were disposed along with industrial wastes for 8 years (DeWall and Chian, 1981). The wastes were initially placed into standing water, and when the site was full, it was covered with sand. Chloride was detected in the groundwater as far as 800 meters downgradient from the fill. The volatile organic compounds most frequently detected included benzene, hexane, dichloroethane, toluene, *p*- and *m*-xylene, bis(2-chloroethyl)-ether, and acetone. The occurrence of acetone, butanol, and pentanol was attributed to microbial activity.

Gordon and Huebner (1983) studied 12 zone-of-saturation landfills in Wisconsin. In one example, permeable deposits were present, and no liner was used. Leachate escaped into the bedrock aquifer, and the groundwater was "severely contaminated." The acceptance of large quantities of liquid waste, and the lack of leachate management were cited as the major factors that led to an excessive leachate head within the site and the subsequent outward gradients away from the landfill. In another example, excessive gas pressures and leachate head may have resulted in outward gradients. Numerous leachate seeps and gas vents developed at the site. The acceptance of large quantities of paper mill sludge, the use of heavy clay soils for daily cover, and an inadequate removal of leachate were attributed to the failure of the site. Gordon and Huebner (1983) summarized the two major problems with zone-of-saturation landfills in Wisconsin. First, the site geology was not suitable for the then prevalent method for constructing landfills. Second, inward groundwater gradients, which are basic to the design of zone-of-saturation sites, did not exist because of the disposal of paper mill sludge and the lack of leachate management systems.

Stecker and Garvin (1983) documented the formation of gas vents and holes in landfill covers in Wisconsin. They recommended the use of a highly permeable soil as a daily cover to allow gases to escape. Sabel and Clark (1983) reported on the occurrence of groundwater contamination by municipal landfills in Minnesota. Groundwater samples were collected at 13 landfills, but no details were given on the spatial distribution of organic contaminants. This sample study may have been biased because inorganic parameters suggested that leachate contamination was present at each sampled site. Seventeen compounds were detected in more than 40% of the sites in the groundwater. The most frequently occurring organic compounds are given in Table 6.

In a very similar study, Nelson and Book (1986) noted that since the 1970s, the Minnesota Pollution Control Agency had issued permits for 131 municipal landfills that were with few exceptions unlined facilities constructed directly in or on native soils. Approximately 54% of these sites were in sand. Nelson and Book (1986) concluded that volatile organic compounds were ubiquitous in groundwater that was downgradient from these landfills. Among volatile organic compounds, 1,1-dichloroethane, 1,1,2-trichloroethylene, benzene, toluene, methylene chloride, and ethylbenzene were the six most commonly occurring compounds, but no concentration data were given.

Duwelius and Greeman (1989) described two unlined landfills in Indiana that were in direct contact with sand and gravel aquifers. Both landfills were closed in the 1970s. Concentrations of dissolved inorganic chemicals in groundwater samples indicated that leachate from both landfills reached the shallow aquifers. The effect on deeper aquifers was regarded as insignificant because of the predominance of horizontal groundwater flow and discharge into streams where the leachate was diluted. The fate of organic chemicals was not discussed.

Battista and Connelly (1989) conducted a follow-up study at 19 unlined landfills in Wisconsin. The results indicated that the traditional indicators of groundwater contamination at landfills corresponded to elevated levels of volatile organic compounds in 20 of 49 wells. In 11 of the 49 wells, inorganic parameters were elevated, but volatile organic compounds were undetected. In three wells, elevated concentrations of volatile organic compounds were noted, with no corresponding increase in inorganic solutes.

In 1980, the USEPA (NSF, 1983) estimated that 0.1 to 0.4% of the usable surface aquifers in the United States had been contaminated by industrial impoundments and landfills. This range was intended to be an order-of-magnitude estimate and was based on the number of sources of contamination and existing information, quantitative and qualitative. The approach used was not referenced to the severity of the contamination nor the potential impacts to human health and the environment. The Illinois Environmental Protection Agency (Illinois Hazardous Waste Task Force, 1983) concluded that only a small percentage of potential groundwater pollution could be attributed to landfill leachate. However, more recent information collected by the USEPA (1988b) indicated that background data on groundwater quality at most (64%) of the active landfills in 1986 were derived from "best estimates" (i.e., there were no data) and that the operators of only 17% of the landfills had site-specific data. Moreover, only between 25 and 35% of the active landfills were believed to have had any type of groundwater monitoring program. The USEPA (1988b) concluded that because the majority of existing municipal landfills did not monitor groundwater quality, the number of landfills that have contaminated the groundwater could not be determined. Some states have only recently started to monitor groundwater quality near landfills. In Iowa, for example, regulations were just adopted in 1989 that require hydrogeologic investigations of landfills.

Aggregate data accumulated up to 1984 for 9,284 landfills indicated 586 groundwater contamination violations, according to the USEPA (1988b). The number of reported violations is an imperfect measure of environmental impacts because violations are defined differently among states and multiple violations can occur at a single facility. Groundwater quality was adversely affected at 146 sites. In 90 of these sites, the on-site groundwater was reported to be contaminated, and 56 sites had contaminated off-site groundwater. The impacts identified ranged in severity from elevated concentrations of various constituents directly beneath the landfill (such as chloride) to the contamination of major aquifers. In 17 of these case studies, alternate water supplies were required. The USEPA (1988b) summarized that most of these problem landfills were located within about 2 meters of the water table, sited in highly permeable soils, or constructed without an effective liner.

Surface-water contamination violations totaled 660. Problems resulted from leachate seeps or run-off control deficiencies or because of the placement of the landfills in wetlands or floodplains. The degree of degradation was limited in most cases, but the USEPA felt that ecological damage occurred more often than documented because of the time and resources needed to make comprehensive assessments.

Minimizing groundwater contamination

Subtitle D of the Resource Conservation and Recovery Act established the framework for coordinating federal, state, and local governments for the management of nonhazardous solid wastes. Subtitle D required the USEPA to promulgate guidelines for the states to develop and implement solid waste disposal plans. The criteria for municipal solid waste landfills (the Subtitle D regulations) were finalized in 1991 (Federal Register, 1991). The regulations were intended to be self-implementing; implementation and enforcement are to be primarily the responsibility of state and local governments. They were also intended to be the minimum requirements for Subtitle D facilities in that the states have the option of adopting requirements more stringent than the federal criteria.

Most relevant to this chapter is the requirement that a facility or practice must not contaminate an underground drinking water source beyond the solid waste boundary or an alternative boundary. The federal criteria to achieve this requirement include the use of soil liners, leachate collection systems, and final covers. Operating criteria include daily cover of refuse, restrictions concerning the placement of liquids, a program for managing the codisposal of hazardous wastes, postclosure care for a minimum of 30 years, and several other requirements concerning groundwater monitoring and the location of landfills.

Because these regulations are relatively recent, very little documentation is available with which to evaluate their efficacy. The most recent survey conducted by the USEPA (1988b) indicated that most of the active landfills in the United States were not using any type of liner. Only about 1% were using a flexible membrane liner and 15 to 27% were using soil or clay liners.

A typical modern landfill in Wisconsin has a soil liner (Figure 4) that is 1.2 to 1.5m thick (Gordon *et al.*, 1984). This thickness was specified to allow a margin of error in constructing liners. In 1976, the first landfill in Wisconsin with a 1.2-m liner and a leachate collection system was constructed. The hydraulic conductivity of the liner reportedly was less than 10^{-7} cm s^{-1}. Groundwater monitoring data collected over an 8-year period indicated that the site had only a minimal impact on the quality of the groundwater. In a second example, a landfill had a soil liner and a leachate collection system, but a leachate seep developed. Remediation revealed that a collection pipe had been crushed during construction. Repairs were made, and during 5 years of operation, groundwater and surface-water monitoring indicated "essentially no significant impact." In a third example, the groundwater beneath a lined landfill with a leachate collection system showed no measurable impacts after 4 years of operation.

Gordon *et al.* (1984) concluded that soil liners can provide an adequate means of leachate containment at municipal landfills and that soil liners is a well-developed technology in Wisconsin, where they have been in use for nearly a decade. If the hydraulic conductivity of the liner is 10^{-7} cm s^{-1} or less, then several decades would elapse before solutes migrate through a 1.2-meter liner and move toward a monitoring well in sufficient concentrations to be detected. In a follow-up study, Huebner *et al.* (1989) collected lysimeter and groundwater samples at a Wisconsin landfill that was constructed in 1982. The test cell had a 1.5-m compacted soil liner with a hydraulic conductivity of less than 10^{-7} cm s^{-1}. Monitoring data indicated that contaminant breakthrough had not occurred. Solute-transport times calculated by advection and diffusion equations suggested that measurable quantities of chloride would not reach the bottom of the liner for approximately 11 years and that it would take about three centuries before significant concentrations of benzene would break through the liner. Huebner *et al.* (1989) argued that the lack of contaminant detection was consistent with the predicted travel times and that these results implied that the soil liner had been properly constructed.

Little research has been done on field-scale soil liners to evaluate their performance (USEPA, 1988c), although they are commonly used. Daniel and Brown (1988) summarized 14 case studies of earthen liners constructed in the United States and Canada. Of the 14 studies, only two met the USEPA hydraulic requirement of 1x 10^{-7} cm s^{-1}: the Keele Valley Landfill in Ontario, and a test liner in California. Recent research conducted at the Illinois State Geological Survey (Krapac *et al.*, 1989) suggests that the low conductivity requirement can be met when field-scale liners are constructed if a rigorous quality-control program is applied. However, this project is ongoing, and definitive results are not yet available.

In recent years, there was a concern that hazardous waste leachate could significantly increase the hydraulic conductivity of a soil liner, in turn reducing its ability to limit leachate movement. Some studies have demonstrated that pure organic solvents will increase hydraulic conductivity (see Griffin and Roy, 1985). However, the concentrations of organic solvents in municipal leachate are too dilute to produce similar increases (Madsen and Mitchell, 1987). Griffin and Shimp (1978) found that landfill leachate lowered the hydraulic conductivity of clay-sand mixtures in laboratory columns. The reductions were attributed to both clay swelling and to microbial growth. Gordon *et al.* (1984) reported that the passage of 0.25 to about seven pore volumes of landfill leachate resulted in no significant changes in the hydraulic conductivity of four clay-soil samples in laboratory tests. Wuellner *et al.* (1985) reported that landfill leachate did not significantly change the permeability of two clay samples after the passage of eight pore volumes. The hydraulic conductivities and dry densities of samples of the subsurface materials beneath three unlined municipal landfills were not found to be significantly different from samples collected outside the landfills (USEPA, 1978). These studies suggest that leachate-induced increases in liner permeability are not a major issue.

Future expectations and research

Landfills are often designated as the least desirable method for the disposal of solid wastes, but they are the common denominator in all waste management approaches. In the future, reliance on landfills will probably be lessened by waste minimization and recycling. The USEPA (1988b) estimated that 25 to 30% of municipal solid wastes are easily recyclable, although markets for this volume of material do not yet exist. Certain wastes will never have a practical use or an alternative to land disposal. Incineration reduces the volume of refuse by as much as 70 to 90% (USEPA, 1987), but the combustion of wastes results in a fly ash that is collected by air pollution control equipment and a bottom ash that accumulates in the combustion chamber. Most generators mix the two ashes and dispose the combined ash in municipal landfills or ash monofills (USEPA, 1987).

In 1983, the National Science Foundation (NSF, 1983) predicted that the relative importance of landfills as sources of groundwater pollution would decline in the future as management practices improve. Morelli (1989) argued that leachate is generated in a landfill and requires management, but that it is not a problem in and of itself. He further concluded that groundwater contamination is a problem when it occurs, but that it is now better described as indicative of an improperly operated landfill or an inadequately designed or constructed leachate containment system. Charnley *et al.* (1988) also concluded that the favorable siting of modern landfills and the use of control technologies for leachate and gas now combine to minimize environmental impacts.

The USEPA (1988a) concluded that if properly located, designed, and operated, future municipal solid waste landfills will be protective of human health and the environment, that proper designs can play a role in minimizing the risks posed by the facility, and that a liner and a leachate collection system will help prevent groundwater and surface-water contamination. However, the USEPA noted that at that time, very few states had yet adopted the federal criteria and that the process continues to be in a state of flux. In 1987, 24 states had requirements for liners, and 27 states proposed requirements for leachate collection systems (USEPA, 1988b). More current information has not been published. Other states were considering liner and collection systems, but not as requirements. Only 19 states had landfill location restrictions with regard to critical habitats and geologically unsuitable areas. Few states had considered standards for wetlands or areas prone to subsidence and earthquakes. The USEPA (1988b) reported that 42 states will require groundwater monitoring in the future. Ten states will also require surface-water monitoring. Because of the variations in state regulations, the extent to which a particular municipal landfill is located, designed, and operated is strongly dependent on politics (USEPA, 1988b). The USEPA further suggested that deficiencies in state regulations were partially responsible for the current lack of groundwater monitoring.

In 1984, the United States Congress required that all new hazardous waste landfills have double liners. A double-liner system has one or more flexible membrane high-density (plastic) liners used in concert with compacted soil liners (see USEPA, 1989). Flexible membrane liners may also be required by some states in the design of municipal landfills. Although the use of flexible membrane liners may be expected in some areas, there is little actual experience with plastic liners, and they are not as easy to install or repair in the field as soil liners. Moreover, no criteria have been established for the thickness of flexible membrane liners, and the significance of solute diffusion through geomembranes remains unresolved.

For some time, those involved in waste management have debated whether landfilling should be managed as a mummification technique. Here, wastes are isolated from the environment and every attempt is made to minimize the influx of infiltration through the wastes to abate leachate formation. Perhaps landfills should be viewed as bioreactors in which leachate is recirculated to accelerate refuse degradation, increase methane production rates, and shorten the time necessary to stabilize the landfill. The USEPA (1988b) reported that the recirculation of leachate can significantly enhance biodegradation and methane production. Morelli (1989) advocated recirculation to accelerate landfill stabilization because he felt that a principal problem with landfills is a lack of realistic design life; leachate recirculation would accelerate the attainment of landfill maturity before the design-life of the landfill is exceeded.

At present, leachate collection systems are rarely used; only about 5 to 11% of the active landfills polled were collecting leachate (USEPA, 1988a). Recirculation by spraying or trucking collected leachate to publicly owned treatment works were the most common management methods. Recirculation has certain benefits, but using it as a treatment method has certain drawbacks. It results in an increased rate of leachate production. The increase in leachate volume may clog the leachate collection system (USEPA, 1988b), and leachate-induced erosion may be a problem when daily cover is required. In 1986, only about 3% of the active landfills polled by the USEPA (1988b) were using recirculation systems. Vasuki (1988) reported that a 5-year study was beginning in Delaware where two test cells were being constructed. One cell will be kept dry, and leachate will be collected from the other cell and recycled through the refuse. After the 5-year interval, the refuse will be exhumed to determine the degree of degradation. The future role of leachate recirculation is uncertain.

Run-on/run-off controls are important in landfill-pollution control because run-on contributes to leachate generation. Surface controls, such as diversion berms, collection/sedimentation ponds, and diversion ditches were being used at about 61% of landfills active in 1986 (USEPA, 1988b). Methane control and recovery represents another method of minimizing the environmental impact of landfills. Only about 11% of all landfills were using such controls. Less than 2% of them were attempting to recover the methane, possibly because several years are necessary before significant methane production becomes apparent. Landfill methane has been identified as a contributor to the greenhouse effect (Augenstein, 1990), can result in explosive situations if it migrates off site, and is now considered as a potential energy resource. Control of methane emissions from landfills is of environmental and economic interest and is receiving increased attention.

The final cover placed on the landfill should minimize water infiltration. During the 1980s, recompacted clay and "topsoil" were most commonly used as the final cover (USEPA, 1988b). However, sand and gravel was used during the closing of 370 sites and was being used at 932 active landfills. The long-term stability of landfill covers is not known. Research is needed on cover failure brought about by differential settlement as the refuse decays.

Although the efficacy of soil liners as physical barriers to leachate has been studied, the concept of soil liners as chemical barriers to the movement of solutes is not a factor in the design of liners. Current Subtitle D regulations do not require that the chemical or mineralogical properties of the soil be characterized. No guidelines state that the adsorption properties of the candidate soil should be measured or used as criteria in screening earth materials for liner construction. Illitic clays may be the most common clay mineral present in surficial materials used to construct soil liners. As discussed previously, the adsorption of some heavy metals correlates to the cation-exchange capacity of the adsorbent. Thus, illite-rich soils would be preferable to kaolinitic soils, which typically have a lower cation-exchange capacity. Moreover, the concept of using travel-time calculations as criteria for estimating the minimum thickness of the liner remains only a concept. For example, Griffin *et al.* (1976) used travel-time calculations to estimate the minimum thickness of a clay layer needed to remove lead from landfill leachate. This general approach was considered by the USEPA in the late 1970s but was abandoned because of uncertainties in measuring solute adsorption, effective porosity, hydraulic conductivity, dispersion, and diffusion. Improved methods for the routine collection of adsorption data now exist (Roy *et al.*, 1992), but it is still difficult to measure properties such as effective porosity, which is still regarded as a "research measurement." In the future, travel-time calculations may be included in liner design.

On a larger scale, previous studies have also concluded that an assessment of the attenuation capacity of a disposal site should be an integral part of the site-selection process and of waste management practices (Landon, 1978; Crutcher and Rovers, 1978; van Genuchten, 1978; Brunner, 1979; Donigian *et al.*, 1983; Roy and Griffin, 1987). Cartwright *et al.* (1977) pointed out that because of the reduction in the number of landfills, municipal refuse is being concentrated in increasing volumes at fewer sites. Thus, large multicounty landfills may generate sufficient leachate such that the attenuation capacity of the site is exceeded regardless of the design engineering in place.

With regard to the vanishing-landfill crisis, short-term remedies include excavating the edges of filled landfills to extend the life of the site (Michaels, 1990), and lateral or vertical expansion where refuse or a new landfill unit is placed directly on top of an old landfill (Johnson, 1985). States that do not have approved landfills have been shipping and will probably continue to ship refuse to states that have active landfills. It has been predicted that railroads will be used to move refuse across the country; by the mid-1990s, American railroads expect to earn $400 million per year by transporting refuse (Bukro, 1989).

Another emerging trend is that of the "megafill," or giant landfill. A megafill covers an area of 405 hectares or more (Bukro, 1989). The largest planned megafill is the 3,360-hectare Eagle Mountain Project in California.

One relatively new approach is landfill mining. On the basis of work by the Collier County Government in Florida, old landfills are exhumed, and the debris is sorted for the purpose of recycling and waste reduction. The feasibility of landfill mining is based largely on the fact that chemical and biological degradation reduces the volume of the buried waste (Spencer, 1990). As the waste is removed, plastics and metals can be concentrated and sold for recycling. Oversized items can be then placed in a new landfill (Watson, 1988), and the exhumed soil can be reused in the operation of the new landfill (Spencer, 1990). A possible scenario for an environmentally sound landfill is where waste is buried and mined over and over again (Watson, 1988). In this approach, the landfill itself is recycled. The major concerns include odors, fires, and explosions (Spencer, 1990) and whether landfill mining is feasible at sites where large quantities of hazardous wastes are present (Watson, 1988). Morelli (1989) concluded that landfill reclamation will become a method of creating new landfill capacity. The possible impacts of landfill mining on groundwater quality are unknown.

Johnson (1985) remarked that public opposition to new landfills seems to be least where the land has already been disturbed by activities such as mining or quarrying or where there has been a long history of land disposal of other materials. However, as the case studies have illustrated, such areas were seldom geologically suited for landfills. Johnson (1985) concluded that the siting of new landfills will depend on the efficacy and support of the local unit of government and the local press, as well as effective public relations.

Summary

Because of public opposition to the siting of landfills and other factors, siting new facilities has become very difficult. About 71% of the nation's landfills have 15 years or less of remaining capacity.

Paper products are the major component of municipal solid waste, but the waste stream may contain some hazardous wastes. Landfill stabilization after closure may be about 15 to 40 years, but the rate by which refuse decays is variable, and no landfill has a single "age." Exhumation studies have indicated that presumably degradable materials may be more persistent than previously assumed.

Landfill leachate can transport inorganic and organic contaminants from a disposal site to groundwater. The pH of leachate samples has varied from 1.5 to 12.5, and the inorganic chemical composition of landfill leachate is quite variable. Leachate may also contain a plethora of organic chemicals, although the past practice of codisposing municipal refuse with industrial wastes has been a major source of organics in landfill leachate. Landfill gas mostly consists of methane and carbon dioxide and, like leachate, can contain a diverse composition of volatile organic compounds. Municipal solid waste generally contains a large microbial population and may contain pathogenic organisms. However, few microbes are transported away from municipal landfills after the waste has been in place for a few months.

The release pathways of contaminants in landfill leachate and concomitant mechanisms of attenuation are diverse and complex. Once contaminants reach groundwater, their concentrations can be reduced by dispersion, dilution, and chemical interactions. About half of the volatile organic compounds reported in landfill leachate have either a medium or low mobility; their rates of migration through a landfill liner and in groundwater would be expected to be dependent on the amount of organic carbon present. A little less than half of the compounds identified in landfill leachate were classified as highly or very highly mobile; adsorption-desorption processes alone may have little effect in retarding their movement. The concentration of some organic contaminants can be reduced by volatilization, biodegradation, hydrolysis, and oxidation. About 19% of the compounds commonly found in landfill leachate are highly volatile; these compounds would be expected to readily volatilize from leachate in the unsaturated zone. Fifteen of the compounds have a medium volatility; volatilization can be an important mechanisms whereby a large fraction of these chemicals are slowly removed from solution. Eleven of the compounds have a low potential to volatilize. Chemical oxidation of volatile organic compounds in landfill leachate is not thought to be significant. Some of the compounds can hydrolyze in solution, but at slow rates. Many organic compounds may be biodegraded by microorganisms to some extent, but 1,2- and 1,4-dichlorobenzene, endrin, and vinyl chloride appear to be resistant to biodegradation.

Few case histories were found in which municipal landfills caused serious groundwater contamination, but only between 25 and 35% of all active landfills had any type of groundwater monitoring program. Because the majority of existing municipal landfills do not monitor groundwater quality, determining the exact number of landfills that have contaminated the groundwater is not possible. Aggregate data accumulated up to 1984 for 9,284 landfills indicated 586 groundwater contamination violations; groundwater quality was adversely affected at

146 sites. The impacts identified ranged in severity from simply elevated concentrations of various constituents directly beneath the landfill to the contamination of major aquifers. In 17 of these case studies, alternate water supplies were required. Most of these problem landfills were located within about 2 meters of the water table, sited in highly permeable soils, or constructed without an effective liner.

Federal regulations from Subtitle D of the Resource Conservation and Recovery Act require the use of soil liners, leachate collection systems, final covers, daily cover of refuse, restrictions concerning the placement of liquids, a program for managing the codisposal of hazardous wastes, postclosure care for a minimum of 30 years, and several other requirements concerning the location of landfills and groundwater monitoring. Because these regulations are relatively recent, very little documentation is available with which to evaluate their efficacy. Most of the active landfills in the United States do not use any type of liner. Researchers in Wisconsin concluded that soil liners can provide an adequate means of leachate containment, but little research has been done on field-scale soil liners to evaluate their performance. Flexible membrane liners may also be required in some states, but there is little actual experience with them, and the synthetic-liner industry is young. The USEPA concluded that if properly located, designed, and operated, future municipal solid waste landfills should be protective of human health and the environment. But few states have adopted the federal criteria, and the process is in a state of flux. Leachate collection systems are rarely used, and the future role of leachate recirculation is uncertain. Less than 2% of the landfills in operation today are attempting to recover the methane. Interest in methane recovery may increase in the future.

Short-term remedies to the landfill crisis include excavating the edges of filled landfills to extend the life of the site, lateral or vertical expansions, and landfill mining. Public opposition to new landfills seems to be least where the land has already been disturbed. The siting of new landfills may depend on the efficacy and support of the local unit of government and the local press and effective public relations. On the basis of numerous case studies and the potential efficacy of Subtitle D guidelines, the public's concern for groundwater contamination by future municipal landfills appears to be unjustified.

Acknowledgments

The author gratefully acknowledges the information provided by the Municipal Solid Waste and Residuals Management Branch of the USEPA in Cincinnati, Ohio: Dr. Michael H. Roulier, Robert Landreth, David Carson, Lynann Hitchens, and Dr. Wendy Hoover. Others who provided helpful information included Marci A. Friedman (Wisconsin Department of Natural Resources), Dr. Thomas C. Voice (Michigan State University), Robert Fahey (Collier County, Florida, County Government), Edward J. Londres (New Jersey Department of Environmental Protection), Prof. Olaf L. Weeks (Waste Management, Inc.), Dr. Richard A. Denison (Environmental Defense Fund), Richard Dweling (US Geological Survey), George

H. Larson (California Waste Management Board), Karen Hill (Texas Environmental Protection Division), Dr. Ralph Piskin (Hydropol, Inc.), William H. Hinkley (Florida Department of Environmental Regulation), Karl Zollner (Michigan Department of Natural Resources), Dr. William F. Pounds (Pennsylvania Department of Environmental Resources), and Drs. Dennis D. Coleman and Keros Cartwright (Illinois State Geological Survey).

References

Andersen, J.R. and Dornbush, J.N. 1967. Influence of sanitary landfill on groundwater quality. *Journal of American Water Works Association,* 59, 457–470.

Anderson, M.A., Ferguson, J.F. and Gavis, J. 1976. Arsenate adsorption on amorphous aluminum hydroxide. *Journal of Colloid and Interface Science,* 54, 391–399.

ATSDR. 1987a. *Toxicological Profile for Methylene Chloride.* Agency for Toxic Substances and Disease Registry (Draft), 120 pp.

ATSDR. 1987b. *Toxicological Profile for Benzene.* Agency for Toxic Substances and Disease Registry (Draft), 182 pp.

ATSDR. 1987c. *Toxicological Profile for Tetrachloroethylene.* Agency for Toxic Substances and Disease Registry (Draft), 112 pp.

Augenstein, D. 1990. Greenhouse effect contributions of United States landfill methane. In: *Proceedings from the GRCDA 13th Annual International Landfill Gas Symposium,* pp.95–125. Governmental Refuse Collection and Disposal Association, Lincolnshire, IL, March 27–29, 1990,

Baedecker, M.J. and Apgar, M. A. 1984. Hydrochemical studies at a landfill in Delaware. In: *Groundwater contamination,* chap.10, pp.127-138. National Academy Press, Washington, DC.

Bagchi, A., Dodge, R.L. and Mitchell, G.R. 1980. Application of two attenuation mechanisms to a sanitary landfill. In: *Proceedings of the Third Annual Madison Waste Conference,* pp.210-213. University of Wisconsin, Madison, September 10–12, 1985.

Battista, J.R. and Connelly, J.1989. *VOC contamination at selected Wisconsin landfills–sampling results and policy implications.* Wisconsin Department of Natural Resources, PUBL-SW-094 89, 74 pp.

Bennett, G.F. 1985. *Fate of solvents in a landfill. Environmental Institute for Waste Management Studies.* Open File Report No. 5, University of Alabama, 82 pp.

Birks, D. 1989. The garbage route: choosing all the options. *Warmer Bulletin,* no. 23, 10–11.

Bonazountas, M. and Wagner, J. 1984. Modeling mobilization and fate of leachates below uncontrolled hazardous wastes sites. In: *The 5th National Conference on Management of Uncontrolled Hazardous Waste Sites,* pp.97-102. Washington, D.C., November 7–9, 1984.

Bukro, C. 1989. Railroads help open the West to garbage. *Chicago Tribune,* December 10.

Burrows, W.D. and Row, R.S. 1975. Ether soluble constituents of landfill leachate. *Journal of the Water Pollution Control Federation,* 47, 921–927.

Brunner, D.R. 1979. Forecasting production of landfill leachate. In: *Proceedings of the Fifth Annual Research Symposium, Orlando, Florida,* pp. 268-282. US Environmental Protection Agency, EPA-600/9-79-023a.

California Waste Management Board. 1988. Landfill gas characterization, 18 pp.

Calvert, C.K. 1932. Contamination of groundwater by impounded garbage waste. *Journal of the American Water Works Association,* 24, 266–270.

Cartwright, K. and McComas, M.R. 1968. Geophysical surveys in the vicinity of sanitary landfills in northeastern Illinois. *Ground Water*, 6, 5–23.

Cartwright, K., R.A. Griffin and Gilkeson, R.H. 1977. Migration of landfill leachate through glacial tills. *Ground Water*, 15, 294–305.

Charnley, G., Crouch, E.A.C., Green, L.C and Lash,T.L. 1988. *Municipal solid waste landfilling: a review of environmental effects.* Meta Systems, Inc., Cambridge, MA, 72pp.

Chian, E.S.K. and DeWalle,F.B. 1977. *Evaluation of leachate treatment.* Vol. I. *Characterization of leachate.* US Environmental Protection Agency, EPA-600/2-77-186a.

Clark, T.P. and Piskin, R. 1977. Chemical quality and indicator parameters for monitoring landfill leachate in Illinois. *Environmental Geology*, 1, 329–339.

Coleman, D.D., Benson, L.J. and Hutchinson, J. 1990. The use of isotopic analysis for identification of landfill gas in the subsurface. In: *Proceedings from the GRCDA 13th Annual International Landfill Gas Symposium, Governmental Refuse Collection and Disposal Association,* pp.213-229, Lincolnshire, IL, March 27–29, 1990.

County of Sonoma. 1973. *Sonoma County refuse stabilization study.* Second Annual Report. Department of Public Works, Santa Rosa, CA.

Crutcher, A.J. and Rovers, F.A. 1978. The design of a natural leachate arrentation system. In: *Proceedings of the First Annual Madison Waste Conference,* pp. 199-223. University of Wisconsin, Madison, September 10–13, 1978.

Daniel, D.E. and Brown, K.W. 1988. Landfill liners: how well do they work and what is their future? In: Cronow, J.R., Schofield, A.N. and Jain, R.K.(eds.), *Land Disposal of Hazardous Wastes,* pp.235-244. Ellis Horwood Limited, West Essex, England.

DeWalle, F.B. and Chian, E.S.K. 1981. Detection of trace organics in well water near a solid waste landfill. *Journal of the American Water Works Association,* 73, 206–210.

Donigian, A.S., Brown, S.M. and Yabusaki, S.B. 1983. *Groundwater modeling of selected hydrogeological settings to determine leachate fate and migration from waste facilities.* US Environmental Protection Agency, Contract Report No. 68-03-3116.

Dunlap, W.J., Shew,D.C., Robertson, J.M. and Toussaint,C.R. 1976. Organic pollutants contributed to groundwater by a landfill. In: *Proceedings of a Research Symposium on Gas and Leachate from Landfills.* US Environmental Protection Agency, New Brunswick, NJ, March 25–26, 1975, EPA-600/9-76-004, 96–110.

Duwelius, R.F. and Greeman, T.K. 1989. *Geohydrology, simulation of groundwater flow and groundwater quality at two landfills.* Marion County, Indiana. US Geological Survey, Water-Resources Investigations Report 89-4100, 135 pp.

Ehrenfeld, J.R., Ong, J.H. Farino, W., Spawn, P., Jasinsli, M., Murphy, B., Dixon,D. and Rissmann,F. 1986. *Controlling Volatile Emissions at Hazardous Wastes Sites. Pollution Technology Review,* No. 126. Noyes Publication, Park Ridge, New Jersey, 412pp.

Eliassen, R. 1942. Decomposition of landfills. *American Journal of Public Health,* 32, 1029–1037.

Engelbrecht, R.S. 1973. *Survival of viruses and bacteria in a simulated sanitary landfill. American Tissue Institute.* NTIS PB-234 589, 50 pp.

Farquhar, G.J. 1977. Leachate treatment by soil methods. In: *Proceedings of the Third Annual Municipal Solid Waste Research Symposium, Management of gas and leachate in landfills,* pp.187-207. US Environmental Protection Agency, St. Louis, Missouri, March 14–16, 1977, EPA-600/9-77-026.

Federal Register. Oct. 9, 1991. *Solid Waste Disposal Facility Criteria;* Final Rule,. 56, 50978-51119.

Franklin Associates, Ltd. 1986. *Characterization of municipal solid waste on the United States, 1960 to 2000.* Report prepared for the US Environmental Protection Agency.

Franklin Associates, Ltd. 1988. *Characterization of municipal solid waste in the United States, 1960 to 2000—Update 1988.* Contract No. 68-01-7310, US Environmental Protection Agency.

Friedman, M.A. 1988. *Volatile organic compounds in groundwater and leachate at Wisconsin landfills.* Wisconsin Department of Natural Resources, PUBL-WR-192, 79 pp.

Fuller, W.H. 1978. *Investigation of landfill leachate pollutant attenuation by soils.* US Environmental Protection Agency, EPA-600/2-78-156.

Gaby, W.L. 1975. Evaluation of the health hazards associated with solid waste/sewage sludge mixtures. US Environmental Protection Agency, EPA-670/2-75-023, 44 pp.

Glynn, W.K. 1985. Treatability of contaminated groundwater from two hazardous waste sites. In: *Proceedings of the Eighth Annual Madison Waste Conference, pp.17-36.* University of Wisconsin, Madison, September 18–19, 1985.

Gominger, D. and Lyon, D.A. 1981. Comprehensive evaluation of the abandoned LiPari Landfill. *Hazardous solid waste testing: First conference, ASTM STP 760, pp.321-328.* American Society for Testing and Materials, Ft. Lauderdale, FL, January 14–15, 1981.

Gordon, M.E. and Huebner, M. 1983. An evaluation of the performance of zone of saturation landfills in Wisconsin. In: *Proceedings of the Sixth Annual Madison Waste Conference, pp.23-53.* University of Wisconsin, Madison, September 14–15, 1983.

Gordon, M.E.,Huebner, P.M. and Kmet, P. 1984. An evaluation of the performance of four clay-lined landfills in Wisconsin. In: *Proceedings of the Seventh Annual Madison Waste Conference*, pp.399-460. University of Wisconsin, Madison, September 11–12, 1984.

Gossett, J.M. 1987. Measurement of Henry's law constants for C_1 and C_2 chlorinated hydrocarbons. *Environmental Science and Technology*, 21, 202–208.

Grossman, D. and Shulman,S. 1990. Down in the dumps. *Discover*, 11, 36–41.

Griffin, R.A. and Roy, W.R. 1985. *Interaction of organic solvents with saturated soil-water systems.* Environmental Institute for Waste Management Studies, Open File Report No. 3, University of Alabama, 86 pp.

Griffin, R.A. and Shim, N.F. 1976. Effect of pH on exchange-adsorption or precipitation of lead from landfill leachates by clay minerals. *Environmental Science and Technology*, 10, 1256–1261.

Griffin, R.A. and Shim, N.F.1978. *Attenuation of pollutants in municipal landfill leachate by clay minerals.* US Environmental Protection Agency, EPA-600/2-78-157.

Griffin, R.A., Au, A.K. and Frost, R.R. 1977. Effect of pH on adsorption of chromium from landfill-leachate by clay minerals. *Journal of Environmental Science and Health*, A12(8), 431–449.

Griffin, R.A., Shimp, N.F., Steele, J.D., Ruch,R.R., White, W.A. and Hughes, G.M. 1976. Attenuation of pollutants in municipal landfill leachate by passage through clay. *Environmental Science and Technology*, 10, 1262–1268.

Ham, R.K. 1979. Predicting gas generation from landfills. *Waste Age*, 10, 50–58.

Hassett, J. J., Banwart, W.L. and Griffin, R.A. 1983. Correlation of compound properties with soil sorption characteristics of nonpolar compounds by soils and sediments: concepts and limitations. In: Francis, C. W. and Auerback, S.I. (eds.), *Environment and Solid Wastes: Characterization, Treatment and Disposal*, chap.15, pp.161-178. Butterworth.

Huebner, P. M., Gordon, M.E. and Miazga, T.J. 1989. An evaluation of the quality of measured seepage from a clay-lined landfill in Wisconsin. In: *Proceedings of the Twelfth Annual Madison Waste Conference*, pp.78-97. University of Wisconsin, Madison, September 20–21, 1989.

Hughes, G. M., Landon, R.A. and Farvolden, R.N. 1971. *Hydrogeology of solid waste* disposal sites in northeastern Illinois. US Environmental Protection Agency, Solid Waste Management Series, Report SW-12d.

Illinois Hazardous Waste Task Force. 1983. *Report to the Committee on Land Disposal*, 42 pp.

Illinois Environmental Protection Agency. 1990. *Environmental Progress*, XV, 8 pp.

Iowa Department of Natural Resources. 1988. *Iowa sanitary landfill survey*: June 1988.

Jacobs, L. W., Syers, J.K. and Keeney,D.R. 1970. Arsenic sorption by soils. *Soil Science Society of America Proceedings*, 34, 750–754.

James, S. C. 1977. Metals in municipal landfill leachate and their health effects. *American Journal of Public Health*, 67, 429–432.

Johnson, C. A. 1985. Success in siting solid waste facilities. In: *Proceedings of the Eighth Annual Madison Waste Conference*, pp.220-235. University of Wisconsin, Madison, September 18–19, 1985.

Kmet, P. and McGinley, P.M. 1982. Chemical characteristics of leachate from municipal solid waste landfills in Wisconsin. In: *Proceedings of the Fifth Annual Madison Waste Conference*, pp.225-254. University of Wisconsin, Madison, September 22–24, 1982.

Krapac, I.G., Panno, S.V., Rehfeldt,K.R., Herzog, B.L., Hensel, B.R. and Cartwright, K. 1989. Hydraulic properties of an experimental soil liner: preliminary results. In: *Proceedings of the Twelfth Annual Madison Waste Conference*, pp.395-411. University of Wisconsin, Madison, September 20–21, 1989.

LaMoreaux, P.E., Moffett, T.B. and Whitaker, L.E. 1988. Landfill capacity: evaluating the estimates. *World Wastes*, July, 1988, pp.64-66.

Landon, R. A. 1978. Pollution prediction techniques for waste disposal siting—a state of the art assessment. In: *Proceedings of the First Annual Madison Waste Conference*, pp.169-198. University of Wisconsin, Madison, September 10–13, 1978.

Lang, R., Herrera, T.A., Chang, D.P.Y., Tchobanoglous, G. and Spicher, R.G. 1987. *Trace organic constituents in landfill gas*. California Waste Management Board, 107 pp.

LaRegina, J., Bozzelli, J.W., Harkov, R. and Gianti, S. 1986. Volatile organic compounds at hazardous waste sites and a sanitary landfill in New Jersey. *Environment Progress*, 5, 18–27.

Lindorff, D. E. and Cartwright, K. 1977. Ground-water contamination: problems and remedial actions. Illinois State Geological Survey, *Environmental Geology Notes* 81, 58 pp.

Lu, J.C.S., Eichenberger, B. and Stearns, R.G. 1985. Leachate from municipal landfills. *Pollution Technology Review* No. 119, Noyes Publications, Park Ridge, NJ.

Mabey, W.R., Smith, J.H., Podoll, R.T., Johnson, H.L., Mill, T., Chou, T.W., Gates, J., Partridge, I.W., Jaber, H. and Vandenberg, D. 1982. *Aquatic fate process data for organic priority pollutants*. US Environmental Protection Agency, EPA 440/4-81-014.

Mac Namara, E.E. 1971. Leachate from landfilling. *Compost Science*, 12, 10–14.

Madsen, F.T. and Mitchell, J.K. 1987. *Chemical effects on clay hydraulic conductivity and their determination*. Environmental Institute for Waste Management Studies, Report No. 13, University of Alabama, Tuscaloosa, AL, 70 pp.

Mao, M.C.M. and Pohland, F.G. 1973. *Continuing investigations and landfill stabilization with leachate recirculation*. Special Research Report, Department of Civil Engineering, Georgia Institute of Technology, Atlanta, GA.

McBride, J.H. 1958. Sanitary landfill was a campaign promise. *American City*, 73, 38.

Michaels, A. 1990. Landfills. *Public Works*, 121, 70–71.

Miller, D.W., DeLuca, F.A. and Tessier, T.L. 1974. *Groundwater contamination in the northeast states*. US Environmental Protection Agency, EPA-660/2-74-056, 280 pp.

Morelli, J. 1989. *Municipal solid waste landfills: optimization, integration and reclamation.* Report by the New York State Energy and Development Authority, Albany, NY.

National Science Foundation. 1983. *Groundwater resources and contamination in the United States.* Policy Research and Analysis Report 83-12, 63 pp.

Nelson, B.R. and Book, R. 1986. Monitoring for volatile organic hydrocarbons at Minnesota sanitary landfills. In: *Proceedings of the Ninth Annual Madison Waste Conference,* pp.72-84 University of Wisconsin, Madison, September 9–10, 1985.

New Jersey Department of Environmental Protection. 1985. *Solid Waste Management Plan. Draft Update: 1985–2000.* 160pp.

Nicholson, R.V., Cherry, J.A. and Reardon, E.J. 1983. Migration of contaminants in groundwater at a landfill: a case study. 6. Hydrogeochemistry. *Journal of Hydrology,* 63, 131–176.

Nirmalakhandan, N.N. and Speece, R.E. 1988. QSAR model for predicting Henry's constant. *Environmental Science and Technology,* 22, 1349–1357.

Olsen, R.L. and Davis, A. 1990. Predicting the fate and transport of organic compounds in groundwater. *Hazardous Materials Control,* 3, 38–64.

Ostenso, N.A. and Gordon, M.E. 1980. Leachate problems and solutions—a case study. In: *Proceedings of the Third Annual Madison Waste Conference,* pp.343-362. University of Wisconsin, Madison, September 10–12, 1980.

Overcast, C.M. and Heintz, E.L. 1990. *Where will the garbage go?* A status report on solid waste management and disposal in New York State. The Legislative Commission on Solid Waste Management, 33 pp.

Pahren, H.R. 1987. Microorganisms in municipal solid waste and public health implications. *CRC Critical Reviews in Environmental Control,* 17, 187–228.

Pettijohn, R.A. 1977. Nature and extent of groundwater quality changes resulting from solid-waste disposal, Marion County, Indiana. *US Geological Survey Water Resources Investigations* 77-40, 119 pp.

Pohland, F.G. and Harper, S.R. 1986. *Critical review and summary of leachate and gas production from landfills.* US Environmental Protection Agency, Cooperative Agreement CR 809 997, NTIS PB 86-240-181.

Rai, D., Zachara, J.M., Schwab, A. P., Schmidt, R.L., Girvin, D.C. and Rogers, J.E. 1986. *Chemical attenuation rates, coefficients and constants in leachate migration.* Vol. 1: A critical review. Electric Power Research Institute, EPRI EA-3356.

Rathje, W.L. 1990. *The solid-waste crisis,* Vol.57, pp.20-21. American Public Works Association Reporter.

Roberts, P.V. and Dandliker, P.P.G.. 1983. Mass transfer of volatile organic contaminants from aqueous solution to the atmosphere during surface aeration. *Environmental Science and Technology,* 17, 484–489.

Roberts, P.V., Schreiner, J.E. and Hopkins, G.D. 1982. Field study of organic water quality changes during groundwater recharge in the Palo Alto baylands. *Water Res.,* 16, 1025–1035.

Robertson, J. M., Toussaint, C.R. and Jorque, M.A. 1974. *Organic compounds entering groundwater from a landfill.* US Environmental Protection Agency, EPA 660/2-74-077, 47 pp.

Rossler, B. 1950. Influence of garbage and rubbish dumps on groundwater. *Vom Vasser,* 18, 43–45.

Rovers, F.A. and Farquhar, G.J. 1973. Infiltration and landfill behavior. *Journal of the Environmental Engineering Division, EE5,American Society of Civil Engineers,* 99, 671-690.

Roy, W.R. and Griffin, R.A. 1985. Mobility of organic solvents in water-saturated soil materials. *Environmental Geology and Water Sciences*, 7, 241–247.

Roy, W.R. and Griffin, R.A. 1987. Estimating threshold values for the land disposal of organic solvent-contaminated wastes. *Journal of Hazardous Materials*, 15, 365–376.

Roy, W.R. and Griffin, R.A. 1990. Vapor-phase interactions and diffusion of organic solvents in the unsaturated zone. *Environmental Geology and Water Sciences*, 15, 101–110.

Roy, W.R., Krapac, I.G., Chou, S.F.J. and Griffin, R.A. 1992. *Batch-type procedures for estimating soil adsorption of chemicals.* US Environmental Protection Agency, Technical Resource Document, EPA/530/SW-87-006-F.

Sabel, G.V. and Clark, T.P. 1983. Volatile organic compounds as indicators of municipal solid waste leachate contamination. In: *Proceedings of the Sixth Annual Madison Waste Conference*, pp.108-125. University of Wisconsin, Madison, September 14–15, 1983.

Scarpino, P.V., Donnelly, J.A. and Brunner, D. 1979. Pathogen content of landfill leachate. In: *Proceedings of the Fifth Annual Research Symposium, Municipal Solid Waste: Land Disposal*, pp.138-167. US Environmental Protection Agency, Orlando, FL, March 26–28, 1979.

Schumacher, M.M. (ed.). 1983. *Landfill methane recovery. Energy Technology Review* No. 84, Noyes Data Corporation, Park Ridge, NJ.

Seitz, H.R., Wallace, A.T. and Williams, R.E. 1972. Investigation of a landfill in granite-loess terrain. *Groundwater*, 10, 35–41.

Solid Wastes Management. 1972. Composition of rubbish in the United States. *Solid Wastes Management*, 15, 74.

Spencer, R. 1990. Landfill space reuse. *Biocycle*, 31, 30–33.

Sridharan, L. and Didier, P.P. 1988. Leachate quality from containment landfills in Wisconsin. In: *Proceedings of the Fifth International Solid Wastes Exhibition and Conference*, pp.133-138. International Solid Waste Association, Denmark, September 11–16, 1988.

Stecker, P.P. and Garvin, J.W. 1983. Control and prevention of landfill leachate seeps. In: *Proceedings of the Sixth Annual Madison Waste Conference*, pp 256-270. University of Wisconsin, Madison, September 14–15, 1983.

Tabak, H.H., Quave, S.A., Mashni, C.I. and Barth, E.F. 1981. Biodegradability studies with organic priority pollutant compounds. *Journal Water Pollution Control Federation*, 53, 1503–1518.

US Environmental Protection Agency. 1978. *Chemical and physical effects of municipal landfills on underlying soils and groundwater.* EPA-600/2-78-096.

US Environmental Protection Agency. 1979. *A 1974 summary report: municipal solid waste generated gas and leachate.* USEPA Contract Report No. 68-03-1339.

US Environmental Protection Agency. 1983. *Treatability Manual*, Vol. 1. *Treatability Data.* EPA-600/2-82-001a.

US Environmental Protection Agency. 1987. *Characterization of MWC ashes and leachates from MSW landfills, monofills and co-disposal sites.* Vol.V. EPA-530-SW-87-28E.

US Environmental Protection Agency. 1988a. *Report to Congress: Solid Waste Disposal in the United States.* Vol. I. EPA/530-SW-88-011.

US Environmental Protection Agency. 1988b. *Report to Congress: Solid Waste Disposal in the United States.* Vol. II. EPA/530-SW-88-011B.

US Environmental Protection Agency. 1988c. *Design, construction and evaluation of clay liners for waste management facilities.* EPA/530-SW-86-007F.

US Environmental Protection Agency. 1989. *Requirements for hazardous waste landfill design, construction and closure.* EPA/625/4-89/022.

van Genuchten, M. Th. 1978. Simulation models and their application to landfill disposal sitting: a review of current technology. In: *Proceedings of the Fourth Annual Research Symposium*, pp.191-214 San Antonio, Texas. US Environmental Protection Agency, EPA-600/9-78-016.

Vasuki, P.E. 1988. Why not recycle the landfill? *Waste Age*, 19, 165–170.

Ware, S. A. 1980. *A survey of pathogen survival during municipal solid waste and manure treatment processes*. US Environmental Protection Agency, EPA-600/8-80-034.

Watson, T. 1988. *Recycling the landfill: the mining of disposal sites. Resource Recycling*, 7, 20–21, 58–59.

Weaver, L. 1961. Refuse disposal, its significance. In: *Proceedings of the 1961 Symposium on Groundwater Contamination*, pp.104-110 US Department of Health, Education and Welfare, Cincinnati, Ohio, April 5–7, 1961.

Wuellner, W.W., Wierman, D.A. and Koch, H. A. 1985. Effect of landfill leachate on the permeability of clay soils. In: *Proceedings of the Eighth Annual Madison Waste Conference*, pp.287-302. University of Wisconsin, Madison, September 18–19, 1985.

Young, P.J. and Heasman, L.A. 1985. An assessment of the odor and toxicity of the trace components of landfill gas. In: *Proceedings of the Eighth International Landfill Gas Symposium, Governmental Refuse Collection and Disposal Association*, pp.175-181. San Antonio, Texas, April 11–13, 1985.

Zanoni, A.E. 1972. Groundwater pollution and sanitary landfills—a critical review. *Groundwater*, 10, 3–13.

Zenone, C., Donaldson, D.E. and Grunwaldt, J.J. 1975. Groundwater quality beneath solid-waste disposal sites at Anchorage, Alaska. *Ground Water*, 13, 182–190.

Zoeteman, B.C.J., De Greef, E. and Brinkman, F.J.J. 1981. Persistency of organic contaminants in groundwater, lessons from soil pollution incidents in the Netherlands. *Science of the Total Environment*, 21, 187–202.

15 Aqueous behavior of elements in a flue gas desulfurization sludge disposal site

Dhanpat Rai[1], John M. Zachara[1], Dean A. Moore[1], and Ishwar P. Murarka[2]

[1]*Battelle, Pacific Northwest Laboratories, Battelle Boulevard, Richland, Washington 99352, USA*
[2]*Electric Power Research Institute, Palo Alto, California, USA*

Abstract
Field and laboratory methods were developed to preserve the chemical integrity of highly reduced utility waste leachate and to measure such important parameters as pH and redox potentials. A combination of field measurements and laboratory experiments with core samples of flue gas desulfurization (FGD) sludge and of soils underlying the sludge were used to identify geochemical reactions controlling elemental concentrations in the sludge pore waters and solute migration.

Pore waters in the sludge were found to be uniform both spatially and with depth in the sludge basin. The pore waters were highly reducing, containing sulfide at parts per million levels. It was found that precipitation/dissolution reactions with sulfate, sulfite, and sulfide solids controlled the concentrations of many elements (Ba, Ca, Cd, Cu, Fe, Ni, Pb, S, Sr, and Zn). Through a combination of techniques, (1)X-ray diffraction, (2)comparison of observed ion activity products with the thermodynamic data for given solid phases, and (3)identification of solids by spiking the pore waters or sludge suspensions with a given element, the solubility-controlling solids for these elements were hypothesized to be $BaSO_4/(Ba,Sr)SO_4$ $CaSO_4 \cdot H_2O$, $CaSO_3 \cdot 0.5H_2O$, $CaCO_3$, CdS, $Cu_2S/Cu_{1.65}S$, FeS, NiS, PbS, $SrSO_4/(Ba,Sr)SO_4$, and ZnS. In addition, the aqueous concentrations of Cr and Se were inferred to be controlled by $Cr(OH)_3$ and $FeSe_2/FeSe$, respectively. The aqueous concentrations of many elements (Cd, Cr, Cu, Ni, Pb, Se, V, and Zn) in sludge pore waters were either at or near the detection limits (parts per billion range). Therefore, these elements were neither expected nor found to migrate into underlying soils over the year disposal period. Migration into underlying soil/subsoil of those elements that were found to be present in measurable concentrations in the sludge pore waters was also minimal because of the low permeability of the soil and its high attenuation capacity. Boron was the most mobile element, and in approximately 8 years it had migrated slightly more than 1 m into the underlying soil, primarily through diffusion. Most other elements that were in detectable concentrations in the sludge pore waters were strongly attenuated by the underlying soil.

Introduction

Flue gas desulfurization (FGD) sludges result from the use of calcium carbonate, calcium oxide (lime), Na_2CO_3, or highly alkaline fly ashes to remove (or scrub) SO_2 from the flue gas stream produced during the burning of coal to generate electricity. According to a 1983 estimate, about 5 million short tons of FGD sludges were produced annually; this figure is expected to quadruple by the year 2000 (Murarka, 1987). The chemical properties of this waste form are highly dependent on the scrubbing agent, the source and composition of the coal, and the degree of entrapment of fly ash (Ainsworth and Rai, 1987). The FGD sludges characteristically contain unused scrubbing agents such as $CaCO_3$, as well as compounds formed by the chemical reactions between gaseous S species and the scrubbing agent, and some entrained fly ash. The calcium compounds reported in the unweathered FGD sludges are $CaSO_3.0.5H_2O$, $CaSO_4.2H_2O$, $CaSO_4$, $CaCO_3$, $Ca(OH)_2$, $CaAl_2(SO_4)_3(OH)_{12}.26H_2O$ (Selmeczi and Knight, 1973; Liem *et al.*, 1982; McCarthy *et al.*, 1983; Groenewold and Manz 1982). To a lesser extent, heavy metals and other chemicals from the flue gas and the fly ash are found in the sludge.

Typically, in the wet-scrubbing process, the FGD sludge is sluiced to disposal ponds. Cost-effective disposal designs that will protect the environment require that the fate of water-soluble constituents in the wastes be accurately predicted. The soluble constituents may be present initially in the sluice waters or they may arise from long-term chemical reactions and changes in the waste materials after disposal. The available literature prior to 1987 on the total chemical composition, mineralogical composition, and leachability of FGD sludges was reviewed by Rai *et al.* (1987). This review showed that only limited data are available for the FGD sludges as compared with the other wastes. The general conclusions that can be drawn from this review are (1) volatile elements (*e.g.*, B, F, and Se) are generally concentrated in fly ash and FGD sludges, (2) except for the major Ca-S minerals as noted above, data on the types of other solid phases present are limited and (3) reliable information on the leachate characteristics or reactions that may control aqueous concentrations in FGD sludges does not exist.

Because of the lack of information on the environmental behavior of FGD sludge, field and laboratory studies were conducted (Rai *et al.*, 1989; and Smith *et al.*, 1991, in press) to ascertain the leaching characteristics of these wastes and to determine the fundamental chemical reactions that may control aqueous concentrations of different elements and that may be used to accurately predict longterm leaching behavior of these wastes. Intuitively, it was expected that flue gas desulfurization sludges and the sites to which they were disposed would differ. Accordingly, the leachate generated at different sites was also expected to vary in its composition and quality. Therefore, extensive studies of the FGD sludges generated at five different sites were conducted (Rai *et al.*, 1989; Smith *et al.*, 1991). These included four sites where $CaCO_3$ is used as the scrubbing agent and one site where Na_2CO_3 is used as the scrubbing agent. The $CaCO_3$ scrubbing sites included both forced air oxidized and non-forced air oxidized sludges. Both the forced air and nonforced air sites from the $CaCO_3$ scrubbers showed similar chemical properties (Smith *et al.*, 1991). The major cation and anion in pore waters from these sites were found to be Ca(~700 ppm) and SO$_4$ (~1,500 ppm), respectively. The major cation and anion in sluice water of the sites using Na$_2$CO$_3$ were found to be Na (18,000 to 78,000 ppm) and SO$_4$ (54,000 to 211,000 ppm), respectively (Smith *et al.*, 1991). Although microbiologically mediated production of H$_2$S was evident in all sites, H$_2$S emissions from the FGD sludge disposal ponds in concentrations high enough for environmental concern were observed only in sites using Na$_2$CO$_3$ as the scrubber agent.

Of all the coal fired power plants in the United States that have SO$_2$ scrubbers, about 85% use Ca-based (CaO, CaCO$_3$) scrubbing agents. Therefore, an eight-year-old field site containing FGD sludge produced from CaCO$_3$ as the scrubbing agent was chosen as an example for this chapter for discussion of field leaching and attenuation behavior. The study successfully identified many of the chemical reactions that control aqueous concentrations of major and minor elements. These reactions are not solely site specific. Rather, the reactions are of generic importance and were in later studies at other FGD sludge sites also found to control pore water chemistry and leaching because of commonality in the chemical characteristics of FGD sludge.

Methods and materials

Description of the field site
The wastes studied originated from a wet limestone flue gas desulfurization (FGD) scrubber. The unit burned a local high-sulfur, high-ash coal, typically containing 5.5% S and 25% ash. A water slurry of finely ground local limestone was used as the scrubbing agent. The scrubber, designed to capture 80% of the S and the flyash, produced FGD solids at a rate of about 3,500 tons per hour. These solids were transported as a slurry to a system of settling ponds covering a total area of 1,820,700 square meters (450 acres), including the study site. Decantate from the sludge ponds was recycled through the scrubbing system.

The deposits studied covered 315,588 m^2 and contained an estimated 1,446,456 m^3 of sludge. Produced between 1973 and 1979, these solids were dredged from their original location in the old waste pond and deposited at their present location during 1976, 1977, and 1979, with the greatest amount being moved during 1979. This suggests that the sludge interfacing with the underlying soils was placed in 1976 or 1977, with placement of the upper portions continuing into 1979.

The FGD waste disposal site is in the central midwestern United States and receives approximately 63.5-76.2 cm of precipitation annually. Some drainage of the sludge has occurred since dredging, allowing most of the sulfites to be oxidized into sulfates in the upper unsaturated zone, which varies in thickness from a few cm up to about 1.2m. Natural and planted grasses, forbs, and trees now cover most of this dredged FGD sludge disposal area. The sludge deposit ranges from 2 to 9 m in depth in the site. The static water table in the sludge ranges from a depth of approximately 20 to 100 cm. The site is located on the flanks of a gradually sloping fluvial basin that drains to the southwest. The downslope sides have been diked with bottom ash to contain the FGD sludge.

The unsaturated surface zone of the sludge was found to be oxidized and brownish-gray in color, and its physical consistency was that of natural soil materials. At depths ranging from 0.2 to 1.0 m, however, the sludge was found to be water saturated, highly reducing, gray to dark gray in color, and gellike in consistency. Thus the physical stability of the FGD sludge varied with its water content and oxidation/reduction status. The reduced sludge supported little weight. The presence of highly reducing sludge near the ground surface attested to the limited oxygen diffusion that apparently occurs through the upper sludge layers.

Collection of cores

Cores of the FGD sludge and underlying soil materials were taken in June 1986 with a hollow-stem rotary auger corer. The core materials were collected in 6 x 76 cm Lexan core barrels. Tight-fitting end caps were placed on the cores immediately after collection and taped carefully to minimize gas exchange. Cores were taken at three locations separated by approximately 100-150 m across the old sludge pond. These locations were designated the north (N), central (C), and southwest (SW) coring points according to their location in the sludge basin. At each coring site, pH and Eh were measured with depth using a specially designed probe (Rai *et al.*, 1989) that was pushed into the sludge. Soil materials from under the sludge were obtained at only the N and SW sampling points. The cores were stored at 4°C and analyzed soon after collection.

Displacement of pore waters and analyses

Different methods were used to displace pore waters from the sludge and the underlying soils because the two types of samples had different physical properties. The sludge solids settled during transportation, and at the time of pore water extraction approximately 20% of the core volume was free water. All pore water extractions and *in situ* measurements were performed on the sludge cores in horizontal position in an N_2-filled anoxic glovebox. A sampling port was cut in the side of the core barrel, and pH and Eh electrodes were inserted in a manner that minimized gas exchange with the glove box. Following *in situ* measurement of pH and Eh, pore waters were removed using a peristaltic pump (Figure 1). The salient characteristics of this sampling procedure included the following: (1) the solution did not come into contact with even the low-O_2 and low-CO_2 glove box air; (2) inline filtration (0.22 m) prevented contact with air during filtration; (3) the aqueous sample was collected in an in-line glass sampler equipped with stopcocks and a rubber septum to withdraw aliquots for analysis without air contact. The extracted pore water was preserved in different ways. The chemical stabilizing agents included sulfur antioxidant buffer (SAOB, NaOH + ascorbic acid + EDTA) for S^2 (1:1 mixture of sample and SAOB), concentrated Ultrex HCl for metals (10µL for 2 mL sample), and formaldehyde (HCOH) for SO_3^{2-} (sample diluted to around 3% with HCOH). Samples for dissolved inorganic carbon were placed in glass septum vials without head space and sealed until analysed. Other samples for

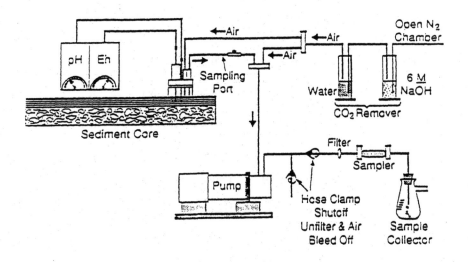

Figure 1. *Schematic illustration of the apparatus used for the extraction of pore water from sludges.*

stable major and minor cations and anions were placed in 3 mL plastic test tubes (with or without acid; see discussion below) and refrigerated until analysed.

In contrast to the sludge cores, the soil cores retained their physical integrity during storage. Visually, the subsoils ranged from saturated at the sludge-soil interface to partially saturated at depths of 2-3 m below the interface. Pore waters were removed by immiscible displacement with the dense organic liquid Freon12 (Kinniburgh and Miles, 1983). The core segments were subsampled in an anoxic glovebox filled with N_2. Subsamples approximately 100 g in size were placed in 250 mL plastic centrifuge tubes, and 150 mL of Freon12 was added to the soil material. The tubes were centrifuged at 2,500 g for 4 h at 25°C. After centrifugation, the extracted pore water, which floats on top of the Freon, was removed in the anoxic glovebox. The pore water was filtered through both 0.22 m and 0.0018 m filters. Samples were preserved and handled in the same way as the pore waters from the sludge.

Major and minor cations (Ba, Ca, K, Li, Mg, Na, and Sr) and metals (Al, Cd, Cr, Cu, Mn, Mo, Ni, Pb, Ti, V, and Zn) were determined on acidified (with Ultrex HCl) and nonacidified filtered samples by inductively coupled argon plasma spectroscopy (ICAP) or, in a few cases, by an atomic absorption spectrometer equipped with a graphite furnace. Silica and B were measured by ICAP. Anions (SO_4, S_2O_3, SCN, PO_4, NO_3, Br, Cl, I), SO_3 (with HCOH preservation), and valence species of iron (Fe^{2+}, Fe^{3+}) were analyzed by ion chromatography. Ammonium (NH_4) and F were analyzed by ion-selective electrode. Sulfide (S^2) was determined by potentiometric titration with Pb. Organic carbon was measured with a carbon analyzer using

Table 1 Standard chemical potentials and ioninteraction parameters for key species.

Species	Standard chemical potentials $\mu°/RT$	Reference
H_2O	–95.664	Harvie *et al.*, 1984
H^+	0.000	Harvie *et al.*, 1984
Ba^{2+}	–226.204	Robie *et al.*, 1978
Ca^{2+}	–223.300	Harvie et al., 1984
Na^+	–105.651	Harvie *et al.*, 1984
Sr^{2+}	–225.679	Robie *et al.*, 1978
Cl	–52.955	Harvie *et al.*, 1984
SO_4^{2-}	–300.386	Harvie *et al.*, 1984
HSO_4^-	–304.942	Harvie *et al.*, 1984
SO_3^{2-}	–196.272	Wagman *et al.*, 1982
HSO_3^-	–212.906	Wagman *et al.*, 1982
$BaSO_{4(aq)}$	–532.850	Felmy *et al.*, 1990
$CaSO_{3(aq)}$	–425.603	Rai *et al.*, 1991
$CaSO_3.O.5H_2O(c)$	–482.704	Rai *et al.*, 1991
$BaSO_{4(c)}$	–549.740	Felmy *et al.*, 1990
$SrSO_{4(c)}$	–541.304	Felmy *et al.*, 1990

persulfate-enhanced UV oxidation, and inorganic carbon was determined with acid sparging and IR detection.

The results of the pore water analyses from sludge and soil were input to the chemical equilibrium code MINTEQ (Felmy *et al.*, 1984) or GMIN (Felmy, 1990) to calculate the activities of aqueous species. The activities of specific combinations of these individual species were used to calculate ion activity products (IAP) that were compared to equilibrium IAPs for mineral solids plausible for the sludge disposal environment. This procedure allowed for the indirect identification of solids that were in equilibrium with the aqueous phase and that were controlling the aqueous concentrations of their constituent ions. The MINTEQ model uses the ion

Table 1 (continued)   Binary ion interaction parameters

	β^0	β^1	β^2	C^-	Reference
H^+-Cl	0.17750	0.29450	0.00000	0.00080	Harvie *et al.*, 1984
H^+-SO_4^{2-}	0.02980	0.000000	0.00000	0.04380	Harvie *et al.*, 1984
Ba^{2+}-Cl$^-$	0.26280	1.49600	0.00000	-0.01938	Pitzer and Mayorga,1973
Ca^{2+}-Cl	0.31590	1.61400	0.00000	-0.00034	Harvie *et al.*, 1984
Ca^{2+}-SO_3^{2-}	0.20000	3.19730	0.00000	0.00000	Rai *et al.*, 1991
Ca^{2+}-SO_4^{2-}	0.20000	3.19730	-54.24000	0.00000	Harvie *et al.*, 1984
Na^+-Cl	0.07650	0.26640	0.00000	0.00127	Harvie *et al.* 1984
Na^+-SO_3^{2-}	0.07890	1.17100	0.00000	-0.00370	Rai *et al.*, 1991
Na^+-SO_4^{2-}	0.01958	1.11300	0.00000	0.00497	Harvie *et al.*, 1984
Sr^{2+}-SO_4^{2-}	0.20000	3.19730	-54.24000	0.00000	Felmy *et al.*,1990

Ternary ion interaction parameters

		Reference
H^+ -Ca^{2+}	0.092	Harvie *et al.* 1984
H^+-Ca^{2+}-Cl	-0.01500	Harvie *et al.*, 1984
H^+ -Na^+	0.03600	Harvie *et al.*, 1984
H^+-Na^+-Cl	-0.00400	Harvie *et al.*, 1984
Ca^{2+}-Na^+	0.07000	Harvie *et al.*, 1984
Ca^{2+}-Na^+-Cl	-0.00700	Harvie *et al.*, 1984
Ca^{2+}-Na^+-SO_4^{2-}	-0.055	Harvie *et al.*, 1984
Cl$^-$-SO_4^{2-}	0.02000	Harvie *et al.*, 1984
Cl$^-$-SO-4^{2-}-Ca^{2+}	-0.01800	Harvie *et al.*, 1984
Cl$^-$-SO_4^{2-}-Na^+	0.00140	Harvie *et al.*, 1984

association approach and the GMIN model uses the ion interaction approach of Pitzer and coworkers, valid to very high ionic strengths. The GMIN model was used for systems in which reliable information for the solid and aqueous species were available. Since the outputs from any modeling calculations depend on the input data, we have listed (Table 1) the key thermodynamic data used in interpreting the results of those aqueous ions that were found to be solubility controlled and were present in measurable concentrations.

Geochemistry of FGD sludges

Bulk chemistry and solid-phase characterization
The sludge disposed of at this site was produced from the use of limestone as a flue gas scrubbing agent. Therefore, the concentrations of all of the elements other than Ca and C in the sludge result from (1) noncarbonate minerals and other impurities in the limestone, (2) adsorption of gases by $CaCO_3$, or (3) the entrapment of fly ash particles during scrubbing. Some variation in total chemical composition of samples from different locations in the disposal basin is therefore expected, depending on the degree of entrapment of fly ash particles.

Given the use of $CaCO_3$ as the scrubbing agent at this plant, it is not surprising that the sludge was very high in Ca (~27%). The other elements present in concentrations greater than 0.38% were Al, Fe, K, Mg, Na, S, Si, and Zn. The concentrations of these elements in this sludge, with the exception of Zn, were within a factor of about 2 of the concentration reported (Ainsworth and Rai, 1987) for sludges from other locations. The average concentrations of Zn in this sludge was about 40 times higher than that in other sludges. Most of the elements of interest, from an environmental standpoint, were present in very low quantities. Arsenic, Cd, Cr, Cu, Ni, and V were present in concentrations less than 0.01%, and Ag, Hg, Mo, Sb, Se, and Sn were present in concentrations less than 0.001%.

Selected samples were analyzed by X-ray diffraction to identify the crystalline components. The samples contained primarily $CaCO_3$ (calcite) and $CaSO_3 \cdot 0.5H_2O$ (hannebachite), with lesser amounts of $CaSO_4 \cdot 2H_2O$ (gypsum) and SiO_2 (quartz).

Redox potentials of pore waters from flue gas desulfurization sludges
The pH and Eh are master variables that affect the aqueous concentrations of elements by controlling their precipitation/dissolution and adsorption/desorption reactions. Many of the elements of interest in the FGD sludges are either redox sensitive (*e.g.*, As, Cr, Fe, S, Se, V) or are affected by the redox reactions (*e.g.*, Cd, Cu, Ni, Pb). To determine the geochemical behavior of an element, it is important to accurately determine both the pH and the Eh of the samples.

Although the field pH-measurement techniques must be appropriately selected (especially for alkaline samples where CO_2 addition or degassing can rapidly change pH values), there is very little controversy regarding the technique and significance of pH measurements in relatively dilute solutions. On the other hand, the significance of Eh measurements in both the field and laboratory has been controversial. Many

Table 2 *Field pH and Eh measurements at various depths in the sludge using an* in situ *probe.*

Depth (cm)	Time (min)	pH	Eh mV
	North Site		
91	6	9.29	-303
	12	9.28	-321
130	7	9.43	-296
	12	9.40	-318
160	5	9.40	-294
	10	9.40	-299
193	5	9.40	-290
	10	9.41	-290
218	10	9.44	-337
	17	9.45	-342
	Central Site		
206	5	8.1	-348
229	7	8.15	-354
	10	8.1	-354
257	5	10.1	-319
	10	10.05	<u>-330</u>
	Average		-320 ± 24

common redox couples in natural waters reportedly exist in apparent disequilibrium (Lindberg and Runnels 1984). Therefore, it is generally difficult to relate the platinum electrode readings to specific redox couples.

In this study, *in situ* measurement techniques and sample preservation and handling techniques were developed to increase the reliability of the measured redox potentials. The Eh values measured using the *in situ* measurement device are listed in Table 2. These data show that the Eh values in most cases reached a steady state rapidly (5 min). The values measured at different locations in the sludge and at various depths at each location show that the sludge is fairly homogeneous with respect to Eh, and the variability in Eh values is relatively small. The average of Eh values measured in the field was -320 ±24 mV. The average of Eh values measured in the laboratory in pore waters from the sludge cores, which had been physically preserved to maintain their chemical integrity, was 306±54mV.

Among the measured redox species (Table 3), Fe(III) and As(V) are in most cases

Table 3 *Analysis of 0.22-μ filtrates from porewater extracts of different sludge samples.*

| | | Aqueous concentration (mg L⁻¹) in samples from different sites and depths (cm) | | | | | | | | |
| | | North | | Central | | | Southwest | | New Pond | |
Species	Method	61-142	231-333	218-254	254-330	396-472	472-549	701-777	NP01	NP02
Al	ICP[a]	<0.3	<0.3	1.03	<0.3	<0.3	<0.3	<0.3	<0.03	<0.03
As(III)	CVAA[b]	0.09	2.66	1.17	1.13	2.61	0.90	2.31	0.02	0.02
AsV	CVAA[c]	0.03	d	0.20	0.34	d	0.11	d	0.45	0.43
B	ICP	12.4	17.9	10.8	10.6	14.8	9.41	21.6	18.9	22.4
Ba	ICP	0.04	0.06	0.1	0.06	0.08	0.10	0.05	0.07	0.06
Br⁻	IC[e]	1.2	2.74	1.8	2.8	2.7	3.1	2.9	4.62	6.1
C inorg	Direct	1.51	0.91	0.74	0.48	0.96	0.55	0.92	4.13	2.81
Ca	ICP	722	788	679	767	755	738	743	721	770
Cd	GFAA[f]	<0.0007	<0.0007	0.002	0.002	0.001	0.001	<0.0007	<0.0007	<0.0007
Cl⁻	IC	194	464	310	321	345	403	482	740	932
Cr	GFAA	<0.001	<0.001	0.003	<0.001	<0.001	<0.001	<0.001	<0.001	<0.0015
Cu	GFAA	<0.01	<0.01	0.015	<0.01	<0.01	<0.01	<0.01	<0.01	<0.01
F⁻	ISE[g]	7.32	7.48	5.81	7.43	7.32	5.66	7.72	6.50	6.59
Fe2+	IC[h]	0.046	0.074	0.1	0.05	0.05	0.046	0.074	0.21	0.005
Fe3+	IC[h]	0.044	0.028	<0.02	<0.01	0.03	<0.018	0.052	0.140	0.11
I⁻	IC	2.25	16.2	4.2	2.1	4	4.05	20.6	29.4	20.9
K	ICP	62	141	172	47	64	92	133	314	431
Li	ICP	0.62	0.55	0.61	0.71	0.76	0.81	0.50	0.82	0.88
Mg	ICP	8.69	1.60	1.73	<0.6	3.33	1.16	7.80	14.70	28.20
Mn	GFAA	0.032	0.0074	0.010	0.0034	0.011	0.0032	0.023	0.245	0.33
Mo	GFAA	0.54	0.48	2.78	2.02	4.09	2.93	8.58	4.36	5.45
Na	ICP	91	79	244	185	137	141	75	123	149
Ni	GFAA	<0.01	<0.01	0.04	<0.01	<0.01	<0.01	<0.01	0.029	0.028
NH4+	ISE[g]	1.28	4.63	NDi	5.97	7.66	10.64	0.62	1.61	2.33
NO2⁻	COLORIM[j]	0.012	0.004	0.01	0.01	0.02	0.005	0.005	15.8	49.7
NO3⁻	IC	0.82	0.24	0.45	0.19	0.32	0.28	0.28	18.9	48.1

Parameter	Method								
Pb	GFAA	<0.01	<0.01	0.028	<0.01	<0.01	<0.01	<0.01	<0.01
PO_4^{3-}	IC	0.46	0.17	<0.3	0.3	0.45	<0.4	0.14	<0.5
Se	GFAA	<0.01	<0.01	<0.01	<0.01	<0.01	<0.01	<0.01	<0.01
Si	ICP	2.4	11.6	7.11	7.54	8.21	6.1	10.2	3.62	4.13
$S_2O_3^{2-}$	IC	7.8	189	92	215	111	196.5	29.9	3	12.8
S_2	ISE^k	14.2	1.2	2.4	2.6	3.0	3.0	0.6	<0.3	<0.04
SO_3^{2-}	IC	17.6	16.1	19.4	20.7	20.5	20.8	17.6	21.4	20.0
SO_4^{2-}	IC	1534	1415	1447	1502	1598	1466	1430	1446	1521
Sr	ICP	9.2	6.9	8.12	7.3	7.15	8.39	8.545	6.82	7.12
V	GFAA	<0.025	0.055	<0.05	0.087	<0.05	<0.05	<0.025	<0.025	<0.025
Zn	ICP	<0.02	<0.02	<0.02	<0.02	<0.02	<0.02	<0.02	<0.02	<0.02
Parameter									
pH		8.07	8.906	9.304	9.66	8.613	9.381	8.63	8.148	8.495
Eh (mV)		-410	-260	-322.5	-318.3	-278.8	-305	-250	-105	-95
EC (mho)		3,250	3,900	4,800	3,750	3,850	3,250	3,750	5,550	3,500

[a] ICP = inductively coupled plasma-atomic emission spectroscopy.
[b] CVAA = analysis by atomic adsorption with cold vapor.
[c] By difference, total As minus As(III).
[d] Total As concentration in the sample was lower than the measured As(III) concentration. Therefore, As(V) concentration cannot be estimated.
[e] IC = analysis by ion chromatography.
[f] GFAA = analysis by atomic adsorption with graphite furnace.
[g] ISE = analysis by ion-selective electrode, with standard addition.
[h] IC with peak height.
[i] ND = not determined. Values for Fe^{2+} are in most cases at near the detection limit and therefore cannot be considered reliable.
[j] COLORIM = analysis by spectrophotometric determination.
[k] ISE with titration.

Table 4 *Thermochemical data in MINTEQ code for S species used in calculating redox potentials.*

Reaction		log K°
$SO_4^{2-} + 8e- + 9H^+$	$HS^- + 4H_2O$	33.66
$0.5S_2O_3^{2-} + 1.5H_2O$	$SO_3^{2-} + 2e^- + 3H^+$	-22.10
$0.5S_2O_3^{2-} + 4H^+ + 4e^-$	$HS^- + 1.5H_2O$	15.21
$SO_4^{2-} + 4e^- + 5H^+$	$0.5S_2O_3^{2-} + 2.5H_2O$	18.45
$HS^- + 3H_2O$	$SO_3^{2-} + 7H^+ + 6e^-$	-37.31

at or near the detection limit. Therefore, redox potentials from Fe(II)/Fe(III) and As(III)/As(V) cannot be accurately calculated. Because potentials calculated from the nitrogen couples do not generally correspond to the equilibrium potentials, S species were used to calculate the redox potentials from the activities of the species obtained from geochemical pore water modeling and from the reactions given in Table 4. The average calculated values (in mV) for the SO_3^{2-}/HS^-, HS^-/SO_4^{2-}, $HS^-/S_2O_3^{2-}$, $SO_3^{2-}/S_2O_3^{2-}$, and $S_2O_3^{2-}/SO_4^{2-}$ couples are -239 ± 38, -327 ± 36, -257± 30, -215 ± 63, and -397 ± 41, respectively, and are within the range of the average measured values (306 ± 54). Even though there is variability in the calculated values, the measured and calculated Eh values are similar and are very low. These data, in addition to the presence of sulfate reducing bacteria that require 100 mV redox potentials for active growth and metabolism (Postgate, 1984), show that the FGD sludge is chemically very reducing.

Pore water composition and reactions that control aqueous concentrations of elements
The pore water concentrations from selected core samples are shown in Table 3. These pore water compositions were used as an input to a geochemical code, MINTEQ, to determine activities for different species. These activities were then used in comparison with the activities calculated to be in equilibrium with different solids under the environmental conditions imposed by the sludge. These results, along with those of additional studies conducted to ascertain (1) whether the aqueous concentrations of a given element are controlled by a solubility phenomenon and (2) the possible identity of the solubility-controlling solid, are discussed below.

Calcium and sulfate are the most dominant soluble ions. The X-ray diffraction analyses of the sludge samples show that the Ca in the solid phase exists as $CaCO_3$ (calcite), $CaSO_4 \cdot 2H_2O$ (gypsum), and $CaSO_3 \cdot 0.5H_2O$ in all of the samples except two where $CaSO_4 \cdot 2H_2O$ was not detected. The presence of these solids in other FGD sludges has also been reported (Selmeczi and Knight, 1973; Liem *et al.*, 1982; McCarthy *et al.*, 1983).

All of these Ca solids ($CaCO_3$, $CaSO_3 \cdot 0.5H_2O$, and $CaSO_4 \cdot 2H_2O$) are known to have rapid precipitation/dissolution kinetics. Therefore, it is likely that the aqueous concentrations of Ca are controlled by the solubility of these solids. The plot of

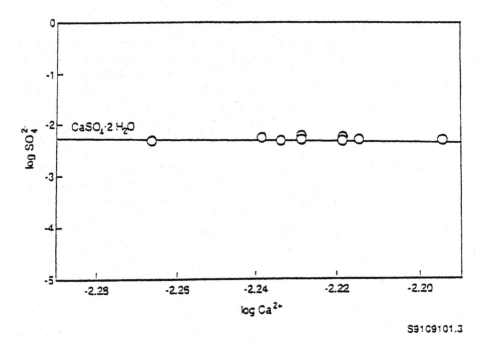

Figure 2 *Activities of Ca^{2+} and SO_4^{2-} in pore water extracts from different core sections from different locations in the sludge. The solid line represents predicted values in equilibrium with $CaSO_4 \cdot 2H_2O$.*

activity of Ca^{2+} versus SO_4^{2-} and the calculated Ca^{2+} and SO_4^{2-} activities in equilibrium with $CaSO_4 \cdot 2H_2O$ show that the observed and the calculated activities are in excellent agreement, indicating that the $CaSO_4 \cdot 2H_2O$ is the solubility-controlling solid (Figure 2). Gypsum was identified as present in all sludge samples except in one core sample from the north and one from the southwest coring sites. The fact that all of the samples appear to be in equilibrium with gypsum suggests that gypsum is also present in those samples, although it was not detected by X-ray diffraction analyses.

To check whether the observed Ca^{2+} activities under the chemical conditions exhibited by the solutions are also in equilibrium with $CaSO_3 \cdot 0.5H_2O$, the ion activities of Ca^{2+} and SO_3^{2-} were calculated using the equilibrium constants for the solubility product of $CaSO_3 \cdot 0.5H_2O$ and for the formation of $CaSO_3^0$ recently reported by Rai *et al.* (1991). These data were necessary to accurately model the sulfite system. The excellent agreement between the calculated and observed ion activities (Figure 3) combined with the reported rapid precipitation/dissolution kinetics of $CaSO_3 \cdot 0.5H_2O$ (Rai *et al,.* 1991) indicates that $CaSO_4 \cdot 2H_2O$ and $CaSO_3 \cdot 0.5H_2O$ are in equilibrium with each other .

Although the comparison of ion activity products for Ca^{2+} and CO_3^{2-} observed in these solutions with those calculated from the thermodynamic data suggests that the solutions are undersaturated with respect to $CaCO_3(c)$, the presence of large quantities of CaCO$_3$(c) and its known rapid precipitation/dissolution kinetics indicates that CaCO$_3$ is also one of the equilibrating phases. The results also show that the aqueous CO$_3^{2-}$ activities are far below those expected, had the solutions been in equilibrium with air, indicating that the gellike structure of the sludge forms an effective barrier against diffusion of air into the sludge. Such a barrier is also conducive to maintaining low redox potentials in the sludge.

Among the trace metals Cd, Cu, Ni, Pb, and Zn, only Zn is present in the sludge with total concentrations as great as a few tenths of a percent (~0.4). Total Pb concentrations are lower than the Zn concentrations by about a factor of 10. All of the other trace elements (Cd, Cu, Ni) are present at concentrations less than 0.01%. However, in pore waters extracted from cores and in pore waters equilibrated with sludge samples adjusted to a range in pH values, the aqueous concentrations of all of these elements are at or near the detection limits (Table 3). Under the low redox potentials encountered in the sludge, Cu is expected to be present as Cu(I) and all of these other trace elements are expected to be present in the divalent state. Because S^{2-} is present in measurable amounts in the aqueous phase, and because these metals readily form compounds with S^{2-} that have very low solubilities, it is probable that the low aqueous concentrations of Cd, Cu, Ni, Pb, and Zn are governed by a

Table 5 *Chemical composition and color of precipitates resulting from making the sludge pore-water. Extracts 10^{-4} M in Cd^{2+}, Cu^{2+}, or Zn^{2+}*

Precipitating element	Composition of precipitates[a]		
	Element	%	Color
Cd	Cd	71	Orangeish yellow
	S	25	
Cu	Cu	93	Brownish black
	S	5	
Zn	Zn	64	White
	S	30	

[a]Determined from energy-dispersive X-ray analysis. Small amounts of Si and Fe were detected in all cases, with a small amount of Pb as well in some cases. In the case of Cu precipitates, the samples were found to decompose when exposed to the electron beam; the result is a higher percentage of Cu and inaccurate analysis of the overall chemical composition. However, the color of the Cu precipitate, the X-ray diffraction patterns, and the formation of Cd and Zn sulfides suggest that Cu precipitates are of copper sulfides.

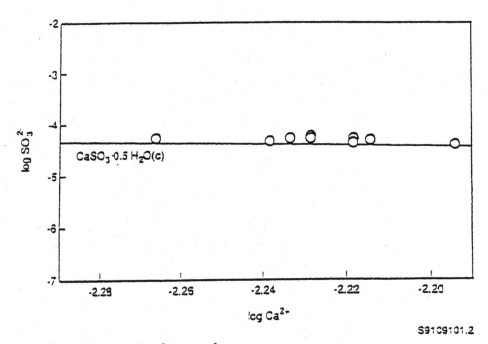

Figure 3 *Activities of SO₃²⁻ and Ca²⁺ in pore water extracts from different core sections from different locations in the sludge. The solid line represents predicted values in equilibrium with CaSO₃O5H₂O(c).*

solubility phenomenon and that the metal sulfides are the potential solubility-controlling solids.

Several approaches were taken to determine (1) whether the aqueous concentrations of Cd, Cu, Ni, Pb and Zn are controlled by a solubility phenomenon and (2) the identity of the solubility-controlling solid. These approaches included (1) observing the change in aqueous concentrations after spiking the sludge suspensions with metal ions, (2) comparing the observed aqueous metal concentrations with those predicted as being in equilibrium with the metal sulfides under the aqueous environmental conditions encountered in the sludge, and (3) characterizing the solid phases resulting from spiking the aqueous extracts from the equilibrated sludge suspensions. The results obtained from these analyses are discussed below. To determine whether the aqueous concentrations are controlled by a solubility phenomenon, different FGD sludge suspensions in water were spiked to an initial concentration of 10^{-4} M for Cd^{2+}, Cu^{2+}, or Ni^{2+}. Comparison of the aqueous concentrations obtained after spiking to those originally present (Figures 4 to 6) shows that the total aqueous concentrations in both cases were similar and were at the detection limits for each of the elements. Concentrations decreased after spiking by three to four orders of magnitude, down to levels originally maintained in the

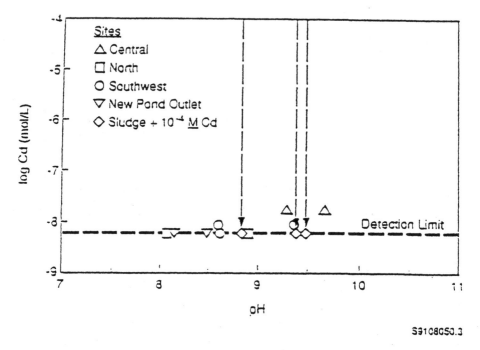

S9108CS0.3

Figure 4 *Concentration of Cd in pore water extracts from different core sections from different locations in the sludge. The top position of the vertical dashed lines represents the concentration of Cd added to the sample, and the bottom position indicates the final concentration.*

solutions, suggesting that the aqueous concentrations are controlled by a solubility phenomenon. To identify the potential solubility-controlling solid, the aqueous Cd^{2+}, Cu^+, Ni^{2+}, Pb^{2+}, and Zn^{2+} activities predicted from the available thermochemical data for different solids of these elements can be compared with the observed aqueous activities (Figure 7). Because the observed aqueous activities of these elements are constrained by the detection limits and are not reliably measurable, it is not possible to determine whether metal sulfides are indeed the solubility-controlling solids. The measured aqueous activities (assuming to be constrained by detection limits) are, in most cases, several orders of magnitude lower than the activities maintained by metal carbonates (Figure 7), a fact that indicates that metal carbonates are not the solubility-controlling solids. However, because the activities predicted to be in equilibrium with metal sulfides (except for Cu) are lower than the measured aqueous activities and the latter are upper limits rather than absolute, these results are consistent with the metal sulfides being potential solubility-controlling solids. The Cu^+ activities in pore water extracts are similar to those in equilibrium with Cu_2S, indicating that Cu_2S may be the solubility-controlling phase.

To identify the potential solubility-controlling solids of Cd, Cu, and Zn, different

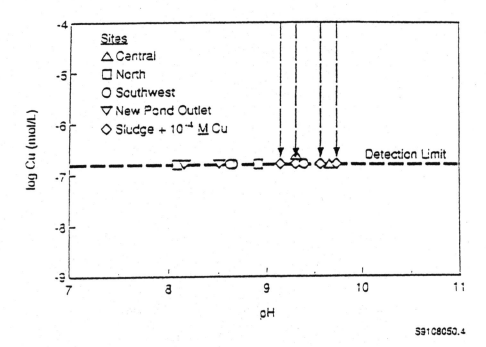

Figure 5. *Concentration of Cu in pore water extracts from different core sections from different locations in the sludge. The top position of the vertical dashed line represents the concentration of Cu added to the sample, and the bottom position indicates the final concentration.*

samples of pore waters extracted from equilibrated sludge suspensions were spiked to obtain initial concentrations of 10^4 M for Cd, Cu, or Zn. The addition of these elements to pore waters resulted in observable precipitates that were analyzed for chemical composition and crystallinity. The chemical analysis of the precipitates shows that they are mostly metal sulfides. Among the Cd, Cu, and Zn solids, Zn precipitates provided the best X-ray diffraction patterns (Figure 8). The Zn precipitates gave a strong reflection at 3.12, a moderately weak and broad reflection at 1.89, and a weak and very broad reflection at about 1.6. These peaks correspond to the three strongest peaks for ß-ZnS (sphalerite) and also to those listed for ZnS (wurtzite-10H,8H). Sphalerite is noted to be white to yellowish in color, whereas wurtzite is noted to be brownish-black (JCPDS). Because the Zn precipitates (Table 5) obtained in this study are white, there is little doubt that the precipitates are poorly crystalline sphalerite.

Cadmium precipitates appear to be primarily amorphous, but there are indications of weak and broad peaks at d spacings that correspond to the three strongest peaks for CdS (hawleyite). Hawleyite is reported to be yellow (JCPDS), which is about the color of these Cd precipitates (Table 5). Given the combination of evidence from

Table 6 *Summary of solubility-limiting solids.*

Element	Solubility-limiting solid
Ba,Sr	$(Ba,Sr)SO_4$[a], $SrSO_4$[a], $BaSO_4$[a]
Ca,S	$CaSO_4 \cdot 2H_2O$[b]
	$CaSO_3 \cdot 0.5H_2O$[b]
Cd	CdS[a]
Cr	$Cr(OH_3)/(Fe,Cr)(OH)_3$[a]
Cu	$Cu_2S/Cu_{1.65}S$[a]
Fe	FeS[c]
Ni	NiS[c]
Pb	PbS[c]
Se	$FeSe_2/FeSe$[d]
Zn	ZnS[a]

[a]Identified using indirect methods.
[b]Physically identified using X-ray diffraction and by indirect method
[c]Identified by analogy with similar compounds identified using indirect method.
[d]Postulated.

Xray diffraction, chemical composition, and color, it appears that poorly crystalline or amorphous hawleyite was the compound precipitated. Like those of Cd, Cu precipitates appeared to be primarily amorphous. However, indications from Xray diffraction and the color of the precipitates suggest that the precipitates were of copper sulfide and that they may be of poorly crystalline or amorphous Cu_2S (or more probably $Cu_{1.65}S$). The brownish black color of the precipitates indicates that they cannot be CuS, which is reported to be indigo blue or darker, often iridescent brass yellow and dark red. In addition, the presence of reduced Cu sulfides such as $Cu_{1.65}S$ or Cu_2S would be consistent with the measured low redox potentials (Table 1) where Cu(I) is the stable oxidation state.

Given that (1) Cd, Cu, and Zn sulfides were identified in the precipitates obtained from spiking the pore water extracts from FGD sludge, (2) Cd, Cu, and Zn sulfides have very rapid precipitation kinetics (a few minutes, as noted in the precipitation experiments described in the methods and materials section), and (3) Cd, Cu, and Zn sulfides have very low solubilities (Figure 7), it appears that the very low observed aqueous concentrations of Cd, Cu, and Zn are controlled by precipitation/dissolution of sulfides of these metals. Although no precipitation experiments similar to those conducted with Cd, Cu, and Zn were conducted with Ni and Pb, it is known that Ni and Pb also form sulfides with low solubilities. Therefore, it is expected that aqueous concentrations of Ni and Pb are also controlled by the solubility of their respective sulfides.

All of the soluble iron in the pore water extracts was found to be present as Fe(II).

Table 7 *Summary of ion-migration depths in soil below sludge..*

Parameter	Estimated migration depth in m below interface	
	SW Site	N Site
Total B	1.27	0.9
Aqueous B	1.1	0.9
Total Pb, Zn, Se	0	0
Aqueous Pb, Zn, Se	0	0
Total Ca^{2+}, Sr^{2+}	0.2	0.2
Aqueous Ca^{2+}, Sr^{2+}	0.4	0.3
Aqueous Na	0.4	0.3
K	0.2	0.2
Eh	~1.0	0.3
S_2O_3	0.8	0.25
NH_4	0.8	0.3
DOC	0.4	0.2

Like Cd, Cu, Ni, Pb, and Zn, Fe(II) also readily forms fairly insoluble compounds with S. The measured Fe^{2+} activities (Figure 9) are about an order of magnitude higher than those in equilibrium with FeS(c) but are several orders of magnitude lower than the Fe^{2+} activities in equilibrium with other known Fe(II) solids such as $Fe(OH)_2(c)$ and $FeCO_3(c)$. It is therefore postulated that the aqueous Fe(II) concentrations are controlled by the precipitation/dissolution of possibly amorphous Fe(II) sulfide.

The measured aqueous Cr concentrations were found to be at the analytical detection limits (Table 3). Therefore, it was not possible to accurately quantify total soluble Cr or the oxidation state of Cr. However, the measured pH and Eh values are such that Cr in the sludge and in the aqueous phase will be present as Cr(III). Extensive studies on the behavior of Cr(III) in the pure systems (Rai *et al.*, 1987; Sass and Rai, 1987), in soils (Rai *et al.*, 1988) and in fly ashes (Rai *et al.*, 1990), show that Cr(III) readily forms compounds, such as $Cr(OH)_3$ and $(Fe,Cr)(OH)_3$, that have fast precipitation/dissolution kinetics. At the pH values of the samples, both of these compounds are expected to control aqueous Cr concentrations near the detection limits. Because of the low total Cr content of the sludge and the amorphous nature of these compounds, direct methods cannot be used to determine whether these compounds are present in the sludge. However, it can definitely be stated that, because of the low uppersolubility limits for $Cr(OH)_3$, aqueous Cr concentrations at the measured pH and Eh values will not exceed detection limits, as was indeed observed in the pore water samples. To further confirm this conclusion,

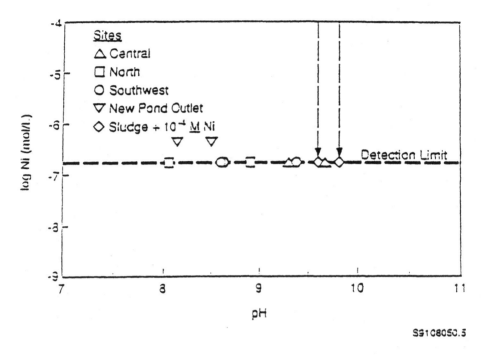

Figure 6 *Concentration of Ni in pore water extracts from different core sections from different locations in the sludge. The top position of the vertical dashed line represents the concentration of Ni added to the sample, and the bottom position indicates the final concentration.*

the pore waters or the suspensions of FGD sludge were spiked with Cr(III) to obtain Cr(III) concentrations of 10^{-4} M. The formation of green $Cr(OH)_3$ was observed in pore waters from sludge that had been spiked with Cr(III). The aqueous Cr(III) concentrations in these spiked sludge suspensions at the end of a 3 day equilibration period were found to be at the detection limit and about four orders of magnitude lower than the level to which they had been spiked (Figure 10). These results confirm that aqueous Cr concentrations in these sludges will not exceed the detection limits because of the formation of the Cr(III) compounds.

Barium and Sr are present in the sludge only in small quantities (<0.02 % and <0.08 % on a total mass basis, respectively), and as a result, such direct identification techniques as Xray diffraction cannot be used to identify crystalline Ba and Sr compounds, even if these compounds are present. However, Ba and Sr in fly ashes have been reported to be controlled by a solubility phenomenon (Ainsworth and Rai, 1987; Fruchter *et al.*, 1988). To check whether Ba concentrations in the sludge are controlled by a solubility phenomenon, the sludge suspensions were spiked with Ba^{2+} to obtain initial Ba^{2+} concentrations of 10^{-4} M. Final Ba concentrations decreased rapidly by about a factor of 100, to levels originally found in FGD extracts

(Figure 11). This finding suggests that Ba concentrations are also controlled by a solubility phenomenon.

Both Ba and Sr rapidly form sparingly soluble compounds with carbonates and

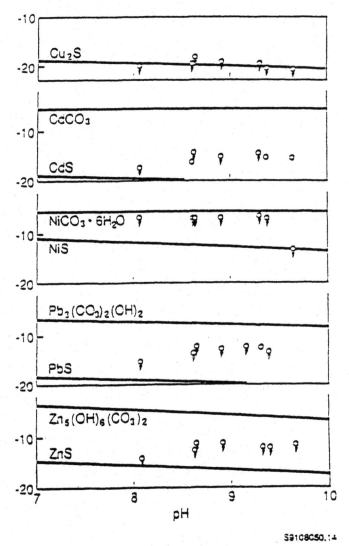

S3108CS0.:4

Figure 7 *Activities of $M^{2+}(Cd^{2+}, Ni^{2+}, Pb^{2+}, and Zn^{2+})$ and $M^+(Cu^+)$ in pore water extracts from different core sections from different locations in the sludge. Because the measured concentrations of M^{2+} and M^+ were at detection limits, the aqueous activity values in pore water extracts are upper limits (shown by downward arrows) rather than absolute values. The solid lines represent values in equilibrium with different solids at the average CO_3^{2-} and HS^- activities (6.8 and 4.2, respectively) in pore water extracts.*

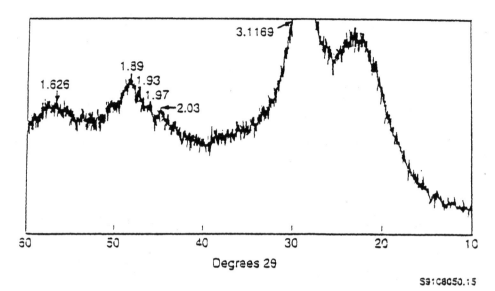

S91C8C5O.15

Figure 8 *X-ray diffraction pattern for the precipitates resulting from spiking the pore water extracts from sludge with Zn.*

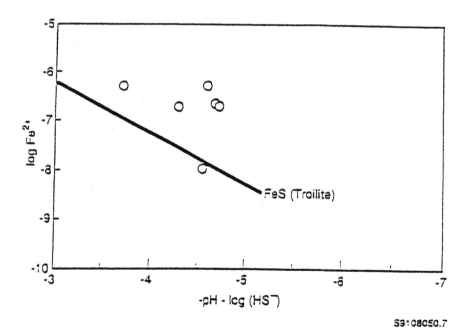

S91C8C5O.7

Figure 9 *Activities of Fe^{2+} plotted as a function of the quantity [-pH -logHS⁻] in pore water extracts from different core sections from different locations in the sludge. The solid line represents predicted values in equilibrium with FeS (troilite).*

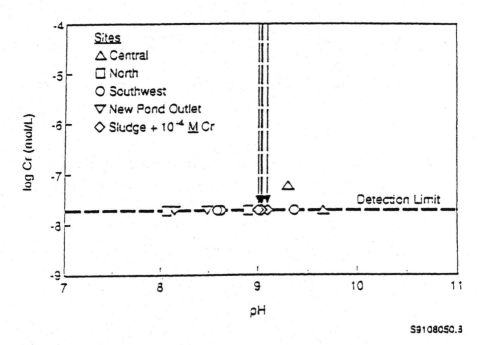

Figure 10 *Concentrations of Cr in pore water extracts from different core sections from different locations in the sludge. The top position of the vertical dashed lines represents the concentration of Cr added to the sample, and the bottom position indicates the final concentration.*

sulfates. To check whether carbonate and/or sulfate solids control aqueous Ba and Sr concentrations, the activities of Ba^{2+} and Sr^{2+} in aqueous extracts were compared with the activities calculated to be in equilibrium with different solid phases at the CO_3^{2-} and SO_4^{2-} activities observed in the sludge extracts. The results indicate that the aqueous extracts are undersaturated with respect to $BaCO_3$ and $SrCO_3$ by several orders of magnitude. Therefore, it is concluded that $BaCO_3$ and $SrCO_3$ are not the solubilitycontrolling solids for Ba and Sr, respectively, and that they are not stable under the environmental conditions encountered in the sludge. The observed Sr^{2+} activities are similar, although slightly lower, than those in equilibrium with $SrSO_4(c)$, indicating that $SrSO_4$ may be the solubility-controlling solid. The observed Ba^{2+} activities are about 0.5 orders of magnitude higher than those in equilibrium with $BaSO_4(c)$ (Figure 12). Our published data (Felmy, *et al.*, 1993) show that Ba and Sr can coprecipitate as a $(Ba,Sr)SO_4$ solid. This $(Ba,Sr)SO_4$ solid has been observed to show slightly higher ion activity products than those in equilibrium with $BaSO_4(c)$, and slightly lower ion activity products (depending on the mole fractions) than those in equilibrium with $SrSO_4(c)$, indicating that the observed aqueous concentrations may reflect equilibrium with $(Ba,Sr)SO_4$.

Among the trace elements of interest in the FGD sludge, total concentrations of Se are the lowest (<0.0005%). The measured aqueous Se concentrations are at the

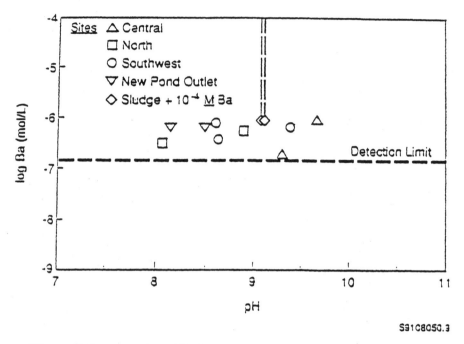

S91C8050.3

Figure 11 *Concentration of Ba in pore water extracts from different core sections from different locations in the sludge. The top position of the vertical dashed lines represents the concentration of Ba added to the sample, and the bottom position indicates the final concentration.*

detection limit (Figure 13). Whether the aqueous Se concentrations are low because there is very little total Se in the sludge, or because low-solubility Se compounds are present cannot be ascertained at this time. However, under the reducing conditions encountered in the sludge, the calculated solubilities of $FeSe_2$ and $FeSe$ are lower than analytically measured aqueous Se levels and may be the concentration-limiting compounds for Se.

The aqueous concentrations of B were observed to range from $10^{-3.01}$ to $10^{-2.68}$ M (Figure 13). The measured concentrations are well above the detection limits. The observed aqueous concentrations of B were essentially independent of pH and varied over a narrow concentration range (Figure 13). The estimated thermochemical data for the borate minerals pinnoite, inderite, inyoite, colemanite, inderborite, hungchaoite, borax, sborgite, McAllisterie, Kaliborite, and nobleite reported by Mattigod (1983) and the experimental data for searlesite (unpublished data of Rosenberg and Kittrick, Washington State University, Pullman, Washington) show that most of these B minerals are very soluble and not likely to control aqueous B concentrations.

Aqueous concentrations of As were observed to range from $10^{-5.81}$ to $10^{-4.45}$ M (Figure 13). These measured concentrations are well above the detection limits.

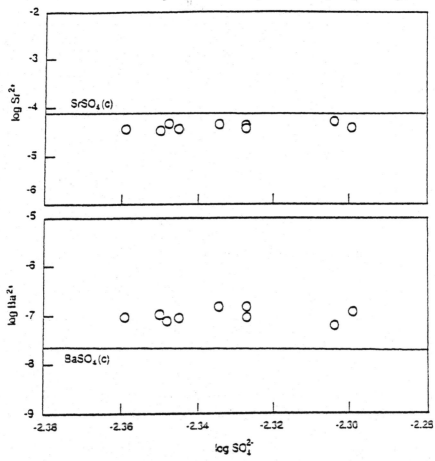

Figure 12. *Activities of SO_4^{2-}, Sr^{2+}, and Ba^{2+} in pore water extracts from different core sections from different locations in the sludge. The solid lines represent equilibrium with the specified solid phases.*

Arsenic was found to be present primarily as As(III) (Table 3), which is the species expected at the low redox potentials measured in the samples. The solubility of As_2S_3 (an As(III) compound that might control As solubility under the chemical conditions encountered in the sludge) is considerably higher than the observed aqueous As(III). Thus, As_2S_3 is not likely to be the solubility-controlling solid. No As(III) compound is currently known that would maintain As concentrations similar to those observed.

As discussed above, the concentrations of many elements (Ba, Ca, Cd, Cr, Cu, Fe, Ni, Pb, S, Se, Sr, and Zn) observed in sludge pore waters are inferred to be controlled by solubility phenomena. Regardless of the locations and depths in the FGD sludge sites at which the samples were taken, the aqueous concentrations of a given element

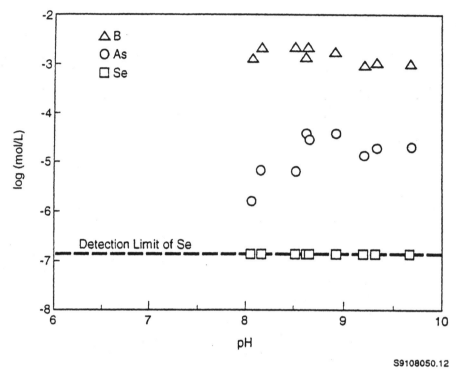

S9108050.12

Figure 13 *Concentrations of As, B, and Se in pore water extracts from different core sections from different locations in the sludge.*

in all of these samples can be explained by a single solubility-controlling solid. The hypothesized and observed solubility controlling solids for different elements are listed in Table 6. Although Se concentrations are expected to be limited by solubility phenomena (with $FeSe_2$ and/or $FeSe$ as the solubility-controlling solid), the solubility-controlling solid cannot be ascertained from the results of this study. The concentrations of As and B cannot currently be interpreted on the basis of known solubility equilibria.

Integrated assessment of solute migration from sludge to soil

The total chemical and pore water analyses and the moisture content determinations have shown that solutes and pore waters from the sludge have only migrated short distances into the underlying clayey soil (Table 7). Boron, the best tracer of sludge water, has moved to a maximum depth of 1.3 m below the sludge. All other solutes migrated less because of chemical attenuation. Solute trends at both coring sites were consistent, but migration was deeper at the southwest site, which was overlain by a greater depth of sludge for a longer time. A definitive analysis of the solute concentration profiles was hampered by a lack of data on the background chemical characteristics of the soils onto which the sludge was disposed. Field observations combined with laboratory measurements suggest that solute migration has been limited in part by (1) the low permeability of the underlying soil and subsoil, which contains over 30% clay and (2) the moisture-retention characteristics of the sludge. Measurements of moisture content at approximate 0.15 to 0.25 m increments indicate that sludge water infiltration into the soil at this site has been minimal, ranging between 0.75 and 2.75 m. Solute transport calculations indicate that the distribution of sludge-derived solutes in the underlying soil (*i.e.*, B) is consistent with transport by molecular diffusion only (*i.e.*, no convective water flux).

Chemical attenuation in the soil and precipitation reactions in the sludge have also been important in limiting solute movement into and through the soil. The formation of metal sulfides in the sludge limit the release of Pb, Zn, and other metals from the waste, while ion exchange reactions on the clayey soil were found to retard the migration of more soluble cations (K^+, Na^+, and Sr^{2+}) in the sludge leachate. Boron was also absorbed by the soil materials (probably edge sites on clays), but it was found to be the best indicator and tracer of sludge leachate. For this reason, B was used in several calculations to estimate the sludge-leachate infiltration depth and the importance of diffusive transport. Boron was adsorbed less than other solutes originating in the sludge, yielding Kd values of 0.3-0.75 mL g^{-1} retardation factors of up to 3.7 in the clay soil.

The investigation of solute migration into soil at this site has allowed an evaluation of the relative merits of total chemical analysis versus pore water extraction and analysis. For many solutes, the total chemical analyses were equally as effective as the pore water analyses in defining the extent of solute migration (Table 7). This was particularly the case for elements that showed a large concentration difference between sludge and soil, such as Zn, and that did not migrate appreciably into soil. The total chemical analyses were also effective for constituents like B that were not originally present in the soil but migrated from sludge into soil in high concentrations. In general, however, the pore water measurements provide a better assessment of ion migration than the total chemical analyses because the instrumental techniques for aqueous-phase analyses are more precise and sensitive than those used for the solid-phase analyses. As a result, mobile sludge species can be identified more readily. In addition, the attenuation phenomena responsible (*e.g.*, precipitation versus adsorption) can be evaluated more rigorously by analyzing the *in situ* pore water without dilution because aqueous-phase concentration data can be used in geochemical model calculations to identify geochemical reactions that may be occurring and controlling aqueous concentrations. Experience here suggests that both types of analyses should be performed because they can corroborate one another, and each can provide unique information.

References

Ainsworth, C,C. and, Rai, D. 1987. *Chemical Characterization of Fossil Fuel Combustion Wastes*. EPRI EA-5321. Electric Power Research Institute, Palo Alto, California.

Felmy, A.R., Girvin, D.C. and Jenne, E.A. 1984. MINTEQ - *A Computer Program for*

Calculating Aqueous Geochemical Equilibria. EPA 600-3-84-032. Athens, Georgia: U.S. Environmental Protection Agency, Office of Research and Development.

Felmy, A.R. 1990. GMIN: *A Computerized Chemical Equilibrium Model Using a Constrained Minimization of the Gibbs Free Energy.* PNL-7281. Pacific Northwest Laboratory, Richland, Washington.

Felmy, A.R., Rai, D. and Amonette, J.E. 1990. The solubility of barite and celestite in sodium sulfate: evaluation of thermodynamic data. *J. Soln. Chem.,* 19(2), 175-185.

Fruchter, J.S., Rai, D., Zachara, J.M. and Schmidt, R.L. 1988. *Leachate Chemistry at the Montour Fly Ash Test Cell.* EPRI EA-5922. Electric Power Research Institute, Palo Alto, California.

Groenewold, G.H. and Manz, O.E. 1982. *Disposal of Fly Ash and Fly Ash Alkali FGD Waste in a Western Decoaled Strip Mine – Interim Report.* Grand Forks, North Dakota: Engineering Experiment Station, University of North Dakota.

Harvie. C.E., Møller, N. and Weare, J.H. 1984. The prediction of mineral solubilities in natural waters: the $Na-K-Mg-Ca-H-Cl-SO_4-OH-HCO_3-CO_2-H_2O$ system to high ionic strengths at 25°C. *Geochim. Cosmochim. Acta,* 48, 723-751.

Kinniburgh, D.G. and Miles, D.L. 1983. Extraction and chemical analysis of interstitial water from soils and rocks. *Environ. Sci. Technol.,* 17, 362-368.

Liem, H.M., Sandstroem, M., Wallin, T., Carne, A., Rydevik, U., Thurenius, B. and Moberg, P.O. 1982. Studies on the Leaching and Weathering Processes of Coal Ashes. In: *International Conference on Coal Fired Power Plants and the Aquatic Environment,* CONF-8208123, 338-366. Water Quality Institute, Hoersholm, Denmark.

Lindberg, R.D. and Runnels, D.D. 1984. A critical analysis of equilibrium of redox reactions in ground waters, applied to Eh measurements and to geochemical modeling. *Science,* 225, 925-927.

Mattigod, S. 1983. An improved method for estimating the standard free energy of formation of borate minerals. *Soil Sci. Soc. Am. J.,* 47, 654-655.

McCarthy, G.J., Swanson, K.D., Schields, P.J. and Groenewold, G.H. 1983. *Mineralogical Controls on Toxic Element Contamination of Groundwater from Buried Electrical Utility Solid Wastes. 1. Solid Waste Mineralogy. 2. Literature Review of Fly Ash Mineralogy.* PB-83-265116. Fargo, North Dakota: North Dakota State University.

Murarka, I.P. 1987. *Solid Waste Disposal and Reuse in the United States.* CRC Press, Inc., Boca Raton, Florida.

Pitzer, K.S. and Mayorga, G. 1973. Thermodynamics of electrolytes. II. activity and osmotic coefficients for strong electrolytes with one or both ions univalent. *J. Phys. Chem.,* 77,2300-2308.

Postgate, J.R. 1984. *The Sulphate-Reducing Bacteria,* 2nd edn. Cambridge University Press, London.

Rai, D., Sass, B.M. and Moore, D.A. 1987. Chromium III. hydrolysis constants and solubility of chromium(III). hydroxide. *Inorg. Chem.,* 26, 345-349.

Rai, D., Ainsworth, C.C., Eary, L.E., Mattigod, S.V. and Jackson, D.R. 1987. *Inorganic and Organic Constituents in Fossil Fuel Combustion Residues.* Vol. 1: *A Critical Review.* EPRI EA-5176, Electric Power Research Institute, Palo Alto, California

Rai, D., Zachara, J.M., Eary, L.E., Ainsworth, C.C., Amonette, J.E., Cowan, C.E., Szelmeczka, R.W., Resch, C.T., Schmidt, R.L., Girvin, D.C. and Smith, S.C. 1988. *Chromium Reactions in Geologic Materials.* EPRI EA-5741. Electric Power Research Institute, Palo Alto, California.

Rai, D., Zachara, J.M., Moore, D.A., McFadden, K.M. and Resch, C.T. 1989. *Field*

Investigation of a Flue Gas Desulfurization FGD. Sludge Disposal Site. EPRI EA-5923, Electric Power Research Institute, Palo Alto, California.

Rai, D. and Szelmeczka, R.W. 1990. Aqueous behavior of chromium in coal fly ash. *J. Environ. Qual.,* 19, 378-382.

Rai, D., Felmy, A.R., Fulton, R.W. and Moore, D.A. 1991. An aqueous thermodynamic model for Ca^{2+}-SO^- ion interactions and the solubility product of crystalline $CaSO_3$-$0.5H_2O$. *J. Soln. Chem.,* 20, 623-632.

Robie, R.A., Hemingway, B.S. and Fisher, J.R. 1978. *Thermodynamic Properties of Minerals and Related Substances at 298.15K and 1 Bar 10^5 Pascals. Pressure and at Higher Temperatures.* Geological Survey Bulletin 1452. United States Government Printing Office, Washington, DC.

Sass, B.M. and Rai, D. 1987. Solubility of amorphous chromium(III)-iron(III) hydroxide solid solutions. *Inorg. Chem.,* 26, 2228-2232.

Selmeczi, J,G, and Knight, R.G. 1973. Properties of Powerplant Waste Sludges. In: *Ash Utilization: Third International Symposium,* pp.123-138. Bureau of Mines Information Circular IC 8640. Washington, DC: U.S. Government Printing Office.

Smith *et al.* *(*in press). Field investigation of calcium and sodium-based flue gas desulfurization FGD. sludge disposal sites. PNL-XXXX. Pacific Northwest Laboratory, Richland, Washington.

Wagman, D.P., Evans, W.H., Parker, V.B., Schumm, R.H., Halow, I., Bailey ,S.M., Churney, K.L. and Nuttall, R.L. 1982. The NBS tables of chemical thermodynamic properties. Selected values for inorganic and C_1 and C_2 organic substances in SI units. *J. Phys. Chem. Ref. Data,* Vol. 11, Supplement 2.

16 Subsurface movement of polychlorinated biphenyls

Thomas E. Hemminger[1] and Benjamin J. Mason[2]

[1]*Environmental Services Manager, Commonwealth Edison Company, P. O. Box 767, Chicago, Illinois 60690, USA*
[2] *Ethura, 9671 Monument Drive, Grants Pass, Oregon 97526, USA*

Abstract

Polychlorinated Biphenyls (PCB) are a family of chemicals manufactured in the United States from 1930 to 1975. PCB were used in a number of throw-away applications and also extensively employed as an electrical insulating fluid. Environmental concerns have lead to strict controls on the use of PCB and standards for cleanup of PCB discharges. The purpose of this chapter is to present information from several typical PCB discharges and data on chemical and physical characteristics of these chemicals. From this, light is shed on the mechanisms of their movement in the subsurface. PCB are relatively insoluble, viscous, and display a strong tendency toward adsorption on soil particles. Their movement through the subsurface is limited by these characteristics such that penetration occurs to a limited depth in formations which have cracks or other connected void spaces. Examination of these histories of discharges of PCB from electrical equipment and manufacturers support this thesis.

Introduction

Manufacturers normally marketed PCB's as mixtures of biphenyls. The combination of the various biphenyls in the mixture controlled the properties of the mixture. Material in this section gives information on the properties of the individual isomers and mixtures used in various industrial applications.

PCB's are especially attractive for industrial applications because of their stability and their dielectric properties (Gustafson, 1970; Mackay, 1982). Figure 1 shows the structure of the biphenyl molecule along with examples of chlorination that can occur at any of the positions on the rings. The physical and chemical properties of both isomers and mixtures used in industrial applications depend upon the degree and position of the chlorine atoms (Mieure *et al.*, 1976; Girvin *et al.*, 1990).

There are 209 possible chlorobiphenyl isomers (Mieure *et al.*, 1976). Table 1 lists the number of isomers for various degrees of substitution. Many of these isomers do not appear in significant levels in commercial products, however. Mieure *et al.* (1976) note that "isomers with four or five chlorine atoms on one ring but none on the other ring are not detectable in PCB containing products".

The five largest uses for PCB's prior to 1970 were dielectric fluids in capacitors, plasticizers, lubricants, transformer fluids, and hydraulic fluids (Hesse, 1976). They

Table 1 *Listing number of possible isomers.*

Degree of substitution	No. of isomers
Mono	3
Di	12
Tri	24
Tetra	42
Penta	46
Hexa	42
Hepta	24
Octa	12
Nona	3
Deca	1

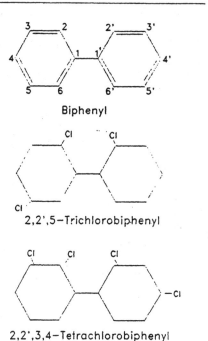

Biphenyl

2,2',5–Trichlorobiphenyl

2,2',3,4–Tetrachlorobiphenyl

Figure 1 *Biphenyl molecule and its numbering system.*

Table 2 *Askarel components in weight percent.*

Component				Type				
	A	B	C	D	E	F	G	H[f]
Hexachlorobiphenyl[a]	60	45						
Pentachlorobiphenyl[b]				70	45	60		
Trichlorobiphenyl[c]			80		100			
Trichlorobenzene[d]				30	40			
Tri-Tetra blend[e]		55	20				55	100

[a] Bipheny chlorinated to a chlorine content of 60 weight percent.
[b] Biphenyl chlorinated to a chlorine content of 54 weight percent.
[c] Biphenyl chlorinated to a chlorine content of 42 weight percent.
[d] A mixture of isomers of trichlorobenzene.
[e] A mixture of isomers of tri- and tetrachlorobenzene.
[f] Non-PCB. Contains no chlorinated biphenyls.
(After ASTM, 1986)

Table 3 *Composition of AROCLORS manufactured for commercial use.*

No. of Cl atoms	MW (g mol^{-1})	Cl (wt %)	AROCLOR 1221	1232	1242	1248	1254	1260	1016
0	154	0	11	<0.1	<0.1	-	<0.1	-	<0.1
1	189	18.8	51	31	1	-	<0.1	-	1
2	223	31.8	32	24	16	2	0.5	-	20
3	258	41.3	4	28	49	18	1	-	57
4	292	48.6	2	12	25	40	21	-	21
5	326	54.3	<0.5	4	8	36	48	12	<0.1
6	361	58.9	-	<0.1	-	4	23	38	-
7	395	62.8	-	-	<0.1	-	6	41	-
8'	430	66.0	-	-	-	-	-	8	-
9	464	68.7	-	-	-	-	-	1	-
Average MW			201	232	267	300	328	376	258

(After Mieure *et al.*,1976 and Pal *et al.*,1980)

were also used widely in protective coatings, sealers, putty, grinding fluids, printing inks, pattern waxes, carbonless paper, *etc*. Because of this widespread use of these chemicals they are found throughout the environment (Risebrough *et al.* 1968; Griffin and Chian, 1980).

A number of important properties of the chemicals are discussed below along with information on their distribution and persistence in the environment.

Table 4. Physical and chemical properties of selected PCB isomers.

CHLORINE PATTERN	#CI ATOMS	MW	WEIGHT % OF ISOMERS			SOLUBILITY	VAPOR PRESSURE	HENRY'S LAW CONSTANT	Log Kow	Log Koc
			1242	1254	1260					
		(g/mol)				(ug/l)	(mm Hg)	atm-m³/mol		
2-	1	188.7	0.0	0.0	0.0	5900 b	1.51E-02 c	(6.35E-04) d	3.9+/-0.01 e	3.2
3-	1	188.7	0.0	0.0	0.0	3500 b	7.14E-03 c	(5.07E-04) d	4.4+/-0.02 e	3.7
4-	1	188.7	0.0	0.0	0.0	1910 b	1.73E-03 c	(2.25E-04) d	4.6+/-0.03 e	3.9
2,2'-	2	223.1	0.0	0.0	0.0	1500 b	1.32E-03 c	2.30E-04 p	4.9+/-0.5	4.3
2,4'-	2	223.1	10.7	0.0	0.0	637+/-7 g	(9.57E-04) h	(3.52+/-0.27)E-04 e	5.1+/-0.4	4.5
2,2',3-	3	257.5	6.5	0.0	0.0	(231) h	(2.05E-04) h	2.00E-04 p	5.6+/-0.4	5.0
2,2',4-	3	257.5	7.6	0.0	0.0	(231) h	(2.05E-04) h	(3.01E-04) p	5.6+/-0.1	5.0
2,2',5-	3	257.5	11.9	0.0	0.0	248+/-4 g	(2.05E-04) h	2.50E-04 p	5.6+/-0.1	5.0
2,3,4'-	3	257.5	3.1	0.0	0.0	(231) h	(2.05E-04) h	(3.01E-04) d	5.6 i	4.9
2,4,4'-	3	257.5	10.3	0.0	0.0	258 j	(2.05E-04) h	2.00E-04 p	5.8+/-0.2	5.2
2,4',5-	3	257.5	10.1	0.0	0.0	(231) h	(2.05E-04) h	1.90E-04 p	5.7+/-0.2	5.1
2,3,4-	3	257.5	7.6	0.0	0.0	78 b	4.00E-04 c	(5.76E-03) d	5.8+/-0.3	5.2
2,2',3,3'-	4	292.0	0.5	0.0	0.0	34 b	1.36E-04 c	1.00E-04 p	5.6+/-0.3	5.0
2,2',3,4-	4	292.0	3.6	0.0	0.0	(70) h	(4.38E-05) h	1.40E-04 p	6.0+/-0.3	5.4
2,2',3,4'-	4	292.0	3.1	0.0	0.0	(70) h	(4.38E-05) h	1.40E-04 p	5.8 i	5.3
2,2',3,5'-	4	292.0	3.9	0.0	0.0	170 b	(4.38E-05) h	(9.90E-05) p	6.0+/-0.3	5.4
2,2',4,4'-	4	292.0	1.9	0.0	0.0	68 b	(4.38E-05) h	1.90E-04 p	5.9+/-0.3	5.3
2,2',4,5'-	4	292.0	2.9	0.0	0.0	(70) h	(4.38E-05) h	2.10E-04 p	6.1+/-0.2	5.5
2,2',5,5'-	4	292.0	4.2	3.2	0.0	26.5+/-0.8 g	4.90E-05 c	(3.25+/-0.34)E-04 e	6.1+/-0.2	5.5
2,3,4,4'-	4	292.0	3.9	2.1	0.0	98 h	(4.38E-05) h	(2.10E-03) p	5.9+/-0.3	5.3
2,3',4,4'-	4	292.0	3.9	7.0	3.4	(70) h	(4.38E-05) h	(2.40E-04) d	5.8+/-0.3	5.2
2,3',4',5-	4	292.0	1.8	7.6	0.0	41 h	(4.38E-05) h	1.00E-04 p	5.9+/-0.3	5.3
2,4,4',5-	4	292.0	0.0	1.6	4.3	(70) h	(4.38E-05) h	1.00E-04 p	6.1+/-0.2 i	5.5
2,2',3,3',4-	5	326.4	0.0	0.6	0.0	(21) h	(9.35E-06) h	6.60E-05 p	6.2 i	5.8
2,2',3,4,4'-	5	326.4	0.0	1.4	0.0	(21) h	(9.35E-06) h	7.40E-05 p	6.2+/-0.4	5.7
2,2',3,4,5'-	5	326.4	0.0	4.4	0.0	22 b	(9.35E-06) h	7.40E-05 p	6.5+/-0.4	6.0
2,2',3,4',5	5	326.4	0.0	2.4	0.0	(21) h	(9.35E-06) h	7.80E-05 p	6.6+/-0.4	6.1
2,2',4,4',5-	5	326.4	0.0	4.3	0.0	(21) h	(9.35E-06) h	(4.59E-04) d	6.4 i	5.8
2,2',4,5,5'-	5	326.4	0.0	11.3	4.5	10.3+/-0.2 g	1.10E-05 c	(1.91E-04) d	6.4+/-0.5	5.7
2,3,3',4',5-	5	326.4	0.0	11.9	0.0	(21) h	(9.35E-06) h	(1.91E-04) d	6.5 i	5.8
2,3',4,4',5-	5	326.4	0.0	15.6	0.0	(21) h	(9.35E-06) h	1.30E-05 p	6.4+/-0.3	5.7
2,2',3,3',4,4'-	6	360.9	0.0	1.3	0.0	(6) h	(2.00E-06) h	2.10E-05 p	7.0+/-0.3	6.5
2,2',3,4,4',5'-	6	360.9	0.0	9.5	11.7	(6) h	(2.00E-06) h	2.30E-05 p	7.0+/-0.5	6.5
2,2',3,4,5,5'-	6	360.9	0.0	0.0	2.2	(6) h	(2.00E-06) h	(1.58E-04) d	6.8 i	6.2
2,2',3,4',5,6'-	6	360.9	0.0	0.0	14.5	(6) h	(2.00E-06) h	2.50E-05 p	6.7 i	6.2
2,2',3,5,5',6-	6	360.9	0.0	4.9	1.6	(6) h	(2.00E-06) h	5.90E-05 p	6.9 i	6.2
2,2',4,4',5,5'-	6	360.9	0.0	8.1	19.0	8.8 b	(2.00E-06) h	(1.38+/-0.15)E-04 e	6.9+/-0.2	6.4

CHLORINE PATTERN	# Cl ATOMS	MW (g/mol)	WEIGHT % OF ISOMERS 1242	WEIGHT % OF ISOMERS 1254	WEIGHT % OF ISOMERS 1260	SOLUBILITY (ug/l)	VAPOR PRESSURE (mm Hg)	HENRY'S LAW CONSTANT (atm-m^3/mol)	Log Kow	a	Log Koc	a
2,2',3,3',4,4',5,-	7	395.3	0.0	0.0	3.8	(2) [h]	(4.27E-07) [h]	9.00E-06 [p]	7.3	i	6.6	
2,2',3,3',4,5,6,-	7	395.3	0.0	0.0	2.1	(2) [h]	(4.27E-07) [h]	1.40E-05 [p]	7.1	i	6.6	
2,2',3,3',4,5',6,-	7	395.3	0.0	0.0	7.7	(2) [h]	(4.27E-07) [h]	(1.11E-4) [d]	7.2	i	6.6	
2,2',3,3',4',5,6,-	7	395.3	0.0	0.5	0.3	(2) [h]	(4.27E-07) [h]	(1.11E-4) [d]	(7.1)	h	6.6	
2,2',3,3',5,6,6',-	7	395.3	0.0	0.0	2.6	(2) [h]	(4.27E-07) [h]	2.40E-05 [p]	6.7	i	6.6	
2,2',3,4,4',5,5',-	7	395.3	0.0	0.0	14.5	(2) [h]	(4.27E-07) [h]	1.00E-05 [p]	7.4	i	6.6	
2,2',3,4,5,6,6',-	7	395.3	0.0	0.0	5.8	(2) [h]	(4.27E-07) [h]	1.60E-05 [p]	7.0+/-0.5		6.5	
2,2',3,4',5,6,6,-	7	395.3	0.0	0.0	2.0	(2) [h]	(4.27E-07) [h]	(1.11E-4) [d]	7.2	i	6.6	
2,2',3,3',4,4',5,5',-	8	429.8	0.0	0.0	0.8	7 [b]	(9.14E-08) [h]	1.00E-05 [p]	7.8	i	7.3	
2,2',3,3',4,4',5',6,-	8	429.8	0.0	0.0	1.4	(0.574) [h]	(9.14E-08) [h]	1.10E-05 [p]	(7.1)	h	7.3	
2,2',3,3',4,5,5',6,-	8	429.8	0.0	0.0	1.5	(0.574) [h]	(9.14E-08) [h]	(9.01E-05) [d]	(7.1)	h	6.6	[f]
2,2',3,3',4,4',5,5',6,6',-	10	498.8	0.0	0.0	0.0	7.3E-03 [l]	5.23E-10 [c]	(4.60E-05) [d]	8.2+/-0.27	e	7.8	[f]
AROCLOR MIXTURES												
1221		201				15000 [m]			2.8	m	2.0	m
1232		232				1450 [m]			3.2	m	2.4	m
1242		267				240 [m]	4.06E-04 [o]	2.28E-04 [n]	3.5	m	2.8	m
1248		300				54 [m]	4.94E-04 [o]	3.43E-04 [n]	3.8	m	3.1	m
1254		328				12 [m]	7.71E-05 [o]	4.40E-04 [n]	4.1	m	3.4	m
1260		376				2.7 [m]	4.05E-05 [o]	2.83E-04 [n]	4.3	m	3.6	m
1016		258				340 [m]		3.36E-04 [n]	3.5	m	2.8	m

[a] Girvin et al.,1990 unless otherwise stated. [b] Kalmaz and Kalmaz,1979. [c] Burkhard et al.,1985. [d] Estimated using the relationship H=VP/S and the data in this report. [e] Woodburn et al.,1984. [f] Calculated from equation 3.9 in Girvin et al.,1990. [g] Haque and Schmeeding,1975. [h] Estimated using regression equations devolped from data presented in this report. [i] Hawker and Connell, 1988. [j] Abdul et al., 1987. [k] Kilzer et al., 1979. [l] Miller et al.,1985. [m] Pal et al.,1980. [n] Burkhard et al., 1985. [o] Mackay and Leinonen,1975. [p] Brunner et al.,1990.

Composition of industrial products

Monsanto Chemical Company was the sole producer of PCB's in the United States marketing them under the trade name AROCLOR. A four-digit number identified the mixture of biphenyls found in a particular product. The first two digits (usually "12") shows that the mixture contained polychlorinated biphenyls. The second two numbers indicate the percentage of chlorine in the mixture. For example: the name AROCLOR 1254 indicates a PCB mixture with 54% chlorine. The only exception to this numbering system was AROCLOR 1016 that contained 41% chlorine. This AROCLOR, although similar to AROCLOR 1242, contains more lower chlorinated biphenyls than AROCLOR 1242 (Griffin and Chian, 1980).

PCB's were also marketed as KANECHLOR and SANTOTHERM in Japan; as PHENOCLOR and PYRALENE in France; as FENCLOR in Italy; as CLOPEN in Germany; as CHEMKO in Czechoslovakia; and as SOVOL in Russia (Griffin and Chian, 1980; Pal *et al.*, 1980).

Transformer fluids containing PCB's are of two types:
(1) Oil filled transformers with a relatively low concentration of PCB's
(2) Askarel filled transformers that contained a significant percentage of PCB's combined with other fluidizers.

ASTM standard D2283-86 defines Askarel mixtures (Table 2) used by the utility industry. The presence of PCB's in many pieces of oil filled equipment is the result of retrofilling of older Askarel transformers. McGraw (1983) notes that about two to 4% of the oil originally placed in the transformer remains within the coil and core structure after draining. This residual PCB contaminated the mineral oil after retrofilling.

Table 3 is a list of the AROCLOR's that are common in commercial use (taken from Mieure *et al.*, 1976 and Pal *et al.*, 1980). Table 3 also contains the chlorinated biphenyls, molecular weights and percentages of chlorine in each mixture. Table 4 lists the specific isomers found in three of the major AROCLOR's used by the utility industry. This table also provides a listing of key environmental parameters used to evaluate the fate and transport of PCB's.

Pal *et al.* (1980) note that the patterns of biphenyls detected in various environmental media have different characteristics. The composition of atmospheric samples contains lighter weight chlorobiphenyls than are found in water or soil. The reason for this phenomenon is a direct result of the chemical characteristics of the individual chlorinated biphenyl compounds.

Physical and chemical properties of PCB's

Mieure *et al.* (1976) notes that it is not possible to carry out an effective assessment of the environmental impact of PCB's without considering the individual isomers that make up the PCB mixtures. This opinion is supported by Girvin *et al.* (1990), Boyd and Sun (1990), Sun and Boyd (1991), Pal *et al.* (1980), Griffin and Chian (1980) and Tucker *et al.* (1975). On the other hand, Lee *et al.* (1979) indicate that there were "No significant differences in the aqueous solubility or the distribution of

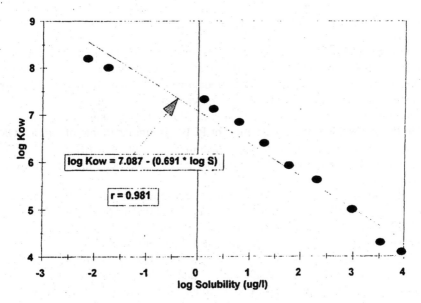

Figure 2 *PCB solubility – K_{ow} relationship. Based on data in Table 4.*

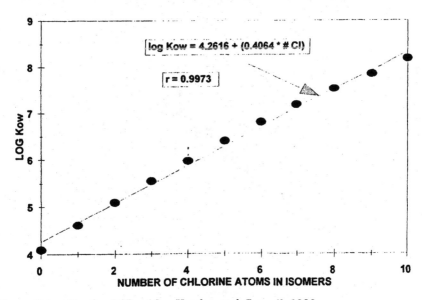

Figure 3 *log K_{ow} for PCBs. After Hawker and Connell, 1988.*

Figure 3 *log K_{ow} for PCBs. After Hawker and Connell, 1988.*

isomers...". This later opinion is not supported by the preponderance of current literature.

Data presented by Mieure *et al.* (1976) show the influence of the environment upon the various biphenyls. In general, the lower chlorinated isomers are more soluble, more readily vaporized, and biodegrade more rapidly than the highly chlorinated isomers. Partitioning on the other hand is stronger with the highly chlorinated isomers. These effects led Pal *et al.* (1980) to conclude that:

"In the atmosphere, the composition of PCB's is similar to that of AROCLOR 1242, while PCB's in surface waters approach the composition of AROCLOR 1254. PCB's in the terrestrial environment are expected to be heavier still, approximating AROCLOR 1260.".

Neely and Moy (1987) noted a similar pattern in that they found that a mixture similar to AROCLOR 1232 occurs in the vapors emitted from soils contaminated with AROCLOR 1242 and 1254 at a chemical processing plant. Brown *et al.* (1988) concur with this pattern of loss of the lower chlorination isomers.

Table 4 lists the molecular weight, solubility, vapor pressure, Henry's Law constants, log K_{ow} and log K_{oc} of the various biphenyls and AROCLOR's at 25° C. It was not practical to include all 209 isomers in Table 4. Isomers present in significant percentages in the AROCLOR's used by the utility industry are included. Decachlorobiphenyl is included to provide an example for the highest weight isomer. The selection of isomers is based on information presented in Lee *et al.* (1979), Griffin and Chian (1980), Petti *et al.* (1977), and Girvin *et al.* (1990).

Several properties listed in Table 4 were not readily available in the literature. It

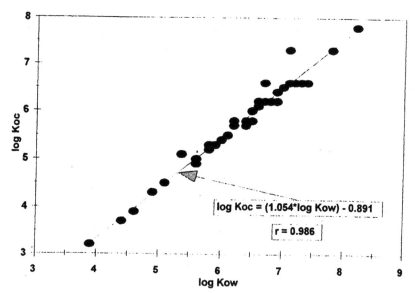

Figure 4 *log K_{ow} – K_{oc} relationship. Based on data in Table 4.*

Figure 4 *log K_{ow} – K_{oc} relationship. Based on data in Table 4.*

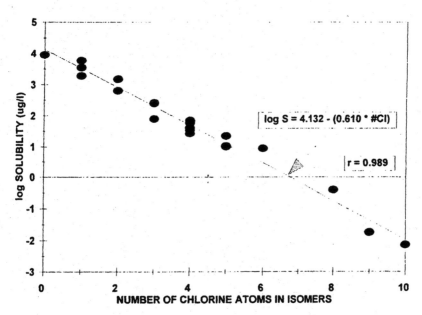

Figure 5 *Solubility of PCB isomers. Based on data in Table 4.*

was necessary to estimate the property for the particular isomer from data either included in this report or found in other literature references. The estimates made are based upon the regression equations shown in the various figures provided in this report. These equations can be used to estimate the property as long as the user understands that measured values are likely to be different from the estimates.

Partitioning

Partition coefficients

Partitioning of PCB's into other chemicals found in the environment alters environmental parameters used to estimate the fate and transport of PCB's. For example, organic material dissolved from the soil can increase the apparent solubility of organic chemicals (Carter and Suffet, 1982; Chiou *et al.*, 1986; Chiou *et al.*, 1987; Chiou, 1989; Gschwend and Wu, 1985; Hassett and Millcic, 1985; Landrum *et al.*, 1984; McCarthy and Jimenez, 1985; and Voice and Weber, 1985). The literature often confuses partitioning and adsorption in the case of the soil-water partitioning. Chiou (1989) defines terms as they apply to the soil system as follows:

Sorption denotes uptake of a solute by soil without reference to a specific mechanism. Adsorption refers to condensation of vapors or solutes on surfaces or interior pores of the solid by physical or chemical binding forces. Adsorption to soil materials is very low in aqueous systems. This is the result of the strong dipole

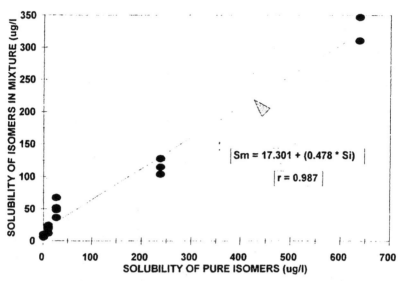

Figure 6 *Effects of PCB mixtures on solubility of PCB isomers. After Dexter and Pavloy 1978.*

interaction of minerals with water, which excludes the organic compounds from the soil mineral surface.

Partitioning denotes uptake in which the sorbed organic chemical permeates into the network of an organic medium by forces common to solution. Partition uptake is analogous to solvent extraction.

Partitioning can be defined by equations with the following form:

$$K_{ij} = C_i / C_j$$

where: K_{ij} = partition coefficient C_i = concentration in one compartment (organic matter) C_j = concentration in second compartment (water)

The most common partition coefficient encountered in environmental work is the octanol-water partition coefficient (K_{ow}) and the soil carbon-water partition coefficient (K_{oC}). A partition coefficient for dissolved organic matter-water ($K_{do}m$) or dissolved organic carbon-water (K_{doC}) occasionally appears in the literature. Boyd and Sun (1989) defined a partition coefficient for residual transformer oil and water (K_{oil}) and Sun and Boyd (1991) defined a coefficient for PCB dielectric fluid-water (K_{PCB}). These last two authors identify a total partition coefficient that combines coefficients for several components of the soil-water system. They define this as:

$$K_p = \Sigma f_m K_m$$

Table 5 *Ecological magnification of PCB isomers by several organisms.*

| PCB isomer | log Ecological magnification | | | |
	Algae	Snail	Mosquito	Fish
2,2',5-	3.86	3.76	2.91	3.81
2,2',5,5'-	4.26	4.60	4.02	4.07
2,2',4,5,5'-	3.74	4.78	4.24	4.08

(After Metcalf *et al.*, 1975)

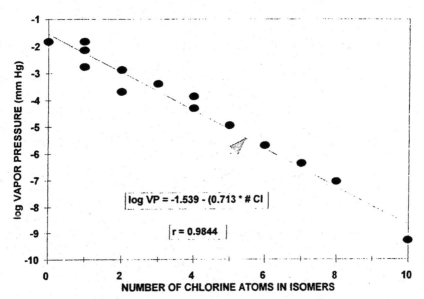

Figure 7 *Vapor pressure of PCB isomers. Based on data in Table 4.*

where: K_p = overall partition coefficient; f_m = fraction of material in medium; and K_m = partition coefficient for medium

Any assessment of PCB's in the environment must consider partitioning into the various media involved in the utility setting. Most research on the PCB's used either pure PCB isomers or AROCLOR's without the fluidizers normally found in utility equipment. The mineral oil, and chlorinated solvents used in the equipment all act as partitioning media for the PCB isomers. Lower weight isomers will partition into the higher weight isomers along with partitioning into the fluidizers. Combining the work of Chiou and his co-workers (Chiou, 1989; Chiou *et al.*, 1977, 1979, 1983, 1986, and 1987) with the work of Boyd and Sun (1990) and Sun and Boyd (1991) the following relationship is defined.

$$K_p = f_{oC} K_{oC} + f_{mo} K_{mo} + f_{PCB} K_{PCB}$$

where: K_p = total partition coefficient f_{oC} = fraction of soil organic carbon K_{oC} = partition coefficient for organic carbon-water f_{mo} = fraction of mineral oil in soil K_{mo} = partition coefficient for mineral oil-water (use K_{ow}) f_{PCB} = fraction of AROCLOR in soil K_{PCB} = partition coefficient for AROCLOR-water

The above equation was evaluated for AROCLOR 1260 using data acquired during the Remedial Investigation/Feasibility Study at the Pepper's Steel and Alloys NPL Site (Duranceau *et al.*, 1985). The PCB concentration found in the groundwater at this site was less than 0.1 μg L^{-1}. Estimates based on K_{oC} alone indicate that the concentration should have been approximately 2 mg L^{-1}. Using the total partition coefficient defined above indicates that the concentration in groundwater should be less than 1 μg L^{-1}.

Octanol-water partition coefficient
Hydrophobic chemicals such as PCB's often partition into oil or fat rather than into water. The octanol-water partition coefficient (K_{ow}) measures this partitioning. K_{ow} is an important parameter for assessing environmental behavior of a chemical. This coefficient provides an indication of the degree to which a chemical accumulates in fatty tissues within organisms as well as the degree to which the chemical partitions into any organic phase. This coefficient is especially useful for determining the release of PCB from mineral oil transformer fluids. Hawker and Connell (1988) provide a listing of the K_{ow} for 180 PCB isomers.

The PCB isomers partition into an oil phase rather than a water phase. Residual oil in soil is approximately ten times more effective for retaining PCB's than is the soil organic matter. Partitioning into an oil phase significantly reduces the mobility of PCB's and other hydrophobic pollutants (Boyd and Sun, 1990). Octanol-water partition coefficients can be estimated from the solubility of the chemical (Miller *et al.*, 1985). The regression equation shown in Figure 2 provides an estimate of K_{ow} for the PCB isomers. The coefficient is also highly correlated with the degree of chlorination of the biphenyl. Figure 3 shows this relationship.

Soil carbon water partition coefficient
Soil organic carbon controls partitioning of hydrophobic chemicals such as the PCB isomers (Karickhoff *et al.*, 1979; Karickhoff, 1981; Chiou *et al.*, 1979 and 1983; Gschwend and Wu, 1985; Means, *et al.* 1970, Hsu and Bartha, 1974; Peck *et al.* 1980). K_{oC} is a measure of this partitioning. K_{oC} can be estimated from either solubility or K_{ow}. Figure 4 shows the relationship between K_{ow} and K_{oC}.

Karickhoff *et al.* (1979) first proposed the relationship between the adsorption of a chemical and the organic carbon in the soil. Research conducted since then shows that the relationship applies to a wide range of chemical compounds. Lee *et al.* (1979) investigated the adsorption of PCB's onto several materials including clays, coal char, and soils. Chiou and his co-workers (Chiou *et al.*, 1977, 1979, 1983, 1986, and 1987) investigated the partitioning of diverse chemicals into soil organic carbon.

DiToro and Horzempa (1982) noted that the adsorption-desorption of PCB's onto

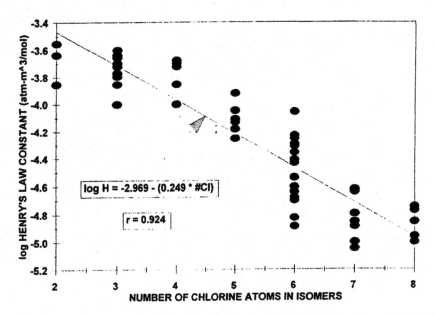

Figure 8 *Henry's law constant for PCB isomers. Based on Brunner* et al., *1990.*

sediments has two components. One component is reversible and the other strongly bound to the sediment. Girvin *et al.* (1990) identifies these two components a labile and a nonlabile fraction. They note that equilibrium between soil and PCB-containing water is a long term process occurring over time scales of months to years. Because of this long equilibration time, they believe that retardation factors developed from short term laboratory studies overestimate the migration of PCB isomers through soil systems.

Several researchers reported that the partition coefficient is dependent upon the type and concentration of particles suspended in water (O'Connor and Connolly, 1980; Nauritter and Wurster, 1983; and Voice *et al.*, 1983). This dependence is the result of organic compounds dissolving from the soil (Garbarini and Lion, 1985; Gauthier *et al.*, 1986; Gschwend and Wu, 1985; Hassett and Millcic, 1985; Landrum *et al.*, 1984; McCarthy and Jimenez, 1985; and Voice and Weber, 1985). Correction for these organic chemicals partitioned into this dissolved organic carbon yields a constant partition coefficient. Therefore, there is no dependence of the partition coefficient upon particle concentration.

Solubility

Aqueous solubility controls the loss of PCB via groundwater migration and is a major factor in understanding the environmental fate of a chemical. The solubility

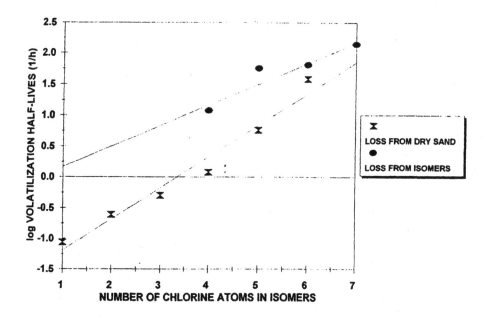

Figure 9 *Vapor loss of PCB isomers. Based on Haque and Kohnert, 1976 and Haque* et al., *1974.*

decreases as the degree of chlorination increases. Figure 5 shows this relationship. Solubility data included in Table 4 and in Figure 5 are based upon pure isomers. When an isomer is part of a mixture such as the AROCLOR's, solubility is reduced. Figure 6 shows the relationship between the solubility of the pure isomer and the same isomer when it is part of AROCLOR 1241 or 1254 (data taken from Dexter and Pavlou, 1978). This difference is due to the partitioning of the isomers into the other biphenyls found in the mixture. Information provided by Boyd and Sun (1990) and Boyd and Sun (1991) shows this effect. Boyd and Sun (1990) noted that 2-chlorophenyl partitioned into an AROCLOR mixture by a factor approximately five times more than in octanol.

Environmental releases of PCB's often accompany releases of carriers from utility equipment. An example would be mineral oil released from oil filled transformers. When PCB's are present in a mineral oil-PCB mixture the aqueous solubility of the PCB's is reduced significantly. Two factors play a role in this reduction -- partitioning of the "oil loving PCB's" into the oil phase and the reduced interaction of the PCB's with precipitation or groundwater caused by the hydrophobic nature of the oil matrix.

Interpretation of aqueous concentrations must consider the presence of dissolved organic carbon. Voice and Weber (1985), Chiou *et al.* (1987), and Gschwend and Wu (1985) researched the partitioning of organic pollutants into organic material dissolved into water.

Vapor pressure

Vapor pressure is a measure of the amount of chemical present in the air at a particular temperature. Figure 7 shows the vapor pressure at 25° C for PCB isomers. Vapor pressure decreases as the degree of chlorination increases. It is one of the factors controlling the vaporization of PCB's into the atmosphere.

Many of the PCB's found in marine environments, in lakes, and in the Arctic and Antarctic snow masses have migrated via atmospheric dispersion of vapors (Westin and Woodcock, 1979; Nelson *et al.*, 1972; Persson, 1971; Pal *et al.*, 1980; Mackay and Wolkoff, 1973). Requejo *et al.* (1977) measured PCB's in the Everglades National Park. These PCB's were the result of open burning of waste plastic mulch on agricultural lands around the Park. Vapor transport of fumes from burning plastic caused the problem.

Vaporization of PCB's from soil decreases as the amount of humic material in the soil increases (Chiou, 1989). Soil cover over the PCB's greatly reduces the loss of PCB's via vaporization (Lewis *et al.*, 1985; and Springer *et al.*, 1983). Griffin and Chian (1980) note that vaporization of PCB's from suspensions of soil or humic acids is reduced by the presence of these materials. Humic acid at 500 mg L^{-1} reduced the volatilization of AROCLOR 1242 from 3.5% to 2.6%. A soil suspension at a concentration of 6,400 mg L^{-1} reduced the loss to 0.74%.

Henry's Law constant

Henry's Law constant (H) expresses the equilibrium relationship between solution concentration of a chemical and air concentration. This constant is a major factor used in estimating the loss of PCB from soil and water. Brunner *et al.* (1990) and Dunnivant *et al.* (1988) measured Henry's Law constants for several isomers. Burkhard *et al.* (1985) estimated H "by calculating the ratio of the vapor pressure of the pure compound to its aqueous solubility". Where the literature did not report on a particular isomer, H was estimated using this relationship given by the following equation.

$$H = VP/S$$

where: H = Henry's Law constant (atm m^3 mol^{-1}); VP = Vapor pressure (atm); S = Solubility (mol m^{-3}).

Henry's Law constant is temperature dependent and must be corrected for environmental conditions. The data and estimates presented in Table 4 are for 25° C. Nicholson *et al.* (1984) outline procedures for altering the constants for temperature effects.

Burkhard *et al.* (1985) notes that H for pure isomers was reduced by two- to threefold when the isomer was in AROCLOR mixtures. The authors attributed this deviation to analytical error. The reduction may be the result of changes in both solubility and vapor pressure resulting from interactions of the isomers with the fluidizers found in the AROCLORS as well as interactions with other isomers. The literature suggests that H does not change with the degree of chlorination alone, but there is variability within the isomers of any chlorination group (Brunner *et al.*, 1990). Figure 8 shows the effects of chlorination on *H* and shows the degree of spread in the values reported for each level of chlorination. Brunner *et al.* (1990) indicate that a more sensitive estimation method is based not only on chlorination of the biphenyl, but also on the degree of chlorination of the ortho-positions on the biphenyl molecule. *H* increases as the degree of substitution on the ortho-position increases.

Environmental fate of PCB's

Each of the properties of the PCB isomers listed above play a role in the environmental distribution of these chemicals. The following information evaluates several routes by which PCB is lost from a particular source, spill or environmental compartment. Pal *et al.* (1980) present one of the best overviews of the environmental fate of the PCB's.

Loss due to vaporisation

As was mentioned above, vapor transport is believed to be one of the major routes of movement of PCB's through the environment. Pal *et al.* (1980) stated that "the low-molecular weight PCB's volatilize more readily than the high-molecular weight species". Because of this tendency there is an atmospheric enrichment of the low-molecular weight isomers. The high-molecular weight species tend to be enriched in the soil environment.

The partitioning exhibited through the Henry's Law constant is used to estimate the vaporization of various chemicals from the soil surface. Hartley (1967) noted that, in the presence of water, pesticides volatilize more rapidly than would be expected based upon vaporization of the pure chemical. This tendency accounts for the presence of low vapor pressure chemicals, such as the PCB's, in the atmosphere at higher concentrations than one would estimate from chemistry of the pure compounds.

Chemicals migrate from surfaces via diffusion (Shen, 1982). This effect also plays a role in the migration of chemicals from and through the soil. Hartley (1967) notes that "the less soluble a substance is in a liquid or in air, the slower its absolute rate of diffusion into previously pure liquid or air". Springer *et al.* (1983) has carried out simulations that indicate that PCB's do volatilize at depth and can potentially migrate through several meters of soil cover.

Lewis *et al.* (1985) measured PCB emissions from several chemical waste landfills. At an uncontrolled site air concentrations ranged up to 18 μg m^{-3}. However, at a landfill designed to meet regulatory standards, the levels were below the detection limit of 0.006 μg m^{-3}. Chromatograms of the air samples indicated that the pattern resembled that of AROCLOR 1242 with a preponderance of the peaks in the low molecular weight region.

Murphy *et al.* (1985) also noted emissions of PCB's from a number of landfills in the Great Lakes Region; however, the air concentrations were lower by several orders of magnitude compared to those seen by Lewis *et al.* (1985). Murphy *et al.* (1985) also measured the PCB concentrations in the stack gases released from sewage sludge and municipal refuse incinerators and found concentrations ranging up to 2.0 g μm^{-3}.

Pal *et al.* (1980) give volatilization half-lives reported in the literature for a number of AROCLOR's. These ranged from 10 days to 12 days for volatilization

Table 6 *Subsurface PCB Levels*

Depth below surface(m)	PCB (ppm)					
	B-1	B-2	B-3	B-4	B-5	B-6
4.61	45	3,000	*	1,580	15	<2
5.23	40	*	NT	NT	NT	
6.77	5	30	32,600	680	5	<2
7.99	25	39,000	5,400	<2	<2	
9.23	15	<2	11,650	4,000	<2	<2
10.46	<2	250	2,200	<2	10	
11.07	<2	5,050	3,800	<2	10	
11.69	<2	180	80	<2	<2	
12.30	<2	<2	45	25	<2	<2

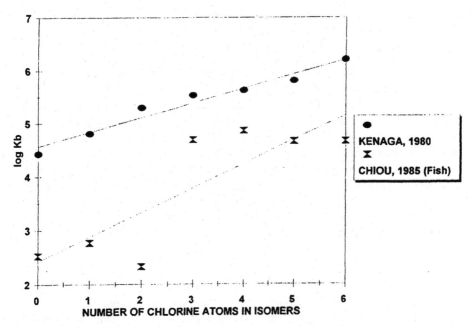

Figure 10 *Biomagnification of PCB isomers. After Kenago, 1980 and Chiou, 1985.*

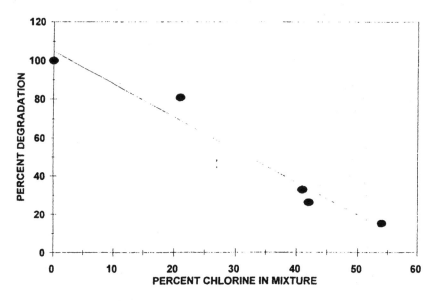

Figure 11 *Degradation of PCB by activated sludge. Based on data from Pal et al., 1980.*

from pure water up to 52 days for volatilization of AROCLOR 1260 from river water. Mackay and Leinonen (1975) give estimated half lives for the AROCLOR's of 12.1 h for AROCLOR 1242, 10.3 h for AROCLOR 1254 and 10.2 h for AROCLOR 1260.

Vaporization of the PCB isomers from pure chemical plated onto a surface generally is dependent upon the degree of chlorination. Haque and Kohnert (1976) reported half-lives for vaporization ranging from 5.2 m for 4-monochlorobiphenyl to 1.6 d for 2,2',4,4'5,5'-hexachlorobiphenyl. Figure 9 shows the effect of chlorination on the rate of PCB volatilization from pure isomer and from a dry sand. The data for this figure were taken from Haque *et al.* (1974) and Haque and Kohnert (1976). Haque *et al.* (1974) note that the vapor loss from Woodburn soil was negligible when compared to planchets or dry sand samples. Girvin and Sklarew (1986) stated that "the mass transfer coefficient is three and one-half orders of magnitude less in soil than in air...". (The mass transfer coefficient is a measure of transfer of a chemical from one phase to another.)

Adsorption, partitioning and retardation
PCB's in soil do not move at the same rate as groundwater because of adsorption onto the surface of the soil particles and partitioning into the organic carbon found in most soils. Pal *et al.* (1980) state that adsorption of PCB's increases with time of contact. Di Toro and Horzempa (1982) show a hysteresis effect during desorption of

one of the hexachlorobiphenyls. They identify this effect with a reversible and a resistant component of PCB adsorption-desorption.

This effect may be explained in part by partitioning of hydrophobic chemicals such as the PCB's into the various fractions of organic matter in the soil (Hsieh, 1992; Parton *et al.*, 1987). Chiou (1989) notes that organic chemicals bind more strongly to the humin than to the humic and fluvic acid fractions of the soil. Garbarini and Lion (1986) showed that toluene and trichloroethylene partitioned the strongest in the most resistant fraction of the soil, the fats, waxes and resins. They quoted work by G. S. Hartley (1986) that showed that the fats, waxes and resins were a solvent for pesticides. This disparity between the partitioning into the various fractions of the soil organic matter may account in part for Di Toro and Horxempa's (1982) observation of the reversible and resistant component of adsorption-desorption of PCB's.

Girvin *et al.* (1990) evaluated the release of PCB's from electrical substation soils contaminated with transformer fluids. They observed that there are two phases to the uptake and release of PCB's with these soils. The initial phase is a rapid, labile phase that is followed by a slower, nonlabile phase. The labile phase occurs at a scale of hours to days while the nonlabile phase releases over weeks and months.

Girvin *et al.* (1988) reviewed the effects of adsorption on the mobility of PCB's and their transport. In an example presented for a hexachlorbiphenyl these authors note that the PCB isomer would have a retardation factor R_f of 1,400 for the particular case given. This means that the groundwater would migrate at a rate 1,400 times faster than the PCB isomer. The retardation factor depends primarily upon the partitioning of the isomer between the soil organic carbon and the aqueous phase.

Retardation can be estimated by the following equation (Girvin *et al.*, 1990):

$$R_f = 1 + (bd/n)\, K_p = v/v'$$

where: R_f = retardation factor; bd = bulk density of soil; n = soil porosity; K_p = soil - water partition coefficient; v = Average velocity of groundwater; v' = rate of advance of the PCB front.

Thus the larger the K_p the more the retardation of the PCB.

Biomagnification

One of the major areas of concern for the hydrophobic chemicals is the area of bioconcentration or biomagnification. Chemicals with a large K_{ow} will concentrate in the lipid and fat fractions of organisms. As the isomers move through the food web of various organisms, biomagnification can become considerable. Bioconcentration factor (BCF) is the ratio of the steady state concentration in the organism (or a part of the organism such as fat, or an organ) with the concentration in water in contact with the organism (Chiou, 1985).

Metcalf *et al.* (1975) report the magnification of three isomers of PCB by several organisms (Table 5). Kenega (1980) developed a general equation for estimating the biomagnification of pesticides and other organic chemicals. He used a log linear relationship with either solubility or K_{oc} to estimate bioconcentration for a number of compounds. Figure 10 is a plot of experimental data obtained during studies carried out by Kenega and his co-workers.

Chiou (1985) measured bioconcentration in fish for a number of compounds and found that the log BCF ranged from 4.4 for 2-chlorobiphenyl to 6.2 for 2,2',4,4',5,5'-hexachlorobiphenyl. The data from Chiou's article are plotted in Figure 10 along with Kenaga's experimental data.

Other bioconcentration patterns reported in the literature are similar to that shown in Figure 10. Biomagnification increases as the degree of chlorination increases. Those compounds with the lowest solubility have the highest K_{ow} and therefore the highest biomagnification.

Travis and Arms (1988) use a different approach for evaluating the potential for accumulation of a chemical in food products such as beef, milk and vegetables. Concentration in the food divided by the daily intake is defined as the biotransfer factor (BTF). The authors report that this term is more meaningful for developing risk assessment information. Travis and Arms (1988) report log BTF values for AROCLOR 1254 of -1.28 in beef, -1.95 in milk and -1.77 in vegetables. BTF values increase as K_{ow} increases for beef and milk but, an inverse relationship holds for vegetables. Higher weight PCB isomers do not accumulate by plants to the same degree as the lower weight isomers (Pal *et al.*, 1980).

Biodegradation
Microorganisms have been shown to degrade PCB's to various degrees depending upon the soil type and other environmental parameters (Pal *et al.*, 1980; Griffin *et al.*, 1980; Hankin and Sawhney, 1984). Figure 11 shows the degradation of the AROCLOR's by microorganisms in activated sewage sludge (Pal *et al.*, 1980.) Baxter and Sutherland, (1984) also observed degradation in soils containing amendments of sewage sludge . The less chlorinated isomers are degraded more readily by soil microorganisms thus contributing to an enrichment of the higher molecular weight compounds (Hankin and Sawhney, 1984).

Photodegradation
PCB's is believed to undergo photodegradation in the environment (Baxter and Sutherland, 1984). No data were available however for evaluating the rate or degree of degradation under actual environmental conditions. Ruzo *et al.* (1974); Herring *et al.* (1972); and Crosby and Moilanen (1973) observed that UV light degrades PCB's dissolved in hexane.

Case histories of PCB spills to the environment

Based on the preceding discussion of the chemical and environmental properties of PCB, it is evident that because of the insolubility of PCB fluids in water, the movement of these substances in the environment can not be predicted from analyses of ground or surface water flow. Depending on the medium in which PCBs are present, they must be treated as settleable solids or as non-aqueous phase liquids (NAPLs).

Discharges of PCB to surface waters

From the foregoing it can be seen that PCB fluids exhibit a strong affinity for solids, such as soil particles. Therefore, one can expect that the distribution of PCB in surface water bodies is a function of the distribution of sediments. Three of the most highly publicized PCB contamination sites show strong evidence of this. These sites are Waukegan Harbor, Bedford Harbor and the Hudson River (Weaver, 1984). In all three cases, PCBs were directly discharged along with process waste waters from manufacturing facilities. All such discharges have been discontinued. However, significant concentrations of PCB are found in the sediments downstream of the wastewater discharge points, where they have remained for a number of years. In all three cases, stream velocities are generally low, so that sedimentation is the prevailing condition for all but the finest solid particles. (Presumably, the particles which have remained in suspension have carried their share of adhered PCB further away from the present regions of high contamination, but their rate of settling has not given rise to easily measured levels of PCB.) All three of these sites are on the National Priorities List and will ultimately be cleaned up by the physical removal and treatment of the contaminated sediments. All three sites are widely publicized due to the large mass of PCB present as well as the measured impacts on aquatic organisms. However, they are of relatively less concern as compared with smaller PCB discharges to the land because they are few in number and are not as likely to be repeated.

Before leaving the subject of the fate of PCB discharged into surface waters, there is one case which illustrates the relationship between PCB distribution and sediment movement. This case involves a facility which has stored and maintained both mineral oil and PCB filled electrical equipment. The facility is located on the banks of the DesPlaines River in Illinois. The DesPlaines River is a small and normally sluggish stream. However, occasional periods of sustained precipitation increase stream velocity in some sections to the point where scouring occurs. The facility, which occupies about ten hectares, has a storm water collection system which discharges into the river at two closely spaced points. Until about 1978 the storm water discharges were untreated. At that time API oil-water separators were installed on both discharges as part of an oil spill control plan. At about the time the oil water separators were being installed, the Illinois Environmental Protection Agency (IEPA) discovered PCB contaminated river sediments in the vicinity of the storm water discharges. Subsequent monitoring of the discharges revealed concentrations of PCB in the range of one to ten parts per million (ppm). Oil and grease levels were in the high tens of parts per million and total suspended solids (TSS) were typically in the low hundreds of parts per million. Concentrations of all parameters varied widely as did flow, as they were all related to precipitation events. In an attempt to reduce discharges of oil, grease, and suspended solids and to attempt to eliminate PCB

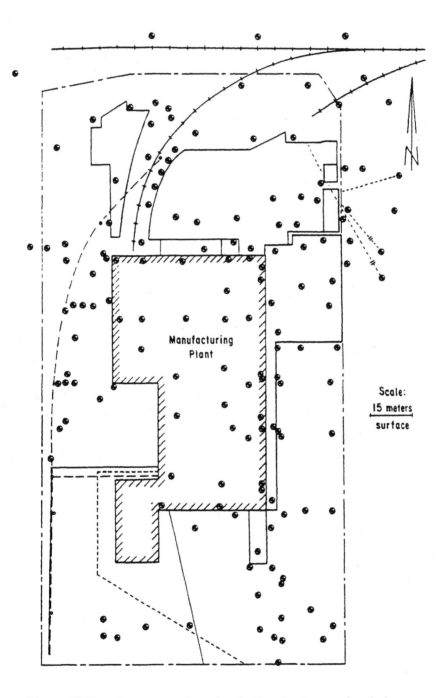

Figure 12 *Transformer manufacturing facility showing test boreholes.*

discharges, the site operator chemically enhanced precipitation in the concrete holding tanks of the oil water separators and improved housekeeping on the site.

After a period of experimentation with treatment chemicals and feed rates, concentrations of TSS are now generally less than 15 ppm, oil and grease less than 1 ppm and PCB from 0 to 5 ppb. Clearly the removal of oil and suspended solids from the discharges resulted in a significant reduction in the levels of PCB entering the river. Moreover, recently measured concentrations of PCB in the river sediments in the vicinity of the discharges were below the detection level of 1 ppm. Sediment cores approximately 12 cm in depth were taken in this study. Periods of river bed scouring that have occurred since discharge treatment began have re-suspended the contaminated sediments and re-distributed them over a broad area downstream. The subsequent spreading of the sediments has resulted in PCB levels below 1 ppm for a considerable distance downstream from the discharge points.

The discharge of PCB to the land, and subsequent potential effects on groundwater are much more difficult to analyze than surface water discharges. Because of this, and because there are many more instances of land contamination, these cases are more interesting. Unfortunately, while a number of such cases have been documented, the amount and type of data available does not allow for a complete understanding of the phenomena involved in PCB movement through the subsurface. In addition to the inherent difficulties in subsurface investigations, data gathering has typically been done with the primary purpose of locating PCB to permit removal down to a level required by environmental standards.

Discharges of PCB to soils

The following are detailed descriptions of incidents and groups of incidents in which PCB fluids have been discharged to soils and thus have had the potential for migration through the subsurface. The cases are arranged in order of decreasing evidence of contaminant transport as well as in decreasing order of detail available in the literature. Even though detailed data may be sparse, tentative conclusions regarding transport mechanisms may be sensed from the following descriptions.

Large spill of transformer fluid at manufacturing plant

Description of incidents

At least one large spill consisting of between 6,800 L and 21,000 L of PCB transformer oil and probably other smaller spills occurred at a transformer manufacturing facility in Regina, Saskatchewan, Canada (Schwartz *et al.*, 1982). The large spill occurred following the rupture of an underground pipe used to transport the transformer fluid within the plant. Other smaller spills are suspected of having occurred at times during the operation of the facility. Evidence for these spills was obtained while investigating the large incident and consisted of near-surface soil contamination that existed remote from the large spill, but at places where PCB fluid had been handled or stored. Because the site investigation focused on the large spill, only its effects are reported here.

Approximately 20,000 to 30,000 L of transformer insulating fluid was used annually at this plant. The fluid used was named as Inerteen 70-30, which is a mixture of Aroclor 1254 (30%) and trichloro and tetrachlorobenzenes (70%).

The actual date of the pipe rupture is unknown. The presence of PCB fluid was first detected during the summer of 1976, at the northeast corner of the plant in the form of seepage into a storm water drainage way. The actual discharge may have occurred over an extended period of time. At the time of detection, a cleanup program was initiated which consisted largely of the removal of an estimated 6,800 to 9,100 L of water and oil mixtures from trenches and sumps around the perimeter

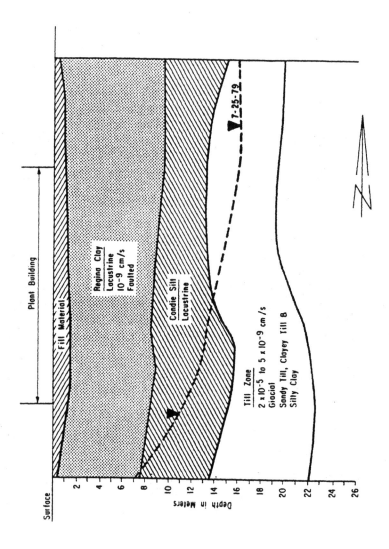

Figure 13 *Geological profile of transformer manufacturing plant.*

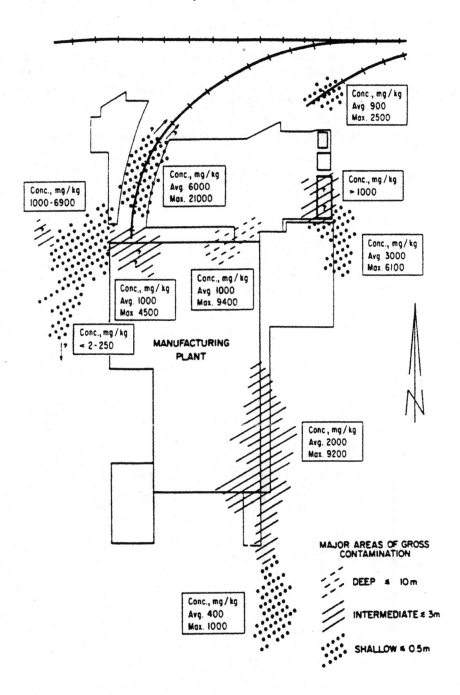

Figure 14 *PCB transformer manufacturing facility site plan showing PCB contamination.*

of the building. In November of 1976, the area around where the PCB oil had been detected was paved with asphalt.

Continuing seepage to the surface drainage ways led to a thorough investigation of the presence and spread of PCB in the fall of 1978. The results of this field study revealed considerable migration of PCB oils downward through the underlying soils. However, the plume of contamination halted before any usable aquifer was encountered. Nevertheless, the study results are interesting in that they reveal the mechanisms by which transformer PCB oil can migrate through the subsurface.

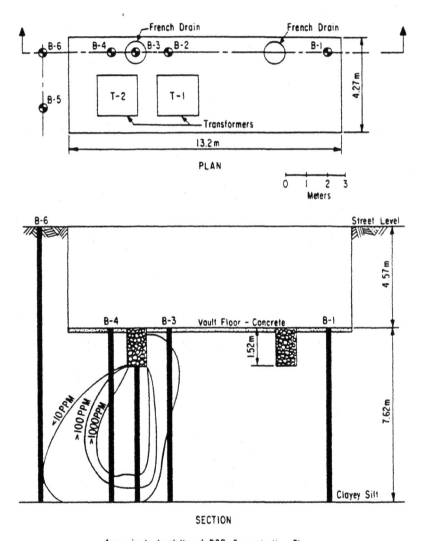

Figure 15 *Plan and section of transformer vault.*

Site investigation

Approximately 100 sampling locations were established in and around the plant structure (Roberts *et al.*, 1982). Boreholes were drilled at these sampling locations and soil samples were examined to characterize the underlying geology. Piezometers were installed in the shallower formations and soil samples were analyzed for PCB oil. The sampling depths varied but in general were at 0.3, 0.6, 0.9, 1.2, 2.4 and 3.0 m in shallow formations, and 5, 6, 8, 9 and 10 m in deeper formations. In all, over 1

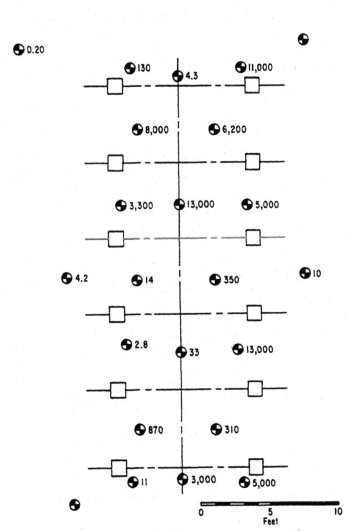

Figure 16 *Substation capacitor bank PCB concentration at surface.*

sil samples were taken and analyzed. A site plan showing approximate borehole locations is shown in Figure 12.

Site geology
The site investigation provided a description of the geology underlying the spill site. A simplified representation is shown in Figure 14 (Roberts *et al.*,1982). The

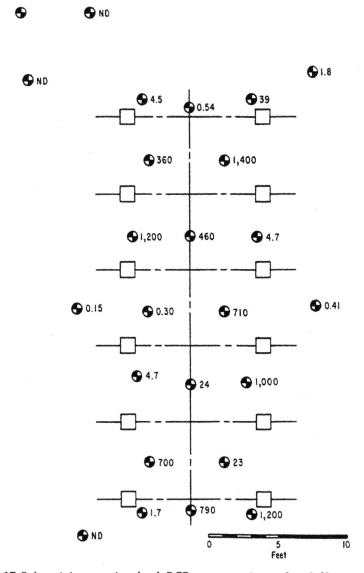

Figure 17 *Substatioin capacitor bank PCB concentration at 0 to 0.61 m*

manufacturing plant was constructed on a porous fill material that averages approximately 1.5 m in thickness. Underlying the fill are 120 – 160 m of Pleistocene deposits consisting of clays, silts, sand and gravel. Below these glacial and lacustrian deposits lies a moderately faulted bedrock formation. The upper 60 m of this series is most relevant to this study and can be described in greater detail. The upper 6-8 m is a lacustrian clay deposit known as the Regina Clay formation. This formation was found to have considerable fracturing. The hydraulic conductivity of unfractured

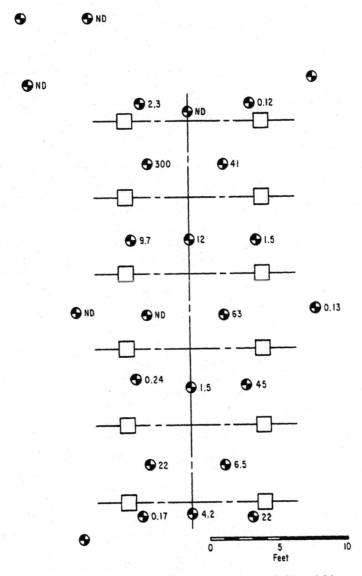

Figure 18 *Substation capacitor bank PCB concentration at 0.61 to 1.22m.*

samples of this clay are estimated to be on the order of 10^{-9} cm s^{-1}. Yet, as will be seen, the transformer oil moved readily through the formation. Beneath the Regina Clay is a silt formation also lacustrian in origin, and known as the Condie Silt. The thickness of this silt formation ranges from 3 to 7 m. Below the Condie Silt is a glacial till formation which ranges from 4 to 7 m in thickness and varies in composition from sandy till through clayey till to silty clay. Hydraulic conductivity ranges from 2 x 10-5 cm s^{-1} to 5 x 10-9 cm s^{-1} with highest hydraulic conductivity in the upper layers.

A saturated zone was discovered in the fill material at the surface. The water table in this zone varied with rainfall and snow melt. In the drier summer months, the saturated areas became discontinuous. Local recharge is indicated as the source of this subsurface water. Lateral movement was observed to be generally southward and no doubt accounts for PCB contaminated water which appeared in ditches and sumps around the plant. Some vertical movement of water into the underlying Regina Clay probably occurs due to the fracturing within this formation. However, some of the piezometers finished with the Regina Clay remained dry, indicating the absence of significant saturated flow through this low hydraulic conductivity clay.

As stated earlier, the Regina Clay is fractured, allowing downward migration of water through what would ordinarily be a very low hydraulic conductivity formation. The Condie Silt in the immediate vicinity of the plant site apparently acts as a barrier to further downward migration. Fracturing in this formation was also discovered. Nevertheless, most of the piezometers finished in the Condie Silt remained dry.

An extensive regional aquifer system occupies part of the till zone under the plant site and, as shown in Figure 13, extends upward into the Condie Silt at the southern tip of the plant. During the study period, no significant changes in water levels were measured in this aquifer. The nearest withdrawal wells finished in this aquifer are located approximately 610 m southwest of the site, while municipal drinking water wells are located approximately 1,910 m to the northwest.

PCB movement

PCB contamination was found in both the shallow fill material and in the underlying Regina Clay. A distribution of PCB is illustrated in Figure 15. Contamination in the fill material is heterogeneous, suggesting several sources of PCB (broken pipe, tank car unloading area, storage tank area) as well as the local variability in hydraulic properties of the fill. This contamination is in the zone of the perched water table where levels and flows were found to be related to rainfall and snow melt, and which discharges locally in drainage ditches around the plant. It is likely that PCB are transported in the liquid phase in this region, as part of the chlorobenzene-PCB mixture. PCB exist in the dissolved phase, but the low solubility level of the transformer fluid probably relegates this phase to one of minor importance in the transport of PCB at the site. Transformer fluid may also be adsorbed to soil particles, and to the extent that such particles are able to move through the coarse fill material, this may account for part of the distribution of PCB. Given the quantity of PCB fluid known to have been discharged at the site, it is probable that flow of transformer fluid accounts for the major transport mechanism. It was noted earlier that PCB contamination was detected in water in the drainage ditches. Very likely, this represents PCB fluid released to the surface in the liquid phase which is quickly adsorbed to soil particles eroded from the surface of the ditch. Dissolved PCB fluid is also present, but again the low solubility of the fluid precludes this as a significant source of PCB in the runoff water.

Levels of contamination in the fill material range from approximately 400 mg kg^{-1} to 3,000 mg kg^{-1}. Some samples in the fill material show no PCB contamination.

Soil sampling results indicated downward and lateral migration of PCB in the Regina Clay formation. The heterogeneous distribution of PCB as well as the presence of fracturing in this clay clearly indicate that the mode of transport into the formation was primarily that of fluid flow or as PCB adsorbed to fine soil particles. An examination of Figure 14 shows that the areas of Regina Clay that were found to be contaminated lie under the highest contaminated levels in the fill material. It appears that liquid phase transformer fluid that was so abundant in the fill material provided a sufficient hydrostatic head to drive fluid into the clay fractures. This downward movement continued until the hydrostatic pressure was balanced by surface tension as the fractures narrowed, or until the relatively unfractured, low hydraulic conductivity silt formation was encountered. Lateral movement in the clay was also discovered. This movement occurred in the upper part of the formation which was found to be porous resulting from calcite and gypsum crystal growth with subsequent dissolving of the crystals. Thus, the same hydrostatic pressures which forced the transformer fluid downward through vertical fractures also accounted for the lateral movement in the porous zone. Despite the presence of fractures in the Condie Silt, no migration of PCB was discovered, either in the initial investigation, or in subsequent sampling performed two years later (Atwater, 1984). This may be due to a lack of continuous fracture pathways or to the presence of water in the silt fractures which would have the effect of reducing the void volume.

Discussion

No groundwater contamination was found at this site due to the low hydraulic conductivity of the silt formation which confines the uppermost aquifer. The apparent lack of channels in this silt can be expected to prevent fluid flow, or even movement of fine soil particles with adsorbed PCB. Therefore, it appears that the only future source of aquifer contamination will be local recharge water which contains dissolved PCB up to the solubility limit. In terms of a contaminant plume, the effect should be small because the recharge flow susceptible to PCB contamination is small with respect to the recharge area of the aquifer as a whole, and the solubility level of PCB is low in any case.

Nevertheless, this case provides interesting insights into the movement of PCB in the subsurface. Large releases of PCB into a small porous formation such as the fill material can lead to saturation and the horizontal movement of PCB in the liquid phase. The presence of chlorobenzene solvents enhances this movement by reducing the viscosity of the pure PCB. Such solvents are nearly always a part of PCB transformer fluids, so their effect is common and should not be overlooked.

Interconnected fractures in an otherwise low hydraulic conductivity formation can have a significant impact on the movement of PCB fluids. However, the phase remains fluid PCB or PCB adsorbed to firm soil particles. The fractures may provide a conduit for contamination migration to lower formations, but do not enhance the spread of PCB fluids to areas outside the fractured zone, or in directions ordinate to the fractures. For example, PCB contamination was found at considerable depths within the fractured Regina Clay, but was not found to move horizontally within this formation from one areal source of contamination to another.

Single occurrence transformer PCB fluid spill

Description of the incident. In this incident, the source of the PCB fluid was a transformer in a subsurface concrete vault located in an eastern city (Molzahn, 1984). The transformer contained about 1,135 L of Pyranol, which is a mixture of about 35% PCB and 65% dichlorobenzene. About 454 L of the insulating fluid was released on June 22, 1982, as a result of a bushing leak. The concrete vault floor was pierced by two 0.92 m diameter drains. Concrete pipes of 0.92 m diameter and 1.69 m length extended into the underlying soil from the floor drains. The pipes were filled with loose, coarse stone, forming a French drain. The bottoms of the pipes were open to the soil, and formed the pathway to the environment for the spilled insulating fluid. The utility which owns the transformer vault estimated that about 378 L of insulating fluid flowed into one of the drains and subsequently found its way into the underlying soil.

Site investigation and geology
A site geological study was made by drilling six borings under and around the vault. The borings are shown in Figure 15 (Molzahn, 1984). In all borings, silty sand was encountered from the elevation of the vault floor down to an elevation nine feet below. The silty sand was underlain by clayey sand, eleven feet thick, which in turn was underlain by semi-cohesive materials classified as clayey silt to silty clay. Borings were completed at this relatively impermeable level.

Water was encountered about fourteen feet below the vault floor. However the geotechnical engineer concluded that, inasmuch as the soils were not saturated, the water indicated the presence of a perched water table. There was no evidence that an aquifer had been reached. It should be noted that the transformer vault is located in the downtown area of a city so there is little opportunity for local groundwater recharge. The perched water table may result from drains such as the two in the vault. Groundwater flow could be expected to be low or non-existent in spite of the relatively permeable formations down to forty feet below the surface.

PCB movement
Soil samples taken at various levels in the test boreholes were analyzed for PCB content. The results showed evidence of rapid vertical migration of PCB fluid down to the level of the impermeable clayey silt. However, the horizontal dispersion of

PCB was small, and the apparent lack of horizontal movement in the perched water table zone indicated both the low solubility of PCB in water and the lack of flow in this formation. The results of these analyses for boreholes B-1, B-2, B-3, B-4, B-5 and B-6 are tabulated in Table 6.

Discussion

The investigators of this event concluded that the estimated 378 L of PCB fluid which was released through the French drain quickly migrated to the full extent to which it was found in the sampling program. It is not known whether there had been previous releases, either sudden or gradual, from this vault. If there had been, then the full volume of soil found to contain PCB might not have been contaminated by the instant event. However a crude mass balance calculation based on the sampling results yielded a fairly close correlation between the quantity of PCB estimated to have been discharged from the vault and the quantity of PCB found in the underlying soils. It seems probable, therefore, that previous discharges were small in quantity, if they existed at all.

The investigators describe the zone through which samples were taken as unsaturated. If this is the case, then this instance does not represent aquifer contamination. The interesting aspect of this case is the relative rapidity with which PCB fluid migrated from the source. Contributing factors in this movement are probably the presence of trichlorobenzene solvent mixed with the PCB, the sandy nature of the soils and the substantial hydrostatic head that existed with approximately 378 L of PCB fluid entering the French drain.

Small capacitor spills at electrical substations

Description of incidents

Electric utilities use thousands of small capacitors installed in banks of up to several hundred units each in substations. Each capacitor unit consists of a metal can containing the operating assembly of the device and from four to twelve liters of insulating fluid. Until the manufacture and distribution of PCB were discontinued in 1970, PCB fluids were most commonly used for this purpose. Many of these PCB-containing capacitors have been replaced with devices containing alternative insulating fluids. However, because there is no requirement in environmental regulations to retire the PCB capacitors before the end of their useful life, many remain in service.

These capacitors have experienced a low failure rate of about 0.02% per year, but because of the large number of units installed, failures at a given location are not uncommon. One large midwestern facility typically experiences five to ten failures per year from a population of somewhat more than one hundred installations. Failures usually result from a break down of the insulating film within the capacitor which gives rise to a short circuit. The resulting electrical energy flow heats and expands and may vaporize the insulating fluid. The pressure thus developed causes the can to bulge and may result in rupture. The insulating fluid is disbursed to the environment as either a spray of droplets which may travel several meters before touching ground, or as a more or less continuous stream downward, or both. The total quantity of fluid discharged varies from a fraction of one liter to nearly eleven liters. While clean up of new substation capacitor spills is now required by federal regulation, there are cases of spills which occurred before the regulations were passed which were not cleaned, giving rise to the possibility of an accumulation of fluid.

Substations are typically constructed with a layer of crushed stone or gravel approximately 10 – 20 cm deep over native soils which have been filled and graded. The stone or gravel provides support for construction and maintenance vehicles as well as a medium to contain fluids that may be discharged from electrical equipment. Since the promulgation of capacitor spill clean up rules, several electrical utilities have adopted the practice of installing an impermeable membrane barrier under all or part of the stone under PCB containing capacitor banks. This practice facilitates future clean-ups by confining PCB to the top gravel cover.

Given the sporadic nature of such capacitor spills and the limited mass of PCB materials released, there is little opportunity to develop a significant hydrostatic head which might force the downward migration of fluid. Soil sampling data at substation capacitor spill sites bear out the obvious conclusion that downward migration of PCB into the subsurface soils is limited. An extensive investigation at a single such substation is illustrative.

A midwestern utility conducted a detailed study of PCB capacitor failures and the fate of discharged fluid at one of its substations (CH2M Hill, Decision Focus, 1990). The substation had a capacitor bank consisting of one hundred and twenty individual capacitor cans, each one containing about 12 L of PCB fluid. Between 1979 and 1986 twenty-two capacitor ruptures were recorded. (After 1986 all capacitors contained non-PCB insulating fluid.) Accurate records do not exist for the years prior to 1979, but the average failure rate of three per year is typical for such installations. It is probable that the same failure rate prevailed prior to 1979. Each capacitor contained 9.46 L of insulating fluid. The amount of fluid released to the environment as a result of a capacitor can rupture is variable, depending to a large extent on the location of the failure on the can and other physical factors. However, some residual fluid always remains in the capacitor, both embedded in the foil windings and as a residue within the can. Assuming that half of the fluid was discharged with each of the recorded failures, a total of approximately 104 L of PCB fluid was discharged over a seven year period ending in 1986. It is reasonable to assume a similar quantity was discharged between 1972 and 1979.

Site geology

The substation is located in an agricultural area whose soils are classified as silt loams that are deep, well drained, and of moderate permeability. Underlying these soils are two formations, the upper consisting of very fine crystalline limestone interbedded with dolomite, and the lower consisting of very fine to medium crystalline dolomite. Joint and solution-channel openings in the upper formation provide low to medium secondary porosity. The lower formation has moderate to high secondary porosity, and provides the capability of high well yields (757 – 1,892

L min^{-1}). Several domestic wells in the vicinity of the substation are finished at 24–37 m, and likely indicate the depth to groundwater.

The substation was constructed on fill material which was placed over the native soils. Normal practice would be to grade and compact this fill material before placing the final cover. At this site, final cover consists of 5 – 10 cm of crushed limestone.

Average annual precipitation at the site is 99 cm and ranges from 6.9 cm in February to 9.4 cm in May. Average annual temperature ranges from -2°C in January to 24°C in July. These conditions can be expected to give rise to a net infiltration rate on the order of 10 cm yr^{-1}. It is reasonable to expect that given the moderately permeable soils in place, that local recharge of the limestone-dolomite aquifers takes place at and around the substation site.

Site investigation

A detailed surface and subsurface investigation was conducted at the site. Twenty sampling points were located at the intersections of a grid more or less centered on the capacitor bank. Surface soil samples were obtained and analyzed at each of the sample points. All locations were also sampled at a 0.6 – 1.2 m depth and at a 1.2 – 1.8 m depth. (Note that 1.8 meters marks the extreme limit of PCB migration into the soil. In fact, most sample locations show a depth of penetration of 1.2 m or less.) The results of this sampling program are shown on Figures 16, 17 and 18 (CH2M Hill, Decision Focus, 1990).

In spite of the apparent downward migration of water at the substation, there has been very little movement of PCB fluid below the surface. The results at this substation are not unique. Another midwestern utility sampled and analyzed soils under substation capacitor banks with similar histories of spills. The climate and geological settings at this company's substations are similar to the case described above, especially with respect to annual precipitation and evaporation rates and the siting of capacitor banks on deep clay formations with crushed rock top covers. Table VII shows the results of surface and subsurface sampling for a composite of data sets for three substations. Subsurface samples were taken at depths of approximately 0.3 m and 0.4 m. The detection limit used was 1.0 ppm for PCB. In these cases, core sampling was discontinued when values of PCB dropped below 5 ppm, as that was the clean up standard employed. It was, therefore, not possible to define the absolute lowest penetration of PCB, but, taken as a whole, the data set revealed the quick reduction in concentration with depth.

Discussion

Given the low hydraulic conductivity of soils underlying the capacitor banks in the above cases, it may seem surprising that there is any penetration of PCB at all, especially in light of the small amount of fluid, hence low fluid head pressure, that is present at any time. In order to understand the mechanisms at work for PCB migration, it is helpful to visualize simple migration models. One such model is transport of dissolved PCB fluid by percolating water. However, if this were the mechanism involved, one would expect PCB concentrations that were much lower than those measured in the upper subsurface zones, but a much less pronounced decrease in PCBs with depth. During late summer and early fall it is normal for the soils in the above mentioned climatological settings to become quite desiccated, often to a depth of nearly 1 m. This being the case, the PCB concentrations, if measured during these drier periods, would very likely be close to zero immediately below the surface, and increase with depth. Such results have not been seen. Another model for PCB transport is simple flow of fluid PCB. We have seen evidence in a previous case that such flow can exist when interconnected voids of sufficient size to overcome the surface tension of PCB fluids exist in soils. Given the annual drying, and concomitant shrinking, of subsurface clays at these sites, it seems probable that voids and cracks down to a depth of about 1 m may develop which would allow the fluid flow of small quantities of PCB fluid. As a corollary, such penetration can be expected to stop at the point that clays are not subject to drying during the climatological cycle. Beyond this depth, PCB migration would be limited to dissolved PCB in recharge water.

Small individual capacitor spills

Description of incidents
Probably the most common release of PCB fluids from electrical equipment resulted from ruptures of pole-mounted PCB capacitors located on the overhead distribution systems of electric utility systems. These units had the same characteristics as the PCB capacitors located in substations which were discussed in the preceding section. They also were subject to the same failure rate. However, instead of being concentrated in a few large installations in substations they were disbursed throughout the overhead distribution system in thousands of small installations. The result was that small quantity spills occurred (up to eleven liters) at many locations. It should be noted that the use of PCB fluids in such capacitors was banned by the United States Environmental Protection Agency (EPA) after October 1, 1988. Therefore, PCB spills from pole mounted capacitors should no longer occur. Also, EPA has required clean-up efforts to various levels of effectiveness at pole-mounted capacitor spill sites since at least 1983 and in some regions earlier.

Pole-mounted capacitor spills have occurred in every conceivable setting including hard pavements, structures, lawns, agricultural lands and undeveloped land. For purposes of this chapter, the settings of greatest interest are spills onto soils.

Site investigations
One large utility collected data on the extent of PCB contamination from pole mounted capacitor spills at several hundred sites. This data was collected for the most part at fresh spill sites while PCB capacitors on poles were still in service. The distribution of PCB took two forms. Most commonly part of the fluid in the capacitor was atomized by the force of the energy released when the capacitor failed. Droplets of PCB fluid, observed to range in size up to approximately one mm in

diameter, were found to be distributed over an area as large as 200 m^2. These droplets could easily be seen on soil and vegetation if viewed within a day or two of the spill. Soil and vegetation samples were taken on a 1.52 m x 1.52 m grid sampling pattern to confirm the extent of contamination. The second form of release consisted of a discharge of a mass of fluid to the base of the pole of sufficient quantity to yield a continuous layer of PCB fluid over a variable area usually on the order of 0.5 m^2. At times, both mechanisms existed at the same spill. When soil samples were taken near the base of the pole, PCB concentrations in the tens of thousands of parts per million were recorded. Analysis of soil samples in the area of spray contamination varied from a few hundred mg/L to the level of detection of 1 mg/L. Concentrations of PCB dropped off with increasing distance from the pole, although not in a predictable way. Soil samples consisted of a surface area of about 0.15 m^2 and a depth of about 5 mm.

Nearly all sampling was performed at a depth no greater than 5 mm on soils. However, a limited number of samples at several spill sites were taken at depths down to about 0.3 m. There was a universal lack of detectable PCB (at a 1 ppm detection level) at all depths below the surface. The only exceptions to this finding were occasional levels of PCB, often in the hundreds of ppm, that were found where a heavy mass of PCB fluid ran down the side of the pole and entered the subsurface via the separation that is sometimes found between the surface of the pole and the soil. In these cases there is a direct conduit for fluids down to a depth of the pole bottom, usually about 2 m. At distances from the pole, where a spray pattern of PCB distribution predominated, there was an insufficient mass of PCB to force a downward migration of fluid, even though soils might exhibit cracking during dry weather. Moreover, solubility of PCB was too low to allow for transport of detectable levels of PCB in solution.

Conclusions

Movement of PCB fluids through the subsurface is dependent upon four site and fluid characteristics: solubility; viscosity; adsorption–desorbtion rates; physical and chemical soil characteristics.

Groundwater flow in both the saturated and unsaturated zones plays a very minor role in the transport of PCB fluids. In the available literature, there are no cases reported in which PCB fluid distribution was influenced by groundwater flow. Indeed, evidence of PCB contamination of groundwater appears to be lacking. Therefore, principles of solute groundwater transport cannot be applied to predict the movement of PCB fluids. However, NAPL and DNAPL transport principles apply to the movement of PCB fluids. In the case of a NAPL or DNAPL spill at or near the ground surface, the contaminant, in this case PCB fluids, can be expected to percolate downward through the unsaturated zone. In this zone water wets the solid soil particles and restricts the flow of non-wetting PCB until a PCB concentration greater than its characteristic residual saturation is reached. Only loosely held pore water and gas is displaced by PCB fluids as they move from pore to pore. In the case of small volume, one-time spills, the migration of PCB fluids downward ceases in the unsaturated zone at the point where the contaminant reaches residual saturation. (Domenico and Schwartz, 1990). In the cases described above the general lack of evidence of PCB fluid flow by advection through the unsaturated zone is likely the result of the small volume of fluid spilled, the low-hydraulic conductivity of the soils into which spills occurred, and perhaps the time between the spill occurrence and investigation. In fact the only observable transport mechanism for the PCB fluids has been conduit flow through observable cracks and voids rather than advective flow.

It should be pointed out that the cases described are typical of most PCB spills in terms of volumes.

With the discontinuance of PCB manufacturing, the incidence of cases wherein large masses of PCB are spilled to the earth or discharged to surface waters will decrease. However, there are still many electrical devices in service which contain PCB which are susceptible to failures resulting in the discharge of these fluids in small quantities. Probably the greatest number of such releases in the future will be from PCB-filled capacitors in substations. The evidence to date indicates that the PCB discharged from such devices remain in the local area of the spill and penetrate soils near the capacitor installation to limited depths, usually of 2 m or less. As such, the PCB contamination is easily removed through minor soil excavation and surface clean-ups.

References

Abdul, A.S., Gibson, T.L. and Rai, D.N. 1987. Statistical Correlations for Predicting the Partition Coefficient for Nonpolar Organic Contaminants Between Aquifer Organic Carbon and Water. *Hazardous Waste and Hazardous Materials*, 4(3), 211-222.

ASTM. 1986. Chlorinated Aromatic Hydrocarbons (ASKARELS) for Transformers. D228386. *Annual Book of ASTM Standards*. Vol 10.03. American Society for the Testing of Materials. Philadelphia, PA 19103.

Baxter, R.M. and Sutherland, D.A.. 1984. Biochemical and Photochemical Processes in the Degredation of Chlorinated Biphenyls. *Environmental Science and Technology*, 18(8), 608-610.

Boyd, S. A. and Sun, S.. 1990. Residual Petroleum and Polychlorinated Biphenyl Oils as Sorptive Phases in Organic Contaminants in Soils. *Environmental Science and Technology*, . 24, 142-144.

Brown, J. F Jr., Wagner, R.E., Bedard, D.L. Carnahan, J.C. and Unterman, R. 1988. The Fate of PCB's in Soil and Water. In: *Proc. 1987 EPRI PCB Seminar. EA/EL5612*. Electric Power Research Institute. Palo Alto, CA 94304.

Brunner, S., Hornung, E. Santi, H. Wolff, E. Piringer, O.G. Altschuh, J.and Bruggemann, R. 1990. Henry's Law Constants for Polychlorinated Biphenyls: Experimental Determination and Structure Property Relationships. *Environmental Science and Technology*, 24(11), 1751-1754.

Burkhard, L.P., Armstrong, D.E. and Andren, A.W. 1985. Henry's Law Constants for the Polychlorinated Biphenyls. *Environmental Science and Technology*, 19(7), 590-596.

Carter, C. W. and Suffet, I.H. 1982. Binding of DDT to Dissolved Humic Materials. *Environmental Science and Technology*, 16(11), 735-740.

Chiou, C. T., Freed, V.H., Schmedding, D.W. and Kohnert, R.L. 1977. Partition Coefficient and Bioaccumulation of Selected Organic Chemicals. *Environmental Science and Technology*, 11(5), 475-478.

Chiou, C.T., Peters, L.J. and Freed, V.H.. 1979. A Physical Concept of Soil-Water Equilibria for Nonionic Organic Compounds. Science, 206, 831-832.

Chiou, C. T., Porter, P.E. and Schmedding, D.W. 1983. Partition Equilibria of Nonionic Organic Compounds between Soil Organic Matter and Water. *Environmental Science and Technology*, 17(4), 227-231.

Chiou, C. T., Malcolm, R.L., Brinton, T.I. and Kile, D.E. 1986. Water Solubility Enhancement of Some Organic Pollutants and Pesticides by Dissolved Humic and Fulvic Acids. *Environmental Science and Technology*, 20(5), 502-508.

Chiou, C. T., Kile, D.E., Brinton, T.I., Malcolm, R.L., Leenheer, J.A. and MacCarthy, P. 1987. A Comparison of Water Solubility Enhancements of Organic Solutes by Aquatic Humic Materials and Commercial Humic Acids. *Environmental Science and Technology*, 21(12), 1231-1234.

Chiou, C. T. 1989. Theoretical Considerations of the Partition Uptake of Nonionic Organic Compounds by Soil Organic Matter. In: Sawhney, B.L. and Brown, K. (eds), *Reaction and Movement of Organic Chemicals in Soils*. Soil Science Society of America Special Publication No. 22, pp. 1-29. Social Science Society of America. Madison, Wisconsin 53711.

Crosby, D.G. and Moilanen, K. W. 1973. Photodecomposition of Chlorinated Biphenyls and Dibenzofurans. *Bulletin of Environmental Contamination and Toxicology*, 10(6), 372-377.

Dexter, R. N. and Pavlou, S. P.. 1978. Mass Solubility and Aqueous Activity Coefficients of Stable Organic Chemicals in the Marine Environment, Polychlorinated Biphenyls. *Marine Chemistry*, 6, 41-53.

Domenico, P. A. and Schwartz, F. A. 1990. *Physical and Chemical Hydrogeology*. John Wiley & Sons, New York.

DiToro, D. M. and Horzempa, L.M.. 1982. Reversible and Resistant Components of PCB AdsorptionDesorption, Isotherms. *Environmental Science and Technology*, 16(9), 594-602.

Dunnivant, F.M., Coates, J.T. and Elzerman, A.W. 1988. Experimentally Determined Henry's Law Constants for 17 Polychlorobiphenyl Congeners. *Environmental Science and Technology*, 22(4), 448-453.

Duranceau, P. E., Tickanen, L.D., Juszczyk, T. and Kunes, T.P. 1985. Laboratory Testing Program, Laboratory Methods, Procedures, Results, and Qualtiy Control -Pepper's Steel and Alloys Site. in Fixation/Stabilization, Final Report for Pepper's Steel and Alloys Site. Medley, Florida. Florida Power and Light Company. Filed with U.S. Federal Court, South Florida District.

Garbarini, D. R. and Lion, L.W. 1985. Evaluation of Sorptive Partitioning of Nonionic Pollutants in Closed Systems by Headspace Analysis. *Environmental Science and Technology*, 19(11), 1122-1128.

Garbarini, D. R. and Lion, L.W. 1986. Influence of the Nature of Soil Organics on the Sorption of Toluene and Trichloroethylene. *Environmental Science and Technology*, 20 (12), 1263-1269.

Gauthier, T. D., Shane, E.C., Guerin, W.F. Seitz, W.R. and Grant, C.L. 1986. Fluorexcence Quenching Method for Determining Equilibrium Constants for Polycyclic Aromatic Hydrocarbons Binding to Dissolved Humic Materials. *Environmental Science and Technology*, 20(11), 1162-1166.

Girvin, D. C. and Sklarew, D. S.. 1986. Attenuation of Polychlorinated Biophenyls,

Polychlorinated Dibenzofurans, and Pol;ychlorinated Dibensodioxins in Soils, Literature Review. Electric Power Research Institute. Palo Alto, CA 94304.

Girvin, D. C., Sklarew, D. S. and Scott, A. J. 1988. Attenuation of Polychlorinated Biphenyls in Soils. In, Proceedings, 1987 EPRI PCB Seminar. EA/EL 5612. Electric Power Research Institute. Palo Alto, CA 94304.

Girvin, D. C., Sklarew, D. S., Scott, A. J. and Zipperer, J. P. 1990. *Release and Attenuation of PCB Congeners, Measurement of Desorption Kinetics and Equilibrium Sorption Partition Coefficients.* GS6875. Electric Power Research Institute. Palo Alto, CA 94304.

Griffin, R. A. and Chian, E. S. K. 1980. *Attenuation of WaterSoluble Polychlorinated Biphenyls by Earth Materials.* EPA600/280027. US Environmental Protection Agency. MERL. Cincinnatii, OH 45268.

Gschwend, P. M. and Wu, S-c. 1985. On the Constancy of Sediment-Water Partition Coefficients of Hydrophobic Organic Pollutants. *Environmental Science and Technology,* 19(1), 90-96.

Gustafson, C. G. 1970. PCB's Prevalent and Persistent. *Environmental Science and Technology,* 4, 814-819.

Hankin, L. and Sawhney, B.L. 1984. Microbial Degradation of Polychlorinated Biphenyls in Soil. *Soil Science,* 137(6), 401-407.

Haque, R. and Kohnert, R. 1976. Studies on the Vapor Behavior of Selected Polychlorinated Biphenyls. *Journal of Environmental Science and Health,* B11(3), 253-264.

Haque, R. and Schmedding, D. 1975. A Method of Measuring the Water Solubility of Hydrophobic Chemicals, Solubility of Five Polychlorinated Biphenyls. *Bulletin of Environmental Contamination and Toxicology,* 14(1), 13-18.

Haque, R., Schmedding, D.W. and Freed, V.H. 1974. Aqueous Solubility, Adsorption, and Vapor Behavior of Polychlorinated Biphenyl Aroclor 1254. *Environmental Science and Technology,* 8(2), 139-142

Hartley, G. S. 1969. Evaporation of Pesticides. In: Van Valkenburg, J. W. (ed.) *Pesticidal Formulations Research, Physical and Colloidal Chemical Aspects,* pp 115-134. American Chemical Society. Washington, D. C.

Hassett, J. P. and Millcic, E. 1985. Determination of Equilibrium and Rate Constants for Binding of a Polychlorinated Biphenyl Congener by Dissolved Humic Substances. *Environmental Science and Technology,* 19(7), 638-643.

Hawker, D.W. and Connell, D.W. 1988. Octanol-Water Partition Coefficients for Polychlorinated Biphenyl Congeners. *Environmental Science and Technology,* 22(4), 382-387.

Herring, J.L., Hannan, E.J. and Bills, D.D. 1972. UV Irradiation of Aroclor 1254. *Bulletin of Environmental Contamination and Toxicology,* 8(3), 153-157.

Hesse, J. L. 1976. Polychlorinated Biphenyl Usage and Sources of Loss to the Environment in Michigan. In: *Proc.National Conference on Polychlorinated Biphenyls.* Electric Power Research Institute. Palo Alto, CA 94304.

Horzempa, L. M. and DiToro, D. M. 1983. PCB Partitioning in Sediment-Water Systems, The Effect of Sediment Concentration. *Journal of Enviornmental Quality,* 12(3), 373-380.

Hsu, T.-S., and Bartha, R. 1974. Interaction of Pesticide-Derived Chloroaniline Residues with Soil Organic Matter. *Soil Science,* 116(6), 444-452.

Hsieh, Y.-P.. 1992. Pool Size and Mean Age of Stable Soil Organic Carbon in Cropland. *Soil Science Society of America Journal.* 56, 460-464.

Kalmaz, E. V. and Kalmaz, D. 1979. Transport, Distribution, and Toxic Effects of Polychlorinated Biphenyls in Ecosystems, Review. *Ecological Modeling*, 6, 223-251

Karickhoff, S.W., Brown, D.S. and Scott, T.A.. 1979. Sorption of Hydro-phobic Pollutants on Natural Sediments. *Water Research*, 13, 241-248.

Karickhoff, S. W. 1981. Semi-Empirical Estimation of Sorption of Hydrophobic Pollutants on Natural Sediments and Soils. *Chemosphere*, 10(8), 833-846.

Kenega, E.. 1980. Predicted Bioconcentration Factors and Soil Sorption Coeffic-ients of Pesticides and Other Chemicals. *Ecotoxicology and Environmental Safety*, 4, 26-38.

Landrum, P.F., Nihart, S.R., Eadie, B.J. and Gardner, W.S.. 1984. Reverse-Phase Separation Method for Determining Pollutant Binding to Aldrich Humic Acid and Dissolved Organic Carbon of Natural Waters. *Environmental Science and Technology*, 18(3), 187-192.

Lee, M. C., Griffin, R. A., Miller, M. L. and Chian, E. S. K. 1979. Adsorption of WaterSoluble Polychlorinated Biphenyl Aroclor 1242 and Used Capacitor Fluid by Soil Materials and Coal Chars. *Journal of Envionmental Science and Health*, A14(5), 415-442.

Lewis, R. G., Martin, B.E. Sgontz, D.L. and Howes, J.E. Jr. 1985. Measurement of Fugitive Atmospheric Emissions of Polychlorinated Biphenyls from Hazardous Waste Landfills. *Environmental Science and Technology*, 19(10), 986-991.

McCarthy, J. F. and Jimenez, B.D. 1985. Interactions between Polycyclic Aromatic Hydrocarbons and Dissolved Humic Material, Binding and Dissociation. *Environmental Science and Technology*, 19(11), 1072-1076.

Mackay, D. 1982. Comments and Studies on the Use of Polychlorinated Biphenyls in Response to an Order of the United States Court of Appeals for the District of Columbia Circuit. University of Toronto. Toronto, Ontario. Submitted to USEPA by The Utility Solid Waste Activities Group, The Edison Electric Institute and The National Rural Electric Cooperative Association.

Mackay, D. and Leinonen, P.J. 1975. Rate of Evaportation of Low-Solubility Contaminants from Water Bodies to Atmosphere. *Environmental Science and Technology*, 9(13), 1178-1180.

Mackay, D. and Wolkoff, A.W. 1973. Rate of Evaporation of LowSolubility Contaminants from Water Bodies to the Atmosphere. *Environmental Science and Technology*, 7(7), 611-614.

McGraw, M. G. 1983. The PCB Problem, Separating Fact from Fiction. *Electrical World*. February, 1983, pp 49-72.

Means, J. C., Wood, S.G., Hassett, J.G. and Banwart, W.L. Sorption of Polynuclear Aromatic Hydrocarbons by Sediments and Soils. *Environmental Science and Technology*, 4(12), 1524-1528.

Metcalf, R. L., Sanborn, J. R., PoYung Lu and Nye, D. 1975. Laboratory Model Ecosystem Studies of the Degredation and Fate of Radiolabled Tri, Tetra, and Pentachlorobiphenyl Compared with DDE. *Archives of Environmental Contamination*, 3(2), 151-165.

Mieure, J. P., Hicks, O. Kaley, R. G. and Saeger, V. W. 1976. Characterization of Polychlorinated Biphenyls. *Proc.National Conference on Polychlorinated Biphenyls*. Electric Power Research Institute. Palo Alto, CA 94304.

Miller, M. M., Wasik, S.P., Huang, G.L., Shiu, W.-Y. and Mackay, D. 1985. Relationships between Octanol-Water Partition Coefficient and Aqueous Solubility. *Environmental Science and Technology*, 19(6), 522-529.

Murphy, T. J., Formanski, L.J. Brownawell, B. and Meyer, J.A. 1985. Polychlorinated

Biphenyl Emissions to the Atmosphere in the Great Lakes Region. Municipal Landfills and Incinerators. *Environmental Science and Technology,* 19(10), 942-946.

NauRitter, G. M. and Wurster, C.F. 1983. Sorption of Polychlorinated Biphenyls (PCB) to Clay Particles and Effects of Desorption on Phytoplankton. *Water Research,* 17(4), 383-387.

Neely, J. and Moy, C.. 1988. Comparison of Estimated and Actual PCB Vapor Exposure During a Soil Excavation Project. In: *Proc. 1987 EPRI PCB Seminar.* EPRI EA/EL-5612. Electric Power Research Institute. Palo Alto, CA 94304.

Nelson, N., Hammond, P. B. Nisbet, I.C.T. , Sarofim, A.F. and Drury, W.H. 1972. Polychlorinated Biphenyls, Environmental Impact. *Environmental Research,* 5, 249 - 362.

Nicholson, B.C., Maquire, B.P. and Bursill, D.B. 1984. Henry's Law Constants for the Trihalomethanes, Effects of Water Composition and Temperature. *Environmental Science and Technology,* 18 (7), 518-521.

O'Connor, D.J. and Connolly, J.P. 1980. The Effect of Concentration of Adsorbing Solids on the Partition Coefficient. *Water Research,* 14, 1517-1523.

Pal, D., Weber, J.B. and Overcash, M. R. 1980. Fate of Polychlorinated Biphenyls (PCB's) in SoilPlant Systems. Residue Reviews. Vol. 74, 46-98.

Parton, W.J., Schimel, D.S., Cole, C.V. and Ojima, D.S. 1987. Analysis of Factors Controlling Soil Organic Matter Levels in Great Plains Grasslands. *Soil Science Society of America Journal,* 51, 1173-1179.

Peck, D. E., Corwin, D. L. and Farmer, W. J. 1980. Adsorption-Desorption of Diuron by Freshwater Sediments. *Journal of Environmental Quality,* 9(1), 101-106.

Persson, B. 1971. Uptake of Chlorinated Hydrocarbons by Whitethroats Sylvia communis. oathl. in Areas Sprayed with DDT. *Ornis Scandinavica,* 2, 127-135.

Petti, R. W., Schmidt, D.J. Rowley, B.B. and Pelton, D.J. 1977. *Polychlorinated Biphenyls.* U. S. Environmental Protection Agency. Research Triangle Park, NC 27711. 24 pp.

Pignatello, J. J. 1989. Sorption Dynamics of Organic Compounds in Soils and Sediments. In: Sawhney, B. L. and Brown, K. (eds.), *Reactions and Movement of Organic Chemicals in Soils,* SSSA Special Publication 22.Chap.3, pp. 45-80. Soil Science Society of America. Madison, Wisconsin 53711.

Requejo, A. G., West, R.H., Harvey, G.R., Hatcher, P.G. and McGillivary, P.A. 1977. Polychlorinated Biphenyls, Chlorinated Pesticides and Trace Metals in Soils of the Everglades National Park and Adjacent Agricultural Areas. ERL AOML32. Environmental Research Laboratories. National Oceanic and Atmospheric Administration. Miami, FL

Risebrough, R.W., Reiche, P., Peakall, D. B. Herman, S. G. and Kirven, M. N. 1968. Polychlorinated Biphenyls in the Global Ecosystem. *Nature,* 220, 1098.

Ruzo, L.O., Zabik, M.J. and Schuetz, R.D. 1972. Polychlorinated Biphenyls, Photolysis of 3, 4, 3', 4'-tetrachlorobiphenyl and 4, 4'-dichlorobiphenyl in Solution. *Bulletin of Environmental Contamination and Toxicoloty,* 8(4), 217-218.

Schwarzenbach, R. P. and Westall, J. 1981. Transport of Nonpolar Organic Compounds from Surface water to Groundwater. Laboratory Sorption Studies. *Environmental Science and Technology,* 15(11), 1360-1367.

Shen, T. 1982. Estimation of Organic Compound Emissions from Waste Lagoons. *Journal of the Air Pollution Control Association,* 32(1), 79-82.

Springer, C., Thibodeaux, L.J. and Chatrathi, S. 1983. Simulation Study of the Volatilization of Polychlorinated Biphenyls from Landfill Disposal Sites. In: Francis, C.

W. and Auerbach, S.I. (eds.) *Environment and Solid Wastes, Characterization, Treatment, and Disposal,* pp 209-222. Butterworths. Boston, MA

Sun, S. and Boyd, S.A. 1991. Sorption of Polychlorobiphenyl (PCB) Congeners by Residual PCB-Oil Phases in Soils. *Journal of Environmental Quality,* 20, 557-561.

Travis, C. C. and Arms, A.D. 1988. Bioconcentration of Organics in Beef, Milk and Vegetation. *Environmental Science and Technology,* 22(3), 271-273.

Tucker, E. S., Litschgi, W. J.and Mees, W. M. 1975. Migration of Polychlorinated Biphenyls in Soil Induced by Percolating Water. *Bulletin of Enviornmental Contamination and Toxicology,* 13(1), 86-93.

Voice, T. C., Rice, C.P. and Weber, W.J. Jr. 1983. Effect of Solids Concentrations on the Sorptive Partitioning of Hydrophobic Pollutants in Aquatic Systems. *Environmental Science and Technology,* 17(9), 513-518.

Voice, T. C. and Weber, W.J.Jr. 1985. Sorbent Concentration Effects in Liquid/Solid Partitioning. *Environmental Science and Technology,* 19(9), 789-796.

Westin, R. and Woodcock, B. 1979. Support Document/Voluntary Environmental Impact Statement and PCB Manufacturing, Processing, Distribution in Commerce, and Use Ban Regulation, Economic Impact Analysis. TS799. Office of Toxic Substances. U. S. Environmental Protection Agency. Washington, DC 10460.

Woodburn, K. B., Doucette, W.J. and Andren, A.W. 1984. Generator Column Determination of Octanol/Water Partition Coefficients for Selected Polychlorinated Biphenyl Congeners. *Environmental Science and Technology,* 18 (6), 457-459.

Index

A

B

C

D

Milton Keynes UK
Ingram Content Group UK Ltd.
UKHW021924071024
449327UK00022B/1701